QUANTUM COMMUNICATION, MEASUREMENT AND COMPUTING (QCMC)

QUANTUM COMMUNICATION, MEASUREMENT AND COMPUTING (QCMC)

The Ninth International Conference

Calgary, Canada *19 - 24 August 2008*

QCMC
Calgary, 2008

EDITOR
Alexander Lvovsky
University of Calgary
Canada

SPONSORING ORGANIZATIONS
National Institute of Information and Communications Technology, Japan
Canadian Institute for Advanced Research
QuantumWorks
Tamagawa University
Pacific Institute for the Mathematical Sciences
Research Laboratory of Electronics at MIT
Informatics Circle of Research Excellence
University of Calgary, Faculty of Science, Department of Computer Science, and Department of Physics and Astronomy
Perimeter Institute
id Quantique
Quantum Communications Victoria
Institute of Physics Publishing
Southern Alberta Institute of Technology

AMERICAN INSTITUTE OF PHYSICS

Melville, New York, 2009
AIP CONFERENCE PROCEEDINGS ■ 1110

Editor

Alexander Lvovsky

Department of Physics and Astronomy
University of Calgary
Calgary AB T2N 1N4
Canada

E-mail: LVOV@ucalgary.ca

L.C. Catalog Card No. 2009923301

ISBN 978-0-7354-0647-6
ISSN 0094-243X
Printed in the United States of America

CONTENTS

CHAPTER 1
THEORETICAL PHYSICS

CHAPTER 2
QUANTUM INFORMATION THEORY

CHAPTER 3
POPESCU-ROHRLICH BOXES

CHAPTER 4
QUANTUM CHANNELS

CHAPTER 5
CLUSTER STATES

CHAPTER 6
CONTINUOUS-VARIABLE QUANTUM OPTICS

CHAPTER 7
PHOTONS

CHAPTER 8
INTERFACING QUANTUM INFORMATION BETWEEN LIGHT AND MATTER

CHAPTER 9
EXPERIMENTAL TOOLS

CHAPTER 10
QUANTUM INFORMATION PROCESSING IN SOLIDS AND LIQUIDS

CHAPTER 11
QUANTUM KEY DISTRIBUTION

CHAPTER 12
QUANTUM COMMUNICATION PROTOCOLS (OTHER THAN CRYPTOGRAPHY)

CHAPTER 13
QUANTUM COMPUTATION AND ALGORITHMS

CHAPTER 14
QUANTUM METROLOGY AND IMAGING

CHAPTER 15
QUANTUM PROCESSES

Preface

The articles in this volume are related to presentations at the Ninth International Conference on Quantum Communication, Measurement and Computing (QCMC). which took place on August 19-24, 2008, on the campus of the University of Calgary. This bi-annual meeting has been the leading conference in the field for more than two decades and has a long standing tradition of bringing together top researchers and students from the mathematics, computer science, engineering, experimental and theoretical physics communities.

The conference featured a set of 46 invited talks delivered by the field's leading researchers. Additionally, we had 16 contributed talks and 248 posters. Altogether, the meeting was attended by 312 delegates from 29 countries, of which 112 were students.

An integral part of QCMC is the International Quantum Communication Award for outstanding achievements in quantum communication research. This award is sponsored by the Tamagawa University and is presented at each QCMC meeting since 1996. This year, the winners were Jeffrey Shapiro (MIT), Akira Furusawa (University of Tokyo), and Anton Zeilinger (University of Vienna).

In addition to the scientific program, we had many special events, such as an outing to Heritage Park Historical Village, a banquet (which featured a performance by the native Canadian Kehewin Dance Theatre), and a reception "Quantum Information in Canada". A definite highlight was a soccer tournament among four teams composed of conference delegates from different part of the world. The winner, the team of Europe, was awarded medals and a challenge trophy.

Additional information about the meeting can be viewed at http://www.qcmc2008.org.

QCMC conferences

Scope of the meeting

- Quantum Cryptography and Quantum Communications
- Quantum Measurement and Quantum Metrology
- Quantum Computing and Quantum Information Theory
- Implementations of Quantum Information Processing
- Quantum Control

Past meetings of the series

- Quantum Aspects of Optical Communication, Paris, France, 1990
- Quantum Communication and Measurement, Nottingham, United Kingdom, 1994
- Quantum Communication and Measurement, Fuji-Hakone, Japan, 1996
- Quantum Communication, Measurement and Computing, Evanston, IL, USA, 1998
- Quantum Communication, Measurement and Computing, Capri, Italy, 2000
- Quantum Communication, Measurement and Computing, Cambridge, MA, USA 2002
- Quantum Communication, Measurement and Computing, Glasgow, United Kingdom, 2004
- Quantum Communication, Measurement and Computing, Tsukuba, Japan, 2006

Committees

Local Organizing Committee

- Alexander Lvovsky (Principal Organizer)
- David Feder
- Gilad Gour
- Peter Høyer
- Christopher Healey
- Karl-Peter Marzlin
- Barry Sanders
- Wolfgang Tittel
- Robert Thompson
- Jeffrey Shapiro, Massachusetts Institute of Technology (ex officio)

Program Committee

- Ignacio Cirac, Max Planck Institute for Quantum Optics, Germany
- Richard Cleve, University of Waterloo, Canada
- Akira Furusawa, University of Tokyo, Japan
- Atac Imamoğlu, ETH Zürich, Switzerland
- Alexander Lvovsky, University of Calgary, Canada
- Christopher Monroe, University of Michigan, USA
- Eugene Polzik, Niels Bohr Insititute, Denmark
- Howard Wiseman, Griffith University, Australia
- Frank Verstraete, California Institute of Technology, USA

QCMC Steering Committee

- Osamu Hirota, Tamagawa Unversity, Japan
- Jeffrey Shapiro, Massachusetts Institute of Technology, USA
- Prem Kumar, Northwestern University, USA
- Gerard Milburn, University of Queensland, Australia
- Raymond Laflamme, Institute for Quantum Computing, Canada
- Philippe Grangier, Institut d'Optique, France
- Giacomo Mauro D'Ariano, Universita' di Pavia, Italy
- Masahide Sasaki, National Institute of Information and Communications Technology, Japan

Quantum Communication Award

The International Quantum Communication Award for outstanding achievements in quantum communication research is presented at each QCMC. That award is sponsored by the Tamagawa University.

This year, the award winners are:

- Jeffrey Shapiro (Massachusetts Institute of Technology), for seminal contributions to the communication theory of systems with quantum effects.
- Akira Furusawa (University of Tokyo), for pioneering experiments on quantum teleportation and macroscopic superposition states.
- Anton Zeilinger (University of Vienna), for pioneering experiments on quantum teleportation and multi-partite entanglement phenomena.

In the past, the award winners have been as follows:

1996: Charles Bennett, Carl Helstrom, Alexander Holevo, Horace Yuen
1998: Jeffrey Kimble, Peter Shor
2000: Paul Benioff, Christopher Monroe, David Wineland
2002: David Deutsch, Serge Haroche, Benjamin Schumacher
2004: Richard Jozsa, Prem Kumar
2006: Ignacio Cirac, Philippe Grangier, William Wootters, Peter Zoller

Acknowledgments

The local committee would like to thank the conference sponsors for their generous support:

- National Institute of Information and Communications Technology, Japan (*http://www.nict.go.jp*)
- Canadian Institute for Advanced Research (*http://cifar.ca*)
- QuantumWorks (*http://www.quantumworks.ca*)
- Tamagawa University (*http://www.tamagawa.ac.jp*)
- Pacific Institute for the Mathematical Sciences (*http://www.pims.math.ca*)
- Research Laboratory of Electronics at MIT (*http://www.rle.mit.edu*)
- Informatics Circle of Research Excellence (*http://www.icore.ca*)
- University of Calgary
    - Faculty of Science (*http://www.ucalgary.ca/science*)
    - Department of Computer Science (*http://www.cpsc.ucalgary.ca*)
    - Department of Physics and Astronomy (*http://www.ucalgary.ca/phas*)
- Perimeter Institute (*http://www.perimeterinstitute.ca*)
- id Quantique (*http://www.idquantique.com*)
- Quantum Communications Victoria (*http://qcvictoria.com*)
- Institute of Physics Publishing (*http://publishing.iop.org*)
- Southern Alberta Institute of Technology (*http://www.sait.ab.ca*)

We are also grateful to

- the conference managers (Rhonda McNeil, Tim Fukami and Erica Borgstrom) for their help in organizing the event;
- Boris Braverman, Anna Ordog, and Travis Brannan for preparing this book;
- Dr. Eden Figueroa, Andrew MacRae and Connor Kupchak for organizing the soccer tournament;
- numerous student volunteers for their help during the meeting.

Thank you very much!

CHAPTER 1

THEORETICAL PHYSICS

Generation of Entanglement Outside of the Light Cone

J.D. Franson

Physics Dept., University of Maryland, Baltimore County, Baltimore, MD 21250

Abstract. The Feynman propagator has nonzero components outside of the light cone. As a result, entanglement and mutual information can be generated between two space-like separated points. Two possible interpretations of these effects are discussed.

Keywords: Entanglement, Feynman, propagator.
PACS: 42.50.-p, 42.50.Xa, 03.65.-w.

INTRODUCTION

The probability amplitude to emit a photon at a location $\mathbf{r}_1, t_1$ and annihilate it at $\mathbf{r}_2, t_2$ is proportional to the Feynman propagator $D_F(\mathbf{x}_2 - \mathbf{x}_1, t_2 - t_1)$ [1]. It is well known that the Feynman propagator has nonzero values outside of the forward light cone, as illustrated in Fig. 1. It should be emphasized that this does not allow messages to be transmitted faster than the speed of light, as will be shown below. But it does allow entanglement and mutual information to be generated outside the light cone, which raises some fundamental issues regarding electromagnetic interactions. These effects will be reviewed here, along with two very different interpretations. A more detailed discussion can be found in Refs. [2,3].

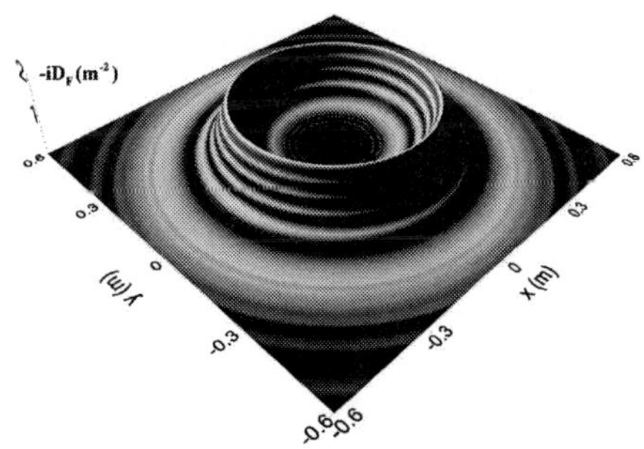

FIGURE 1. A plot of the Feynman propagator in the x-y plane, corresponding to a 1 ns delay after a photon was emitted at the origin. The yellow ring corresponds to the light cone propagating outwards. It can be seen that there are nonzero components outside the light cone as well as retarded components inside the light cone [3].

CP1110, *Quantum Communication, Measurement and Computing (QCMC), The Ninth International Conference,* edited by A. Lvovsky

ENTANGLEMENT OF TWO DISTANT ATOMS

The effects of the Feynman propagator can be illustrated by considering two distant atoms as shown in Fig. 2. Atom 1 is initially assumed to be in its first excited state while atom 2 is in its ground state. Atom 1 can emit a photon which can propagate to atom 2, as described by the Feynman propagator, and atom 2 can then absorb the photon and make a transition to its excited state. The question is whether or not this process can occur after a time interval Δt that is less than the time required for a signal to propagate between the two atoms at the speed of light.

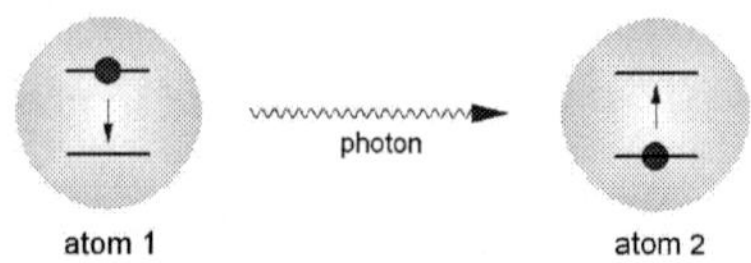

FIGURE 2. Two distant atoms, one in an excited state and the other in its ground state. The emission of a photon by atom 1 and its absorption by atom 2 can reverse the states of the two atoms [3].

This problem was first considered by Fermi, who made several unwarranted assumptions and concluded that there were no effects at all outside the light cone. Several subsequent authors concluded that the state of atom 2 could be changed in an arbitrarily short amount of time, in apparent violation of causality. It was later shown that the expectation value of atom 2 cannot change, but that correlations between the two atoms can be produced outside the light cone [4]. I generalized these results to show that entanglement can be produced between the two atoms in an arbitrarily short time interval [2,3].

The state of the system after time interval Δt can be found using second-order perturbation theory:

$$|\psi(\Delta t)\rangle = \frac{1}{(i\hbar)^2} \int_0^{\Delta t} dt' \int_0^{t'} dt'' \hat{H}'(t') \hat{H}'(t'') |\psi_0\rangle. \tag{1}$$

Within the dipole approximation, the probability amplitude b to find atom 1 in its ground state and atom 2 in its excited state, with no photons present, becomes

$$b = \frac{1}{(i\hbar)^2} \left(\frac{edE_A}{\hbar c} \right)^2 \int_0^{\Delta t} dt' \int_0^{t'} dt'' e^{-iE_A(t''-t')/\hbar} \langle 0 | \hat{A}_x(\mathbf{x}_2, t') \hat{A}_x(\mathbf{x}_1, t'') | 0 \rangle. \tag{2}$$

The matrix elements of the vector potential operator can be evaluated using commutator techniques:

$$
\begin{aligned}
&\langle 0|\hat{A}_x(\mathbf{x}_2,t')\hat{A}_x(\mathbf{x}_1,t'')|0\rangle = \langle 0|\hat{A}_x^{(+)}(\mathbf{x}_2,t')\hat{A}_x^{(-)}(\mathbf{x}_1,t'')|0\rangle \\
&= \langle 0|\hat{A}_x^{(-)}(\mathbf{x}_1,t'')\hat{A}_x^{(+)}(\mathbf{x}_2,t') + [\hat{A}_x^{(+)}(\mathbf{x}_2,t'), \hat{A}_x^{(-)}(\mathbf{x}_1,t'')]|0\rangle \\
&= \langle 0|[\hat{A}_x^{(+)}(\mathbf{x}_2,t'), \hat{A}_x^{(-)}(\mathbf{x}_1,t'')]|0\rangle \\
&= -ic\hbar D_F(\mathbf{x}_2 - \mathbf{x}_1, t' - t'').
\end{aligned}
\tag{3}
$$

Eq. (3) relates the probability amplitude for this process to the Feynman propagator.

For a massless particle and $c\Delta t < r$, the Feynman propagator is given by

$$
D_F(x_2 - x_1) = -\frac{1}{4i\pi^2}\frac{1}{|\mathbf{x}_2 - \mathbf{x}_1|^2 - c^2(t' - t'')^2 - i\varepsilon}. \tag{4}
$$

We will consider the limit of $c\Delta t << r$, where r is the distance between the atoms, in which case the propagator is inversely proportional to r^2. Inserting this into Eqs. (2) and (3) shows that the state vector after time Δt is given by

$$
|\psi\rangle = a|E_1\rangle|G_2\rangle + b|G_1\rangle|E_2\rangle + \gamma|\phi_\perp\rangle. \tag{5}
$$

Here a is a coefficient on the order of unity and $|\phi_\perp\rangle$ is a state that is orthogonal to the other two and reflects the fact that atom 1 may emit a photon that is not absorbed by atom 2. The coefficient γ is of no interest, while b reduces to

$$
b = -\frac{\alpha}{4\pi^2}\frac{d^2}{r^2}\left(i\omega_A\Delta t + 1 - e^{i\omega_A\Delta t}\right). \tag{6}
$$

It is important to include all possible processes in calculations of this kind, including a virtual process in which atom 2 emits a photon that is absorbed by atom 1. With the inclusion of that process, the probability amplitude for the exchange of a photon becomes

$$
b = -\frac{\alpha}{2\pi^2}\frac{d^2}{r^2}[1 - \cos(\omega_0\Delta t)]. \tag{7}
$$

Eq. (7) shows that there is a small but nonzero probability amplitude for the atoms to have exchanged a photon outside of the light cone.

Ref. [3] shows that the overall system consisting of the two atoms and the field is entangled:

$$
|\psi\rangle \neq |\Psi_1\rangle|\Psi_2\rangle. \tag{8}
$$

In most applications in quantum information processing, it would be desirable to have an entangled state of the atoms independent of the state of the field. In principle, this can be obtained by using an array of detectors to post-select results in which there are no photons in the final state. That eliminates the last term in Eq. (5) and gives a state of the form

$$|\psi\rangle = a|E_1\rangle|G_2\rangle + b|G_1\rangle|E_2\rangle \qquad (\gamma = 0) \tag{9}$$

which is an entangled state of the atoms alone. Finally, entanglement concentration can be used as shown in Fig. 3 to produce a maximally entangled state of the two atoms. It should be noted that, although this process depends on post-selection, there is no correction applied to atom 2 and the process can be completed in a time interval much smaller than r/c. It should also be noted that post-selection would have no effect if there had not been a physical interaction between the two atoms.

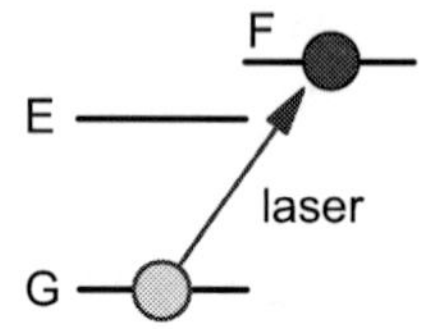

FIGURE 3. Use of entanglement concentration to produce a maximally entangled state of the two atoms. A laser beam is applied to atom 2 to transfer an appropriate amount of probability amplitude to state F. If a measurement shows that state F is not populated, then states E and G will be left with equal probability amplitudes [3].

These results show that an entangled state of the field and atoms can be generated without post-selection in less time than it would take for light to travel between the two atoms. Post-selection can be used to generate a maximally entangled state of the atoms alone. In addition, it has previously been shown that correlations between the two atoms can be generated without any post-selection [4]. All of these results suggest that a physical interaction must have occurred between the two atoms outside the light cone, as will be discussed in more detail below.

MUTUAL INFORMATION AND CAUSALITY

If we simply measure the states of the two atoms, an atom in the ground state can be taken to represent a classical bit value of 0 and an atom in the excited state can represent a bit value of 1. Given two sequences X and Y of classical bits, the mutual information is defined as

$$I(X;Y) = \sum_{y\in Y}\sum_{x\in X} p(x,y)\log_2 \frac{p(x,y)}{p(x)p(y)}. \tag{10}$$

Since the states of the two atoms are correlated, this process corresponds to the generation of mutual information outside of the light cone. No post-selection is required for this to occur.

Although entanglement and correlations can be generated outside the light cone, it is not difficult to show that no messages can be transmitted for $c\Delta t < r$. In analogy with ordinary perturbation theory for the state vector, the expectation value of an operator $\hat{Q}$ is given [5] by

$$
\langle\psi(t)|\hat{Q}|\psi(t)\rangle = \langle\psi_0|\hat{Q}|\psi_0\rangle + \frac{1}{i\hbar}\int_0^t dt'\langle\psi_0|\left[\hat{Q},\hat{H}'(t')\right]|\psi_0\rangle
$$

$$
+\frac{1}{(i\hbar)^2}\int_0^t dt'\int_0^{t'} dt''\langle\psi_0|\left[\left[\hat{Q},\hat{H}'(t')\right],\hat{H}'(t'')\right]|\psi_0\rangle + \ldots \tag{11}
$$

The probability of finding atom 2 in its excited state is equal to the expectation value of the projection operator $\hat{p}_2 = |E_2\rangle\langle E_2|$. Since this operator commutes with the Hamiltonian at atom 1, it follows immediately that no messages can be transmitted outside the light cone. But this argument does not apply to correlations corresponding to operators such as $\hat{p}_{12} = |E_2\rangle|G_1\rangle\langle G_1|\langle E_2|$, which explains why correlations and entanglement can be generated outside the light cone while messages cannot be transmitted [3].

DETECTION PROBABILITIES AND OPTICAL COHERENCE

The usual expression [6] for the probability of detecting a photon is given by

$$
P_d = \eta\left\langle \hat{E}^{(-)}(\mathbf{x},t)\hat{E}^{(+)}(\mathbf{x},t)\right\rangle \tag{12}
$$

where η is a constant related to the detector efficiency. It can be shown using the techniques described above that the use of this formula would allow messages to be transmitted outside the light cone [7]. These difficulties are due to the fact that Eq. (12) is based on the rotating-wave approximation.

A similar problem occurs in the usual definition [6] of optical coherence, where the function $g^{(2)}$ is defined as

$$
g^{(2)}(\mathbf{x}_1,t_1;\mathbf{x}_2,t_2) = \frac{\left\langle \hat{E}^{(-)}(\mathbf{x}_1,t_1)\hat{E}^{(-)}(\mathbf{x}_2,t_2)\hat{E}^{(+)}(\mathbf{x}_2,t_2)\hat{E}^{(+)}(\mathbf{x}_1,t_1)\right\rangle}{\left\langle \hat{E}^{(-)}(\mathbf{x}_1,t_1)\hat{E}^{(+)}(\mathbf{x}_1,t_1)\right\rangle\left\langle \hat{E}^{(-)}(\mathbf{x}_2,t_2)\hat{E}^{(+)}(\mathbf{x}_2,t_2)\right\rangle}. \tag{13}
$$

This expression is also based on the rotating-wave approximation, and it can be shown that Eq. (13) would also allow messages to be transmitted outside the light cone. Approximate solutions to this problem have been previously discussed [7,8], but it may be that the definition of optical coherence is inherently approximate. These difficulties are related to the fact that the quantized electric field cannot be measured at a point.

WHAT IS THE "CORRECT" INTERPRETATION?

It has been shown that entanglement and mutual information can be generated outside the light cone. The usual view of electromagnetic interactions is that they are due to the exchange of virtual photons. In that case, these effects could be interpreted as being due to the exchange of virtual photons that propagate outside the light cone, as suggested by the plot of the Feynman propagator in Fig. 1.

Feynman himself said that "possible trajectories are not limited to regions within the light cone" but "in reality, *not much* of the $t-|\mathbf{x}|$ space outside the light cone is accessible" [9]. But the axes in Fig. 1 are in units of meters, and it is not apparent that these effects are limited to microscopic distances.

An alternative interpretation has been suggested based on algebraic quantum field theory. Within that context, it has been shown that the quantum vacuum is entangled [10] and effects of this kind for spin-1/2 particles have been interpreted as being due to the transfer of entanglement from the quantum vacuum to the particles [11]. This interpretation does not explain what physically causes the entanglement, and it seems to suggest that we are to abandon the idea of virtual photons.

REMAINING QUESTIONS

Regardless of the interpretation, physics is an experimental science and it would be highly desirable to test for the effects discussed here. Experiments of this kind may be difficult, and one might ask whether or not there are any feasible experiments that can test for effects of this kind outside of the light cone. Quantum mechanics and quantum field theory predict these effects, but one might also ask whether or not they would be observed if such an experiment were to be performed. And perhaps the most interesting question is whether or not there are any alternative theories that agree with all of the other predictions of quantum electrodynamics but do not have this property. Further research of this kind would be a natural extension of the earlier work on Bell's inequality.

REFERENCES

1. R.P. Feynman, Phys. Rev. **76**, 167 (1949).
2. J.D. Franson, Proceedings of the 9th Rochester Conference on Coherence and Quantum Optics, Rochester, NY, June 11, 2007 (N.P. Bigelow, J.H. Eberly, and C.R. Stroud, Jr., eds., American Institute of Physics, 2008).
3. J.D. Franson, J. of Mod. Optics **55**, 2117 (2008).
4. E.A. Power and T. Thirunamachandran, Phys. Rev. A **56**, 3395 (1997).
5. J.D. Franson and M.M. Donegan, Phys. Rev. A **65**, 052107 (2002).
6. R.J. Glauber, Phys. Rev. **130**, 2529 (1963).
7. V.P. Bykov and V.I. Tatarskii, Phys. Lett. A **136**, 77 (1989).
8. P.W. Milonni, D.F.V. James, and H. Fearn, Phys. Rev. A **52**, 1525 (1995).
9. R.P. Feynman, *Quantum Electrodynamics* (Benjamin, 1962).
10. S.J. Summers and R. Werner, Phys. Lett. A **110**, 257 (1985).
11. B. Reznik, Found. of Phys. **33**, 167 (2003).

Weak values and entanglement concentration

David Menzies and Natalia Korolkova

School of Physics and Astronomy, University of St. Andrews, North Haugh, St. Andrews KY16 9SS, UK.

Abstract. Weak measurements are a special class of indirect measurements where the signal system is both pre and post-selected in particular quantum states. The accompanying probe system records information about the associated weak value of the interaction between the pre and post-selection and is modified by it. We provide a review of the operational application of weak measurement formalism to entanglement concentration protocols.

Keywords: Weak values, weak measurements, entanglement concentration
PACS: 03.65.Ud, 03.65.Ta, 03.67.Ac

INTRODUCTION

Any physical theory makes contact with empirical observations through the observable numbers it predicts. In quantum theories, there are three types of observable numbers: eigenvalues or measurement results, expectation values and weak values [1]. While the former two are established components of the theory of measurement in standard quantum mechanics, the latter weak values only arise in a particular class of indirect measurements known as weak measurements [1, 2]. Such measurements typically involve the interaction between two quantum systems, the signal and probe, where the signal system is subjected to independent pre and post-selection processes [1]. In the standard presentation [1], the probe system has an infinite dimensional Hilbert space with canonical observables $\hat{Q},\hat{P}$ where $[\hat{Q},\hat{P}] = i\hbar$ and the interaction between signal and probe is mediated by the Hamiltonian $\hat{H}_I = \hbar\kappa(t)\hat{A}\otimes\hat{P}$, where $\hat{A}$ is the observable of interest. For simplicity, it is assumed that $\kappa(t) = \text{const}$ and that interaction persists for $T = t_f - t_i$ seconds. In addition, it is assumed that all of the systems have vanishing free Hamiltonians. The initial state is further assumed to be Gaussian superposition of pointer eigenstates $|\Psi_i\rangle \propto \int dQ\, e^{-Q^2/4\Delta^2 Q}|Q\rangle$, where $\Delta^2 Q$ is the uncertainty in pointer observable and so the global initial state is $|\Phi_1,\Psi_i\rangle$. This then evolves in accordance with $\hat{U}_I = \exp\left(-i\kappa\int_{t_i}^{t_f}\hat{A}\otimes\hat{P}\,dt\right) = e^{-i\kappa T\hat{A}\hat{P}}$ before the signal is post-selected in $|\Phi_2\rangle$. The final probe state is then

$$|\Psi_f\rangle \propto \sum_{n=0}^{\infty}\frac{(-i\kappa T)^n}{n!}(A_W)^n\hat{P}^n|\Psi_i\rangle + \sum_{m=2}^{\infty}\frac{(-i\kappa T)^m}{m!}\left(\frac{\langle\Phi_2|\hat{A}^m|\Phi_1\rangle}{\langle\Phi_2|\Phi_1\rangle} - (A_W)^m\right)\hat{P}^m|\Psi_i\rangle, \quad (1)$$

where the the weak value $\hat{A}$ on a system which is pre-selected in the state $|\Phi_1\rangle$ and post-selected in $|\Phi_2\rangle$ is defined as $A_W = \langle\Phi_2|\hat{A}|\Phi_1\rangle/\langle\Phi_2|\Phi_1\rangle$ and need not be a real number. This concept is valid provided that $\{|\Phi_1\rangle,|\Phi_2\rangle\}$ are not orthogonal. From a physical point of view, such weak values are regarded as the possible values of the observable at

CP1110, *Quantum Communication, Measurement and Computing (QCMC), The Ninth International Conference,* edited by A. Lvovsky

intermediate times between the pre and post-selection [1, 3]. The state of the measuring device records the so-called weak value only if the second term in (1) vanishes, which constrains both the magnitude of the coupling constant and the noise in the initial state of the measuring device. This follows from considering the momentum wavefunction of $\Psi_i(Q) = e^{-Q^2/4\Delta^2 Q}$ as $\tilde{\Psi}_i(P) = e^{-P^2\Delta^2 Q}$ and so

$$e^{-P^2\Delta^2 Q}\left(\frac{\langle\Phi_2|e^{-i\kappa_T\hat{A}P}|\Phi_1\rangle}{\langle\Phi_2|\Phi_1\rangle} - e^{-i\kappa_T A_W P}\right) \approx 0 \;\forall P \in (-\infty,\infty). \tag{2}$$

If the coupling constant is small ($\kappa_T << 1$) then (2) are satisfied for small P. For large P, the above is can only be satisfied if the noise in the initial state of the measuring device is large i.e. $\Delta^2 Q >> 1$ and so the superposition is broad. Thus, the weak value is only encoded provided the coupling is weak and the Gaussian distribution is broad enough.

Since their original conception, weak values have enjoyed a great deal of theoretical investigation and eventual experimental confirmation. The former has revealed the great versatility of weak values and weak measurements in providing an explanatory basis for a number of physical phenomena. Indeed, weak values and weak measurements explain or appear in: fast and slow light effects [4], superluminal quantum tunneling effects [5] and optical telecommunications networks [6]. In the context of quantum information theory, weak values have found application in communication protocols [7] and in the control of transitions between different states [8]. Weak values have also been experimentally verified in the case of quantum optical systems [9, 10, 11].

WEAK VALUES AND ENTANGLEMENT CONCENTRATION

Entanglement concentration protocols are designed to augment the entanglement content of a particular entangled state. For example, Procrustean entanglement concentration [12] is where the initial entangled state is transformed into another entangled state with the Schmidt basis is preserved $|\psi_i\rangle = \sum_{k=1}^{d} s_k|k,k\rangle \rightarrow |\psi_f\rangle = \sum_{k=1}^{d} t_k|k,k\rangle$, with the output state being more entangled than the input. The fact that weak measurements can have an operational use in procrustean entanglement concentration was demonstrated in [13]. In [13], the probe is initially prepared in an entangled state $|\psi_i\rangle \in \mathscr{H}^{\otimes 2}$ with *Schmidt* decomposition $|\psi_i\rangle = \sum_{k=1}^{K} s_k|a_k\rangle|a_k\rangle$. The usual properties are assumed with $\{s_k\}_{k=1}^{K}$ obeying $\sum_{k=1}^{K} s_k^2 = 1$, $s_k \geq 0\,\forall k \in [1,K]$ and $\langle a_k|a_j\rangle = \delta_{jk}$. One subsystem of the probe interacts with the system initially prepared in $|i\rangle$. This mixing is provided by the interaction Hamiltonian $\hat{H}_I = \hbar\kappa\hat{K}\otimes\hat{O}$, with, for simplicity, $\kappa = \text{const}$. In addition, the observable $\hat{K}$ is required to admit the *Schmidt* basis as an eigen-basis with $\hat{K}|a_k\rangle = \lambda_k|a_k\rangle$. Consequently, assuming the pre and post-selected states on the ancilla are $|i\rangle$ and $|f\rangle$, respectively, meant that the final probe state was

$$|\psi_f\rangle = \frac{(\hat{I} - i\varphi\, O_W(\hat{I}\otimes\hat{K}))\,|\psi_i\rangle}{\sqrt{1 + 2\varphi\,\Im(O_W)\langle\psi_i|(\hat{I}\otimes\hat{K})|\psi_i\rangle}}, \tag{3}$$

where $\varphi = \kappa t$. In contrast to the usual weak condition, this result is obtained by assuming that

$$s_k \frac{\langle f|e^{-i\kappa_T \lambda_k \hat{O}}|i\rangle}{\langle f|i\rangle} \approx s_k(1 - i\kappa_T \lambda_k O_W)\ \forall k \in [1, K], \tag{4}$$

this restriction is required to show an increase of entanglement of $|\psi_i\rangle$ in the absence of knowledge of the actual s_k [13].

Entanglement concentration could only be verified by comparing the entanglement entropies before and after the weak measurement [13] $\mathscr{E}(|\Psi\rangle) = \{S \circ \mathrm{Tr}_j\}|\Psi\rangle\langle\Psi| = -\mathrm{Tr}\left(\hat{\rho}_j \ln \hat{\rho}_j\right)$. After some algebra [13], the result becomes

$$\frac{S(\hat{\sigma}_f)}{S(\hat{\sigma}_i)} = \frac{1 + 2\varphi\,\Im(O_W)\mathrm{Tr}(\hat{K}\hat{\omega})}{1 + 2\varphi\,\Im(O_W)\mathrm{Tr}(\hat{K}\hat{\sigma}_i)}, \tag{5}$$

where $\hat{\sigma}_i = \mathrm{Tr}_1(|\psi_i\rangle\langle\psi_i|)$ and $\hat{\sigma}_f = \mathrm{Tr}_1(|\psi_f\rangle\langle\psi_f|)$. Furthermore, note that $\hat{\omega} = \hat{\sigma}_i \ln \hat{\sigma}_i / \mathrm{Tr}(\hat{\sigma}_i \ln \hat{\sigma}_i)$ is identified as a density operator in its own right. Thus, the entanglement content of the probe is altered if and only if $S_f \neq S_i$ and so $S_f/S_i \neq 1$, which is true if both $\Im(O_W)$ and $\mathrm{Tr}(\hat{K}(\hat{\sigma}_i - \hat{\omega}))$ cannot be zero. The requirement that $\Im(O_W) \neq 0$ is obvious from (3) as it accompanies a non-unitary transformation of the probe state. This follows from the properties of entanglement measures which are designed to be non-increasing under local unitary operations. Furthermore, the second simultaneous requirement that $\mathrm{Tr}(\hat{K}(\hat{\sigma}_i - \hat{\omega})) \neq 0$ means that that $\hat{K}$ must be able to distinguish states of different entropies. This is because the entropy of $\hat{\sigma}_i$ is identical to that of $\hat{\omega}$ only when the global probe state is either separable or maximally entangled [13].

Thus, entanglement concentration would occur in this regime provided $S_f/S_i > 1$ and so $\Im(O_W)\mathrm{Tr}(\hat{K}(\hat{\omega} - \hat{\sigma}_i)) > 0$. This condition acts to constrain the possible ancilla ingredients employed in the entanglement concentration protocol [13]. Thus, the weak value formalism can be used in a quantum information context to single out individual examples of entanglement concentration protocols. Consequently, one can view $\Im(O_W)$ as a calculational aid allowing one to pick out suitable ancilla ingredients. In addition, the constraint $\Im(O_W) > 0$ also provides a measurement-based success condition for the protocol thereby allowing Alice and Bob to decide whether they keep the shared entangled state or not. To see this, let $|i\rangle$ be a fixed pre-selected state of the ancilla and assume that the interaction Hamiltonian is fixed with $\hat{H}_I = \hbar\kappa\hat{K}\hat{O}$. We further assume that the post-selected state in each run of the protocol is one of the eigenstates of an operator $\hat{F}$ i.e. $|f\rangle$. Different runs of the protocol result in different measurement outcomes f (from a set F) and imprint a different weak value $O_W(f) = \langle f|\hat{O}|i\rangle/\langle f|i\rangle$. Thus, we can establish a measurement-based success condition from

$$\langle i|\hat{O}|i\rangle = \sum_{f\in F}\langle i|f\rangle\langle f|\hat{O}|i\rangle = \sum_{f\in F} P(f|i)O_W(f), \tag{6}$$

where $P(f|i) = |\langle f|i\rangle|^2$ is the probability of finding $|f\rangle$ given $|i\rangle$. Now, given $\Im(\langle i|\hat{O}|i\rangle) = \sum_{f\in F} P(f|i)\Im(O_W(f)) = 0$ and since $P(f|i) \geq 0$ for all possible f, then either $\Im(O_W(f)) = 0$ for all f or $\Im(O_W(f)) > 0 \forall f \in F_+,\ \Im(O_W(f)) < 0 \forall f \in F_-$.

That is, there is a subset of measurement results that correspond to an increase in shared entanglement and a subset corresponding to a decrease in entanglement. Thus, the selection criterion is equivalent to the existence of a measurement-based success condition. Accordingly, to each fixed configuration we can calculate a success probability of the protocol. This is the probability of obtaining any $f \in F_+$ and it can be obtained from the probability (or probability density if the measurement results form a continuum) of obtaining a particular post-selected state $|f\rangle$ denoted as $P(f)$. Then, the probability of success, i.e. imprinting a weak value and increasing the shared entanglement is $P_S = \sum_{f \in F_+} P(f)$ and the corresponding probability of failure is then $P_{Fail} = 1 - P_S$. The capacity of $\Im(O_W)$ as a calculational tool in entanglement concentration protocols has been previous noted in [14] in the context of Gaussian continuous-variable entangled states.

CONCLUDING REMARKS AND FUTURE WORK

In conclusion, we have provided a review of the operational application of weak measurement formalism to entanglement concentration protocols. The utility of the weak measurement formalism to these quantum information protocols motivates a larger comprehensive investigation into their application to quantum information. For example, can weak values be considered a quantum resource for state engineering? In particular, can one establish a rigorous resource theory related to weak values and measurements in a similar vein to that of entanglement? Exploring these questions is the goal of future research.

ACKNOWLEDGEMENTS

This work was supported by the *Engineering and Physical Sciences Research Council* and by the EU project FP7-212008 COMPAS.

REFERENCES

1. Y. Aharonov and L. Vaidman. *Phys. Rev. A.* **41**, **1**, *11*, 1990.
2. Y Aharonov and L Vaidman. *Physica Scripta* **T76**, *85*, 1998.
3. Y. Aharonov and A. Botero. *Phys. Rev. A.* **72**, *052111*, 2005.
4. D. R. Solli, C. F. McCormick, R. Y. Chiao, S. Popescu, and J. M. Hickmann. *Phys. Rev. Lett.* **92**, **4**, *043601*, 2004.
5. Y Aharonov, N Erez, and B Reznik. *Phys. Rev. A.* **65**, *052124*, 2002.
6. N Brunner, A Acín, D Collins, N Gisin, and V Scarani. *Phys. Rev. Lett.* **91**, **18**, *180402*, 2003.
7. A Botero and B Reznik. *Phys. Rev. A.* **61**, *050301*, 2000.
8. A D Parks. *J. Phys. A: Math. Gen.* **33**, *2555*, 2000.
9. G. J. Pryde, J. L. O'Brien, A. G. White, T. C. Ralph, and H. M. Wiseman. *Phys. Rev. Lett.* **94**, **22**, *220405*, 2005.
10. T. C. Ralph. *Rep. Prog. Phys.* **69**, *853*, 2006.
11. G. S. Agarwal and P. K. Pathak. *Phys. Rev. A* **75**, *032108*, 2007.
12. C. H. Bennett, H. J. Bernstein, S. Popescu, and B. Schumacher. *Phys. Rev. A* **53**, *4*, *2046*, 1996.
13. D. Menzies and N. Korolkova. *Phys. Rev. A* **77**, *062105*, 2008.
14. D. Menzies and N. Korolkova. *Phys. Rev. A* **76**, *062310*, 2007.

A unified bound on the rate of quantum dynamics

Lev B. Levitin and Tommaso Toffoli

Boston University, ECE Dept., 8 Saint Mary's St., Boston MA 02215

Abstract. Two different bounds on the least time required for a quantum state to evolve to an orthogonal one are reconsidered and unified. It is shown that these bounds can be attained only when they coincide. Otherwise, they still remain tight, but can be approached only asymptotically.

Keywords: Quantum dynamics, Mandelstam–Tamm bound, Margolus–Levitin bound
PACS: 03.65.-w, 03.67.-a

The question of how fast a quantum state can evolve inot a state orthogonal to it has attracted a considerable attention in connection with quantum measurements and quantum information processing. Indeed, since only orthogonal states can be unambiguously distinguished, a transition from a state to another state orthogonal to the previous one can be considered as an elementary step in a computational process[10].

Starting with the classical result of Mandelstam and Tamm[16], it was later shown by Anandan and Aharonov[1] and Vaidman[21] that the minimum time τ required for arriving to an orthogonal state is bounded by

$$\tau \geq h/4\Delta E, \tag{1}$$

where $(\Delta E)^2 = \langle\psi|H^2|\psi\rangle - (\langle\psi|H|\psi\rangle)^2$, and $|\psi\rangle$ is the wavefunction of the system. A different bound was obtained in [17], namely,

$$\tau \geq h/4E. \tag{2}$$

Here, $E = \langle\psi|H|\psi\rangle$ is the quantum-mechanical average energy of the system (the energy of the ground state is taken to be zero). Both bounds (1) and (2) are tight, and achieved for a quantum state such that $\Delta E = E$.

Since then, a vast literature has been devoted to various aspects of this problem (e.g., [2–9,11–15,18–20,22,23]). However, what remained unnoticed is the paradoxical situation of the existence of two bounds based on two different characteristics of the quantum state, seemingly independent of one another. Since the average energy E and the energy uncertainty ΔE play the most determinative role in quantum evolution, it is important to have a unified bound that would take into account both of these characteristics.

In all known cases where bounds (1) and (2) can be exactly attained, the ratio $\alpha = \frac{\Delta E}{E}$ equals 1. A question arises: what happens if $\alpha \neq 1$? Some authors just assumed without justification that the minimum orthogonalization time is

$$\tau_{\min} = \max\left(\frac{h}{4E}, \frac{h}{4\Delta E}\right) = \frac{h}{2(E+\Delta E-|E-\Delta E|}. \tag{3}$$

CP1110, *Quantum Communication, Measurement and Computing (QCMC), The Ninth International Conference,* edited by A. Lvovsky

In fact, the situation is not so simple. Bound (3), indeed, can only be achieved for $\alpha = 1$. However, this bound remains tight when $\alpha \neq 1$ as well, though in this case it is only asymptotically attainable.

Theorem 1 *Under the assumption that the ground energy of a quantum system is zero,*
1. The only pure state that attains bound (1) is the two-level state

$$|\psi\rangle = \tfrac{1}{\sqrt{2}}(|\psi_0\rangle + |\psi_1\rangle), \quad (where \quad H|\psi_k\rangle = kE_1 \; for\, k = 0,1); \tag{4}$$

2. The only pure state that attains bound (2) is likewise state (4).

Proof. To prove the first statement we shall use the trigonometric inequality

$$\cos x \geq 1 - \tfrac{4}{\pi^2} x \sin x - \tfrac{2}{\pi^2} x^2, \tag{5}$$

which is valid for any real x. Note that (5) turns into an equality iff $x = 0$ or $x = \pm\pi$.

Let the initial state be $|\psi(0)\rangle = \sum_{n=0}^{\infty} c_n |E_n\rangle$, where the $|E_n\rangle$ are energy eigenstates of the system and $\sum_{n=0}^{\infty} |c_n|^2 = 1$. Then

$$|S(t)|^2 = |\langle\psi(0)|\psi(t)\rangle|^2 = \sum_{n,n'=0}^{\infty} |c_n|^2 |c_{n'}|^2 e^{-i\frac{E_n - E_{n'}}{\hbar/t}} = \sum_{n,n'=0}^{\infty} |c_n|^2 |c_{n'}|^2 \cos \tfrac{E_n - E_{n'}}{\hbar/t}. \tag{6}$$

Using inequality (5), we obtain

$$\begin{aligned} |S(t)|^2 &\geq 1 - \tfrac{4}{\pi^2} \sum_{n,n'=0}^{\infty} |c_n|^2 |c_{n'}|^2 \tfrac{E_n - E_{n'}}{\hbar/t} \sin \tfrac{E_n - E_{n'}}{\hbar/t} - \frac{2}{\pi^2} \sum_{n,n'=0}^{\infty} |c_n|^2 |c_{n'}|^2 \left(\frac{E_n - E_{n'}}{\hbar/t}\right)^2 \\ &= 1 + \frac{4t}{\pi^2} \frac{d|S(t)|^2}{dt} - \left(\frac{\Delta E}{\hbar/2t}\right)^2. \end{aligned} \tag{7}$$

Note that, if $S(t) = 0$, then $\frac{d|S(t)|^2}{dt} = S(t)\frac{dS^*(t)}{dt} + S^*t)\frac{dS(t)}{dt} = 0$. Thus, at a time τ such that $S(\tau) = 0$, the second term in (7) vanishes, and we obtain $0 \geq 1 - \frac{4\tau^2}{\pi^2\hbar^2}(\Delta E)^2$, which yields inequality (1), so that this is just another way to derive that bound. However, for (1) to turn into an equality, inequality (5) must turn into an equality for every term of the double summation (6). Hence, either $x_{nn'} = (E_n - E_{n'})\tau/\hbar = 0$ or $x_{nn'} = (E_n - E_{n'})\tau/\hbar = \pm\pi$ for all n, n' such that $c_n \neq 0, c_{n'} \neq 0$. It follows that, to attain bound (1), $|\psi(0)\rangle$ must be a superposition of only two energy eigenstates with energies $E_0 = 0$ and E_1.

To prove the second statement, we repeat briefly the derivation given in [17]. This time we use the trigonometric inequality

$$\cos x \geq 1 - \tfrac{2}{\pi}(x + \sin x), \tag{8}$$

valid for all $x \geq 0$. Again, (8) turns into an equality only for $x = 0$ or $x = \pi$. Then

$$\mathrm{Re}\, S(t) = \sum_{n=0}^{\infty} |c_n|^2 \cos \tfrac{E_n t}{\hbar} \geq \sum_{n=0}^{\infty} |c_n|^2 \Big[1 - \tfrac{2}{\pi}\big(\tfrac{E_n t}{\hbar} + \sin \tfrac{E_n t}{\hbar}\big)\Big] = 1 - \tfrac{2Et}{\pi\hbar} + \tfrac{2}{\pi} \mathrm{Im}\, S(t). \tag{9}$$

At time τ, $\mathrm{Re}\,S(\tau) = \mathrm{Im}\,S(\tau) = 0$. Hence $0 \geq 1 - \frac{2Et}{\pi\hbar}$, which results in bound (2). However, to actually attain this bound, inequality (8) must become an equality for every term of the sum (9); that is, for any n, either $x_n = E_n\tau/\hbar = 0$ or $x_n = E_n\tau/\hbar = \pi$. This is possible iff $|\psi(0)\rangle$ has the form (4). ■

Results of Theorem 1 also follow from the analyses presented in [9, 20, 23].

Theorem 1 shows that bounds (1) and (2) can only be attained by a state for which $\alpha = \frac{\Delta E}{E} = 1$. Such a state is unique up to degeneracy of the energy level E_1 and arbitrary phase factors for $|\psi_0\rangle$ and $|\psi_1\rangle$. It follows from that theorem that there exists no initial state $|\psi(0)\rangle$ with $\alpha \neq 1$ that would attain bound (3), i.e., no state can evolve into an orthogonal state in the minimum time given by (1) or (2). The question to be answered is how close it is possible to approach the larger of those bounds. Let us rewrite (3) in a different form,

$$\tau_{\min} = \max\left(\frac{h}{4E}, \frac{h}{4\Delta E}\right) = \frac{h(1+e^{|\ln\alpha|})}{4E(1+\alpha)}. \tag{10}$$

Theorem 2 *For any $\alpha \neq 1$ and any $\varepsilon > 0$, there exists a state $|\psi(0)\rangle$ such that $\langle\psi(0)|\psi(\tau)\rangle = 0$ at time*

$$\tau = \frac{h(1+e^{|\ln\alpha|})}{4E(1+\alpha)}(1+\varepsilon) \tag{11}$$

Proof. We will show that in both cases, $\alpha < 1$ and $\alpha > 1$, there exist families of initial states that approach limit (10) arbitrarily close.

1. Let $\alpha < 1$. Consider a family of states

$$|\psi(0)\rangle = c_0|0\rangle + c_1|E_1\rangle + c_2|E_2\rangle. \tag{12}$$

Let $p_i = |c_i|^2$, and introduce dimensionless variables (rotation angles of the states $|E_i\rangle$)

$$x_i(t) = 2\pi E_i\, t/h, \quad i = 0, 1, 2. \tag{13}$$

Then $S(\tau) = \langle\psi(0)|\psi(\tau)\rangle = 0$ iff $x_i(\tau) = x_i$ such that

$$p_1 \sin x_1 + p_2 \sin x_2 = 0 \text{ and } p_0 + p_1\cos x_1 + p_2 \cos x_2 = 0, \text{ (with } p_0 + p_1 + p_2 = 1). \tag{14}$$

Let $0 < p_0 \ll 1$ be a parameter of the family of states (12) (one cannot set $p_0 = 0$, since that would change the ground energy level). When p_0 is small the values of x_1 and x_2 differ almost exactly by π. Let $x_2 = \pi + x_1 - \delta\sin x_1$, with $\delta \ll 1$. Then, from (14),

$$p_1 = \tfrac{1}{2} - \tfrac{\delta}{4}(1+\cos x_1) + O(\delta^2),\; p_2 = \tfrac{1}{2} - \tfrac{\delta}{4}(1-\cos x_1) + O(\delta^2),\; p_0 = \tfrac{\delta}{2} + O(\delta^2). \tag{15}$$

Using (15), we can calculate the average value $\langle x\rangle = \frac{2\pi\tau}{h}E$, namely,

$$\langle x\rangle = p_1x_1 + p_2x_2 = \tfrac{\pi}{2} + x_1 - \tfrac{\delta}{4}[2x_1 - 2\sin x_1 + \pi(1-\cos x_1)] + O(\delta^2). \tag{16}$$

The standard deviation $\Delta x = \frac{2\pi\tau}{h}\Delta E$ is obtained as

$$\Delta x = \sqrt{p_1x_1^2 + p_2x_2^2 - (p_1x_1 + p_2x_2)^2} = \tfrac{\pi}{2} + \tfrac{\delta}{2\pi}[x_1(\pi + x_1) - \pi\sin x_1] + O(\delta^2). \tag{17}$$

On the other hand, $\alpha = \frac{\Delta E}{E} = \frac{\Delta x}{\langle x\rangle} = \frac{\pi}{\pi+2x_1} + O(\delta)$. Hence, $x_1 = \frac{\pi}{2}\left(\frac{1}{\alpha}-1\right) + O(\delta)$. Now, using (17) and taking into account the last expression from (15) we get
$\Delta x = \frac{\pi}{2} + p_0\left[\frac{\pi}{4}\left(\frac{1}{\alpha^2}-1\right) - \sin\frac{\pi}{2}\left(\frac{1}{\alpha}-1\right)\right] + O(p_0^2)$, which results in
$\tau = \frac{h\Delta x}{2\pi\Delta E} = \frac{h}{4\Delta E}\left[1 + \frac{p_0}{2}\left(\frac{1}{\alpha^2} - 1 - \frac{4}{\pi}\sin 2\pi\left(\frac{1}{\alpha}-1\right)\right) + O(p_0^2)\right]$.
Finally, choosing $p_0 < 2\varepsilon\left[\frac{1}{\alpha^2} - 1 - \frac{2}{\pi}\sin\frac{\pi}{2}\left(\frac{1}{\alpha}-1\right)\right]^{-1}$, we obtain $\frac{h}{4\Delta E} < \tau \leq \frac{h}{4\Delta E}(1+\varepsilon)$.

2. Let $\alpha > 1$. Consider a family of states

$$|\psi(0)\rangle = c_0|0\rangle + c_1|E_1\rangle + c_{2k+1}|E_{2k+1}\rangle, \tag{18}$$

where $k = 1, 2, \ldots;\ E_{2k+1} = (2k+1)E_1$;

$$|c_0|^2 = p_0 = \tfrac{1}{2};\ |c_1|^2 = p_1 + \tfrac{1}{2}\left(1 - \tfrac{\beta}{k^2}\right);\ |c_{2k+1}|^2 = p_{2k+1} = \tfrac{\beta}{2k^2}. \tag{19}$$

Using the dimensionless variables (13), it is readily seen from (18) and (19) that $S(\tau) = 0$ at the least time τ for which

$$x_1(\tau) = x_1 = \tfrac{2\pi\tau}{h}E_1 = \pi \quad \text{and} \quad x_{2k+1} = \tfrac{2\pi\tau}{h}(2k+1)E_1 = \pi(2k+1). \tag{20}$$

Then $\langle x\rangle = p_1x_1 + p_{2k+1}x_{2k+1} = \frac{\pi}{2}\left(1 + \frac{2\beta}{k}\right)$ and
$\Delta x = \left[p_1x_1^2 + p_{2k+1}x_{2k+1}^2 - (p_1x_1 + p_{2k+1}x_{2k+1})^2\right]^{1/2} = \frac{\pi}{2}\left(1 + 4\beta + \frac{2\beta}{k}\right) + O\left(\frac{1}{k^2}\right)$.
Hence, $\alpha = \frac{\Delta x}{\langle x\rangle} = 1 + 4\beta + O\left(\frac{1}{\alpha}\right)$ and $\beta = \frac{\alpha-1}{4} + O\left(\frac{1}{k}\right)$.

Since $\langle x\rangle = \frac{2\pi\tau}{h}E$, it follows that $\tau = \frac{h}{4E}\left[1 + \frac{\alpha-1}{2k} + O\left(\frac{1}{k^2}\right)\right]$. Thus, choosing a sufficiently large $k > \frac{\alpha-1}{2\varepsilon}$ guarantees that $\frac{h}{4E} < \tau \leq \frac{h}{4E}(1+\varepsilon)$. ∎

REFERENCES

1. Anandan, J, Y Aharonov, *Phys. Rev. Lett.* **65** (1990), 1967–1760.
2. Andrecut, M, MK Ali, *Int. J Theor. Phys.* **43** (2004), 969–974.
3. Borras, A, C Zander, AR Plastino, M Casas, A Plastino, *EPL* **81** 30007 (2007).
4. Brody, DC, DW Hook, *J Phys. A* **39** (2006), L167–L170.
5. Carlini, A, A Hosoya, T Koike, Y Okudaira, *Phys. Rev. Lett.* **96** 060503 (2006).
6. Carlini, A, A Hosoya, T Koike, Y Okudaira, `arXiv:quant-ph/06080039v4`, 11 April 2007.
7. Giovannetti, V, S Lloyd, L Maccone, *Phys. Review A* **67** 052109 (2003).
8. Giovannetti, V, S Lloyd, L Maccone, *Europhys. Lett* **62** (2003), 615–621.
9. Kosiński, P, M Zych, *Phys. Review A* **73** 024303 (2006).
10. Levitin, Lev B, *Int. J Theor. Phys.* **21** (1982), 299-309.
11. Levitin, LB, T Toffoli, Z Walton, in *Quantum Commmunication, Measurement and Computing* (J Schapiro O Hirota, eds.), Rinton 2003, 457–459.
12. Levitin, L, T Toffoli, Z Walton, *Int. J Theor. Phys.* **44** (2005), 965–970.
13. Lloyd, S, *Nature* **406** (2000), 1047–1054.
14. Lloyd, *Phys. Rev. Lett.* **88** 237901 (2002).
15. Luo, S, *Physica D* **109** (2004), 1–7.
16. Mandelstam, L, I Tamm, *J Physics* **9**, Moscow (1945), 249–254.
17. Margolus, N, LB Levitin, *Physica D* **120** (1998), 188–195.
18. Pati, AK, *Phys. Lett. A* **262** (1999), 296–301.
19. Pfeifer, P, *Phys. Rev. Lett.* **70** (1993), 3365–3368.
20. Söderholm, J, B Gunnar, T Tedros, and A Trifonov, *Phys. Review A* **59** (1999), 1788–1790.
21. Vaidman, L, *Am. J Phys.* **60** 1992, 182–183.
22. Zander, C, AR Plastino, A Plastino, M Casas, *J Phys. A* **40** (2007), 2861–2872.
23. Zielinski, B, M Zych, *Phys. Review A* **74** 034301 (2006).

Storage and transmission of entanglement in a spin chain

F. Plastina* and T. J. G. Apollaro†

*Dip. Fisica, Università della Calabria, & INFN - Gruppo collegato di Cosenza, 87036 Arcavacata di Rende (CS) Italy
†Dipartimento di Fisica, Università di Firenze, Via G. Sansone 1, I-50019 Sesto Fiorentino (FI), Italy

Abstract. We show that a local control of the effective magnetic field in a spin chain allows for an efficient storage of bipartite entanglement and for its transmission "on demand" towards a selected receiving site.

Keywords: Quantum communication, entanglement, spin systems, Anderson localization
PACS: 03.67.Hk, 03.67.Mn, 05.50.+q

INTRODUCTION

Spin systems provide ideal models to study the static and dynamical properties of entanglement in many body systems as they can be naturally thought of as qubit registers and exploited as quantum channels (or coherent data bus) [1]. They also constitute paradigmatic models in the study of magnetism and magnetic ordering at quantum phase transitions, so that they also form a playground to explore the relationship between entanglement and criticality [2]. We consider an XX $S = 1/2$ spin chain in transverse field. This model-system displays a quantum phase transition separating a disordered (paramagnetic) phase from a topologically quasi-long-range-ordered one. We will work in the disordered phase only, and study the effect of local magnetic impurities on the behavior of the system. We show that such defects give rise to an Anderson-like localization of entanglement [3], as they induce the appearance of spatially localized levels, whose presence can be exploited to store quantum information. Furthermore, we show that the ability to control the local effective magnetic field allows one to manipulate the dynamics of entanglement in order to achieve its transfer on demand along the chain, towards a selected site [4].

THE MODEL

We consider a 1-D XX spin-$\frac{1}{2}$ system, with periodic boundary conditions. This spin ring is placed in an effective external magnetic field which is homogeneous everywhere but for at most two sites (defects) l_1 and l_2:

$$H = -\frac{\omega_0}{2}\sum_{i=-\frac{N}{2}}^{\frac{N}{2}}\sigma_z^i - J\sum_{i=-\frac{N}{2}}^{\frac{N}{2}}(\sigma_x^i\sigma_x^{i+1}+\sigma_y^i\sigma_y^{i+1}) - \sum_{l=l_1,l_2}\frac{\alpha_l}{2}\sigma_z^l \tag{1}$$

CP1110, *Quantum Communication, Measurement and Computing (QCMC), The Ninth International Conference,* edited by A. Lvovsky

where σ_i are the Pauli matrices, while ω_0 is the level spacing for every qubit, except for those residing at l_i, which are given by $\omega_0 + \alpha_{l_i}, i = 1,2$. Setting the exchange constant as our energy unit ($J = 1$), we stay in the disordered phase $\omega_0 > 1$, in which (for $\alpha_i = 0$) the ground state of the chain is the purely factorized state $|0\rangle^{\otimes N}$.

We consider the general problem of sending entanglement along the chain. In particular, we assume that the sender, residing at the site s of the chain, prepares a singlet state between its qubit and an external (uncoupled) qubit. The aim is to swap the states of the sender qubit and of a receiving one, located at site r. The quality of entanglement transmission is given by the concurrence between the r-th and the external qubits, denoted by $C_r(t)$, which has been shown [1] to be related to the transmission amplitude to send an excitation from s to r: $C_r(t) = |\langle r|e^{-iHt}|s\rangle|$, where we denote by $|n\rangle = |1\rangle_n \otimes |0\rangle^{\otimes(N-1)}$ the state with one reversed spin at site n.

ENTANGLEMENT LOCALIZATION AT ONE DEFECT

We first assume that there is only one defect at site l_1 (while $\alpha_{l_2} = 0$) and show that this can be used to store entanglement. Indeed, the defect produces the appearance of a new eigen-state, given by, $|\Psi_{l_1}\rangle = \sum_n b_n |n\rangle$, with amplitudes

$$b_n = \left(1+\alpha_{l_1}^2\right)^{-1/4} \sqrt{\alpha_{l_1}} \exp[-\xi|n-l_1|] \tag{2}$$

where $\xi = -\ln\left(\sqrt{1+\alpha_{l_1}^2} - |\alpha_{l_1}|\right)$. This expression implies that this state is spatially localized (à la Anderson) with a localization length given by ξ^{-1}, which characterizes the spatial extension of the wave function around the site l_1 (and which goes below the lattice spacing, $\xi^{-1} < 1$, already for $\alpha_{l_1} \geq 1.175$).

Once the sender prepares its singlet state involving the s-th spin in order to store the entanglement, what it's really (approximately) prepared is, in fact, precisely this localized state, which, being an eigen-state of H, does not evolve in time. This intuition can be confirmed rigorously as, in the presence of the defect, the concurrence can be written (to second order in $1/\alpha$) as

$$C_r(t) \simeq 1 - \frac{1}{2\alpha^2} \quad \text{if } r = l_1\,, \qquad C_r(t) \simeq \frac{r}{\alpha t} J_r(t) \quad \text{if } r \neq l_1 \tag{3}$$

where J_r is the Bessel function of order r. These results are easily interpreted: entanglement is stored at the defect site. Indeed, the concurrence is very high when the receiver coincides with the sender (and the defect); while, if one tries to retrieve the entanglement at a different site, the result is very poor. We presented here only an approximate expression for $C_r(t)$, but it is easy to obtain the exact result for any value of α_l, see [3]. Fig. 1 show that the concurrence is extremely peaked at the sender site. The secondary V-shaped propagation lines are essentially a reminiscence of the $\alpha_{l_1} = 0$ case, in which these would be the only existing lines, describing the distribution of information along the chain as a result of the spin wave propagation. In our case, they signal that the storage is not perfect since some information is carried away and, thus, lost. To characterize the

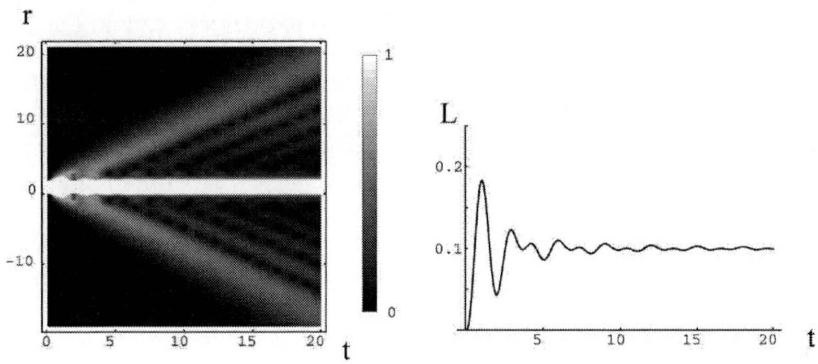

FIGURE 1. Left: space-time evolution of the concurrence C_r for $\alpha_{l_1} = 3$ with $s = l_1 = 0$. Right: time evolution of the leakage, as defined in Eq. (4)

ability to store entanglement at the defect, a leakage coefficient can be evaluated which quantifies how much can be obtained from other sites. This can be defined as

$$L := \sum_{r \neq l_1} C_r^2 \equiv 1 - C_{l_1}^2, \tag{4}$$

where the second equality comes from the monogamy relation for the pairwise tangles, which in our case gives $\sum_r C_r^2 = 1$, [6]. As shown in the right panel of Figure 1, after a transient regime, leakage does not change with time anymore and the amount of stored entanglement is quite large even for a moderate value of α_1.

ENTANGLEMENT TRANSMISSION BETWEEN TWO DEFECTS

To transmit entanglement along the chain towards a selected site, we need to single out the receiving spin by applying an extra magnetic field there. So we go back to the Hamiltonian of Eq. (1) including both of the extra fields. It is possible to show that with the two local fields present (and for positive α_{l_i}), the ground state of H can be written in the form of a "bi-localized" state; that is, a coherent superposition of two exponentially decaying parts, localized around the two defects:

$$|gnd\rangle = \sum_n b_n |n\rangle, \qquad \text{with} \quad b_n = \text{const}\left(K_1 e^{-\xi|n-l_1|} + K_2 e^{-\xi|n-l_2|}\right). \tag{5}$$

Here, the inverse localization length, $\xi = -\ln[-x_{gnd} - \sqrt{x_{gnd}^2 - 1}]$ is written in terms of the re-scaled ground state energy $x_{gnd} = E_{gnd} - \omega_0$, while the weighting factors for the two components are

$$K_i = \left(\frac{\alpha_{l_i}}{2\sqrt{x_{gnd}^2 - 1} - \alpha_{l_i}} \right)^{1/2}.$$

Their ratio measures the relative weights of the two localization region. For $\alpha_{l_1} \gg \alpha_{l_2}$ one gets $|K_2| \ll |K_1|$ and $|gnd\rangle$ is localized around l_1; similarly, for $\alpha_1 \ll \alpha_2$ the localization center is l_2. For equal defect strengths, $\alpha_{l_1} = \alpha_{l_2}$, one finds $K_2 = K_1$.

The form of the ground state immediately suggests a possibility to send quantum information between the two selected sites: if one is able to adiabatically control the two local fields α_{l_i}, the state of the two spins can be exchanged by exploiting the adiabatic passage technique [7]. The idea is to take $s = l_1$ as sender site and $r = l_2$ as target receiving site, and then to change the defect strengths adiabatically, so that the ground state of the system, having the general form given in Eq. (5), is adiabatically changed from the initial state $|gnd(i)\rangle \simeq |l_1\rangle$ localized at l_1, to the final state $|gnd(f)\rangle \simeq |l_2\rangle$, localized at l_2. This can be done by modifying the defect fields from the initial values

$$\alpha_{l_1}(i) \gg 1, \quad \alpha_{l_2}(i) \ll 1 \qquad \Rightarrow \quad b_n(i) \simeq \delta_{n,l_1},$$

to the final (reversed) ones

$$\alpha_{l_1}(f) \ll 1, \quad \alpha_{l_2}(f) \gg 1 \qquad \Rightarrow \quad b_n(f) \simeq \delta_{n,l_2}.$$

If the local fields are modified very slowly, the entire chain always remains in its instantaneous ground state, thus realizing the desired entanglement transfer. To ensure adiabaticity, the rates of change of the α's have to be much smaller than any other energy scale in the system. One can show that the most dangerous point occurs at the crossing of the values of the local fields, i.e. when $\alpha_{l_1}(t) = \alpha_{l_2}(t)$. However, if the adiabatic procedure is designed such that this crossing occurs for a very small α, than the adiabaticity can be preserved even for reasonably short transfer times.

To summarize, we have shown that entanglement storage and transmission can be achieved in a spin chain provided one is able to control and adiabatically manipulate the local Zeeman splitting of the spins. Since effective spin models can be realized with a variety of physical systems, including coupled micro-cavities [8] and atomic condensates in optical lattices [9], our results can be applied to all of these cases.

REFERENCES

1. S. Bose, Phys. Rev. Lett. **91**, 207901 (2003); S. Bose, Contemp. Phys. **48**, 13 (2007); V. Giovannetti and D. Burgarth, Phys. Rev. Lett. **96**, 030501 (2006); L. Amico *et al.*, Phys. Rev. A, **69**, 022304 (2004).
2. L. Amico *et al.*, Rev. Mod. Phys. **80** 517 (2008).
3. L.F. Santos, G. Rigolin, and C.O. Escobar, Phys. Rev A **69**, 042304 (2004); L.F. Santos, and G. Rigolin, Phys. Rev. A **71**, 032321 (2005); T. J. G. Apollaro and F.Plastina, Phys. Rev. A **74**, 062316 (2006); T. J. G. Apollaro and F. Plastina, Open Sys. & Information Dyn. **14**, 41 (2007).
4. F. Plastina and T. J. G. Apollaro, Phys. Rev. Lett. **99**, 177210 (2007).
5. T. J. G. Apollaro *et al.*, Phys. Rev. A **77**, 062314 (2008).
6. V. Coffman *et al.*, Phys. Rev. A **61**, 052306 (2001).
7. K. Bergmann, H. Theuer, and B. W. Shore , Rev. Mod. Phys. **70**, 1003 (1998); P. Král, I. Thanopulos, and M. Shapiro, Rev. Mod. Phys. **79**, 53 (2007).
8. M.J. Hartmann, F.G.S.L. Brandao, and M.B. Plenio, Phys. Rev. Lett. **99**, 160501 (2007).
9. M.Lewenstein *et al.*, Adv. Phys. **56**, 243 (2007) and references therein.

Second Law of Thermodynamics with Discrete Quantum Feedback Control

Takahiro Sagawa* and Masahito Ueda*,†

*Department of Physics, The University of Tokyo, 7-3-1, Hongo, Bunkyo-ku, Tokyo, 113-8654, Japan
†ERATO Macroscopic Quantum Control Project, JST, 2-11-16 Yayoi, Bunkyo-ku, Tokyo 113-8656, Japan

Abstract. We propose a new thermodynamic inequality which leads to the maximum work that can be extracted from multi-heat baths with the assistance of discrete quantum feedback control by "Maxwell's demon". The maximum work is determined by the free-energy difference and a generalized mutual information content between the thermodynamic system and the feedback controller.

Keywords: Quantum feedback control, Maxwell's demon, Nonequilibrium statistical mechanics
PACS: 03.67.-a, 05.30.-d, 05.70.Ln,

INTRODUCTION

Maxwell's demon, which has the capability to perform information processing at a microscopic level, has sparked numerous discussions concerning the foundation of the second law of thermodynamics [1]. It is also attracting renewed interest in the light of quantum information theory.

In a different context, quantum feedback control has attracted considerable attention for controlling and stabilizing a quantum system. It can be applied, for example, to squeezing an electromagnetic field, spin squeezing, and stabilizing macroscopic coherence. While the theoretical framework of quantum feedback control as a stochastic dynamic system is well developed, the possible thermodynamic gain of quantum feedback control has yet to be fully understood.

In the present study [2], we derive a new thermodynamic inequality which sets the fundamental limit on the work that can be extracted from a thermodynamic engine with the assistance of the demon. To derive the inequality, we combine nonequilibrium quantum-statistical mechanics and quantum measurement theory. The demon plays the role of a discrete quantum feedback controller, which performs a quantum measurement on a thermodynamic system and next performs a mechanical operation on the system depending on the measurement outcome.

SETUP

We consider a thermodynamic process for system S which can contact heat baths B_1, B_2, $\cdots$, B_n at respective temperatures T_1, T_2, $\cdots$, T_n. We assume that system S is in thermodynamic equilibrium in the initial and final states. For simplicity, we also assume

CP1110, *Quantum Communication, Measurement and Computing (QCMC), The Ninth International Conference,* edited by A. Lvovsky

that the initial and final temperature of S is given by $T \equiv (k_B\beta)^{-1}$. This can be realized by contacting S with, for example, B_1 in the preparation of the initial state and during equilibration to the final state; in this case $T = T_1$. We do not, however, assume that the system is in thermodynamic equilibrium between the initial and final states.

We assume that system S and heat baths B_m are as a whole isolated and that they only come into contact with some external mechanical systems and the feedback controller. Apart from the feedback controller, the total Hamiltonian can be written as

$$\hat{H}(t) = \hat{H}^{S}(t) + \sum_{m=1}^{n} (\hat{H}^{SB_m}(t) + \hat{H}^{B_m}), \tag{1}$$

where $\hat{H}^{SB_m}(t)$ is the interaction Hamiltonian between system S and heat bath B_m. The Hamiltonian $\hat{H}^{S}(t)$ describes a mechanical operation on S through such external parameters as an applied magnetic field or volume of the gas, and the Hamiltonian $\hat{H}^{SB_m}(t)$ describes, for example, the attachment (detachment) of an adiabatic wall or B_m to (from) S. We consider a time evolution from t_i to t_f, assume $\hat{H}^{SB_m}(t_i) = \hat{H}^{SB_m}(t_f) = 0$ for all m, and write $\hat{H}^{S}(t_i) = \hat{H}^{S}_{i}$ and $\hat{H}^{S}(t_f) = \hat{H}^{S}_{f}$. The time evolution of the total system with discrete quantum feedback control can be divided into the following five stages:

Stage 1 (*Initial state*) At time t_i, the initial state of S and that of B_m are in thermodynamic equilibrium at temperatures T and T_m, respectively. We assume that the density operator of the entire state is given by the canonical distribution $\hat{\rho}_i = \frac{\exp(-\beta\hat{H}^{S}_{i})}{Z^{S}_{i}} \otimes \frac{\exp(-\beta_1\hat{H}^{B_1})}{Z^{B_1}} \otimes \cdots \otimes \frac{\exp(-\beta_n\hat{H}^{B_n})}{Z^{B_n}}$, where $\beta_m \equiv (k_B T_m)^{-1}$ $(m = 1, 2, \cdots, n)$, $Z^{S}_{i} \equiv \mathrm{tr}\{\exp(-\beta\hat{H}^{S}_{i})\}$, and $Z^{B_m} \equiv \mathrm{tr}\{\exp(-\beta_m\hat{H}^{B_m})\}$. We denote the Helmholtz free energy of system S as $F^{S}_{i} \equiv -k_B T \ln Z^{S}_{i}$.

Stage 2 (*Unitary evolution*) From t_i to t_1, the entire system undergoes unitary evolution $\hat{U}_i = \mathrm{T}\exp\left(\int_{t_i}^{t_1} \hat{H}(t)dt/i\hbar\right)$. Let $\hat{\rho}_1 \equiv \hat{U}_i\hat{\rho}_i\hat{U}_i^{\dagger}$.

Stage 3 (*Measurement*) From t_1 to t_2, the feedback controller performs quantum measurement on S described by measurement operators $\{\hat{M}_k\}$ and obtains each outcome k with probability p_k. Let X be the set of outcomes k's, and $\{\hat{D}_k\}$ be POVM as defined by $\hat{D}_k \equiv \hat{M}_k^{\dagger}\hat{M}_k$; we then have $p_k = \mathrm{tr}(\hat{D}_k\hat{\rho}_1)$. The post-measurement density operator with outcome k as $\hat{\rho}_2^{(k)} \equiv \hat{M}_k\hat{\rho}_1\hat{M}_k^{\dagger}/p_k$, and define $\hat{\rho}_2 \equiv \sum_k p_k\hat{\rho}_2^{(k)}$. Note that our scheme can be applied not only to a quantum measurement, but also to a classical measurement which can be described by setting $[\hat{\rho}_1, \hat{D}_k] = 0$ for all k.

Stage 4 (*Feedback control*) From t_2 to t_3, the feedback controller performs a mechanical operation on S depending on outcome k. Let $\hat{U}_k$ be the corresponding unitary operator on the entire system, and $\hat{\rho}_3^{(k)} \equiv \hat{U}_k\hat{\rho}_2^{(k)}\hat{U}_k^{\dagger}$ be the density operator of the entire system at t_3 corresponding to outcome k. We define $\hat{\rho}_3 \equiv \sum_k p_k\hat{\rho}_3^{(k)}$. Note that the feedback control is characterized by $\{\hat{M}_k\}$ and $\{\hat{U}_k\}$.

Stage 5 (*Equilibration and final state*) From t_3 to t_f, the entire system evolves according to unitary operator $\hat{U}_f$ which is independent of outcome k. We assume that by t_f system S and heat bath B_m will have reached thermodynamic equilibrium at temperatures T and T_m, respectively. We denote as $\hat{\rho}_f$ the density operator of the final state of the entire system, which is related to the initial state as

$$\hat{\rho}_{\rm f} = \mathscr{E}(\hat{\rho}_{\rm i}) \equiv \sum_k \hat{U}_{\rm f}\hat{U}_k\hat{M}_k\hat{U}_{\rm i}\hat{\rho}_{\rm i}\hat{U}_{\rm i}^{\dagger}\hat{M}_k^{\dagger}\hat{U}_k^{\dagger}\hat{U}_{\rm f}^{\dagger}. \tag{2}$$

We emphasize that $\hat{\rho}_{\rm f}$ need not equal the rigorous canonical distribution $\hat{\rho}_{\rm f}^{\rm can}$, as given by $\hat{\rho}_{\rm f}^{\rm can} = \frac{\exp(-\beta\hat{H}_{\rm f}^{\rm S})}{Z_{\rm f}^{\rm S}} \otimes \frac{\exp(-\beta_1\hat{H}^{\rm B_1})}{Z^{\rm B_1}} \otimes \cdots \otimes \frac{\exp(-\beta_n\hat{H}^{\rm B_n})}{Z^{\rm B_n}}$, where $Z_{\rm f}^{\rm S} \equiv {\rm tr}\{\exp(-\beta\hat{H}_{\rm f}^{\rm S})\}$.

MAIN RESULTS

Our main result is given by

$$-\Delta U^{\rm S} + \sum_{m=1}^{n} \frac{T}{T_m} Q_m \leq -\Delta F^{\rm S} + k_{\rm B} T I(\hat{\rho}_1 : X), \tag{3}$$

where $W_{\rm ext}$ is the work extracted from the heat bath, $\Delta F^{\rm S}$ is the difference of the Helmholtz free energy between the initial and final thermodynamic equilibrium states, and I is a generalized mutual information content between the demon (i.e. the feedback controller) and the thermodynamic system. The equality in (3) holds if and only if $\hat{\rho}_3^{(k)}$ is independent of measurement outcome k (i.e. the feedback control is perfect), and $\hat{\rho}_{\rm f}$ coincides with $\hat{\rho}_{\rm f}^{\rm can}$.

The generalized mutual information is defined as

$$I(\hat{\rho}_1 : X) \equiv S(\hat{\rho}_1) + H(\{p_k\}) + \sum_k {\rm tr}[\hat{M}_k\hat{\rho}_1\hat{M}_k^{\dagger} \ln(\hat{M}_k\hat{\rho}_1\hat{M}_k^{\dagger})], \tag{4}$$

where $S(\hat{\rho}_1) \equiv -{\rm tr}(\hat{\rho}_1 \ln \hat{\rho}_1)$ is the von Neumann entropy, and $H(\{p_k\}) \equiv -\sum_k p_k \ln p_k$ is the Shannon information of the measurement outcomes. We shall refer to this quantity as the QC-mutual information content, where QC indicates that the measured system is quantal and that the measurement outcome is classical. The QC-mutual information content reduces to the usual mutual information content in the case of a classical measurement. It can be shown that the QC-mutual information content satisfies

$$0 \leq I(\hat{\rho}_1 : X) \leq H(\{p_k\}). \tag{5}$$

In the limit of $I(\hat{\rho}_1 : X) \to 0$, inequality (3) reduces to the conventional second law of thermodynamics. In particular, for a thermodynamic heat cycle, inequality (3) reduces to the Clausius inequality: $\sum_{m=1}^{n} \frac{Q_m}{T_m} \leq 0$.

We will discuss two important cases for inequality. Let us first consider a situation in which the system undergoes an isothermal process in contact with single heat bath B at temperature T. In this case, (3) reduces to

$$W_{\rm ext} \leq -\Delta F^{\rm S} + k_{\rm B} T I(\hat{\rho}_1 : X), \tag{6}$$

where the first law of thermodynamics, $W_{\rm ext} = \sum_{m=1}^{n} Q_m - \Delta U^{\rm S}$, is used. Inequality (6) implies that we can extract work greater than $-\Delta F^{\rm S}$ from a single heat bath with feedback control, but that we cannot extract work larger than $-\Delta F^{\mathrm{S}} + k_{\mathrm{B}} T I(\hat{\rho}_1 : X)$. If we do not get any information, (6) reduces to $W_{\mathrm{ext}} \leq -\Delta F^{\mathrm{S}}$. On the other hand, in the case of classical and error-free measurement, (6) becomes $W_{\mathrm{ext}} \leq -\Delta F^{\mathrm{S}} + k_{\mathrm{B}} T H(\{p_k\})$.

The upper bound of inequality (6) can be achieved with the Szilard engine which is described as follows. A molecule is initially in thermal equilibrium in a box in contact with a heat bath at temperature T. We quasi-statically partition the box into two smaller boxes of equal volume, and perform a measurement on the system to find out in which box the molecule is. When the molecule is found in the right one, we remove the left one and move the right one to the left position, which is the feedback control. We then expand the box quasi-statically and isothermally so that the final state of the entire system returns to the initial state from a macroscopic point of view. During the entire process, we obtain $\ln 2$ of information and extract $k_{\mathrm{B}} T \ln 2$ of work from the system.

We next consider a heat cycle which contacts two heat baths: $\mathrm{B_H}$ at temperature T_{H} and $\mathrm{B_L}$ at T_{L} with $T_{\mathrm{H}} > T_{\mathrm{L}}$. We assume that $\hat{H}_{\mathrm{i}}^{\mathrm{S}} = \hat{H}_{\mathrm{f}}^{\mathrm{S}}$, $\Delta U^{\mathrm{S}} = 0$, and $\Delta F^{\mathrm{S}} = 0$. Noting that $W_{\mathrm{ext}} = Q_{\mathrm{H}} + Q_{\mathrm{L}}$, we can obtain

$$W_{\mathrm{ext}} \leq \left(1 - \frac{T_{\mathrm{L}}}{T_{\mathrm{H}}}\right) Q_{\mathrm{H}} + k_{\mathrm{B}} T_{\mathrm{L}} I(\hat{\rho}_1 : X). \tag{7}$$

Without a feedback control, (7) shows that the upper bound for the efficiency of heat cycles is given by that of the Carnot cycle: $W_{\mathrm{ext}}/Q_{\mathrm{H}} \leq 1 - T_{\mathrm{L}}/T_{\mathrm{H}}$. With feedback control, (7) implies that the upper bound for the efficiency of heat cycles becomes larger than that of the Carnot cycle.

We note that the consistency of our result with the second law of thermodynamics is ensured by the fact that work is needed for information processing of the feedback controller [1, 3].

ACKNOWLEDGMENTS

This work was supported by a Grant-in-Aid for Scientific Research (Grant No. 17071005).

REFERENCES

1. *"Maxwell's demon 2: Entropy, Classical and Quantum Information, Computing"*, H. S. Leff and A. F. Rex (eds.), (Princeton University Press, New Jersey, 2003).
2. T. Sagawa and M. Ueda, Phys. Rev. Lett. **100**, 080403 (2008).
3. T. Sagawa and M. Ueda, arXiv:0809.4098 (2008).

Optimal Covariant Measurement of Momentum on a Half Line

Yutaka Shikano and Akio Hosoya

Department of Physics, Tokyo Institute of Technology, Tokyo, Japan, 152-8551

Abstract. We investigate an optimal covariant measurement to measure a momentum of a particle. Given an instantaneous interaction and an infinite measurement time, we explicitly construct a measurement model to satisfy the optimal covariant positive operator valued measure by Holevo and physically consider this measurement model. Furthermore, we apply this model to measure a momentum of a particle on a half line.

Keywords: Measurement Theory, Quantum Mechanics on a Half Line, Mathematical Physics
PACS: 03.65.-w, 03.65.Db, 03.65.Ta, 03.67.-a

In quantum information, measurement is often described by a positive operator valued measure (POVM) since we often discuss a state change and a probability distribution about the measured system. When a particular POVM is given, we do not know how to realize this process using the probe system. An aim of our work is to construct a measurement model to obtain an optimal covariant POVM initiated by Holevo [1, 2, 3, 4]. Furthermore, we cannot perform the projective measurement of a momentum on a half line since it is not a self-adjoint operator. Nevertheless, we would like to obtain some physical information of the momentum on a half line. In our work [5], we explicitly construct the optimal covariant measurement model to measure a momentum of a particle on a half line.

Let us introduce a convenient notation $[\cdot]$ [2]. By von Neumann's spectral theorem, any Hilbert space $\mathscr{H}$ can be formally described as the direct integral of a Hilbert space $\mathscr{H}_x$, $\mathscr{H} = \int \oplus \mathscr{H}_x dx$, so that any state vector $\psi \in \mathscr{H}$ is described by the vector-valued function $\psi = [\psi_x]$ with $\psi_x \in \mathscr{H}_x$. There, a position operator $\hat{x}$ acts as multiplication operators $\hat{x}\psi = [x\psi_x]$ in this notation. The same notation $[\cdot]$ is used for an operator-valued function. A kernel $[K(x,x')]$, where $K(x,x')$ is a mapping from $\mathscr{H}_{x'}$ to $\mathscr{H}_x$ for all x and x', defines an operator $\hat{K}$ on $\mathscr{H}$. We can write

$$\hat{K}\psi = [K(x,x')]\,[\psi_{x'}] = \left[\int K(x,x')\psi_{x'}dx'\right]. \tag{1}$$

Also we express the norm in $\mathscr{H}_x$ as $\|\cdot\|_x$.

We define an optimality for measurement as minimizing the variance between an inferred outcome of the measured system before a measuring process and a measurement outcome of the probe system after the measuring process, restricting our attention to the covariant measurement. Holevo derived the optimal covariant POVM [2] as

$$M_0(dp) = \left[\frac{\psi_x \cdot \psi_{x'}^{\dagger}}{\|\psi_x\|_x \|\psi_{x'}\|_{x'}} e^{i(x-x')p}\frac{dp}{2\pi}\right]. \tag{2}$$

CP1110, *Quantum Communication, Measurement and Computing (QCMC), The Ninth International Conference,* edited by A. Lvovsky

This POVM corresponds to the projective measurement, $M_0(dp) = |p\rangle\langle p|dp$, expressing this POVM in the bracket notation via the Fourier transformation. Also note that (2) is independent of a domain of x. In our work, we explicitly construct a model Hamiltonian to reproduce the POVM (2). We first consider the whole line system and the projective measurement model.

Let us consider a model Hamiltonian,

$$\hat{H}_{com} = \frac{1}{2m}\hat{p}^2 + \frac{1}{2M}\hat{P}^2 + g\hat{P}\hat{x}\delta(t) + \frac{m\omega^2}{2}\hat{x}^2, \tag{3}$$

where a pair $(\hat{x}, \hat{p})$ are the position and the momentum operators of the measured system, a pair $(\hat{X}, \hat{P})$ are those of the probe system and $\delta(t)$ is the Dirac δ-function. This Hamiltonian is modeled from the following consideration. We take the potential of the measured system as a harmonic oscillator for simplicity and the probe system is assumed to be a free particle system. Furthermore, the interaction is assumed to be instantaneous with a coupling constant g. We calculate the evolution operator,

$$\hat{U} = \mathrm{T}\exp\left(-i\int_{0-}^{t_f} \hat{H}_{com}dt\right), \tag{4}$$

where T stands for the time-ordered product, from the model Hamiltonian (3) to the Kraus operator. Taking the measuring time t_f infinity and given the initial probe state $|X=0\rangle = \int |P\rangle dP$, an eigen state of the momentum $\hat{X}$ of the probe system[6], we calculate the POVM from the Kraus operator as

$$M = \left[\psi_{x'}^{\dagger} \cdot \psi_x \exp\left(-ig\tilde{P}\{x(0) - x'(0)\}\right)\right], \tag{5}$$

following the iε prescription [7]. We identify $g\tilde{P}$ with the measurement outcome p itself of the probe observable to reproduce the optimal covariant POVM (2) except for the normalization of the wave functions and the measure $dp/2\pi$.

Now, we physically describe how we optimally infer the momentum of the measured system just before the measuring process. First, we instantaneously couple the measured system to the probe system. Second, we keep the measured system in contact with the bulk system at zero temperature and wait for a sufficiently long time. Since the energy of the measured system is dissipated to the bulk system, the state of the measured system settles down to the ground state. If we let the energy of the ground state zero, i.e., $\omega \to 0$ of the interaction Hamiltonian (3), the momentum of the measured system $p_{sys,\infty}$ becomes zero at $t_f = \infty$. According to the momentum conservation law, we obtain

$$p_{sys,0} + p_{p,0} = p_{sys,\infty} + p_{p,\infty} = p_{p,\infty}, \tag{6}$$

where $p_{sys,t}$ and $p_{p,t}$ are the momenta of the measured system and the probe system at a time t. Since we can control the probe system, we can precisely infer the "measurement" value $p_{sys,0}$ of the momentum of the measured system at the beginning of the measuring process from the measurement outcome $p_{p,\infty}$, which we measure in the probe system at $t_f = \infty$ (See Fig. 1). If ω of the Hamiltonian (3) were finite, the variance of the

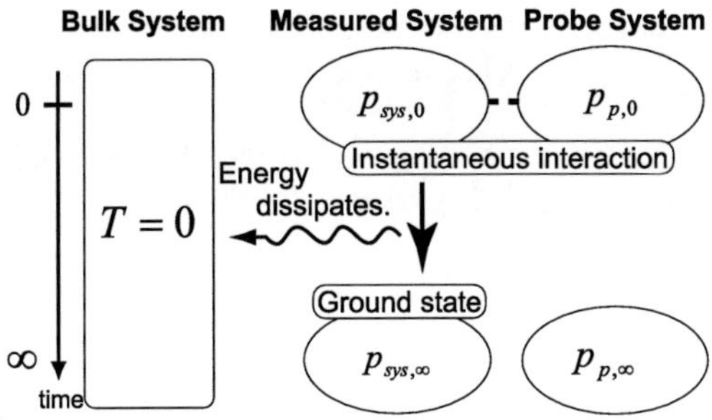

FIGURE 1. An optimal covariant measurement model. By the instantaneous interaction between the measured and probe systems, the measured system is entangled with the probe system. On the other hand, the measured system is coupled with the bulk system at zero temperature to dissipate the energy of the measured system. Thus we optimally evaluate the system observable at $t = 0$ inferred from the outcome of the probe system at $t = \infty$ by the momentum conservation law.

momentum of the measured system would remain finite due to the zero point oscillation and Eq. (6) would be modified.

Our proposed model is different from the standard von Neumann model[8], which we only consider the interaction Hamiltonian following the coupling constant is large. However, our proposed model can also be realized for the weak coupling constant. Furthermore, it is very curious to point out that this model can physically realize the post-selected state picking up the ground state of the measured system. It may be analogous to weak measurement initiated and developed by Aharonov *et al.* [9, 10, 11].

Let us apply the optimal measurement model to the half line system. Extending the domain of the momentum operator on a half line by introducing a two dimensional Hilbert space $\mathscr{H}_2$ with the orthonormal bases $|0\rangle$ and $|1\rangle$ to be tensored as $\mathscr{H} = \mathscr{H}_+ \otimes \mathscr{H}_2$, we make it self-adjoint as $\hat{p} \equiv \hat{p}_+ \otimes |0\rangle\langle 0| - \hat{p}_+ \otimes |1\rangle\langle 1|$ *á la* Naimark. Taking the unitary transformation Π_1, which is the space inversion around the zero point only for the spin state $|1\rangle$, the extended Hilbert space becomes the whole line system except for the origin. In the extended system, we consider the model Hamiltonian (3) and obtain the POVM as

$$\Pi_1 M_0 \Pi_1^\dagger = \left[\psi_{x_+} \cdot \psi_{x'_+}^\dagger e^{i(x_+ - x'_+)p_+}\right] \otimes |0\rangle\langle 0| + \left[\psi_{x_-} \cdot \psi_{x'_-}^\dagger e^{i(x_- - x'_-)p_-}\right] \otimes |1\rangle\langle 1|. \quad (7)$$

By taking the partial trace over $\mathscr{H}_2$, we obtain the reduced POVM as

$$\tilde{M}_0 \equiv \mathrm{Tr}_2 M_0 = \left[\psi_{x_+} \cdot \psi_{x'_+}^\dagger e^{i(x_+ - x'_+)p_+}\right],$$

up to a normalization constant. We have reproduced the optimal covariant POVM (2) restricted to positive parameters x and x'. We physically describe the measuring process to optimally evaluate the momentum of a particle on a half line.

We calculate the probability distribution of the momentum on a half line in the optimal case. As an example, let us assume the pure state $\rho = \left[\phi_{x_+} \cdot \phi_{x'_+}^\dagger\right]$, which is a plane wave with a momentum p_{true}, $\phi_{x_+} = Ae^{ip_{true}x_+}$, for the measured system before the measuring process. We assume that the state is properly localized to be an element of the Hilbert space $\mathscr{H}_+$. The state, $[\phi_{x_+}]$, is relaxed by the measuring process to the ground state $\psi_{x_+} \in \mathscr{H}_+$ given by $\psi_{x_+} = 2\left(\frac{(m\omega)^3}{\pi}\right)^{\frac{1}{4}} x_+ \exp\left(-\frac{m\omega}{2}x_+^2\right)$. Then we obtain the probability distribution of the momentum as

$$\begin{aligned}\mathrm{Tr}(\rho\tilde{M}_0) &= \mathrm{Tr}\left(\left[\phi_{x''_+}\cdot\phi^\dagger_{x'_+}\right]\left[\psi_{x_+}\cdot\psi^\dagger_{x''_+}e^{i(x_+-x''_+)p}\right]\right)\\ &= 16\sqrt{\frac{\pi}{(m\omega)^3}}|A|^2(p-p_{true})^2\exp\left(-\frac{1}{m\omega}(p-p_{true})^2\right),\end{aligned} \tag{8}$$

which has two peaks at $p = p_{true} \pm \sqrt{m\omega}$ and vanishes at $p = p_{true}$. If we take $\omega \to 0$, i.e., the measured system is a free particle system, we can precisely evaluate the momentum of the plane wave since we obtain $\mathrm{Tr}(\rho\tilde{M}_0) = \delta(p - p_{true})$. Otherwise there remains uncertainty by quantum zero point oscillation and the momentum with the maximum probability deviates by $\sqrt{m\omega}$ from the precise momentum p_{true}. When the potential of the measured system is a general convex function, the probability distribution for the momentum becomes the modulus square of the Fourier transformation of the odd parity ground state wave function.

In conclusion, we have discussed the optimal covariant measurement model to measure a momentum of a particle. By applying Holevo's works [1, 2, 3, 4], we have obtained the optimal covariant POVM in the optimal sense to minimize the variance between the "measurement" outcome of the measured system before the interaction and the measurement outcome of the probe system after the interaction. To realize physical systems, we have explicitly constructed the model Hamiltonian for the measured and probe systems and coupled the measured system to the bulk system at zero temperature for infinitely long time. We have shown that the optimal covariant POVM coincides with the calculated POVM from the model Hamiltonian. As a result, we have presented the optimal covariant measurement model. Then we have physically explained the optimal covariant measuring process.

REFERENCES

1. A. S. Holevo, Rep. Math. Phys. **13**, 379-399 (1978).
2. A. S. Holevo, Rep. Math. Phys. **16**, 385-400 (1979).
3. A. S. Holevo, *Probabilistic and statistical aspects of quantum theory* (North-Holland, Amsterdam, 1982).
4. A. S. Holevo, *Statistical Structure of Quantum Theory* (Springer, Berlin, 2001).
5. Y. Shikano and A. Hosoya, J. Math. Phys. **49**, 052104 (2008).
6. In our paper [5], we have taken the mistake about the setting the initial state of the probe system. However, we confirm the later discussion to calculate the POVM is right.
7. E. S. Abers and B. W. Lee, Phys. Rep. **9**, 1-141 (1973).
8. J. von Neumann, *Mathematische Grundlagen der Quantumechanik* (Springer, Berlin, 1932), [*Mathematical Foundations of Quantum Mechanics* (Princeton University Press, Princeton, 1955).]
9. Y. Aharonov, D. Z. Albert and L. Vaidman, Phys. Rev. Lett. **60**, 1351-1354 (1988).
10. Y. Aharonov and L. Vaidman, in *Time in Quantum Mechanics*, Vol. 1, edited by J. G. Muga, R. Sala Mayato, and I. L. Egusquiza, (Springer, Berlin Heidelberg, 2008) pp. 399-447.
11. R. Jozsa, arXiv: 0706.4207 (quant-ph).

Quantum Theory of Optical Temporal Phase in the Continuous Time Limit

Mankei Tsang, Jeffrey H. Shapiro, and Seth Lloyd

Research Laboratory of Electronics, Massachusetts Institute of Technology, Cambridge, Massachusetts 02139, USA

Abstract. We consider our recently proposed quantum theory of optical temporal phase in the continuous time limit.

We have recently introduced a quantum theory of optical temporal phase and instantaneous frequency for slowly varying signals in the discrete time domain [1]. In this paper, we consider the theory in the continuous time limit. The frequency-domain field operators satisfy the bosonic commutation relation:

$$[\hat{a}(f),\hat{a}^\dagger(f')] = \delta(f-f'). \tag{1}$$

The time-domain envelope operators are defined as

$$\hat{A}(t) \equiv \int_{-\infty}^{\infty} df\, \hat{a}(f)\exp(-i2\pi ft), \qquad [\hat{A}(t),\hat{A}^\dagger(t')] = \delta(t-t'). \tag{2}$$

The Hilbert space can be spanned by Fock states. Let $dn(t)$ be a continuous-time discrete-photon-number process, and t_j be the times at which $dn(t_j) > 0$. The photon-number flux is then

$$I(t) \equiv \frac{dn(t)}{dt} = \sum_j dn(t_j)\delta(t-t_j). \tag{3}$$

A Fock state is defined as

$$|dn(t)\rangle \equiv \left\{\prod_j \frac{1}{\sqrt{dn(t_j)!}}\left[\hat{A}^\dagger(t_j)\sqrt{dt}\right]^{dn(t_j)}\right\}|0\rangle, \qquad \hat{1} = \sum_{dn(t)} |dn(t)\rangle\langle dn(t)|. \tag{4}$$

For pure states, the photon-number representation is $C[dn(t)] \equiv \langle dn(t)|\Psi\rangle$, related to the N-photon temporal wavefunction by

$$\psi_N(\tau_1,\ldots,\tau_N) = \sum_{dn(t)} C[dn(t)]\Phi_{dn(t)}(\tau_1,\ldots,\tau_N), \quad N = \int_{-\infty}^{\infty} dt\, I(t) = \sum_j dn(t_j),$$

$$\Phi_{dn(t)}(\tau_1,\ldots,\tau_N) \equiv \left[\frac{\prod_j dn(t_j)!}{N!}\right]^{1/2} \sum_P \prod_j \delta^{\frac{1}{2}}(\tau_{P(j,1)}-t_j)\times\cdots\times\delta^{\frac{1}{2}}(\tau_{P(j,dn(t_j))}-t_j), \tag{5}$$

CP1110, *Quantum Communication, Measurement and Computing (QCMC), The Ninth International Conference,* edited by A. Lvovsky

where the summation over all permutations symmetrizes $\Phi_{dn(t)}$. The inverse relation is

$$C[dn(t)] = \left[\frac{N!dt^N}{\prod_t dn(t)!}\right]^{1/2} \psi_N(\ldots, \underbrace{t,\ldots,t}_{dn(t)\text{ terms}}, \ldots). \tag{6}$$

For example, a multimode coherent state can be written as

$$|\alpha(t)\rangle = \exp\left[-\frac{\bar{N}}{2} + \int_{-\infty}^{\infty} dt\, \alpha(t)\hat{A}^\dagger(t)\right]|0\rangle, \quad \bar{N} \equiv \int_{-\infty}^{\infty} dt\, |\alpha(t)|^2, \tag{7}$$

$$C[dn(t)] = \prod_t \exp\left[-\frac{1}{2}dt|\alpha(t)|^2\right] \frac{\left[\alpha(t)\sqrt{dt}\right]^{dn(t)}}{\sqrt{dn(t)!}}, \tag{8}$$

and the photon-number probability density $|C[dn(t)]|^2$ describes a Poisson process,

$$[dn(t)]^2 = dn(t), \qquad \langle dn(t)\rangle = |\alpha(t)|^2 dt, \tag{9}$$

$$\langle I(t)I(\tau)\rangle = \langle I(t)\rangle\, \delta(t-\tau), \qquad \langle I(t)\rangle = |\alpha(t)|^2 = \frac{\mathscr{P}(t)}{hf_0}, \tag{10}$$

$\mathscr{P}(t)$ being the optical power and f_0 the carrier frequency.

The Susskind-Glogower phase state is defined as the functional Fourier transform of Fock states,

$$|\phi(t)\rangle \equiv \sum_{dn(t)} \exp\left[i\int_{-\infty}^{\infty} dt\, I(t)\phi(t)\right]|dn(t)\rangle = \sum_{dn(t)} \exp\left[i\sum_j dn(t_j)\phi(t_j)\right]|dn(t)\rangle. \tag{11}$$

A temporal-phase POVM can then be written as

$$\hat{\Pi}[\phi(t)] \equiv |\phi(t)\rangle\langle\phi(t)|, \quad \int_{\phi_0(t)}^{\phi_0(t)+2\pi} D\phi(t)\, \hat{\Pi}[\phi(t)] = \hat{1}, \quad D\phi(t) \equiv \prod_t \frac{d\phi(t)}{2\pi}. \tag{12}$$

To measure the temporal phase in practice, in Ref. [1] we have designed homodyne phase-locked loops (PLLs) using the Wigner distribution. In the continuous time domain, the Wigner distribution of a squeezed state is

$$\Delta X_1(t) \equiv A(t) + A^*(t) - \langle A(t) + A^*(t)\rangle, \tag{13}$$

$$\Delta X_2(t) \equiv -i[A(t) - A^*(t)] - \langle -i[A(t) - A^*(t)]\rangle, \tag{14}$$

$$W_0[\Delta X_1(t), \Delta X_2(t)] \propto \exp\left[-\frac{1}{2}\sum_{k=1,2}\int_{-\infty}^{\infty} dt \int_{-\infty}^{\infty} d\tau\, \Delta X_k(t) K_k^{-1}(t,\tau)\Delta X_k(\tau)\right], \tag{15}$$

$$K_{1,2}(t,\tau) \equiv \left\langle \Delta X_{1,2}(t)\Delta X_{1,2}(\tau)\right\rangle, \quad K_{1,2}(t,\tau) = K_{2,1}^{-1}(t,\tau), \tag{16}$$

$$\int_{-\infty}^{\infty} d\tau\, K_{1,2}(t,\tau) K_{1,2}^{-1}(\tau,u) = \delta(t-u). \tag{17}$$

Assuming that the squeezed state has time-invariant statistics, $K_{1,2}(t,\tau) = K_{1,2}(t-\tau)$, we can define the quadrature power spectral densities as

$$S_{1,2}(f) \equiv \int_{-\infty}^{\infty} dt\, K_{1,2}(t) \exp(i2\pi f t), \qquad S_1(f)S_2(f) = 1. \tag{18}$$

For example, the coherent state has the covariance functions $K_{1,2}(t,\tau) = \delta(t-\tau)$, which means that it can be regarded as a classical signal with additive white Gaussian noise upon homodyne detection.

Assume that a message $m(t)$ is linearly encoded in the optical phase,

$$\bar{\phi}(t) = \int_{-\infty}^{\infty} d\tau\, h(t-\tau) m(\tau). \tag{19}$$

For example, for phase modulation $h(t-\tau) = \beta\delta(t-\tau)$ and for frequency modulation $h(t-\tau) = -2\pi\mathscr{F}\int_{-\infty}^{t} du\, \delta(u-\tau)$. The a priori probability density for the message is assumed to be

$$P[m(t)] \propto \exp\left[-\frac{1}{2}\int_{-\infty}^{\infty} dt \int_{-\infty}^{\infty} d\tau\, m(t) K_m^{-1}(t-\tau) m(\tau)\right], \quad K_m(t-\tau) \equiv \langle m(t) m(\tau)\rangle. \tag{20}$$

The conditional Wigner distribution, upon phase modulation, is

$$W[X_1(t), X_2(t)|m(t)] = W_0[Y_1(t), Y_2(t)], \tag{21}$$

$$Y_1(t) = X_1(t)\cos\bar{\phi}(t) + X_2(t)\sin\bar{\phi}(t) - 2|\alpha(t)|, \tag{22}$$

$$Y_2(t) = -X_1(t)\sin\bar{\phi}(t) + X_2(t)\cos\bar{\phi}(t). \tag{23}$$

The maximum a posteriori (MAP) estimate equation becomes

$$\frac{\delta}{\delta m(t)} \ln P[m(t)|X_1(t), X_2(t)] = \frac{\delta}{\delta m(t)}\left\{\ln W[X_1(t), X_2(t)|m(t)] + \ln P[m(t)]\right\} = 0. \tag{24}$$

The solution of this equation gives the MAP estimate $\widetilde{m}(t)$. For coherent states the MAP equation is

$$\widetilde{m}(t) = 4|\alpha|^2 \int_{-\infty}^{\infty} d\tau \int_{-\infty}^{\infty} du\, K_m(t-\tau) h(u-\tau)\left\{\sin[\bar{\phi}(u) - \widetilde{\phi}(u)] + Z(u)\right\}, \tag{25}$$

$$\langle Z(t) Z(\tau)\rangle = \frac{1}{4|\alpha|^2}\delta(t-\tau), \quad \widetilde{\phi}(t) \equiv \int_{-\infty}^{\infty} d\tau\, h(t-\tau)\widetilde{m}(\tau). \tag{26}$$

Linearizing the MAP equation by assuming $\sin(\bar{\phi} - \widetilde{\phi}) \approx \bar{\phi} - \widetilde{\phi}$,

$$\widetilde{m}(t) = \int_{-\infty}^{\infty} d\tau\, g(t-\tau)\phi(\tau), \quad \phi(t) \equiv \bar{\phi}(t) + Z(t), \tag{27}$$

where $g(t-\tau)$ is the optimal linear filter,

$$\int_{-\infty}^{\infty} d\tau\, g(t-\tau)\left[K_{\bar{\phi}}(\tau-u)+\frac{1}{4|\alpha|^2}\delta(\tau-u)\right]=\int_{-\infty}^{\infty} dv\, K_m(t-v)h(u-v), \qquad (28)$$

and $K_{\bar{\phi}}$ is

$$K_{\bar{\phi}}(t-\tau)\equiv\left\langle\bar{\phi}(t)\bar{\phi}(\tau)\right\rangle=\int_{-\infty}^{\infty} du\int_{-\infty}^{\infty} dv\, h(t-u)K_m(u-v)h(\tau-v). \qquad (29)$$

In the frequency domain, defining

$$G(f)\equiv\int_{-\infty}^{\infty} dt\, g(t)\exp(i2\pi ft), \qquad S_m(f)\equiv\int_{-\infty}^{\infty} dt\, K_m(t)\exp(i2\pi ft),$$
$$H(f)\equiv\int_{-\infty}^{\infty} dt\, h(t)\exp(i2\pi ft), \qquad (30)$$

we can derive the irreducible error,

$$G(f)=\frac{S_m(f)H^*(f)}{S_m(f)|H(f)|^2+1/4|\alpha|^2}, \quad \left\langle[m(t)-\tilde{m}(t)]^2\right\rangle=\int_{-\infty}^{\infty} df\,\frac{S_m(f)}{4|\alpha|^2S_m(f)|H(f)|^2+1}. \qquad (31)$$

To implement the PLLs, we also need the realizable estimate of the phase in terms of the Wiener-Hopf filter,

$$\phi'(t)=\int_{-\infty}^{t} d\tau\, g'(t-\tau)\phi(\tau), \quad \int_{0}^{\infty} d\tau\, g'(\tau)\left[K_{\bar{\phi}}(t-\tau)+\frac{1}{4|\alpha|^2}\delta(t-\tau)\right]=K_{\bar{\phi}}(t). \qquad (32)$$

The error of Wiener filtering is

$$\left\langle[\bar{\phi}(t)-\phi'(t)]^2\right\rangle=\frac{1}{4|\alpha|^2}\int_{-\infty}^{\infty} df\,\ln\left[1+4|\alpha|^2S_m(f)|H(f)|^2\right], \qquad (33)$$

which is required to be much smaller than 1 for the linear analysis to be valid. Under this constraint, the quantum limits to temporal phase and instantaneous frequency measurements can be derived and are given in [1]. For broadband phase-squeezed states, we can replace $|\alpha|^2$ with $|\alpha|^2\exp(2r)$ if the phase-locking is tight enough so that the anti-squeezed quadrature can be neglected.

This work is financially supported by the W. M. Keck Foundation Center for Extreme Quantum Information Theory.

REFERENCES

1. M. Tsang, J. H. Shapiro, and S. Lloyd, e-print arXiv:0804.0463 (Phys. Rev. A, in press).

Incoherently generated coherences

Raisa I. Karasik*, Karl-Peter Marzlin†,**, Barry C. Sanders** and K. Birgitta Whaley*,‡

*Applied Science & Technology and Berkeley Quantum Information Center, University of California, Berkeley, California 94720, USA
†Department of Physics, St Francis Xavier University, Antigonish, Nova Scotia B2G 2W5, Canada
**Institute for Quantum Information Science, University of Calgary, Calgary, Alberta T2N 1N4, Canada
‡Department of Chemistry, University of California, Berkeley, California 94720, USA

Abstract. We have discovered that, in specific cases, decoherence due to the interaction with the environment can be avoided by just introducing a special driving field for an open quantum system. We show that this phenomenon constitutes a new class of decoherence-free subspaces which are useful for passive preservation of quantum information and discuss a physical example.

Keywords: open systems, decoherence free subspaces
PACS: 03.67.Pp, 42.50.Ct

INTRODUCTION

Encoding into decoherence-free subspaces (DFS) [1, 2] protects quantum states from the effects of decoherence by exploiting certain symmetries in the decoherence processes. This encoding can be useful for quantum memory and may reduce the need for quantum error correction and has been subject to experimental tests [3, 4, 5, 6, 7].

In this paper, we discuss a new type of DFS that is created through "incoherently generated coherences" (IGC). Such subspaces exist for certain open systems, where the dynamics of a quantum system can connive together with the interactions between the system and its environment in a special way to reduce decoherence. We provide an example of a physical system that supports such states.

DECOHERENCE-FREE SUBSPACES

We study DFS in open quantum systems whose evolution is given by a Markovian semigroup master equation. In this approach, an open quantum system S is coupled to the environment R. Tracing over the degrees of freedom of the environment leads to a description of the system's dynamics in terms of the reduced density matrix ρ which describes the state of the system S. The evolution of ρ is given by the Markovian master equation

$$\dot{\rho} = -i[\hat{H}_{\text{eff}}, \rho] + L_D[\rho], \tag{1}$$

where $\hat{H}_{\text{eff}}$ is the effective-system Hamiltonian that also includes energy level shifts due to the interaction with the environment. $L_D[\rho]$ is often called dissipative or decoherence

CP1110, *Quantum Communication, Measurement and Computing (QCMC), The Ninth International Conference,* edited by A. Lvovsky

part of the Markovian master equation and is given by the following expression

$$L_{\mathrm{D}}[\rho] = \frac{1}{2}\sum_{l=1}^{M} \lambda_l \left([\hat{J}_l, \rho \hat{J}_l^\dagger] + [\hat{J}_l \rho, \hat{J}_l^\dagger] \right), \tag{2}$$

where $\lambda_l > 0$ and $\hat{J}_l$ are Lindblad operators which describe all possible decoherence processes in a quantum system S.

DFS have been proposed as a collection of states that undergo pure unitary evolution [1, 2]. Early works on DFS [2] interpreted Markovian master equation by saying that the first term in the equation ($[\hat{H}_{\mathrm{eff}}, \rho]$) generates unitary evolution, whereas $L_D[\rho]$ generates non-unitary decoherening dynamics. With this interpretation, it would be natural to define DFS as follows:
DFS definition 1: a DFS corresponds to a collection of states that satisfy $L_D[\rho(t)] = 0$ for all times t. Then the evolution of ρ will be generated by $-i[\hat{H}_{\mathrm{eff}}, \rho]$ alone, which is unitary.

INCOHERENTLY GENERATED COHERENCES

In Ref. [8] we have shown that $L_D[\rho] \neq 0$ is not a necessary condition generating unitary dynamics. For systems described by Markovian master equations, pure states undergo unitary dynamics when purity $\mathrm{Tr}[\rho^2(t)]$ is preserved throughout the evolution. This suggests a different definition for decoherence-free subspaces in Markovian systems:
DFS definition 2: A DFS $\mathscr{H}_{\mathrm{DFS}} \subset \mathscr{H}_{\mathrm{S}}$ is a collection of pure states $\rho(t)$ generated by $\mathscr{H}_{\mathrm{DFS}}$ that fulfill $\partial_t \mathrm{Tr}[\rho^2(t)] = 0$ for all t.

DFS definitions 1 and 2 are different in a subtle way. To shed light on this we introduce a new class of DFS states:
Definition of IGC states: States with incoherent generation of coherences (IGC) are states $\rho(t)$ that satisfy DFS definition 2 but not DFS definition 1, so that $\partial_t \mathrm{Tr}[\rho^2(t)] = 0$ with $L_D[\rho(t)] \neq 0$.

Hence, IGC states evolve coherently, but for the - generally decohering - effect of the dissipative part of the master equation cannot be neglected during their evolution. As we have shown in Ref. [8] this is only possible if the dissipative part does generate a coherent, instead of decohering, effect on IGC states. This is not possible for all decoherence models but can occur under special circumstances.

Example

One example of a system that can support IGC states is given by a two-level atom with ground state $|1\rangle$ and excited state $|2\rangle$ that interacts with the quantized radiation field. In this example the environment corresponds to the radiation field which is initially prepared in a squeezed vacuum state. The time evolution of this system is given by the Markovian master equation [9]

$$\dot{\rho} = -i[\hat{H}_D, \rho] + \frac{\gamma}{2}[2\hat{J}\rho\hat{J}^\dagger - \hat{J}^\dagger\hat{J}\rho - \rho\hat{J}^\dagger\hat{J}], \tag{3}$$

where the Lindblad operator take the form $\hat{J} = c\sigma_- + s\sigma_+$ with $c \equiv \cosh(r)$, $s \equiv \sinh(r)$ and r is the real squeezing parameter. σ_+ and σ_- are, respectively, the raising and lowering operators for the two-level atom, and γ is the decay rate of the two-level atom in vacuum. The reduced density matrix of the two-level corresponds to a Hermitean 2×2 matrix with matrix elements $\rho_{ij}(t)$, $i,j = 1,2$.

If the system Hamiltonian takes the form

$$\hat{H}_D = \frac{\gamma}{2}\sqrt{sc}(s-c)\hat{\sigma}_y = \frac{\gamma_0}{2}\sqrt{sc}(s-c)\begin{pmatrix} 0 & -i \\ i & 0 \end{pmatrix}, \tag{4}$$

then the general solution of Eq. (3) is given by

$$\begin{aligned}
\rho_{11}(t) =& \frac{s}{s+c} + \frac{\sqrt{sc}(s-c)(\rho_{12}(0)+\rho_{21}(0))}{s^2-6sc+c^2}\left(e^{-4(s-c)^2 t} - e^{-2(s+c)^2 t}\right) \\
& - \frac{s}{s+c}\left(\frac{s^2-c^2}{s^2-6sc+c^2}e^{-4(s-c)^2 t} + \frac{2c(c-3s)}{s^2-6sc+c^2}e^{-2(s+c)^2 t}\right) \\
& + \left(\frac{(s-c)^2}{s^2-6sc+c^2}e^{-4(s-c)^2 t} - \frac{4sc}{s^2-6sc+c^2}e^{-2(s+c)^2 t}\right)\rho_{11}(0) && (5) \\
\rho_{12}(t) =& \rho_{21}^*(t) && (6) \\
\rho_{21}(t) =& \frac{\sqrt{sc}}{s+c} - \frac{\sqrt{sc}}{s+c}\left(\frac{(s-c)(3s-c)}{s^2-6sc+c^2}e^{-2(s+c)^2 t} - \frac{2s(s+c)}{s^2-6sc+c^2}e^{-4(s-c)^2 t}\right) \\
& + \frac{2sc}{s^2-6sc+c^2}\left(e^{-2(s+c)^2 t} - e^{-4(s-c)^2 t}\right)\rho_{12}(0) \\
& + \left(\frac{s^2-4sc+c^2}{s^2-6sc+c^2}e^{-2(s+b)^2 t} - \frac{2sc}{s^2-6sc+c^2}e^{-4(s-c)^2 t}\right)\rho_{21}(0) \\
& + \frac{2\sqrt{sc}(s-c)}{s^2-4sc+c^2}\left(e^{-2(s+c)^2 t} - e^{-4(s-c)^2 t}\right)\rho_{11}(0) && (7) \\
\rho_{22}(t) =& 1-\rho_{11}(t). && (8)
\end{aligned}$$

This implies that the pure state $|\psi(t)\rangle = \sqrt{s}|2\rangle + \sqrt{c}|1\rangle$, with corresponding density matrix

$$\rho(t) = |\psi\rangle\langle\psi| = \frac{1}{s+c}\begin{pmatrix} s & \sqrt{cs} \\ \sqrt{cs} & c \end{pmatrix}, \tag{9}$$

solves Eqs. (5-8) and thus indeed undergoes a (trivial) unitary evolution. Furthermore,

$$L_D[\rho(t)] = \frac{\gamma}{2}\frac{s-c}{s+c}\begin{pmatrix} 2cs & \sqrt{cs}(s-c) \\ \sqrt{cs}(s-c) & 2cs \end{pmatrix} \neq 0, \tag{10}$$

so that $|\psi(t)\rangle$ corresponds to an IGC state: it undergoes unitary evolution even though the decoherence term of the Markovian master equation is not zero.

The interplay between the dissipative part and the coherent part of the master equation is critical for the existence of IGC states. To demonstrate this, we consider the case when the driving Hamiltonian is absent from the master equation, $\hat{H}_D = 0$. The general solution to Eq. (3) is then given by

$$\rho_{11}(t) = \frac{s^2}{s^2+c^2} + \left(\rho_{11}(0) - \frac{s^2}{s^2+c^2}\right) \mathrm{e}^{-4(s^2+c^2)t} \tag{11}$$

$$\rho_{12}(t) = \frac{\rho_{12}(0)+\rho_{21}(0)}{2}\mathrm{e}^{-2(s-c)^2 t} + \frac{\rho_{12}(0)-\rho_{21}(0)}{2}\mathrm{e}^{-2(s+c)^2 t} \tag{12}$$

$$\rho_{21}(t) = \frac{\rho_{12}(0)+\rho_{21}(0)}{2}\mathrm{e}^{-2(s-c)^2 t} - \frac{\rho_{12}(0)-\rho_{21}(0)}{2}\mathrm{e}^{-2(s+c)^2 t} \tag{13}$$

$$\rho_{22}(t) = \frac{c^2}{s^2+c^2} + \left(\rho_{22}(0) - \frac{c^2}{s^2+c^2}\right) \mathrm{e}^{-4(s^2+c^2)t}, \tag{14}$$

so that all states decay to

$$\rho(t \to \infty) = \begin{pmatrix} \frac{s^2}{s^2+c^2} & 0 \\ 0 & \frac{c^2}{s^2+c^2} \end{pmatrix}. \tag{15}$$

Thus, without driving all pure states decay into incoherent mixture of ground and excited state. We see that in this system the application of a specific driving Hamiltonian can eliminate decoherence for IGC states.

In conclusion, we have introduced a new class of decoherence-free states for which the effect of the dissipative part of the master equation is crucial to achieve coherence. In the example presented above in Eqs. (5-8), one has $L_D[\rho(t)] = i[\hat{H}_D, \rho(t)]$, so that the driving field negates the decohering effects of $L_D[\rho]$ and allows the state to be preserved.

ACKNOWLEDGMENTS

B.C.S. is a CIFAR Associate and acknowledges financial support from iCORE and NSERC for this project. R.I.K. and K.B.W. thank the NSF for financial support under ITR Grant No. EIA-0205641, and the Defense Advanced Research Projects Agency (DARPA) and the Air Force Laboratory, Air Force Material Command, USAF, under Contract No. F30602-01-2-0524.

REFERENCES

1. P. Zanardi and M. Rasetti, *Phys. Rev. Lett.* **79**, 3306 (1997).
2. D. A. Lidar, I. L. Chuang and K. B. Whaley, *Phys. Rev. Lett.* **81**, 2594 (1998).
3. P. G. Kwiat, A. J. Berglund, J. B. Altepeter, and A. G. White, *Science* **290**, 498 (2000).
4. D. Kielpinski, *et al.*, *Science* **291**, 1013 (2001).
5. L. Viola, *et al.*, *Science* **293**, 2059 (2001).
6. M. Mohseni, J. S. Lundeen, K. J. Resch, and A. M. Steinberg, *Phys. Rev. Lett.* **91**, 187903 (2003).
7. C. Langer, *et al.*, *Phys. Rev. Lett.* **95**, 060502 (2005).
8. R. Karasik, K.-P. Marzlin, B. C. Sanders and K. B. Whaley, *Phys. Rev. A* **77**(5), 052301 (2008).
9. H.-P. Breuer and F. Petruccione, *The Theory of Open Quantum Systems* (Oxford University Press, Oxford, 2002).

The Cost of Erasing Information

J.A. Vaccaro* and S.M. Barnett†

*Centre for Quantum Dynamics, Griffith University, Brisbane, Queensland 4111 Australia
†Department of Physics, University of Strathclyde, Glasgow G4 0NG, United Kingdom

Abstract. Landauer has argued that the process of erasing 1 bit of information requires a minimum amount of work equal to $kT\ln(2)$ where T is the temperature of a thermal reservoir that consumes the entropy of the bit. We show that, in principle, the thermal reservoir can be replaced by a reservoir of N spin-1/2 particles which is subject to the conservation of spin angular momentum. The cost of erasing 1 bit of information in this case is found to be at least $\beta^{-1}\ln(2)$ where $e^{-\hbar\beta} = (N\hbar + 2\langle J_z\rangle)/(N\hbar - 2\langle J_z\rangle)$ and $\langle J_z\rangle$ is the z component of the total spin angular momentum of the spin reservoir.

Keywords: information erasure, thermodynamics, canonical ensemble, spin system
PACS: 03.65.Ta, 89.70.-a, 05.70.-a

INTRODUCTION

Entropy has different incarnations in thermodynamics and information theory. For example, one of the essential elements of thermodynamics is the conservation of energy. Indeed without energy conservation, the maximization of the entropy would lead to the trivial situation of completely disordered systems. In contrast, the Shannon entropy associated with information is independent of energy. In fact, almost all processing of quantum information could take place within a manifold of degenerate eigenstates of energy. As classical information is a special case of quantum information, this energy independence also applies to classical case. The only exception to the energy independence appears to be the processing involved in the erasure of information. Landauer's erasure principle [1] states that the erasure of information incurs an energy cost. Here we show that the physical cost of the erasure of information is directly related to the type of statistical ensemble that is used to absorb the entropy of the unwanted information. In particular, the energy cost in Landauer's erasure principle is incurred because of the use of a thermal reservoir. We show that the thermal reservoir can be replaced by a statistical ensemble of energy degenerate spin-$\frac{1}{2}$ particles. With this replacement the erasure of information incurs a physical cost in terms of spin angular momentum only.

ERASURE MODEL WITH ENERGY COST

First we describe an erasure model based on a thermal reservoir which incurs an energy cost in accord with Landauer's erasure principle. Let a qubit with a degenerate energy spectrum be in the state of maximum entropy $(|0\rangle\langle 0| + |1\rangle\langle 1|)/2$ where $|0\rangle$ and $|1\rangle$ represent states in the computational basis. The qubit represents a 1 bit memory element. In order to reset the qubit state to the logical state $|0\rangle$, it is placed in contact with a

CP1110, *Quantum Communication, Measurement and Computing (QCMC), The Ninth International Conference,* edited by A. Lvovsky

thermal reservoir at temperature T. Since the qubit states are degenerate the qubit is already in thermal equilibrium with the reservoir. Next a splitting is slowly induced in the energy spectrum of the qubit such that the energy of logical state $|1\rangle$ is E higher than $|0\rangle$. This splitting is induced sufficiently slowly that the qubit can always be regarded as being in equilibrium with the thermal reservoir. The state of the qubit when the energy splitting is E is given by the density operator

$$\frac{|0\rangle\langle 0| + e^{-E/kT}|1\rangle\langle 1|}{1+e^{-E/kT}} \tag{1}$$

in accordance with the Boltzmann distribution. The work dW required to increase the splitting from E to $E + dE$ while in contact with the reservoir is given by the probability of occupation of the state $|1\rangle$ times dE, i.e. $dW = e^{-E/kT}(1+e^{-E/kT})^{-1}dE$. The total work needed to increase the splitting slowly from zero to infinity in this way is $W = \int dW = kT\ln(2)$. At the end of this process the state of the qubit is the $|0\rangle$ as required. The qubit is removed from the reservoir and the energy degeneracy is restored at no cost in energy. The erasure here is given by maximizing the entropy subject to the conservation of energy while increasing the energy splitting of the memory qubit.

ERASURE MODEL WITH SPIN COST

We now consider a reservoir consisting of a collection of N spin-$\frac{1}{2}$ particles which are in thermal equilibrium with a heat bath. The spatial degrees of freedom of the particles have a probability distribution which depends on the temperature of the heat bath and the confining potential. We assume that the Hamiltonian of the system has no spin-dependent term, and so the probability distributions of the spatial and spin degrees of freedom are decoupled. In this case the thermal motion can be taken to induce random spin exchanges in the particles through collisions. We do not need to consider the heat bath nor the spatial degrees of freedom further. We shall label the eigenstates of the z component of spin angular momentum of each particle as the logical states $|0\rangle$ and $|1\rangle$.

The state space of the spin reservoir can be described as follows. Consider the set of eigenstates corresponding to the eigenvalue $(n - N/2)\hbar$ of the z component of the spin angular momentum operator J_z . There are $\binom{N}{n}$ such states, where $\binom{\cdot}{\cdot}$ is the usual binomial coefficient. We label each element of this set as the collective state $|n,\lambda\rangle\rangle$ where $\lambda = 1,2,\cdots,\binom{N}{n}$ uniquely identifies the state within the set. As an example, the collective state $|N,1\rangle\rangle$ corresponds to the product of N single particle states $|1\rangle|1\rangle\cdots$. The full basis set of the spin reservoir is given by $\{|n,\lambda_n\rangle\rangle : n = 0,\ldots,N;\ \lambda_n = 1,\ldots,\binom{N}{n}\}$.

In order to model the equilibrium state of the spin reservoir, we treat it as a canonical ensemble but instead of energy being exchanged between it and a larger bath, the two systems randomly exchange the z component of spin. For this we require the bath to also consist of spin-$\frac{1}{2}$ particles. We shall refer to this exchange interaction as "spin-exchange contact". The spin bath maintains the spin reservoir in a state such that the z component of the total spin in the reservoir is given by

$$\langle J_z\rangle = (\alpha - \frac{1}{2})N\hbar \tag{2}$$

for $0 \leq \alpha \leq 1$. At equilibrium, the probability $P_{n,\lambda}$ of finding the spin reservoir in the state $|n,\lambda\rangle\rangle$ is given by maximizing the entropy $-\sum_{n,\lambda} P_{n,\lambda} \ln P_{n,\lambda}$ with respect to $P_{n,\lambda}$ subject to the constraints $\sum_{n,\lambda} nP_{n,\lambda} = \alpha N$ and $\sum_{n,\lambda} P_{n,\lambda} = 1$. The optimum probability distribution is found to be

$$P_{n,\lambda} = \frac{e^{-n\hbar\beta}}{(1+e^{-\hbar\beta})^N} \tag{3}$$

where $\beta = \frac{1}{\hbar}\ln(\frac{1-\alpha}{\alpha})$ or, on using Eq. (2), where

$$e^{-\hbar\beta} = \frac{N\hbar + 2\langle J_z\rangle}{N\hbar - 2\langle J_z\rangle}\,. \tag{4}$$

Now imagine another spin-$\frac{1}{2}$ particle is our binary memory element. It begins in the maximally mixed state $\frac{1}{2}(|0\rangle\langle 0| + |1\rangle\langle 1|)$ and we wish to erase its "memory" and leave it in the logical zero state $|0\rangle\langle 0|$. Let there be a large collection of ancillary spin-$\frac{1}{2}$ particles in the state $|0\rangle\langle 0|$ for our use. The first stage of the erasure protocol entails putting the memory spin in spin-exchange contact with the spin reservoir, letting the combined reservoir-memory spin system come to equilibrium, and then separating the memory spin from the spin reservoir. At this point the state of the memory spin would be

$$p_0|0\rangle\langle 0| + p_1|1\rangle\langle 1| \tag{5}$$

with $p_1 = e^{-\hbar\beta}/(1+e^{-\hbar\beta}) = 1 - p_0$. We have assumed that the value of $\langle J_z\rangle = (\alpha - \frac{1}{2})N\hbar$ is maintained by the spin bath and so the value of β remains fixed despite the contact with the memory spin. A controlled-not (CNOT) operation is them performed on the memory spin and one ancilla spin, with the former being the control qubit; this yields the state $p_0|00\rangle\langle 00| + p_1|11\rangle\langle 11|$ where $|xy\rangle$ represents the state $|x\rangle$ of the memory spin and $|y\rangle$ of the ancilla spin. The spin cost of this operation is $\hbar e^{-\hbar\beta}/(1+e^{-\hbar\beta})$. The memory-ancilla system is placed in random spin-exchange contact with the spin reservoir while the reservoir-bath is undergoing random spin exchange as before. The spin exchange between the reservoir and the memory-ancilla system is constructed to leave all states unchanged except for the following mapping

$$|2,1\rangle\rangle|00\rangle \leftrightarrow |0,1\rangle\rangle|11\rangle \tag{6}$$

where $|n,\lambda\rangle\rangle|ij\rangle$ represents the reservoir collective state $|n,\lambda\rangle\rangle$ and memory-ancilla system state $|ij\rangle$. The random spin exchange continues for a sufficient time for the reservoir and memory-ancilla system to equilibrate. The state of the memory-ancilla spin system is then given by

$$p'_0|00\rangle\langle 00| + p'_1|11\rangle\langle 11| \tag{7}$$

where now $p'_1 = e^{-2\hbar\beta}/(1+e^{-2\hbar\beta}) = 1 - p'_0$. Another ancilla spin is added and a CNOT operation is performed as before to yield the state $p'_0|000\rangle\langle 000| + p'_1|111\rangle\langle 111|$ with a spin cost of $\hbar e^{-2\hbar\beta}/(1+e^{-2\hbar\beta})$. The combined memory-ancilla spin system put in spin-exchange contact with the reservoir with the mapping

$$|3,1\rangle\rangle|000\rangle \leftrightarrow |0,1\rangle\rangle|111\rangle\,, \tag{8}$$

and so on. After n cycles of this process, the memory-ancilla spins are all in either the logical zero state or all in the logical 1 state with probabilities p_0'' and p_1'', respectively, where $p_1'' = e^{-n\hbar\beta}/(1+e^{-n\hbar\beta}) = 1 - p_0''$. In the limit of many repetitions and large N, the memory-ancilla spin system approaches a pure state where each spin is in the logical zero state.

The spin cost of the nth cycle is $\hbar$ times the probability p_1'', that is, $\hbar e^{-n\hbar\beta}/(1+e^{-n\hbar\beta})$. The total spin cost of the whole process is therefore given by

$$\Delta J_z = \sum_{n=1}^{\infty} \hbar \frac{e^{-n\hbar\beta}}{1+e^{-n\hbar\beta}} \, . \tag{9}$$

This sum is approximated by

$$\beta^{-1}\ln(1+e^{-\hbar\beta}) < \Delta J_z < \beta^{-1}\ln(2) \, . \tag{10}$$

If we include the spin of the initial state, then the cost is

$$\Delta J_z' = \sum_{n=0}^{\infty} \frac{e^{-n\hbar\beta}}{1+e^{-n\hbar\beta}} = \Delta J_z + \frac{1}{2}\hbar \, , \tag{11}$$

$$\beta^{-1}\ln(2) < \Delta J_z' < \beta^{-1}\ln(1+e^{\hbar\beta}) \, . \tag{12}$$

Noting that β has the same units as $\hbar$, we see that the physical cost of erasure is in terms of spin angular momentum.

CONCLUSION

Our analysis shows that the physical cost associated with the erasure of information depends on the physical nature of statistical reservoir that is used to absorb the entropy of the memory bit. This cost need not be in terms of energy. We demonstrated this using an example of an erasure protocol based on a statistical ensemble of spin-$\frac{1}{2}$ particles. While the operations needed for the protocol are expected to be difficult to realize in practice, they are nevertheless possible in principle. The cost of the erasure in this case is shown to be in terms of spin angular momentum.

REFERENCES

1. R. Landauer, *IBM J. Res. Lett.,* **67** 661 (1961).

Generalized spin squeezing criteria: Entanglement detection with collective measurements

Géza Tóth[*,†], Christian Knapp[**], Otfried Gühne[**,‡] and Hans J. Briegel[**,‡]

[*]*Department of Theoretical Physics and History of Science, The University of the Basque Country, P.O. Box 644, E-48080 Bilbao, Spain*
[†]*ICFO-The Institute of Photonic Sciences, E-08860 Castelldefels (Barcelona), Spain*
[**]*Institute for Theoretical Physics, University of Innsbruck, A-6020 Innsbruck, Austria*
[‡]*Institute for Quantum Optics and Quantum Information, Austrian Academy of Sciences, A-6020 Innsbruck, Austria*

Abstract. What is the relation between spin squeezing and entanglement? To clarify this, we derive the full set of generalized spin squeezing inequalities for the detection of entanglement. These are inequalities for the mean values and variances of the collective angular momentum components. They can be used for the experimental detection of entanglement in a system of spin-$\frac{1}{2}$ particles in which the spins cannot be individually addressed. We also show how to obtain the complete set of entanglement conditions if the mean spin and the spin correlation matrix are known. Finally, we apply our criteria for entanglement detection in spin models, showing that they can be used to detect bound entanglement in the thermal state of these systems.

Keywords: spin squeezing, bound entanglement, spin chain
PACS: 03.65.Ud, 03.67.Mn, 05.50.+q, 42.50.Dv

INTRODUCTION

Entanglement lies at the heart of many problems in quantum mechanics and has attracted increasing attention in recent years. In physical systems such as ensembles of cold atoms or trapped ions, spin squeezing [1, 2] is one of the most successful approaches for creating large scale quantum entanglement. Since the variance of a spin component is small, spin squeezed states can be used for reducing spectroscopic noise or to improve the accuracy of atomic clocks. Moreover, if an N-qubit state violates the inequality [3]

$$\frac{(\Delta J_x)^2}{\langle J_y\rangle^2 + \langle J_z\rangle^2} \geq \frac{1}{N},$$

where $J_l := \frac{1}{2}\sum_{k=1}^{N} \sigma_l^{(k)}$ for $l = x, y, z$ are the collective angular momentum components and $\sigma_l^{(k)}$ are Pauli matrices, then the state is entangled (i.e., not fully separable), which is necessary for using it in quantum information processing applications.

Recently, several generalized spin squeezing criteria for the detection of entanglement appeared in the literature [4] and have been used experimentally [5]. Such criteria need the measurement of collective spin components and their variances. They have a large practical importance since in many quantum control experiments the spins cannot be individually addressed, and only collective quantities can be measured.

CP1110, *Quantum Communication, Measurement and Computing (QCMC), The Ninth International Conference,* edited by A. Lvovsky

In this paper we explain the ideas of Ref. [6]. We present the full set of generalized spin squeezing inequalities for many-qubit systems. They detect all entangled states that can be detected knowing the expectation values and variances of spin components. We also show, how this approach can be extended to a set of criteria that detect entanglement based on the correlation matrix. We also discuss that our conditions are able to detect bound entanglement in the ground state of spin models. Surprisingly, these bound entangled states have a positive partial transpose with respect to all bipartitions [7].

GENERALIZED SPIN SQUEEZING CRITERIA

Next, we describe the complete set of generalized spin squeezing entanglement criteria in a system of N spin-$\frac{1}{2}$ particles [6]:

Observation 1. *Let us assume that for a physical system the values of* $\vec{J} := (\langle J_x\rangle, \langle J_y\rangle, \langle J_z\rangle)$ *and* $\vec{K} := (\langle J_x^2\rangle, \langle J_y^2\rangle, \langle J_z^2\rangle)$ *are known. For separable states the following inequalities hold:*

$$\langle J_x^2\rangle + \langle J_y^2\rangle + \langle J_z^2\rangle \leq \tfrac{N(N+2)}{4}, \tag{1a}$$

$$(\Delta J_x)^2 + (\Delta J_y)^2 + (\Delta J_z)^2 \geq \tfrac{N}{2}, \tag{1b}$$

$$\langle J_k^2\rangle + \langle J_l^2\rangle - \tfrac{N}{2} \leq (N-1)(\Delta J_m)^2, \tag{1c}$$

$$(N-1)\left[(\Delta J_k)^2 + (\Delta J_l)^2\right] \geq \langle J_m^2\rangle + \tfrac{N(N-2)}{4}, \tag{1d}$$

where i, j, k take all the possible permutations of x, y, z. The first equation of Eq. (1a) is valid for all quantum states, while if one of the other three inequalities are violated then we know that the state is entangled.

The quantum states fulfilling inequalities Eqs. (1) correspond to a convex object in the six dimensional space of the components of $\vec{J}$ and $\vec{K}$. In order to understand better these inequalities, let us examine what happens if we fix three of these variables. If we fix $\vec{J}$, the points corresponding to separable states in the $(\langle J_x^2\rangle, \langle J_y^2\rangle, \langle J_z^2\rangle)-$space are in a polytope as can be seen in Fig. 1.

It can be proven that for large N the criteria of Observation 1 are complete. Thus, if the inequalities are satisfied, then the first and second moments of J_k do not suffice to prove entanglement. In other words, it is not possible to find criteria detecting more entangled states based on these moments.

CRITERIA WITH THE CORRELATION MATRIX

What happens if we can measure the collective spin in any direction, not only along the x, y and z axes? Knowledge of $\langle J_i\rangle$ and $\langle J_i^2\rangle$ in arbitrary directions is equivalent to the knowledge of the vector $\vec{J}$ and the correlation matrix C defined as

$$C_{kl} := \frac{1}{2}\langle J_k J_l + J_l J_k\rangle \tag{2}$$

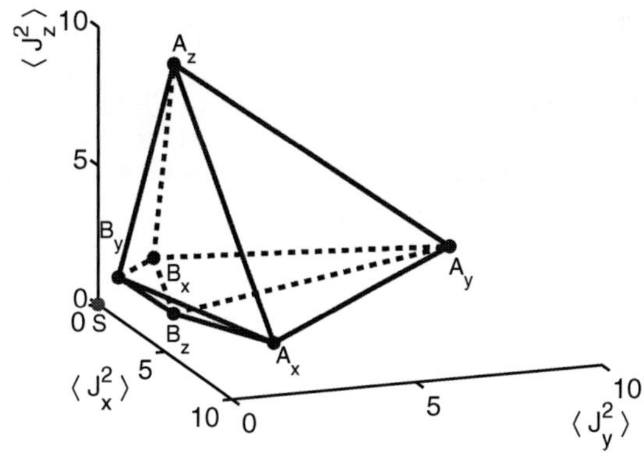

FIGURE 1. The polytope of separable states corresponding to Observation 1 for $N = 6$ and $\vec{J} = 0$. S corresponds to a many body singlet state.

for $k, l = x, y, z$. For convenience, we also define the covariance matrix γ with elements

$$\gamma_{kl} \;:=\; C_{kl} - \langle J_k \rangle \langle J_l \rangle. \tag{3}$$

When changing the coordinate system to x', y', z', vector $\vec{J}$ and the matrices C and γ transform as $\vec{J} \mapsto O\vec{J}$, $C \mapsto OCO^T$ and $\gamma \mapsto O\gamma O^T$ where O is an orthogonal 3×3-matrix. Looking at the inequalities of Observation 1 one finds that the first two inequalities are invariant under a change of the coordinate system. Concerning Eq. (1c), we can reformulate it as

$$\langle J_i^2 \rangle + \langle J_j^2 \rangle + \langle J_k^2 \rangle - \tfrac{N}{2} \leq (N-1)(\Delta J_k)^2 + \langle J_k^2 \rangle. \tag{4}$$

Then, the left hand side is again invariant under rotations, and we find a violation of Eq. (1c) in some direction if the minimal eigenvalue of

$$\mathfrak{X} := (N-1)\gamma + C \tag{5}$$

is smaller than $\mathrm{Tr}(C) - \frac{N}{2}$. Similarly, we find a violation of Eq. (1d) if the largest eigenvalue of $\mathfrak{X}$ exceeds $(N-1)\mathrm{Tr}(\gamma) - N(N-2)/4$.

Observation 2. *We can rewrite our conditions Eqs. (1) in a coordinate system independent way as [6]*

$$\begin{aligned} \mathrm{Tr}(C) &\leq \tfrac{N(N+2)}{4}, && \text{(6a)} \\ \mathrm{Tr}(\gamma) &\geq \tfrac{N}{2}, && \text{(6b)} \\ \lambda_{\min}(\mathfrak{X}) &\geq \mathrm{Tr}(C) - \tfrac{N}{2}, && \text{(6c)} \\ \lambda_{\max}(\mathfrak{X}) &\leq (N-1)\mathrm{Tr}(\gamma) - \tfrac{N(N-2)}{4}, && \text{(6d)} \end{aligned}$$

where $\lambda_{\min}(A)$ and $\lambda_{\max}(A)$ are the smallest and largest eigenvalue of matrix A, respectively. If Eqs. (1) are violated by a quantum state for any choice of coordinate axes x, y, and z then Eqs. (6) are also violated [8].

BOUND ENTANGLEMENT IN SPIN SYSTEMS

One can now ask, what kind of entanglement is detected by our inequalities? Our conditions need only the measurement of two-body correlations and do not have information on higher order correlations. Do they detect only states for which the reduced two-qubit state is entangled? It can be shown that our inequalities can detect entangled states that have a separable two-qubit density matrix, and even states that have a positive partial transpose with respect to all bipartitons. Such entangled states cannot be detected by the Peres-Horodecki criterion [7]. Our criteria can be used to show that such states appear in the thermal state of many spin models, such as for example the Heisenberg and XY models. This way we show that fully PPT bound entanglement appears under natural circumstances.

SUMMARY

We presented the full set of generalized spin squeezing inequalities for multi-qubit systems. These detect all entangled states that can be detected knowing the collective spin components and their variances. We also presented another set of criteria that detect all entanglement that can be detected knowing the expectation values of the collective spin components and the correlation matrix. Our conditions detect multi-qubit bound entanglement in the thermal state of several well-known spin models.

ACKNOWLEDGMENTS

We thank A. Acín, A. Cabello, M. Christandl, J.I. Cirac, S.R. de Echaniz, K. Hammerer, S. Iblisdir, M. Koschorreck, J. Korbicz, J.I. Latorre, M. Lewenstein, M.W. Mitchell, L. Tagliacozzo, and R.F. Werner for fruitful discussions. We also thank the support of the EU (OLAQUI, SCALA, QICS), the National Research Fund of Hungary OTKA (Contract No. T049234), and the Hungarian Academy of Sciences (Bolyai Programme). This work was further supported by the FWF and the Spanish MEC (Ramon y Cajal Programme, Consolider-Ingenio 2010 project "QOIT"). G.T. thanks Ikerbasque, the Basque Foundation for Science.

REFERENCES

1. M. Kitagawa and M. Ueda, Phys. Rev. A **47**, 5138 (1993).
2. D.J. Wineland , J.J. Bollinger, and W.M. Itano, and D.J. Heinzen, Phys. Rev. A **50**, 67 (1994).
3. A. Sørensen, L.-M. Duan, J.I. Cirac, and P. Zoller, Nature **409**, 63 (2001).
4. G. Tóth, Phys. Rev. A **69**, 052327 (2004); J. Korbicz, J.I. Cirac, and M. Lewenstein, Phys. Rev. Lett. **95**, 120502 (2005); G. Tóth, J. Opt. Soc. Am. B **24**, 275 (2007).
5. J. Korbicz *et al.*, Phys. Rev. A **74**, 052319 (2006); N. Kiesel *et al.*, Phys. Rev. Lett. **98**, 063604 (2007).
6. G. Tóth, C. Knapp, O. Gühne, and H.J. Briegel, Phys. Rev. Lett. **99**, 250405 (2007); arxiv:0806.1048.
7. A. Peres, Phys. Rev. Lett. **77**, 1413 (1996); M. Horodecki, P. Horodecki, R. Horodecki, Phys. Lett. A **223**, 1 (1996).
8. Command `optspinsq.m` of the QUBIT4MATLAB V3.0 package computes the violations of Eqs. (6) for a quantum state given as input. See G. Tóth, Comput. Phys. Comm. **179**, 430 (2008).

CHAPTER 2

QUANTUM INFORMATION THEORY

Optimal covariant quantum networks

Giulio Chiribella, Giacomo Mauro D'Ariano, Paolo Perinotti

QUIT *Group, Dipartimento di Fisica "A. Volta", via Bassi 6, 27100 Pavia, Italy*

Abstract. A sequential network of quantum operations is efficiently described by its *quantum comb* [1], a non-negative operator with suitable normalization constraints. Here we analyze the case of networks enjoying symmetry with respect to the action of a given group of physical transformations, introducing the notion of covariant combs and testers, and proving the basic structure theorems for these objects. As an application, we discuss the optimal alignment of reference frames (without pre-established common references) with multiple rounds of quantum communication, showing that *i)* allowing an arbitrary amount of classical communication does not improve the alignment, and *ii)* a single round of quantum communication is sufficient.

A quantum comb [1] describes a quantum network with N open slots in which an ordered sequence of variable quantum devices can be inserted, thus programming the quantum operation of the resulting circuit. Mathematically, a comb implements an *admissible supermap* [2, 3], that transforms an input network of N quantum operations into an output quantum operation. Having at disposal a suitable formalism opens the possibility of optimizing the architecture of quantum circuits for a large number of computational, cryptographic, and game-theoretical tasks, such as discrimination of single-party strategies, cloning of quantum transformations, and storing of quantum algorithms into quantum memories [1, 4, 5, 6]. For example, quantum combs allow one to find the optimal networks for the estimation of an unknown group transformation with N uses at disposal, a problem that has been solved in the past only in the particular case of phase estimation [7]. Using combs and supermaps one can prove in full generality that a parallel disposition of the N black boxes is sufficient to achieve the optimal estimation of the unknown group element [5], thus reducing the problem to the optimal parallel estimation of group transformations already solved in Ref. [8].

In this paper we summarize the main concepts and methods developed so far in the optimization of quantum networks, with focus on the case of networks with symmetry properties, and we present an original result on multi-round protocols for reference frame alignment.

1. BASIC NOTIONS OF QUANTUM CIRCUITS ARCHITECTURE

1.1. Quantum N-combs

Consider a sequential network of N quantum operations (QOs) with memory, as in Fig. 1. Due to the presence of internal memories, there can be other networks that are indistinguishable from it in all experiments that involve only the incoming and outgoing quantum systems. The quantum comb is the equivalence class of all networks having

CP1110, *Quantum Communication, Measurement and Computing (QCMC), The Ninth International Conference,* edited by A. Lvovsky

the same input/output relations, irrespectively of what happens inside. The equivalence

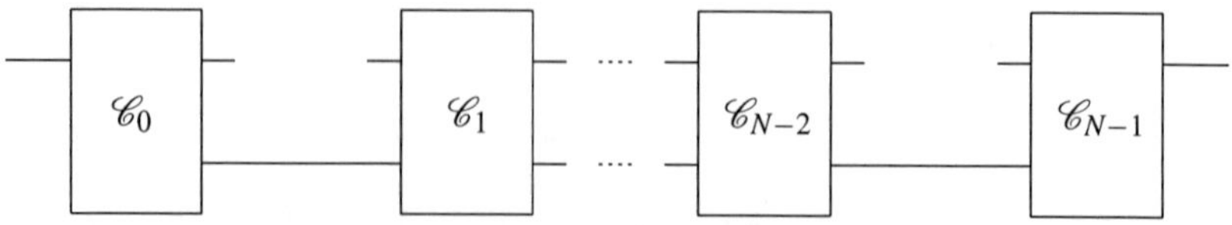

FIGURE 1. N-comb: sequential network of N quantum operations with memory. The network contains input and output systems (free wires in the diagram), as well as internal memories (wires connecting the boxes).

class is in one-to-one correspondence with the Choi operator of the network, which can be computed as the *link product* [1] of the Choi operators of the QOs $(\mathscr{C}_i)_{i=0}^{N-1}$. Here we adopt the convention that the input (output) spaces for the QO $\mathscr{C}_j$ are indicated as $\mathscr{H}_{2j}$ ($\mathscr{H}_{2j+1}$). Accordingly, the Choi operator of the network is a non-negative operator $R^{(N)} \in \mathsf{Lin}\left(\bigotimes_{j=0}^{2N-1} \mathscr{H}_j\right)$. The quantum comb can be then identified with such a Choi operator. For networks of channels (trace-preserving QOs) one has the recursive normalization condition

$$\mathrm{Tr}_{2k-1}[R^{(k)}] = I_{2k-2} \otimes R^{(k-1)} \qquad k = 1, \dots, N \tag{1}$$

where $R^{(k)} \in \mathsf{Lin}\left(\bigotimes_{j=0}^{2k-1} \mathscr{H}_j\right)$, and $R^{(0)} = 1$. Eq. (1) is the translation in terms of Choi operators of the fact that the N-partite channel $\mathscr{R}^{(N)} = \mathscr{C}_{N-1} \circ \mathscr{C}_{N-2} \circ \cdots \circ \mathscr{C}_0$, sending states on the even Hilbert spaces $\mathsf{St}(\bigotimes_{k=0}^{N-1} \mathscr{H}_{2k})$ to states on the odd ones $\mathsf{St}(\bigotimes_{k=0}^{N-1} \mathscr{H}_{2k+1})$, is a deterministic *causal automaton* [9, 10], namely a channel where the reduced dynamics of an input state at step k depends only on input states at steps $k' \leq k$, and not at steps $k' > k$. With different motivations from supermaps and causal automata, Eq. (1) also appeared in the work by Gutoski and Watrous toward a general formulation of quantum games [4].

We call $\mathsf{DetComb}\left(\bigotimes_{j=0}^{2N-1} \mathscr{H}_j\right)$ the set of non-negative operators satisfying Eq.(1), and $\mathsf{ProbComb}\left(\bigotimes_{j=0}^{2N-1} \mathscr{H}_j\right)$ the set

$$\mathsf{ProbComb}\left(\bigotimes_{j=0}^{2N-1} \mathscr{H}_j\right) = \left\{ R^{(N)} \geq 0 \mid \exists S^{(N)} \in \mathsf{DetComb}\left(\bigotimes_{j=0}^{2N-1} \mathscr{H}_j\right) : R^{(N)} \leq S^{(N)} \right\}. \tag{2}$$

It is possible to prove that any operator $R^{(N)} \in \mathsf{DetComb}\left(\bigotimes_{j=0}^{2N-1} \mathscr{H}_j\right)$ is the Choi operator of some sequential network of N channels, or, equivalently, of some causal channel $\mathscr{R}^{(N)}$ [9, 3]. The minimal Stinespring dilation of the channel in terms of the Choi operator is given by [11]

$$\mathscr{R}^{(N)}(\rho) = \mathrm{Tr}_A[V \rho V^\dagger] \qquad V = \left(I_{odd} \otimes \sqrt{R^{(N)\tau}}\right) (|I_{odd}\rangle\rangle \otimes I_{even}) \,, \tag{3}$$

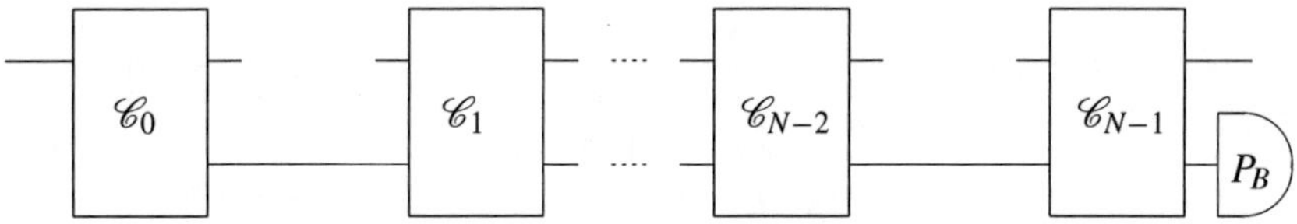

FIGURE 2. N-instrument: sequence of N channels followed with postselection on the last ancilla.

where $\mathscr{H}_{odd} = \bigotimes_{k=0}^{N-1} \mathscr{H}_{2k+1}$, $\mathscr{H}_{even} = \bigotimes_{k=0}^{N-1} \mathscr{H}_{2k}$, τ denotes transposition w.r.t. a fixed orthonormal basis, $\mathscr{H}_A = \mathsf{Supp}\left(R^{(N)\tau}\right)$ is the minimal ancilla space, $|I_{odd}\rangle\rangle$ is the unnormalized maximally entangled state on $\mathscr{H}_{odd}^{\otimes 2}$, and V is an isometry from $\mathscr{H}_{even}$ to $\mathscr{H}_{odd} \otimes \mathscr{H}_A$.

1.2. Quantum N-instruments

Let Ω be a measurable space and $\sigma(\Omega)$ be its σ-algebra of events. A quantum N-instrument $R^{(N)}$ on $\bigotimes_{j=0}^{2N-1} \mathscr{H}_j$ is an operator-valued measure that associates to any event $B \in \sigma(\Omega)$ an N-comb $R_B^{(N)} \in \mathsf{ProbComb}\left(\bigotimes_{j=0}^{2N-1} \mathscr{H}_j\right)$, and satisfies the normalization

$$R_\Omega^{(N)} \in \mathsf{DetComb}\left(\bigotimes_{j=0}^{2N-1} \mathscr{H}_j\right). \tag{4}$$

Theorem 1 (Dilation of N-instruments) *For any N-instrument $R^{(N)}$ on $\bigotimes_{j=0}^{2N-1} \mathscr{H}_j$ there exist a deterministic N-comb $S^{(N)} \in \mathsf{DetComb}\left(\bigotimes_{j=0}^{2N-1} \mathscr{H}'_j\right)$ with $\mathscr{H}'_j = \mathscr{H}_j$ for $j = 0, \ldots, 2N-2$, and $\mathscr{H}'_{2N-1} = \mathscr{H}_{2N-1} \otimes \mathscr{H}_A$, and a POVM P on the ancilla $\mathscr{H}_A$ such that*

$$R_B^{(N)} = \mathrm{Tr}_A\left[S^{(N)}\left(I_0 \otimes \cdots \otimes I_{2N-1} \otimes P_B^\tau\right)\right] \qquad \forall B \in \sigma(\Omega), \tag{5}$$

τ denoting transposition w.r.t. a fixed orthonormal basis.

The meaning of the theorem is that a quantum N-instrument can be always achieved by a network of N channels with postselection induced by the measurement on an ancilla exiting from the N-th channel, as in Fig. 2.

Proof. Diagonalize $R_\Omega^{(N)}$ as $R_\Omega^{(N)} = \sum_{i=1}^r \lambda_i |\phi_i\rangle\langle\phi_i|$, and take $\mathscr{H}_A = \mathsf{Supp}\left(R_\Omega^{(N)\tau}\right) = \mathsf{Span}\{|\phi_i^*\rangle \mid i = 1, \ldots, r\}$, $|\phi_i^*\rangle := \sum_n \langle n|\phi_i\rangle^* |n\rangle$. Consider the purification $S^{(N)} = |R_\Omega^{(N)\frac{1}{2}}\rangle\rangle\langle\langle R_\Omega^{(N)\frac{1}{2}}|$ where $|R_\Omega^{(N)\frac{1}{2}}\rangle\rangle = \sum_i \sqrt{\lambda_i} |\phi_i\rangle |\phi_i^*\rangle \in \left(\bigotimes_{j=0}^{2N-1} \mathscr{H}_j\right) \otimes \mathscr{H}_A$. By construction, $S^{(N)}$ is a deterministic comb in $\mathsf{DetComb}\left(\bigotimes_{j=0}^{2N-1} \mathscr{H}'_j\right)$. Now define the POVM P by $P_B = \left[R_\Omega^{(N)}\right]^{-\frac{1}{2}} R_B^{(N)} \left[R_\Omega^{(N)}\right]^{-\frac{1}{2}}$, where $\left[R_\Omega^{(N)}\right]^{-\frac{1}{2}}$ is the inverse of $R_\Omega^{(N)\frac{1}{2}}$ on its support. It is immediate to check that for any event $B \in \sigma(\Omega)$ we have $R_B^{(N)} = \mathrm{Tr}_A[S^{(N)}\,(I_0 \otimes \cdots \otimes I_{2N-1} \otimes P_B^\tau)]$. ■

This theorem is similar in spirit to Ozawa's dilation theorem for quantum instruments [12]. The important difference here that P is a POVM on a finite-dimensional ancilla space, rather than a von Neumann measurement in infinite dimension.

1.3. Quantum N-testers

An N-tester $T^{(N)}$ is an $(N+1)$-instrument where the first and last Hilbert spaces, $\mathscr{H}_0$ and $\mathscr{H}_{2N+1}$, respectively, are one-dimensional. Accordingly, we can shift back by one unit the numeration of Hilbert spaces, so that, if $B \in \sigma(\Omega)$ is an event, then $T_B^{(N)}$ is an operator on $\bigotimes_{j=0}^{2N-1} \mathscr{H}_j$. With this shifting, the normalization of the tester is given by

$$\begin{aligned} &T_\Omega^{(N)} = I_{2N-1} \otimes \Xi^{(N-1)} \\ &\mathrm{Tr}_{2k-2}[\Xi^{(k)}] = I_{2k-3} \otimes \Xi^{(k-1)} \qquad k = 2, \ldots, N \\ &\mathrm{Tr}_0[\Xi^{(1)}] = 1 \ , \end{aligned} \tag{6}$$

with $\Xi^{(k)} \in \mathsf{Lin}\left(\bigotimes_{j=0}^{2k-2} \mathscr{H}_j\right)$.

A tester represents a quantum network starting with a state preparation and finishing with a measurement on the ancilla. When such a network is connected to a network of N quantum operations as in Fig. 3, the only outputs are measurement outcomes.

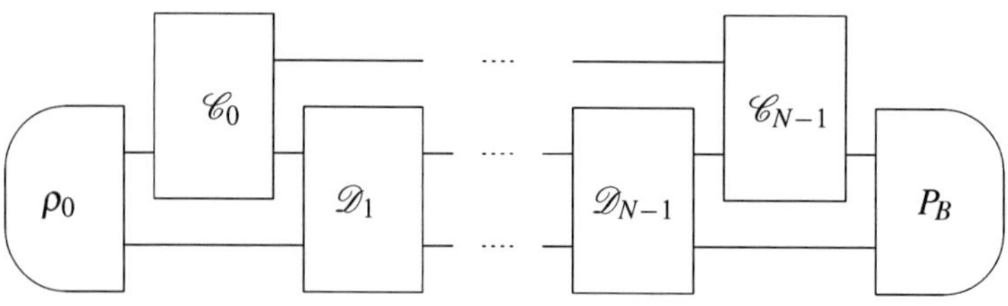

FIGURE 3. Testing a network of N quantum operations $(\mathscr{C}_i)_{i=0}^{N-1}$. The N-tester consists in the preparation of an input state ρ_0, followed by channels $\{\mathscr{D}_1, \ldots, \mathscr{D}_{N-1}\}$, and a final measurement P_B.

Precisely, if the comb of the measured network is $R^{(N)} \in \mathsf{ProbComb}\left(\bigotimes_{j=0}^{2N-1} \mathscr{H}_j\right)$, then the probabilities of any event are given by the generalized Born rule [1, 13]

$$p(B|R^{(N)}) = \mathrm{Tr}[T_B^{(N)\tau}\, R^{(N)}] \qquad \forall B \in \sigma(\Omega) \ . \tag{7}$$

For deterministic combs $R^{(N)} \in \mathsf{DetComb}\left(\bigotimes_{j=0}^{2N-1} \mathscr{H}_j\right)$ the probabilities sum up to one:

$$p(\Omega|R^{(N)}) = \mathrm{Tr}[T_\Omega^{(N)\tau} R^{(N)}] = 1 \ . \tag{8}$$

Clearly, since $T^{(N)\tau}$ is also a tester, the Born rule can be written in the familiar way without the transpose. However, here we preferred to write probabilities in terms of the combs $R^{(N)}$ and $T_B^{(N)}$ of the measured and measuring networks, respectively. In fact, the Born rule is nothing but a particular case of link product [1], and the transpose appears as the signature of the interlinking of the two networks.

Proposition 1 (Decomposition of N-testers [5]) *Let $T^{(N)}$ be a quantum N-tester on $\bigotimes_{j=0}^{2N-1}\mathscr{H}_j$, and consider the ancilla space $\mathscr{H}_A = \mathsf{Supp}\left(T_\Omega^{(N)\tau}\right)$. Let $\mathscr{S}$ be the linear supermap from $\mathsf{ProbComb}\left(\bigotimes_{j=0}^{2N-1}\mathscr{H}_j\right)$ to $\mathsf{St}(\mathscr{H}_A)$ given by*

$$\mathscr{S}(R^{(N)}) = \left[T_\Omega^{(N)\tau}\right]^{\frac{1}{2}} R^{(N)} \left[T_\Omega^{(N)\tau}\right]^{\frac{1}{2}} \tag{9}$$

and P be the POVM on $\mathscr{H}_A$ defined by

$$P_B = \left[T_\Omega^{(N)}\right]^{-\frac{1}{2}} T_B^{(N)} \left[T_\Omega^{(N)}\right]^{-\frac{1}{2}} . \tag{10}$$

The supermap $\mathscr{S}$ transforms deterministic combs into normalized states of the ancilla. The probabilities of events are given by

$$p(B|R^{(N)}) = \mathrm{Tr}[T_B^{(N)\tau} R^{(N)}] = \mathrm{Tr}[P_B^\tau \mathscr{S}(R^{(N)})] . \tag{11}$$

This proposition reduces any measurement on an input quantum network to a measurement on a suitable state, which is obtained by linear transformation of the input comb. As we will see in the following, this simple result has very strong consequences in quantum estimation.

Proof. If $R^{(N)}$ is in $\mathsf{DetComb}\left(\bigotimes_{j=0}^{2N-1}\mathscr{H}_j\right)$, then $\mathrm{Tr}[\mathscr{S}(R^{(N)})] = \mathrm{Tr}[T_\Omega^{(N)\tau} R^{(N)}] = 1$, having used Eq. (8). Eq. (11) is an obvious consequence of the definitions of $\mathscr{S}$ and P. ■

Proposition 1 reduces the discrimination of two networks to the discrimination of two states. This allows us to define an operational notion of distance between networks [5], whose meaning is directly related to minimum error discrimination:

$$\left\| R^{(N)} - R^{(N)'} \right\|_{op} = \max_{T_\Omega^{(N)}} \left\| \left[T_\Omega^{(N)\tau}\right]^{\frac{1}{2}} \left(R^{(N)} - R^{(N)'}\right) \left[T_\Omega^{(N)\tau}\right]^{\frac{1}{2}} \right\|_1 , \tag{12}$$

with $\|A\|_1 = \mathrm{Tr}\,|A|$. Remarkably, the above norm can be strictly greater than the diamond (cb) norm of the difference $\mathscr{R}^{(N)} - \mathscr{R}^{(N)'}$ of the two multipartite channels [5]. This means that a scheme such as in Fig. 3 can achieve a strictly better discrimination than a parallel scheme where a multipartite entangled state is fed in the unknown channel and a multipartite measurement is performed on the output.

2. COVARIANT QUANTUM NETWORKS

2.1. Covariant N-combs

Let $\mathbf{G}$ be a group, acting on the Hilbert space $(\mathscr{H}_j)_{j=0}^{2N-1}$ via the a unitary representation $\{U_{g,j} \mid g \in \mathbf{G}\}$. Denote by $\mathscr{U}_{g,j}$ the map $\mathscr{U}_{g,j}(\rho) = U_{g,j}\rho U_{g,j}^{\dagger}$. Suppose that the causal channel $\mathscr{R}^{(N)}$ from $\mathsf{St}(\bigotimes_{k=0}^{N-1}\mathscr{H}_{2k})$ to $\mathsf{St}(\bigotimes_{k=0}^{N-1}\mathscr{H}_{2k+1})$ is *covariant*, namely

$$\mathscr{R}^{(N)} \circ \left(\bigotimes_{k=0}^{N-1} \mathscr{U}_{g,2k}\right)(\rho) = \left(\bigotimes_{k=0}^{N-1} \mathscr{U}_{g,2k+1}\right) \circ \mathscr{R}^{(N)}(\rho) \,. \tag{13}$$

Then the corresponding comb, which we call *covariant* either, satisfies the commutation property

$$\left[R^{(N)}, \bigotimes_{k=0}^{N-1} (U_{g,2k+1} \otimes U_{g,2k}^{*})\right] = 0 \qquad \forall g \in \mathbf{G} \,. \tag{14}$$

For covariant combs, the minimal dilation of the memory channel $\mathscr{R}^{(N)}$ given by Eq. (3) satisfies the commutation relation

$$\left[\left(\bigotimes_{k=0}^{N-1} U_{g,2k+1}\right) \otimes U_{g,A}\right] V = V \left(\bigotimes_{k=0}^{N-1} U_{g,2k}\right) \,, \tag{15}$$

where $U_{g,A}$ is the compression of $\left(\bigotimes_{k=0}^{N-1}(U_{g,2k+1}^{*} \otimes U_{g,2k})\right)$ to the invariant subspace $\mathscr{H}_A = \mathsf{Supp}\left(R^{(N)\tau}\right)$.

2.2. Covariant N-instruments and testers

Suppose that the group $\mathbf{G}$ acts on the outcome space Ω. For $B \in \sigma(\Omega)$, denote by $gB := \{g\omega \mid \omega \in B\}$. A covariant N-instrument $R^{(N)}$ is defined by the property

$$R_{gB}^{(N)} = \left(\bigotimes_{k=0}^{N-1} (\mathscr{U}_{g,2k+1} \otimes \mathscr{U}_{g,2k}^{*})\right) \left(R_B^{(N)}\right) \,. \tag{16}$$

A covariant tester is simply a covariant N-instrument with one-dimensional $\mathscr{H}_0$ and $\mathscr{H}_{2N-1}$ and with all remaining labels shifted back by unit. We now suppose that $\mathbf{G}$ is compact and Ω is transitive, i.e. for any pair $\omega_1, \omega_2 \in \Omega$ there always exists a group element $g \in \mathbf{G}$ such that $\omega_2 = g\omega_1$.

Theorem 2 (Structure of covariant N-instruments/testers) *Let $\mathbf{G}$ be compact and Ω be transitive, with normalized Haar measure $\mathrm{d}\,\omega$. Let $\omega_0 \in \Omega$ be a point of Ω, and let $\mathbf{G}_0 = \{g \in \mathbf{G} \mid g\omega_0 = \omega_0\}$ be the stabilizer of ω_0. Let $\sigma : \Omega \to \mathbf{G}$ be a measurable section,* *such that $\omega = \sigma_\omega \omega_0$. If $R^{(N)}$ is a covariant instrument, then there exists a non-negative operator $D_0^{(N)}$ such that*

$$
\begin{aligned}
R_B^{(N)} &= \int_B \mathrm{d}\omega\, D_\omega^{(N)} \\
D_\omega^{(N)} &= \left(\bigotimes_{k=0}^{N-1} (U_{\sigma_\omega,2k+1} \otimes U^*_{\sigma_\omega,2k})\right) D_0^{(N)} \left(\bigotimes_{k=0}^{N-1} (U_{\sigma_\omega,2k+1} \otimes U^*_{\sigma_\omega,2k})\right)^\dagger \qquad (17) \\
&\left[D_0^{(N)}, \bigotimes_{k=0}^{N-1} (U_{g_0,2k+1} \otimes U^*_{g_0,2k})\right] = 0 \qquad \forall g_0 \in \mathbf{G}_0 \,.
\end{aligned}
$$

Proof. Simple generalization of the standard proof for covariant POVMs [14].

For a covariant N-instrument/tester, Eq. (16) implies the commutation

$$\left[R_\Omega^{(N)}, \bigotimes_{k=0}^{N-1} (U_{g,2k+1} \otimes U^*_{g,2k})\right] = 0 \qquad \forall g \in \mathbf{G} \,. \tag{18}$$

This implies additional group structure in the results of Theorem 1 and Proposition 1. In particular, for covariant testers, the map $\mathscr{S} : \mathsf{ProbComb}\left(\bigotimes_{j=0}^{2N-1} \mathscr{H}_j\right) \to \mathsf{St}(\mathscr{H}_A)$ is a *covariant supermap*:

$$\mathscr{S} \circ \left(\bigotimes_{k=0}^{N-1} (\mathscr{U}^*_{g,2k+1} \otimes \mathscr{U}_{g,2k})\right) \left(R^{(N)}\right) = U_{g,A}\, \mathscr{S}(R^{(N)})\, U^\dagger_{g,A} \,, \tag{19}$$

where $U_{g,A}$ is the compression of $\bigotimes_{k=0}^{N-1} \left(U^*_{g,2k+1} \otimes U_{g,2k}\right)$ to the invariant subspace $\mathscr{H}_A = \mathsf{Supp}\left(T_\Omega^{(N)\tau}\right) \subseteq \bigotimes_{j=0}^{2N-1} \mathscr{H}_j$.

3. OPTIMAL COVARIANT ESTIMATION OF QUANTUM NETWORKS

Let $\left\{R_\omega^{(N)} \in \mathsf{DetComb}\left(\bigotimes_{j=0}^{2N-1} \mathscr{H}_j\right) \mid \omega \in \Omega\right\}$ be a family of quantum networks parametrized by ω. We now want to find the optimal tester to estimate the parameter ω. For simplicity, we consider here the special case in which $\Omega \equiv \mathbf{G}$, for some compact group $\mathbf{G}$, and $R_g^{(N)}$ has the form

$$R_g^{(N)} = \left(\bigotimes_{k=0}^{N-1} (U_{g,2k+1} \otimes U^*_{g,2k})\right) R_0^{(N)} \left(\bigotimes_{k=0}^{N-1} (U_{g,2k+1} \otimes U^*_{g,2k})\right)^\dagger \tag{20}$$

Let $c(\hat{g}, g)$ be a cost function, penalizing the differences between the estimated parameter $\hat{g}$ and the true one g. Suppose that $c(\hat{g}, g)$ is left-invariant, namely $c(h\hat{g}, hg) =$

$c(\hat{g},g)\ \forall h \in \mathbf{G}$. The optimal estimation is then given by the tester $T^{(N)}$ that minimizes the average cost

$$\langle c \rangle = \int_{\mathbf{G}} \mathrm{d}g \int_{\mathbf{G}} c(\hat{g},g)\ \mathrm{Tr}[T_{\mathrm{d}\hat{g}}^{(N)\tau} R_g^{(N)}] , \tag{21}$$

where $\mathrm{d}g$ is the normalized Haar measure, and $\int_{\mathbf{G}} f(\hat{g})\ \mathrm{Tr}\left[T_{\mathrm{d}\hat{g}}^{(N)\tau} R^{(N)}\right]$ denotes integration of f against the scalar measure $\mu_B = \mathrm{Tr}[T_B^{(N)\tau} R^{(N)}]$. An alternative notion of optimality is the minimization of the worst-case cost

$$c_{wc} = \max_{g \in \mathbf{G}} \left(\int_{\mathbf{G}} c(\hat{g},g)\ \mathrm{Tr}[T_{\mathrm{d}\hat{g}}^{(N)\tau} R_g^{(N)}] \right) . \tag{22}$$

However, it it easy to prove that in the covariant setting it is sufficient to consider covariant testers, for which the average and worst-case cost coincide:

Theorem 3 *There exists a covariant tester $T_B^{(N)} = \int_B \mathrm{d}g\, D_g^{(N)}$, with density*

$$D_g^{(N)} = \left(\bigotimes_{k=0}^{N-1} (U_{g,2k+1}^* \otimes U_{g,2k}) \right) D_0^{(N)} \left(\bigotimes_{k=0}^{N-1} (U_{g,2k+1}^* \otimes U_{g,2k}) \right)^{\dagger} \tag{23}$$

that is optimal both for the average and worst-case cost.

Proof. The standard averaging argument [14]: if $T^{(N)}$ is an optimal tester, then the tester $\overline{T^{(N)}}$ defined by $\overline{T_B^{(N)}} = \int_{\mathbf{G}} \mathrm{d}h \bigotimes_{k=0}^{N-1} (\mathscr{U}_{h,2k+1}^* \otimes \mathscr{U}_{h,2k}) \left(T_{h^{-1}B}^{(N)} \right)$ is covariant and has the same average and worst-case cost as $T^{(N)}$. Moreover, for any covariant tester, the average and worst-case cost coincide.

4. APPLICATIONS

4.1. Optimal estimation of group transformations with N copies

Suppose we have at disposal N uses of a black box performing the unknown group transformation U_g, and that we want to find the optimal network for estimating g. In this case the parametric family of networks is $R_g^{(N)} = (|U_g\rangle\rangle\langle\langle U_g|)^{\otimes N}$, where $|U_g\rangle\rangle := (U_g \otimes I)|I\rangle\rangle$, $|I\rangle\rangle = \sum_{i=1}^d |i\rangle|i\rangle$. Using Proposition 1 and Eq. (19), the optimal estimation on these networks is reduced to the optimal estimation on the ancilla states $\rho_g = \mathscr{S}\left(R_g^{(N)}\right) = U_{g,A} \mathscr{S}\left(R_0^{(N)}\right) U_{g,A}^{\dagger}$, with $R_0^{(N)} = (|I\rangle\rangle\langle\langle I|)^{\otimes N}$. Since the ancilla space is an invariant subspace of $\bigotimes_{j=0}^{2N-1} \mathscr{H}_j$ and the representation $U_{g,A}$ is a sub-representation of $\bigotimes_{k=0}(U_{g,2k+1} \otimes I_{2k})$, it is clear that the minimum cost in the estimation is lower bounded by the minimum cost achievable in a parallel scheme, where the unitary $U_g^{\otimes N} \otimes I_{ref}$ is applied to a multipartite entangled state in $\mathsf{St}(\bigotimes_{k=1}^N \mathscr{H}_k \otimes \mathscr{H}_{ref})$, with $\mathscr{H}_{ref}$ suitable reference space. In this way the optimal estimation is reduced to the optimal parallel estimation of Ref. [8].

4.2. Optimal alignment of reference frames with multi-round protocols

Two distant parties Alice and Bob, who lack a shared reference frame, can try to establish one by sending suitable physical systems, such as clocks and gyroscopes for time and orientation references, respectively. In the quantum scenario, the role of elementary clocks and gyroscopes is played by spin $1/2$ particles, and it has been shown that the optimal protocol using N particles in a single round of quantum communication from Alice to Bob has a r.m.s. error scaling to zero as $1/N$ (with suitable constants) for both for clock synchronization [15] and Cartesian axes alignment [16]. However, the optimal protocol for establishing reference frames with many rounds of quantum communication and arbitrary amount of classical communication has been not analyzed yet. In principle, an adaptive strategy might improve the alignment, if not by changing the scaling with N, at least by improving the constant. With the formalism of covariant combs and testers, however, it is rather straightforward to prove that this is not the case.

Let us consider the general case in which the mismatch between Alice's and Bob's reference frames is represented by an unknown element g of some group of physical transformations $\mathbf{G}$. The unitary (projective) representation in the Hilbert spaces of quantum systems yields the passive transformation of states due to the change from Alice's to Bob's viewpoint: a single-particle state that is $|\psi^{(A)}\rangle$ is Alice's reference frame becomes $|\psi^{(B)}\rangle = U_g|\psi^{(A)}\rangle$ in Bob's one, a single-particle operator $O^{(A)}$ becomes $O^{(B)} = U_g O^{(A)} U_g^\dagger$, and a single-particle operation $\mathscr{C}^{(A)}$ becomes $\mathscr{C}^{(B)} = \mathscr{U}_g \mathscr{C}^{(A)} \mathscr{U}_g^\dagger$. Consider a protocol with $2r$ rounds of quantum communication (r rounds from Alice to Bob and r from Bob to Alice) with q_i quantum particles exchanged per round. We also allow an unbounded amount of classical communication, represented by the exchange of $\mathbf{G}$-invariant systems prepared in classical (diagonal) states. The goal of the protocol is to give the best possible estimate $\hat{g}$ of the mismatch g. Notice that, since Alice and Bob are not restricted in sending classical data, we can imagine without loss of generality that the estimate $\hat{g}$ is produced by Bob (if it were produced by Alice, she could always transmit this classical information to Bob). The protocol is then represented by the interlinking of two networks of quantum operations: *i)* Alice's network is a deterministic r-comb $R^{(r,A)} \in \mathsf{DetComb}(\mathscr{H}_{A\to B} \otimes \mathscr{H}_{B\to A} \otimes \mathscr{H}_C)$, where $\mathscr{H}_{A\to B}$ ($\mathscr{H}_{B\to A}$) is the Hilbert space of all particles sent from Alice to Bob (from Bob to Alice), and $\mathscr{H}_C$ is the Hilbert space of the invariant systems used for classical communication, and *ii)* Bob's network is an r-tester $T^{(r,B)}_{\mathrm{d}\hat{g}}$ on the same Hilbert spaces. When switching to Bob's reference frame, all Alice's operations are conjugated by unitaries, and her comb becomes

$$R_g^{(r,B)} = \left(U_g^{\otimes N_{A\to B}} \otimes U_g^{*\otimes N_{B\to A}} \otimes I_C\right) R^{(r,A)} \left(U_g^{\otimes N_{A\to B}} \otimes U_g^{*\otimes N_{B\to A}} \otimes I_C\right) . \qquad (24)$$

where $N_{A\to B}$ ($N_{B\to A}$) is the number of particles traveling from Alice to Bob (from Bob to Alice). Notice that we have the identity I_C on the classical systems, since classical communication (strings of bits) is invariant under changes of reference frame. Therefore, for any left-invariant cost function $c(\hat{g}, g)$ we are in the case of covariant network estimation treated before. The estimation of g from the networks $R_g^{(r,B)}$ is then reduced to the estimation of g from the states $\rho_g = \mathscr{S}(R_g^{(r,B)}) = U_{g,A}\rho_0 U_{g,A}^\dagger$, where $U_{g,A}$ is a sub-representation of $U_g^{\otimes N_{A\to B}} \otimes U_g^{*\otimes N_{B\to A}} \otimes I_C$. For $\mathbf{G} = U(1)$ and $\mathbf{G} = \mathbb{SU}(2)$ U_g and U_g^* are equivalent representations (up to global phases), hence this is exactly the same estimation that can be achieved by sending $N_{A\to B} + N_{B\to A}$ particles in a single round. Even for groups for which U_g and U_g^* are not equivalent (such as $\mathbb{SU}(d)$), one can achieve the same estimation precision in a single round by sending $N_{A\to B}$ particles and $N_{B\to A}$ charge-conjugate particles from Alice to Bob. This proves that anyway there is no advantage in using more than one round of quantum communication, and that classical communication is completely useless.

This work is supported by the EC through the project CORNER. G. C. acknowledges financial support from the Risk and Security Study Center, IUSS Pavia.

REFERENCES

1. G. Chiribella, G. M. D'Ariano, and P. Perinotti, Phys. Rev. Lett. **101**, 060401 (2008).
2. G. Chiribella, G. M. D'Ariano, and P. Perinotti, Europhys. Lett. **83**, 30004 (2008).
3. G. Chiribella, G. M. D'Ariano, and P. Perinotti, in preparation.
4. G. Gutoski and J. Watrous, in *Proc. of the 39th Annual ACM Symposium on Theory of Computation*, 565 (2007).
5. G. Chiribella, G. M. D'Ariano, and P. Perinotti, Phys. Rev. Lett. **101**, 180501 (2008).
6. G. Chiribella, G. M. D'Ariano, and P. Perinotti, Phys. Rev. Lett. **101**, 180504 (2008).
7. W. van Dam, G. M. D'Ariano, A. Ekert, C. Macchiavello, M. Mosca, Phys. Rev. Lett. **98**, 090501 (2007).
8. G. Chiribella, G. M. D'Ariano, and M. F. Sacchi, Phys. Rev. A **72**, 042338 (2005).
9. D. Kretschmann and R. F. Werner, Phys. Rev. A. **72**, 062323 (2005).
10. T. Eggeling, D. Schlingemann and R. F. Werner, Europhys. Lett. **57**, 782-788 (2002).
11. G. Chiribella, P. Perinotti, and G. M. D'Ariano, arXiv:0810.3211.
12. M. Ozawa, J. Math. Phys. **25** (1984), 79.
13. M. Ziman, Phys. Rev. A **77**, 062112 (2008).
14. A. S. Holevo *Probabilistic and Statistical Aspects of Quantum Theory*, North-Holland, Amsterdam, 1982.
15. V. Bužek, R. Derka, and S. Massar, Phys. Rev. Lett. **82**, 2207 (1999).
16. G. Chiribella, G. M. D'Ariano, P. Perinotti, and M. F. Sacchi, Phys. Rev. Lett. **93**, 180503 (2004).

Quantum Filtering and Control in Example: Unstable Qubits

V P Belavkin

School of Mathematics, University of Nottingham,
E-mail:vpb@maths.nott.ac.uk

Abstract. A brief account of the quantum unstable controlled dynamics admitting continuous demolition measurement is given. The dynamical programming method for optimization of quantum control with convex constraints and concave cost functions of quantum conditional state is presented. Consideration is given to open loop control corresponding to deterministic conditionally-Markov dynamics of a quantum unstable system. A single controlled decaying qubit conditioned by its survival is considered as an example.

Keywords: Quantum Filtering, Quantum Control, Unstable Qubit
PACS: 02.30.-Yy; 45.80.+r; 03.67.-a; 02.50.-r

INTRODUCTION

Quantum Cybernetics (QC) will be dealing with self-organizing nanomechanical, nanoelectrical or nanobiological systems described by *quantum laws* of physics, probability and information. Like the classical cybernetics is essentially the systems theory of classical bits with input controls, output observations and feedback in the description of Norbert Wiener, QC can be defined as quantum systems theory of open to observation and feedback control qubit systems. However, unlike in the classical case, quantum observed systems must decohere and may become even unstable under the demolition measurements. The aim of QC is therefore the dynamical optimization of quantum observation and feedback control on the purpose described by costs and bequest functionals of quantum states during the quantum information processing. The main ingredients of such theory is the theory of optimal quantum filtering and quantum feedback control theory based on the quantum stochastic innovation dynamics which was developed by the author since [1] in a serious cited in the review paper [2].

Here we give a brief account of this theory on an example of a decaying qubit under the continuous watching of the decreasing event of its survival. Such systems can be described as sub-Markov quantum processes governed by nonlinear nonstochastic master equation specified in this paper.

SOME FACTS AND NOTATIONS

Let $\mathbb{A} = \mathfrak{B}(\mathfrak{h})$ be the operator algebra on a Hilbert space $\mathfrak{h}$. The predual space $\mathbb{A}_\star$ can be realized by the densities $\rho \in \mathbb{A}$, $\mathrm{tr}|\rho| < \infty$ with respect to the trace tr such that the linear functional $\rho(\mathtt{A}) = \mathrm{tr}[\rho\mathtt{A}] \equiv \langle \rho, \mathtt{A} \rangle$ describes a quantum state defining the expectations of

CP1110, *Quantum Communication, Measurement and Computing (QCMC), The Ninth International Conference,* edited by A. Lvovsky

all $\mathtt{A} \in \mathbb{A}$ by a Hermitian-positive trace-one operator ρ.

Let $\mathscr{S} \subset \mathbb{A}_\star$ denote the state space realized by such $\rho \geq 0$, $\langle \rho, \mathtt{I} \rangle = 1$ with the tangent space $\mathscr{T}_0 = \left\{ \upsilon = \upsilon^\dagger \in \mathbb{A}_\star : \langle \upsilon, \mathtt{I} \rangle = 0 \right\}$ and the cotangent space $\mathscr{T}_0^\star = \bar{\mathbb{A}}^{\mathrm{h}} / \mathbb{R}\mathtt{I}$. Every state $\rho \in \mathscr{S}$ can be parametrized as $\rho(\mathrm{q}) = \rho_0 - \mathrm{q}$ by a *tangent element* $\mathrm{q} \in \mathscr{T}_0$. Cotangent elements $\mathrm{p} \in \mathscr{T}_0^\star$ are the equivalence classes

$$\mathrm{p}(\mathtt{X}) = \{ \mathtt{A} \in \mathbb{A} : \mathtt{A} = \mathtt{X} + \lambda \mathtt{I}, \text{ for some } \lambda \in \mathbb{R} \}.$$

Example: A single quantum bit is described by the algebra of (2×2)-matrices $\mathtt{A} = \alpha \mathtt{I} + \sigma_{\vec{p}}$, where $\sigma_{\vec{p}} \in \mathscr{T}_0$ is decomposed into Pauli matrices

$$\sigma_{\vec{a}} = a_x \sigma_x + a_y \sigma_y + a_z \sigma_z \equiv \vec{a} \cdot \vec{\sigma}, \qquad \vec{a} \in \mathbb{C}^3.$$

The normalized trace $\langle \mathtt{I}, \mathtt{A} \rangle = \mathrm{tr}\{\mathtt{A}\} = \alpha$ defines the standard pairing $\langle \rho, \mathtt{X} \rangle = -\vec{q} \cdot \vec{p}$ of $\mathtt{X} = \sigma_{\vec{p}}$ with quantum bit states $\rho = 1 - \sigma_{\vec{q}}$ given by the coordinate $\mathrm{q} = \sigma_{\vec{q}}$ parametrized by real vector $\vec{q} \in \mathbb{R}^3$ from the unit ball $|\vec{q}| \leq 1$ with respect to the central state $\rho_0 = \mathtt{I}$ and $\vec{p}(\mathtt{X}) \in \mathbb{R}^3$ identified with $\mathrm{p}(\mathtt{X}) \ni \mathtt{X}$ such that $\langle \mathrm{q}, \mathrm{p} \rangle = \vec{q} \cdot \vec{p}$.

Affine and concave costs

An affine functional $\mathsf{G}(u, \rho) = \langle \rho, G(u) \rangle$ of $\rho \in \mathscr{S}$ given by a function $u \mapsto G(u)$ in Hermitian operators $G(u) \vdash \mathbb{A}$ on a measurable space $\mathscr{U}$ is called *expected cost* of the control $u \in \mathscr{U}$. The *minimal expected cost* $\mathsf{S}[\rho] = \inf\{\langle \rho, G(u) \rangle : u \in \mathscr{U}\}$ is not affine but *concave* functional into $\mathbb{R}_\wedge = [-\infty, \infty[$ on the convex set $\mathscr{S}$ in the sense

$$\mathsf{S}[\lambda \rho_0 + (1-\lambda)\rho_1] \geq \lambda \mathsf{S}[\rho_0] + (1-\lambda)\mathsf{S}[\rho_1]$$

for any $\lambda \in [0,1]$ and $\rho_0, \rho_1 \in \mathscr{S}$. It can have value $-\infty$ for an unbounded from below function $G(u)$ and is continuous in the lower topology on $\mathbb{R}_\wedge$. Such lower continuous concave function will be called *proper concave* if $\mathsf{S}[\rho] > -\infty$ at least for one ρ.

Example: $\mathsf{S}[\rho] = \inf_{\vec{u} \in \mathbb{R}^3} \{ \langle \rho, \sigma_{\vec{u}} \rangle + |\vec{u}| \} = O^-_{\mathscr{B}_1}(\vec{q})$, where $O^-_{\mathscr{B}_1}(\vec{q}) := \left\{ \begin{smallmatrix} 0, & \vec{q} \in \mathscr{B}_1 \\ -\infty, & \vec{q} \notin \mathscr{B}_1 \end{smallmatrix} \right. \equiv -O^+_{\mathscr{B}_1}(\vec{q})$ is the max-plus indicator function of for the ball $\mathscr{B}_1 = \left\{ \vec{q} \in \mathbb{R}^3 : |\vec{q}| \leq 1 \right\}$.

Legendre-Fenchel transform

The above example for an affine G corresponding to $G(u) = |\vec{u}|\, \mathtt{I} + \sigma_{\vec{u}}$ is a special case of the Legendre-Fenchel transform

$$\mathsf{S}[\rho] = \inf_{\mathrm{p} \in \mathscr{T}_0^\star} \{ g(\mathrm{p}) - \langle \mathrm{q}(\rho), \mathrm{p} \rangle \} = \inf_{\mathtt{X} = \mathtt{X}^\dagger} \{ \langle \rho, \mathtt{X} \rangle + \mathsf{G}[\mathtt{X}] \}$$

of the proper convex function $g(\mathrm{p}) = |\vec{p}|$ for $\mathtt{X} = \sigma_{\vec{p}} - \lambda \mathtt{I}$ with $\mathsf{G}[\mathtt{X}] = g(\mathrm{p}) + \lambda$. Every proper concave functional $\mathsf{S}[\rho]$ on $\mathscr{S}$ can be obtained as optimization $\inf_{u \in \mathscr{U}} \{ \mathsf{G}(u, \rho) \}$ of a proper affine function $\rho \mapsto \mathsf{G}(u, \rho)$, e.g. as the LF transform corresponding to $u = \mathrm{p}$ and

$$\mathsf{G}(\mathrm{p}(\mathtt{X}), \rho) = \mathsf{G}[\mathtt{X}] + \langle \rho, \mathtt{X} \rangle = g(\mathrm{p}(\mathtt{X})) - \langle \mathrm{q}(\rho), \mathrm{p}(\mathtt{X}) \rangle.$$

The functional G, uniquely defining $g(\mathrm{p}) = \mathsf{G}(\mathrm{p}, \rho_0)$ on $\mathscr{T}_0^\star$ by the property $\mathsf{G}[\mathtt{X}] = \mathsf{G}[\mathtt{X} + \lambda \mathtt{I}] - \lambda$, can be found as a *proper convex* functional on the representatives X of $\mathrm{p}(\mathtt{X})$ into $\mathbb{R}_\vee =]-\infty, \infty]$ by the inverse Legendre transform $\mathsf{G}[\mathtt{X}] = \sup\{\langle \rho, \mathtt{X}\rangle + \mathsf{S}[\rho]\}$ of the functional $\mathsf{S}[\rho]$ extended as $-\infty$ outside of $\mathscr{S}$. Thus, the *relative entropies* $\mathsf{S}[\rho : \mu] = -\mathrm{tr}\left\{\rho \ln \frac{\rho}{\mu}\right\}$ of (a) $\ln \frac{\rho}{\mu} = \ln \rho - \ln \mu$ and (b) $\ln \frac{\rho}{\mu} = \ln\left(\mu^{-1}\rho\right)$ types are the transforms of some proper convex functionals G_a and G_b, and the *thermodynamic relative entropy* is defined as the transform

$$\mathsf{S}_\mathrm{t}[\rho : \mu] = \inf\left\{\langle \rho, \mathtt{X}\rangle + \ln\left\langle \mu, \mathrm{e}^{-\mathtt{X}}\right\rangle : \mathtt{X} = \mathtt{X}^\dagger\right\}.$$

QUANTUM SUB-MARKOV DYNAMICS

In general quantum controlled systems are governed by sub-Markov dynamics described for each controlling process $y = [u(r) | r \in \mathbb{R}_+]$ in terms of a hemigroup $\left[\tau_r^y(t) | t > r \in \mathbb{R}_+\right]$ of normal completely positive contracting maps $\tau_r^y(t) : \mathbb{A} \to \mathbb{A}$ such that

$$\tau_r^y(t) \circ \tau_t^y(s) = \tau_r^y(s) \quad \forall r < t < s.$$

The dependence on y is assumed to be causal, $\tau_r^{y_1}(t) = \tau_r^{y_2}(t)$ whenever $y_1^t = [u(r) | r < t] = y_2^t$, and usually $\tau_t^y(s)$ are assumed to be independent also on the past y^t of the admissible controls y defined by the cadlag trajectories $u(t) = u(t^+)$ as right continuous in $\mathbb{R}^{n_t}$ having the left limits $u(t^-)$ at each t.

Quantum controlled generator

The continuous feedback controlled sub-Markov dynamics $\left\{\tau_r^y(t) | t > r \in \mathbb{R}_+\right\}$ is usually determined by its generator

$$\mathfrak{L}(u(t), \mathtt{X}) = \lim_{s \searrow 0} \frac{1}{s}\left(\tau_t^y(t+s, \mathtt{X}) - \mathtt{X}\right)$$

given by a completely dissipative map $\mathfrak{L}(u) : \mathbb{A} \to \mathbb{A}$ controlled by the values $u(t) = \left(u^1, \ldots, u^n\right)$ of some parameters $u^j \in \mathbb{R}$. Such dynamics is described by the controlled semi-Lindblad generators $\mathfrak{L}(u, \mathtt{X}) = \mathrm{i}[H(u), \mathtt{X}] + \mathfrak{L}^m(u, \mathtt{X})$ with the dissipation part

$$\mathfrak{L}^m(\mathtt{X}) = \sum_j \mathtt{L}_j \mathtt{X} \mathtt{L}_j^\dagger - \frac{1}{2}(\mathtt{MX} + \mathtt{XM}), \quad \mathtt{M} \geq \sum_j \mathtt{L}_j \mathtt{L}_j^\dagger$$

prepared for a continuous indirect measurement, say, of the observables $\mathtt{L}_j + \mathtt{L}_j^\dagger$ affiliated to $\mathbb{A}$. Here we assumed that the dynamics is controlled only by the Hamiltonian $H(u) = H(u)^\dagger$ and $\mathtt{M} \vdash \mathbb{A}$ (affiliated to $\mathbb{A}$) determines the decay operator $\mathtt{D} = \mathtt{M} - \sum_j \mathtt{L}_j \mathtt{L}_j^\dagger = -\mathfrak{L}(\mathtt{I})$.

Deterministic quantum Master equations

The density operator $\rho^t \in \mathbb{A}_\star$ of a quantum state evolves from a normalized $\rho^0 = \rho_0$ by resolving the *controlled Master equation* $\frac{\mathrm{d}}{\mathrm{d}t}\rho^t = \mathfrak{L}_\star(u,\rho^t)$ with the predual generator $\langle \mathfrak{L}_\star(\rho), \mathtt{A}\rangle = \langle \rho, \mathfrak{L}(\mathtt{A})\rangle$. It is normalized on the decaying probability of the survival $\pi^t = \langle \rho^t, \mathtt{I}\rangle$. The renormalized $\rho_t = \rho^t/\pi^t$ describes the quantum state conditioned by the survival effect up to time t. Its evolution is described by the velocity $\upsilon(\rho) = \frac{\mathrm{d}}{\mathrm{d}t}\mathtt{q}(\rho) \in \mathscr{T}_0$ defining the *deterministic nonlinear filtering equation*

$$\frac{\mathrm{d}}{\mathrm{d}t}\rho_t = \mathfrak{L}_\star(\rho_t) - \langle \rho_t, \mathfrak{L}(\mathtt{I})\rangle \rho_t \equiv -\upsilon(\rho_t).$$

Example: An unstable quantum bit is described by the Hamiltonian $H(\vec{u}) = \frac{1}{2}\sigma_{\vec{u}}$, $\mathtt{L} = \frac{1}{2}\lambda\sigma_z = \mathtt{L}^\dagger$, $\mathtt{M} = \mu\mathtt{I} + \sigma_{\vec{m}}$ with $\mu \geq |\vec{m}| + \lambda^2/4$. Then $\upsilon(\rho) = \sigma_{\vec{u}\times\vec{q}} + \upsilon_m(\rho)$ where

$$\upsilon_m(\rho) = \left(\vec{m} - \frac{\lambda^2}{2}\vec{q}_z^{\perp} - (\vec{m}\cdot\vec{q})\vec{q}\right)\cdot\vec{\sigma} \equiv \vec{v}_m\cdot\vec{\sigma}.$$

QUANTUM DYNAMICAL PROGRAMMING

The *cost to go* of a control $u(t)$ conditioned by survival is

$$\mathsf{J}[\{u(\cdot)\};t,\rho_t] = \int_t^\tau \mathsf{C}(u(r),\rho_r)\,\mathrm{d}r + \mathsf{G}(u(\tau),\rho_\tau).$$

Due to the statistical interpretation of quantum states,

$$\mathsf{C}(u,\rho) = \langle \rho, C(u)\rangle,\ \mathsf{G}(u,\rho) = \langle \rho, G(u)\rangle.$$

The optimal average cost on the interval $(t,\tau]$ to be $\mathsf{S}(t,\rho) := \inf_{\{u(r)\}}[\mathsf{J}[u(\cdot);t,\rho]]$ with $\mathsf{S}(\tau,\rho) = \inf\{\langle\rho, G(u)\rangle : u \in \mathbb{R}^{n_\tau}\} \equiv \mathsf{S}[\rho]$ given by any concave functional $\mathsf{S} : \rho \mapsto \mathsf{S}[\rho] \in \mathbb{R}_\wedge$ of the terminal state $\rho = \rho_\tau$.

Quantum Hamilton-Jacobi equation

Since $\mathsf{J}[\{u(\cdot)\};t,\rho_t] = \int_t^{t'}\mathsf{C}(u(r),\rho_r)\,\mathrm{d}r + \mathsf{J}[\{u\};t',\rho_t]$ at the times $t < t' \leq \tau$, one has

$$\mathsf{S}(t,\rho) = \inf_{\{u\}}\left\{\int_t^{t'}\mathsf{C}(u(r),\rho_r)\,\mathrm{d}r + \mathsf{J}\left[u(\cdot);t',\rho_{t'}\right]\right\}.$$

Suppose that $\{u^{\mathrm{o}}(r,\rho) : r > t\}$ is an optimal control when starting in state ρ at the time t, and denote by $\{\rho_r : r \in (t,\tau]\}$ the corresponding state trajectory starting at a state ρ at t. Bellman's optimality principle observes that

$$-\frac{\partial}{\partial t}\mathsf{S}(t,\rho) = \inf_{u\in\mathscr{U}}\left\{\mathsf{C}(u,\rho) - \left\langle \upsilon(u,\rho), \nabla_\rho \mathsf{S}(t,\rho)\right\rangle\right\},$$

The equation is then to be solved subject to

$$\mathsf{S}(\tau,\rho)=\mathsf{G}(u^{o}(\tau_{+},\rho),\rho)\equiv\mathsf{S}[\rho].$$

Pontryagin's maximum principle

We may rewrite this as the *Hamilton-Jacobi* equation

$$-\frac{\partial}{\partial t}\mathsf{S}(t,\rho(\mathsf{q}))+\mathsf{H}_{\upsilon}(\mathsf{q},\mathsf{p}(\nabla_{\rho}\mathsf{S}(t,\rho))(\mathsf{q}))=0$$

introducing the Pontryagin's Hamiltonian as the transform

$$\mathsf{H}_{\upsilon}(\mathsf{q},\mathsf{p})=\sup_{u\in\mathscr{U}}\{\langle\upsilon(u),\mathsf{p}\rangle-\mathsf{C}(u)\}(\mathsf{q})$$

which is affine in $\mathsf{p}\in\mathscr{T}_0^{\star}$. This leads to the Hamiltonian boundary value problem

$$\begin{cases}\dot{\mathsf{q}}_t-\nabla_{\mathsf{p}}\mathsf{H}_{\upsilon}(\mathsf{q}_t,\mathsf{p}_t)=0, & \mathsf{q}_0=\mathsf{a}\\ \dot{\mathsf{p}}_t+\nabla_{\mathsf{q}}\mathsf{H}_{\upsilon}(\mathsf{q}_t,\mathsf{p}_t)=0, & \mathsf{p}_\tau=\mathsf{b}\end{cases}$$

as the *Hamilton-Pontryagin problem* with the solutions defining for $\mathsf{a}=0$ if $\rho^0=\rho_0$, $\mathsf{b}=\nabla_{\rho}\mathsf{S}[\rho]$ the minimal costs by the path integral in

$$\mathsf{S}(t_0,\rho_0)=\int_{t_0}^{\tau}[\langle\dot{\mathsf{q}}_t,\mathsf{p}_t\rangle-\mathsf{H}(\mathsf{q}_t,\mathsf{p}_t)]\,\mathrm{d}r+\mathsf{S}[\rho(\mathsf{q}_\tau)].$$

Optimal qubit decay control

Let L_0 be $\mathsf{L}=\frac{\lambda}{2}\sigma_z=\mathsf{L}^{\dagger}$ $(\lambda=\overline{\lambda})$, and let us ignore the effect of environment by taking $\mathsf{L}_j=0$ for $j\neq0$. We may also take cost $c(\vec{u})=O^{+}_{\mathscr{B}_1}(\vec{u})$ of constraint $\mathscr{B}_1=\{\vec{u}:|\vec{u}|\leq1\}$ and any concave bequest functions $\mathsf{S}[\rho]$, the qubit entropy say, as conditional entropy of survival $\mathsf{S}[\rho_\tau]=\frac{1}{\pi^\tau}\mathsf{S}[\rho^\tau:2\pi^\tau\mathrm{I}]$:

$$\mathsf{S}[\mathrm{I}+\sigma_{\vec{q}}]=q\ln\sqrt{\frac{1-q}{1+q}}-\ln\sqrt{\frac{1-q^2}{4}},\quad q=|\vec{q}|\ .$$

We have $\langle\upsilon(u,\rho),\mathsf{p}\rangle=\vec{v}(\vec{u},\vec{q})\cdot\vec{p}$ linear in $\vec{u}$:

$$\vec{v}(\vec{u},\vec{q})=\vec{v}_m+\vec{u}\times\vec{q},\quad \vec{v}_m=\vec{m}-(\vec{m}\cdot\vec{q})\vec{q}-\frac{\lambda^2}{2}\vec{q}_z^{\perp}$$

which is maximized by the unit $\vec{u}^{o}=\vec{q}\times\vec{p}/|\vec{q}\times\vec{p}|$ under the constraint $|\vec{u}|\leq1$. This leads to the Hamiltonian function

$$\mathsf{H}_{\upsilon}(\vec{p},\vec{q})=|\vec{q}\times\vec{p}|+\vec{v}_m\cdot\vec{p}.$$

The HJ equation for optimal qubit control under this constraint is

$$-\frac{\partial \mathsf{S}}{\partial t}+\left|\vec{p}\times\vec{\nabla}\mathsf{S}\right|+\left(\vec{m}-(\vec{m}\cdot\vec{r})\vec{r}+\frac{\lambda^2}{2}\vec{r}_z^{\perp}\right)\vec{\nabla}\mathsf{S}=0.$$

The solution $\mathsf{S}(t_0,\rho_0)$ of this equation corresponding to the entropy $\mathsf{S}(\tau,\rho_\tau)=\mathsf{S}[\rho_\tau]$ gives the minimal entropic cost of the controlled unstable qubit under the condition of its survival up to a time $\tau > t_0$.

DISCUSSION

Our analysis is based on the fact that quantum state is a sufficient coordinate not only for closed, but also for open unstable quantum systems under the Markov approximation. However we have to deal with differential equation of high or infinite dimensionality if $\dim\mathfrak{h}=\infty$. Nevertheless, the Bellman principle can then be applied in much the same spirit as for classical states and we are able to derive the corresponding HJB theory for a wider class of cost functionals than traditionally considered.

When restricted to the Bloch sphere for the qubit with the cost being a quantum expectation, we recover the class of Bellman equations encountered so far in quantum feedback purification.

REFERENCES

1. V. P. Belavkin, Theory of the Control of Observable Quantum Systems, Automatica and Remote Control 44 (2) 178–188 (1983), quant-ph/0408003,
2. V. P. Belavkin and S. Edwards, "Quantum Filtering and Optimal Control" in *Quantum Stochastics, Filtering and Control*, edited by V. Belavkin and M. Guta, World Scientific, 2008, pp. 143-206, preprint in quant-ph/0506018.

Covariance matrix criterion for separability

O. Gittsovich*,†, O. Gühne†,*, P. Hyllus** and J. Eisert‡

*Institut für Theoretische Physik, Universität Innsbruck, 6020 Innsbruck, Austria
†Institut für Quantenoptik und Quanteninformation, Österreichische Akademie der Wissenschaften, 6020 Innsbruck, Austria
**Institut für Theoretische Physik, Universität Hannover, 30167 Hannover, Germany
‡Institute for Physics and Astronomy, University of Potsdam, 14476 Potsdam, Germany

Abstract. We review an approach to the separability problem using covariance matrices of locally measurable observables. This approach leads to a strong entanglement criterion which we call covariance matrix criterion (CMC). We consider some corollaries of the CMC that in general allow to detect many bound entangled states. In particular we give a new example which sheds some light on the connections between different entanglement criteria derived from the general CMC.

Keywords: separability problem, covariance matrix, bound entanglement
PACS: 03.67.-a, 03.65.Ud

INTRODUCTION

Characterization of entanglement is an important issue in quantum theory. Entangled states play the central role in quantum information science as a resource for information processing tasks and therefore have received growing attention in the last decade [1]. One of the fundamental questions is how to find good criteria to decide whether a state is entangled or classically correlated in the first place. Formally, one asks the question whether a given state is contained in the convex hull of product states and can hence be written as $\rho = \sum_k p_k |a_k\rangle\langle a_k| \otimes |b_k\rangle\langle b_k|$, where the coefficients p_k form a probability distribution. If so, the state is called classically correlated or separable, otherwise it is an entangled state.

Typically one aims at finding criteria to make a decision based on these criteria about the separability of a given state in the case of some fixed Hilbert space dimension. The first and the most prominent criterion of this sort was derived from the observation that every separable state stays positive under the partial transposition map and that this condition can be used as an entanglement criterion *(PPT Criterion)* that is necessary and sufficient for low dimensional systems (2×2 and 2×3) [2, 3]. The PPT criterion is an instance of the family of criteria *based on positive maps* [4].

Apart from the criteria that are directly related to the PPT criterion, a number of other separability criteria have been suggested where such a connection seemingly does not exist [5, 6, 8, 9]

In [10] we proposed to investigate the separability problem using covariance matrices (CMs) of certain observables

$$\gamma_{i,j} = \frac{\langle M_i M_j\rangle + \langle M_j M_i\rangle}{2} - \langle M_i\rangle\langle M_j\rangle, \tag{1}$$

CP1110, *Quantum Communication, Measurement and Computing (QCMC), The Ninth International Conference,* edited by A. Lvovsky

where the observables M_i are chosen to be orthonormal $\mathrm{tr}(M_iM_j) = \delta_{i,j}$. In this context we have developed a separability criterion in terms of CMs (*covariance matrix criterion* or CMC). We have shown that the CMC can detect states where the PPT criterion fails and that, augmented by proper filtering, the CMC leads to the condition, which is necessary and sufficient for two qubits. In [11] we have shown that the CMC, when evaluated with certain observables, leads to the several other separability conditions, which in fact coincide with the conditions proposed in [5, 6, 7, 8, 9]. In this note we review some of the results of [11] and provide a further example which shows, that the derived criteria are not equivalent, namely some of them detect states where the other fail.

GENERAL CMC FOR BIPARTITE SYSTEMS

We consider the situation where the Hilbert space is a tensor product $\mathscr{H} = \mathscr{H}^A \otimes \mathscr{H}^B$ of Hilbert spaces of two subsystems A and B with dimensions d_A and d_B respectively. We can choose a basis of the observable algebra in A as $A_k : k = 1, \ldots, d_A^2$ and in B as $B_k : k = 1, \ldots, d_B^2$ and consider the set of $d_A^2 + d_B^2$ observables $M_k = \{A_k \otimes \mathbb{1}, \mathbb{1} \otimes B_k\}$. This set of observables can be employed to define a very useful form of CMs.

Definition 1. (Block covariance matrices) *Let ρ be a state of a bipartite system, and let $M_k = \{A_k \otimes \mathbb{1}, \mathbb{1} \otimes B_k\}$ be a set of observables as outlined above. Then, the covariance matrix $\gamma(\rho, \{M_k\})$ has a block structure:*

$$\gamma = \begin{pmatrix} A & C \\ C^T & B \end{pmatrix}, \tag{2}$$

where $A = \gamma(\rho_A, \{A_k\})$ and $B = \gamma(\rho_B, \{B_k\})$ are CMs of the reduced states of systems A and B, and $C_{i,j} = \langle A_i \otimes B_j \rangle - \langle A_i \rangle \langle B_j \rangle$.

CMs in the block form presented in Definition 1 can be now used to formulate the CMC, that is derived from the concavity property for CMs and the fact that the block CM for product states is block diagonal.

Proposition 2. (Covariance matrix criterion) *Let ρ be a separable state and γ be a block CM as in Definition 1. Then there exist pure states $|\psi_k\rangle\langle\psi_k|$ for A and $|\phi_k\rangle\langle\phi_k|$ for B and convex weights p_k such that if we define $\kappa_A = \sum_k p_k \gamma(|\psi_k\rangle\langle\psi_k|)$ and $\kappa_B = \sum_k p_k \gamma(|\phi_k\rangle\langle\phi_k|)$ the inequality*

$$\begin{pmatrix} A & C \\ C^T & B \end{pmatrix} \geq \begin{pmatrix} \kappa_A & 0 \\ 0 & \kappa_B \end{pmatrix} \tag{3}$$

holds. If there are no such $\kappa_{A,B}$ then the state ρ must be entangled.

Note that the CMC, as formulated as in Proposition 2, is not clearly a physically plausible and efficient test for separability and therefore one has to evaluate it in an appropriate way. The main problem is to characterize the possible κ_A and κ_B. The formulation above still contains an optimization over all pure product states. However the appropriate choice of the observables helps to simplify evaluation of the CMC a lot. Notably the CMC is manifestly invariant under a change of observables $\{A_k\}$ and $\{B_k\}$ [11].

EVALUATION OF THE CMC

We consider two strategies to evaluate the CMC. Both strategies are based on matrix invariants such as eigenvalues or singular values. The positive semi-definiteness of the matrix in the Proposition 2 can be characterized in terms of traces of its submatrices.

Proposition 3. (CMC evaluated from traces) *Let γ be the symmetric block CM of a state ρ. Then if ρ is separable, we have $2\cdot\sum_{i=1}^{d_A^2}|C_{i,i}| \leq \left[1-\mathrm{tr}(\rho_A^2)\right]+\left[1-\mathrm{tr}(\rho_B^2)\right]$. If this inequality is violated, then ρ must be entangled.*

Note that the criterion in Proposition 3 detects most states if the observables are chosen in such a way that C is diagonal. For any state there exist a choice of observables such that this can be achieved.

Using the Schmidt decomposition for density matrices in operator space $\rho = \sum_{k=1}^{d_A^2}\lambda_k G_k^A\otimes G_k^B$, where we assumed that $d_A \leq d_B$, and taking $\{G_k^A\}$ and $\{G_k^B\}$ as the new basis observables for the construction of the symmetric block CM, we arrive at the following form of the CMC - *Schmidt CMC*:

Proposition 4. (Schmidt CMC) *If ρ is separable, then*

$$2\sum_i|\lambda_i-\lambda_i^2 g_i^A g_i^B| \leq 2-\sum_i\lambda_i^2\left[(g_i^A)^2+(g_i^B)^2\right] \tag{4}$$

must hold, where we defined $g_i^A=\mathrm{tr}(G_i^A)$ and $g_i^B=\mathrm{tr}(G_i^B)$. If this condition is violated, the state must be entangled.

Let us now consider local filtering operations or SLOCC (stochastic local operations assisted by classical communication) [12] of the form $\rho\mapsto\tilde{\rho}=(F_A\otimes F_B)\rho(F_A\otimes F_B)^\dagger=\frac{1}{d_Ad_B}\left(\mathbb{1}+\sum_k\xi_k J_k^A\otimes J_k^B\right)$, where and $F_A\in\mathrm{SL}(d_A,\mathbb{C})$ and $F_B\subset\mathrm{SL}(d_B,\mathbb{C})$ are invertible matrices on the respective Hilbert spaces and $J_k^{A/B}$ are generators of $su(d_{A/B})$. Clearly, such operations can not map a separable state into an entangled one and vice versa, since they are local and invertible. In Refs. [12, 13] it has been shown, that one can bring any state of the full rank in its *standard form* using such filtering operations. Moreover it has been shown that the standard form of a state maximizes all *SLOCC-entanglement monotones* such concurrence or 3-tangle.

Now one can bring the state into its normal form and then apply the new separability criteria. In the asymmetric case ($d_A<d_B$) we can formulate the following statement:

Proposition 5. (Filter CMC) *If ρ is separable, then the following inequalities hold*

$$\sum_i\xi_i \leq \frac{d_Ad_B}{2}\left[1-\frac{1}{d_A}+(d_A^2-1)\frac{1}{d_B}+\min(0,-(d_B-1)+(d_B^2-d_A^2)\frac{1}{d_B})\right] \tag{5}$$

$$\sum_i\xi_i \leq \left[d_Ad_B(d_A-1)(d_B-1)\right]^{1/2}. \tag{6}$$

If one of these inequalities is violated, the state must be entangled.

EXAMPLES

The Propositions 3, 4 and 5 formulated in the last chapter are not equivalent, although they all follow from the general CMC, as stated in Proposition 2 [11]. To prove this non-equivalence we provide two one parametric families of the bound entangled states introduced by P. Horodecki [14] - $\rho_{3\times3}(a)$ and $\rho_{2\times4}(b)$ for the 3×3 and 2×4 systems respectively.

Testing different criteria on this states shows (FIG. 1), that the criteria are indeed not equivalent. For example the $\rho_{3\times3}$ is not detected by *Filter CMC* for some values of the parameter a, whereas other two criteria detect it for all a. Therefore we conclude that applying filtering operations is not always the best strategy to detect bipartite entanglement, as one could conclude from an example provided in [11].

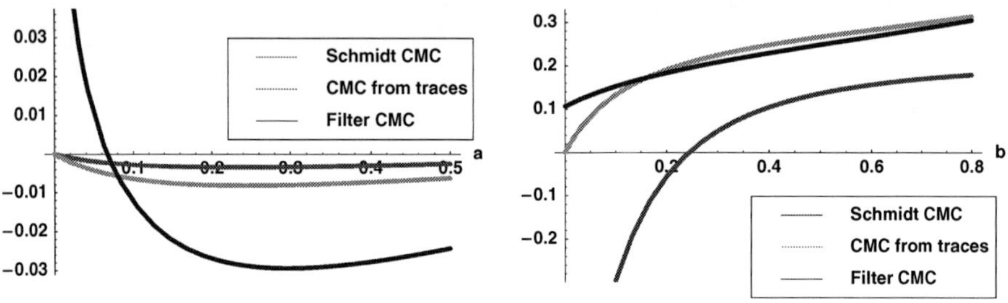

FIGURE 1. (Color online) *Left picture*: Detection of the $\rho_{3\times3}$ depending on the parameter a. For every a the state is detected by the Schmidt CMC and by the CMC evaluated from traces of submatrices of the block covariance matrix. However there is a specific region of the parameter a where the filter CMC fails to detect the state. *Right picture*: Detection of the $\rho_{2\times4}$ depending on the parameter b. The state is detected only for certain parameter values and only by the Proposition 4.

We thank H.J. Briegel, R. Horodecki and M. van den Nest for discussions. This work has been supported by the EU (OLAQUI, QAP, QICS, SCALA, COMPAS, MINOS), the FWF, Microsoft Research, the EPSRC and the EURYI.

REFERENCES

1. R. Horodecki, P. Horodecki, M. Horodecki, and K. Horodecki, quant-ph/0702225.
2. A. Peres, Phys. Rev. Lett **97**, 1413 (1996).
3. M. Horodecki, P. Horodecki, and R. Horodecki, Phys. Lett. A **223**, 1 (1996).
4. B. M. Terhal, Lin. Alg. Appl. **323**, 61 (2000); H.-P. Breuer, Phys. Rev. Lett. **97**, 080501 (2006).
5. O. Rudolph, quant-ph/0202121; K. Chen and L.-A. Wu, Quant. Inf. Comp. **3**, 193 (2003).
6. J.I. de Vicente, Quantum Inf. Comput. **7**, 624 (2007); J. Phys. A **41**, 065309 (2008).
7. H.F. Hofmann and S. Takeuchi, Phys. Rev. A **68**, 032103 (2003).
8. S. Yu and N. Liu, Phys. Rev. Lett. **95**, 150504 (2005).
9. C.-J. Zhang *et. al.* Phys. Rev. A **76**, 012334 (2007); Phys. Rev. A **77**, 060301(R) (2008).
10. O. Gühne, P. Hyllus, O. Gittsovich and J. Eisert, Phys. Rev. Lett. **99**, 130504 (2007).
11. O. Gittsovich, O. Gühne, P. Hyllus and J. Eisert, Phys. Rev. A **78**, 052319 (2008).
12. A. Kent *et. al.* Phys. Rev. Lett. **83**, 2656 (1999); F. Verstraete *et. al.* Phys. Rev. A **65**, 032308 (2002).
13. J.M. Leinaas, J. Myrheim, and E. Ovrum, Phys. Rev. A **74**, 012313 (2006).
14. P. Horodecki, Phys. Lett. **A232**, 333 (1997).

Reverse Coherent Information

Raúl García-Patrón, Stefano Pirandola, Seth Lloyd,
and Jeffrey H. Shapiro

Research Laboratory of Electronics, MIT, Cambridge, MA 02139

Abstract. We define a family of entanglement distribution protocols assisted by classical feedback communication that gives an operational interpretation to reverse coherent information, i.e., the symmetric counterpart of the well-known coherent information. This protocol family leads to the definition of a new entanglement distribution capacity that exceeds the unassisted entanglement distribution capacity for some interesting channels.

Keywords: Quantum Information Theory, Entanglement Distribution, Damping Channel
PACS: 03.67.-a, 03.67.Dd, 03.67.Hk

ENTANGLEMENT DISTRIBUTION

Quantum information theory has attracted a great deal of interest in the last decade, as it offers new potential applications, such as quantum communication and entanglement distribution over quantum channels. Quantum communication allows faithful transmission of quantum states through a noisy quantum channel Λ. The quantum communication capacity, $\mathscr{Q}(\Lambda)$, gives the number of qubits per channel use that can be reliably transmitted over the channel Λ [1]. Entanglement is another important resource for quantum information processing. Therefore, the study of entanglement distribution over quantum channels is of crucial importance. By analogy with the quantum communication scenario, we can define an entanglement distribution capacity $\mathscr{E}(\Lambda)$. As explained in Fig. 1, it is known that $\mathscr{Q}(\Lambda) = \mathscr{E}(\Lambda)$. If they have free access to a classical communication channel, Alice and Bob can improve their quantum communication (entanglement distribution) capacity, by defining three new pairs of capacities depending on the use of that classical channel (see Fig. 1); forward classical communication $\mathscr{Q}_{\rightarrow}$ ($\mathscr{E}_{\rightarrow}$); classical feedback communication $\mathscr{Q}_{\leftarrow}$ ($\mathscr{E}_{\leftarrow}$); and two-way classical communication $\mathscr{Q}_{\leftrightarrow}$ ($\mathscr{E}_{\leftrightarrow}$). It is interesting to study $\mathscr{Q}_{\leftarrow}$ and $\mathscr{E}_{\leftarrow}$, because it is the only example in which quantum communication and entanglement distribution capacities are different, see Fig. 1.

It was shown in [2] that the coherent information $I(\Lambda, \rho_A)$, a function of Alice's input ρ_A on channel Λ, plays a crucial role in proving that $\mathscr{Q}(\Lambda) = \mathscr{E}(\Lambda)$. Only for the restricted class of *degradable channels* [3], $\mathscr{Q}(\Lambda)$ is known to be additive, i.e., $Q(\Lambda) = Q^{(1)}(\Lambda)$, where $Q^{(1)}(\Lambda)$ is the single-letter capacity. The channel Λ is called degradable if there exist a channel $\mathscr{M}$ that maps Bob's output ρ_B into the environment state ρ_E ($\mathscr{M}(\rho_B) = \rho_E$). Similarly, if there is a map $\mathscr{G}$ such that $\mathscr{G}(\rho_E) = \rho_B$, then the channel is called *antidegradable* and $\mathscr{Q}(\Lambda) = 0$.

CP1110, *Quantum Communication, Measurement and Computing (QCMC), The Ninth International Conference,* edited by A. Lvovsky

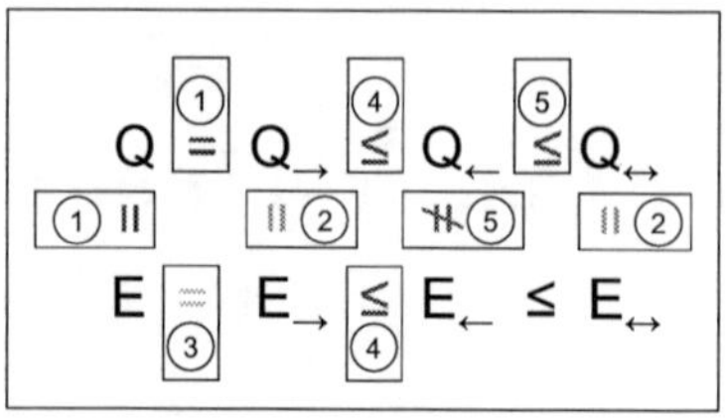

FIGURE 1. Summary of the known relations between the quantum communication and entanglement distribution assisted (unassisted) capacities. Being able to send a noiseless qubit is a stronger resource than distributing units of entanglement (e-bits): $\mathscr{E}_x \geq \mathscr{Q}_x$ for all x. Increasing the complexity of the assistance cannot decrease the capacity: $\mathscr{X} \leq \mathscr{X}_{\leftarrow} \leq \mathscr{X}_{\leftrightarrow}$. (1) The equality $\mathscr{E} = \mathscr{Q} = \mathscr{Q}_{\rightarrow}$ was shown in [4]. (2) Any entanglement distribution protocol can be transformed into a quantum communication protocol by means of teleportation. (3) Results from combining 1 and 2. (4) The erasure channel is an example of strict separation between $\mathscr{Q}$ and $\mathscr{Q}_{\leftarrow}$ ($\mathscr{E}$ and $\mathscr{E}_{\leftarrow}$) [5, 6]. (5) It is easy to prove that $\mathscr{E}_{\leftarrow} = \mathscr{Q}_{\leftrightarrow}$ for the erasure channel. In [6] it was shown that the erasure channel satisfies the strict inequality $\mathscr{Q}_{\leftarrow} < \mathscr{Q}_{\leftrightarrow}$, thus providing an example for which $\mathscr{E}_{\leftarrow} \neq \mathscr{Q}_{\leftarrow}$.

REVERSE ENTANGLEMENT DISTRIBUTION

In a fully general entanglement distribution protocol assisted by classical feedback, Alice has to receive Bob's measurement result before sending the next qubit. This has a big practical disadvantage, in that it greatly decreases the transmission rate. A way of avoiding this problem is simplifying the protocol to a single post-processing round, as shown in Fig. 2. The single post-processing round of this simplified protocol can be seen as an

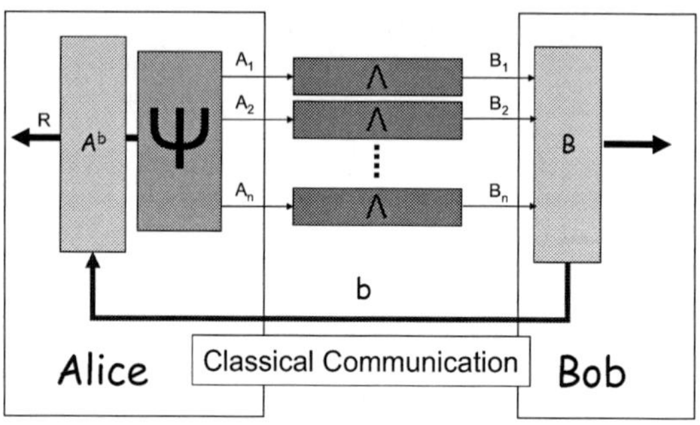

FIGURE 2. After Alice has sent all her qubits ($A_1A_2...A_n$) though the quantum channel Λ, Bob applies a collective incomplete measurement $\mathscr{B}$ among all the qubits $B_1B_2...B_n$ and communicates the classical outcome b to Alice. Finally, conditioned on the message b, Alice applies the quantum operation $\mathscr{A}^b$ on system R.

entanglement distillation protocol applied to the quantum state $\rho_{R|B_1...B_n}$ shared by Alice and Bob after the transmission of qubits $A_1A_2...A_n$. Therefore, we can transform any entanglement distillation protocol into an entanglement distribution version. For example, by exchanging the roles of Alice and Bob in the family of distillation protocols assisted by one-way classical communication defined in [8], we can build a new family of distillation protocols with rate $I_R(\rho_{RB}) = S(R) - S(R,B)$. By analogy with the quantum key distribution scenario [9], we call this family of protocols *reverse reconciliation* entanglement distillation protocols and the quantity $I_R(\rho_{RB})$ the *reverse coherent information*. It is then straightforward to define a family of entanglement distribution protocols assisted by classical feedback with rate $I_R(\Lambda, \rho_A)$. Furthermore, optimizing over ρ_A we obtain the single-letter reverse entanglement distribution capacity $\mathscr{E}_R^{(1)}(\Lambda)$. Surprisingly, this new capacity can be shown to be additive ($\mathscr{E}_R = \mathscr{E}_R^{(1)}$) for all channels. In what follows we detail an example of an important channel for which we obtain a remarkable improvement over the unassisted capacity $\mathscr{E}(\Lambda)$.

AMPLITUDE DAMPING CHANNEL

The amplitude damping channel describes the process of energy dissipation through spontaneous emission in a two-level system. The effect of the channel on the input state ρ is $\mathscr{D}_\eta(\rho) = E_0 \rho E_0^\dagger + E_1 \rho E_1^\dagger$, where

$$E_0 = \begin{bmatrix} 1 & 0 \\ 0 & \sqrt{\eta} \end{bmatrix}, E_1 = \begin{bmatrix} 0 & \sqrt{1-\eta} \\ 0 & 0 \end{bmatrix}, \tag{1}$$

and $1-\eta$ is the probability of spontaneous emission. Generalizing the results of [10], we can restrict the input state to the class $\rho_A = \mathrm{diag}(1-p, p)$ without loss of generality. Optimizing over the input population we obtain $\mathscr{E}^{(1)}(\mathscr{D}_\eta)$ and $\mathscr{E}_R(\mathscr{D}_\eta)$ as functions of the damping parameter η, see Fig. 3(a). Using the concatenation property of the amplitude damping channel ($\mathscr{D}_\eta \circ \mathscr{D}_{\eta'} = \mathscr{D}_{\eta\eta'}$), it is easy to prove that the amplitude damping channel is degradable for $\eta \geq 1/2$ and antidegradable for $\eta \leq 1/2$. We conclude that $\mathscr{E}_R(\mathscr{D}_\eta)$ outperforms $\mathscr{E}(\mathscr{D}_\eta)$ for all η. Even more interestingly, $\mathscr{E}_R(\mathscr{D}_\eta)$ remains positive in the range $\eta \leq 1/2$ where $\mathscr{E}(\mathscr{D}_\eta) = 0$, see Fig. 3 (a).

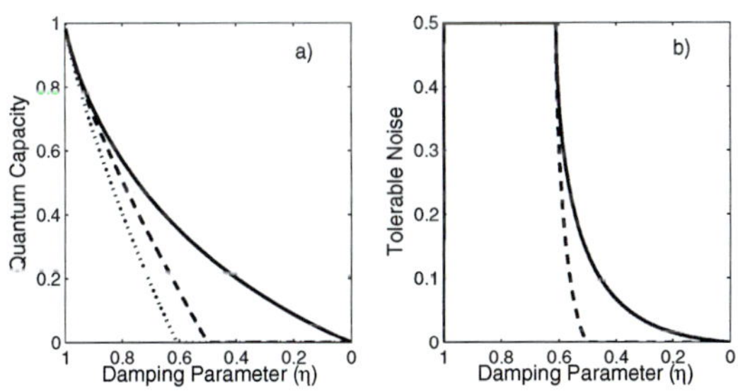

FIGURE 3. (a) Comparison of $\mathscr{E}(\mathscr{D}_\eta)$ (dashed line) and $\mathscr{E}_R(\mathscr{D}_\eta)$ (solid line) as functions of the damping parameter η for the amplitude damping channel, together with the capacity $\mathscr{E}^{(1)}(\mathscr{D}_\eta) = \mathscr{E}_R(\mathscr{D}_\eta)$ of the generalized amplitude damping channel with a maximally-mixed environment ($\alpha = 1/2$) (dotted line). (b) Tolerable thermal noise of the channel $\mathscr{D}_{(\eta,\alpha)}$ (minimum α such that the capacity is zero) as a function of the damping parameter η for: $\mathscr{E}^{(1)}(\mathscr{D}_{(\eta,\alpha)})$ (dashed line), and $\mathscr{E}_R(\mathscr{D}_{(\eta,\alpha)})$ (solid line).

Spontaneous emission to an environment at thermal equilibrium leads to the generalized amplitude damping channel $\mathscr{D}_{(\eta,\alpha)}$. The channel can be regarded as the random mixing $\mathscr{D}_{(\eta,\alpha)} = \alpha \mathscr{D}_{(\eta,0)} + (1-\alpha)\mathscr{D}_{(\eta,1)}$ of two limiting cases: (1) the amplitude damping channel ($\mathscr{D}_{(\eta,0)}$); and (2) a *populating channel* ($\mathscr{D}_{(\eta,1)}$). We restrict the analysis to $0 \le \alpha \le 1/2$ as for any channel $\mathscr{D}_{(\eta,\alpha=1/2+x)}$ with optimal input population p^* there is a companion channel $\mathscr{D}_{(\eta,\alpha=1/2-x)}$ with optimal population $1-p^*$ reaching the same capacity. As before, Alice's input can be restricted to $\rho_A = \mathrm{diag}(1-p,p)$. Optimizing over the input population p we obtain the capacities $\mathscr{E}^{(1)}(\mathscr{D}_{(\eta,\alpha)})$ and $\mathscr{E}_R(\mathscr{D}_{(\eta,\alpha)})$. It is easy to show that $\mathscr{E}_R(\mathscr{D}_{(\eta,\alpha)}) \ge \mathscr{E}^{(1)}(\mathscr{D}_{(\eta,\alpha)})$ for any noise α except $\alpha = 1/2$, where both are equal. Unfortunately, we cannot conclude that $\mathscr{E}_R(\mathscr{D}_{\eta,\alpha}) \ge \mathscr{E}(\mathscr{D}_{(\eta,\alpha)})$ for $\alpha > 0$, as the channel $\mathscr{D}_{(\eta,\alpha)}$ is not degradable for $\alpha > 0$. Nevertheless, it is easy to prove that generalized amplitude damping channels ($\mathscr{D}_{(\eta,\alpha)}$) with $\eta \le 1/2$ are antidegradable ($\rho_B = \mathscr{D}_{(\eta/(1-\eta),\alpha)}(\rho_E)$), which shows that for such channels $\mathscr{E}_R(\mathscr{D}_{(\eta,\alpha)}) \ge \mathscr{E}(\mathscr{D}_{(\eta,\alpha)}) = 0$, see Fig. 3 (b).

CONCLUSIONS

We reviewed the relations between quantum communication and entanglement distribution capacities. By restricting ourselves to realistic entanglement distribution protocols assisted by classical feedback, we defined an interesting subset of protocols that we call reverse entanglement distribution. This new definition gives an operational interpretation for the reverse coherent information, the symmetric counterpart of the well-known coherent information. This, in turn, allowed us to define a new entanglement distribution capacity that is additive and outperforms the unassisted capacity for some important channels, such as the damping channel and its generalization.

ACKNOWLEDGMENTS

We acknowledge financial support from the W. M. Keck Foundation Center for Extreme Quantum Information Theory. S.P. acknowledges financial support from the EU (Marie Curie fellowship).

REFERENCES

1. S. Lloyd, Phys. Rev. A **55,** 1613 (1997); I. Devetak, IEEE Trans. Inf. Theory **51,** 44 (2005).
2. B. Schumacher and M. A. Nielsen, Phys. Rev. A **54,** 2629 (1996).
3. I. Devetak and P. W. Shor, Commun. Math. Phys. **256,** 287 (2005).
4. H. Barnum, E. Knill, and M. A. Nielsen, IEEE Trans.Info.Theor. **46,** 1317 (2000)
5. C. H. Bennett, D. P. DiVincenzo, and J. A. Smolin, Phys. Rev. Lett. **78,** 3217 (1997);
6. D. Leung, J. Lim, and P. W, Shor, quant-ph/0710.5943.
7. A. W. Leung, Phys. Rev. A **77,** 012322 (2005).
8. I. Devetak and A. Winter, Phys. Rev. Lett. **93,** 080501 (2004).
9. F. Grosshans, G. van Assche, J. Wenger, R. Tualle-Brouri, and P. Grangier, Nature (London) **421,** 238 (2003).
10. V. Giovannetti and R. Fazio, Phys. Rev. A **71,** 032314 (2005).

Monogamy of correlations for broadcast copies of entangled states

M. Piani*, M. Christandl†, P. Horodecki**,‡ and C. E. Mora*

*Institute for Quantum Computing, University of Waterloo, 200 University Ave. W., N2L 3G1 Waterloo, Canada
†Arnold Sommerfeld Center for Theoretical Physics, Ludwig-Maximilians-Universität München, Theresienstrasse 37, 80333 München, Germany
**Faculty of Applied Physics and Mathematics, Technical University of Gdańsk, 80-952 Gdańsk, Poland
‡National Quantum Information Centre of Gdańsk, 81-824 Sopot, Poland

Abstract. By combining a measure of total correlations, mutual information, and the notion of broadcast —i.e. generally non-factorized— copies of states of distributed systems, we are able to characterize in a single framework different levels of quantumness of correlations, corresponding to the classification of strictly classical, separable, and entangled states. We find that the broadcast regularization of mutual information —defined as the minimal per-copy mutual information between parties, when they share an infinite amount of broadcast copies— besides having by construction an appealing interpretation, satisfies many of the properties of an entanglement measure. In particular, it vanishes for separable states while it is strictly positive for all entangled states. The latter fact may be considered a signature of monogamy of entangled correlations.

Keywords: Entanglement, Quantumness, Quantum Correlations, Monogamy
PACS: 03.65.Ud,03.67.Mn

INTRODUCTION

Much work has been recently performed to analyze whether and how correlations can be understood, quantified and classified as either classical or quantum [1]. Efforts to address the *quantumness* of correlations are relevant to improve our understanding of the fundamental differences between classical and quantum mechanics. These differences are important from the point of view of quantum computing and quantum information, as we would like to better understand the origin of the quantum advantage for many tasks with respect to the classical scenario.

In studying the quantumness of correlations, *classical-classical* (CC) states $\rho_{AB}^{\mathrm{CC}} = \sum_{ij} p_{ij}^{AB} |i_A\rangle\langle i_A| \otimes |j_B\rangle\langle j_B|$, with $\langle k_X|l_X\rangle = \delta_{kl}$ for $X = A, B$ and p_{ij}^{AB} a classical joint probability distribution, are considered to have no quantum features. Indeed, CC states are a proper subclass of *separable* states $\rho_{AB}^{\mathrm{sep}} = \sum_k p_k \rho_A^k \otimes \rho_B^k$, and, most importantly, they can be seen as the embedding of joint classical distributions in the quantum formalism. On the contrary, states that are *entangled*, i.e. non-separable, are expected to exhibit the strongest quantum behaviour.

In [2] the quantumness of correlations of a bipartite state ρ_{AB} was addressed from an operational point of view, by employing the notion of broadcast copies, and by quantifying total correlations by means of mutual information $I(\rho_{AB}) \equiv I(A:B)_{\rho_{AB}} \equiv$

CP1110, *Quantum Communication, Measurement and Computing (QCMC), The Ninth International Conference,* edited by A. Lvovsky

$S(\rho_A)+S(\rho_B)-S(\rho_{AB})$, with $S(\sigma)=-\mathrm{Tr}\sigma\log_2\sigma$ the von Neumann entropy. Given a state $\sigma=\sigma_X$, we say that a state $\sigma^{(n)}=\sigma^{(n)}_{X_1\ldots X_n}$ is an n-copy broadcast state of σ if $\sigma^{(n)}_{X_k}=\mathrm{Tr}_{\backslash X_k}(\sigma^{(n)})=\sigma$, where $\mathrm{Tr}_{\backslash X_k}$ stands for the partial trace over all subsystems but X_k. Each system X_k may be composed of subsystems itself, and we will actually refer to such a case, with $X_k=A_kB_k$. Combining mutual information and the notion of broadcast copies, we can define the *n-copy broadcast mutual information* of ρ_{AB} as $(I_b)_n(\rho_{AB})\equiv\min_{\rho^{(n)}}I(A_1\ldots A_n:B_1\ldots B_n)_{\rho^{(n)}}$, where the minimum is taken over all n-copy broadcast states of ρ_{AB}.

Results presented in [2] suggest that the quantumness of the correlations present in ρ_{AB} may be revealed by the dependence of $(I_b)_n(\rho_{AB})$ on n, i.e., on the number of broadcast copies. Indeed, it was proved that for non-CC states —even separable ones— $(I_b)_n(\rho_{AB})>I(\rho_{AB})$, for $n\geq 2$. The intuition is that, while classical correlations can be freely shared among the broadcast copies, quantum correlations can not.

MAIN RESULTS

The characterization of different levels of quantumness of correlations is given by the following theorem.

Theorem 1 *The n-copy broadcast mutual information $(I_b)_n$ as a function of n: (i) is constant for CC states; (ii) grows but it is bounded for general separable states; (iii) grows linearly for all entangled states.*

Given any mixed ensemble realization $\{(p_k,\rho^k_{AB})\}$ of ρ_{AB}, such that $\rho_{AB}=\sum_k p_k\rho^k_{AB}$, the convex combination of factorized copies $\rho^{(n)}[\{(p_k,\rho^k_{AB})\}]\equiv\sum_k p_k{\rho^k_{AB}}^{\otimes n}$ is a possible n-copy broadcast state of ρ_{AB}. Therefore, $I(\rho_{AB})\leq(I_b)_n(\rho_{AB})\leq I\big(\rho^{(n)}[\{(p_k,\rho^k_{AB})\}]\big)$. Points (i) and (ii) are then easily checked by choosing: for CC states, $p_k\equiv p_{ij}$ and $\rho^k_{AB}\equiv|i_A\rangle\langle i_A|\otimes|j_B\rangle\langle j_B|$; for separable-but-not-CC states, $\rho^k_{AB}=\rho^k_A\otimes\rho^k_B$. In the latter case, the strict growth of $(I_b)_n$ with n was proved in [2].

Point (iii) of Theorem is seen to be true by studying the *broadcast regularization of mutual information*, defined as the minimal per-copy mutual information between parties, when they share an infinite amount of broadcast copies of a state ρ_{AB}:

$$I_b^{(\infty)}(\rho_{AB})\equiv\lim_n\frac{1}{n}(I_b)_n(\rho_{AB}).$$

We can prove that $I_b^{(\infty)}$ satisfies many of the requirements of an entanglement measure. More in general, we found that

Proposition 1 *The asymptotic broadcast mutual information $I_b^{(\infty)}$:*

(i) vanishes on separable states: $I_b^{(\infty)}(\rho^{\mathrm{sep}}_{AB})=0$;

(ii) is convex: $I_b^{(\infty)}(\sum_i p_i\rho_i^{AB})\leq\sum_i p_i I_b^{(\infty)}(\rho_i^{AB})$;

(iii) does not increase under local operations: $I_b^{(\infty)}((\Lambda_A\otimes\Lambda_B)(\rho_{AB}))\leq I_b^{(\infty)}(\rho_{AB})$ for completely-positive trace-preserving maps Λ_A and Λ_B;

(iv) *is weakly additive:* $I_b^{(\infty)}(\rho_{AB}^{\otimes m}) = mI_b^{(\infty)}(\rho_{AB})$;

(v) *is asymptotically continuous:* $|I_b^{(\infty)}(\rho) - I_b^{(\infty)}(\sigma)| \leq \delta_1(||\rho-\sigma||_1)\log d + \delta_2(||\rho - \sigma||_1)$, *where d is the dimension of AB,* $||X||_1 = \mathrm{Tr}\sqrt{X^\dagger X}$, *and* $\delta_{1,2}(x) \xrightarrow{x\to 0} 0$.

The monotonicity of $I_b^{(\infty)}$ under general Local Operations and Classical Communication (LOCC) is an open problem.

We prove that $I_b^{(\infty)}$ is strictly positive for all entangled states. More precisely,

Lemma 1 $I_b^{(\infty)}(\rho_{AB})$ *is lower bounded by*

$$I_b^{(\infty)}(\rho_{AB}) \geq \inf_{\sigma_{AB}^{\text{sep}}} \sup_{\{M_i\otimes N_j\}} S(\{p_{ij}(\rho_{AB})\}||\{p_{ij}(\sigma_{AB}^{\text{sep}})\}), \tag{1}$$

where: the infimum runs over all separable states; the supremum is taken with respect to all local POVMs $M_i \geq 0$, $N_j \geq 0$, $\sum_i M_i = \mathbb{I}$, $\sum_i N_j = \mathbb{I}$; *we consider the joint classical probabilities* $p_{ij}(\tau_{AB}) \equiv \mathrm{Tr}(\tau_{AB} M_i \otimes N_j)$; $S(\{p_k\}||\{q_k\}) \equiv -\sum_k p_k \log_2 \frac{q_k}{p_k}$ *is the classical relative entropy between two classical probability distributions* $\{p_k\}$ *and* $\{q_k\}$.

The right-hand side of (1) is strictly positive for all entangled states, because there exist informationally complete local POVMs [3] and relative entropy vanishes only for $\{p_k\} = \{q_k\}$.

We relate $I_b^{(\infty)}$ to some entanglement measures known in literature [4, 5], which are important both conceptually and operationally [6] (we adopt here a slightly non-standard convention as regards normalization constants). *Quantum squashed entanglement* is defined as $E_{\text{sq}}^Q(\rho_{AB}) = \inf_{\sigma_{ABE}} \big(I(A:BE)_\sigma - I(A:E))_\sigma\big)$, where the infimum is on states satisfying $\mathrm{Tr}_E(\sigma_{ABE}) = \rho_{AB}$. Similarly, *classical squashed entanglement* is given by $E_{\text{sq}}^C(\rho_{AB}) = \min_{\sigma_{AB\hat{E}}} \big(I(A:B\hat{E})_\sigma - I(A:\hat{E}))_\sigma\big)$, where the minimum is taken with respect to states that are classical on $\hat{E}$ —i.e, states $\sigma_{AB\hat{E}} = \sum_k p_k \rho_{AB}^k \otimes |k_{\hat{E}}\rangle\langle k_{\hat{E}}|$, with $\langle k_{\hat{E}}|l_{\hat{E}}\rangle - \delta_{kl}$— and that satisfy $\sum_k p_k \rho_{AB}^k = \rho_{AB}$. It has recently been proved that these two quantities are different [7, 8, 9]. We find $E_{\text{sq}}^Q \leq I_b^{(\infty)} \leq E_{\text{sq}}^C$.

We further analyze the case where we modify the definition of $I_b^{(\infty)}$ by considering only permutationally invariant broadcast copies of ρ_{AB}, i.e. only states $(\rho_{\text{sym}}^{(n)})_{A_1\ldots A_n B_1 \ldots B_n}$ that satisfy $\Pi \rho_{\text{sym}}^{(n)} \Pi^\dagger = \rho_{\text{sym}}^{(n)}$, where Π permutes the pairs $\{A_i B_i\}$. In this way one may define a *symmetric broadcast regularization of mutual information*:

$$I_{b,\text{sym}}^{(\infty)}(\rho_{AB}) \equiv \lim_n \frac{1}{n} \min_{\rho_{\text{sym}}^{(n)}} I(A_1 \ldots A_n : B_1 \ldots B_n)_{\rho_{\text{sym}}^{(n)}}$$

The following theorem can be proved either by a de Finetti-type argument [11] or by an argument along the lines of [12].

Theorem 2 $I_{b,\text{sym}}^{(\infty)} = E_{\text{sq}}^C$, *i.e., optimal symmetric broadcast copies are of the form* $\rho^{(n)}[\{(p_k, \rho_{AB}^k)\}]$.

CONCLUSIONS

By analyzing the minimal per-copy mutual information as a function of n, when parties share n broadcast copies, we were able to classify strictly classical, separable, and entangled distributed states. That such a quantity grows linearly with n for and only for entangled states may be interpreted as a signature of *monogamy of entanglement* [10]. The standard notion of monogamy of entanglement refers to the impossibility for a system to be strongly entangled a the same time with two or more other systems. Here we find a copy-copy monogamy of correlations, as quantified by mutual information, in the broadcast scenario. Entangled correlations, contrary to the classical correlations of CC states and —up to a certain approximation— to the correlations of separable states, cannot be freely shared among copies: each broadcast copy needs its own amount of correlations.

We defined the asymptotic broadcast regularization of mutual information and studied its properties. We found that it fullfills many requirements of an entanglement measure, and showed that it lies between two known entanglement measures, the quantum and classical versions of squashed entanglement. Furthermore, we proved that when only symmetric broadcast copies are considered, the symmetric broadcast regularization of mutual information coincides with classical squashed entanglement.

ACKNOWLEDGMENTS

We thank Michał Horodecki for useful discussions. This work was initiated when MP was a Lise Meitner fellow at the Institute for Theoretical Physics of the University of Innsbruck, and is partially supported by the Austrian Science Fund (FWF). MP and CM acknowledge support by the Institute for Quantum Computing, QuantumWorks NSERC Innovation Platform, and Ontario Centres of Excellence.

REFERENCES

1. H. Ollivier, and W.H. Zurek, *Phys. Rev. Lett.* **88**, 017901 (2001); L. Henderson and V. Vedral, *J. Phys. A: Math. Gen.* **34**, 6899–6905 (2001); J. Oppenheim et al., *Phys. Rev. Lett.* **89**, 180402 (2002); M. Horodecki et al., *Phys. Rev. A* **71**, 062307 (2005); B. Groisman, S. Popescu, and A. Winter, *Phys. Rev. A* **72**, 032317 (2005); B. Groisman et al., arXiv:quant-ph/0703103.
2. M. Piani, P. Horodecki, and R. Horodecki, *Phys. Rev. Lett.* **100**, 090502 (2008).
3. C. M. Caves et al., *J. Math. Phys.* **43**, 4537 (2002).
4. M. Christandl, and A. Winter, *J. Math. Phys.* **45**, 829–840 (2004).
5. D. Yang et al., arXiv:0704.2236.
6. J. Oppenheim, arXiv:0801.0458.
7. S. Aaronson et al., to appear in *Theory of Computing*, arXiv:0804.0802.
8. F. Brandao, *Entanglement Theory and the Quantum Simulation of Many-Body Physics*, PhD thesis, Imperial College, London, 2008, pp. 121–127, arXiv:0810.0026.
9. M. Christandl and N. Schuch, private communication.
10. V. Coffman, J. Kundu, and W. K. Wootters, *Phys.Rev. A* **61**, 052306 (2000).
11. R. Renner, *Nature Physics* **3**, 645–649 (2007).
12. M. Christandl, A. Harrow, and G. Mitchison, *Comm. Math. Phys.* **270**, 575–585 (2007).

The partial $K-$way negativities and entanglement of N qubit states

S. Shelly Sharma* and N. K. Sharma†

*Depto. de Física, Universidade Estadual de Londrina, Londrina 86051-990, PR Brazil
†Depto. de Matemática, Universidade Estadual de Londrina, Londrina 86051-990 PR, Brazil

Abstract. We define partial K-way negativities for N-qubit states. For the states in canonical form, global and K-way negativities measure the entanglement of the state.

Keywords: Entanglement, Global negativity, Partial k-way negativity
PACS: 03.65.Ud, 03.67.Mn

WHAT IS A PARTIAL K-WAY TRANSPOSE?

The Hilbert space, $C^d = C^{d_1} \otimes C^{d_2} \otimes ... \otimes C^{d_N}$, associated with a quantum system composed of N sub-systems, is spanned by basis vectors of the form $|i_1 i_2 ... i_N\rangle$, where $i_m = 0$ to $(d_m - 1)$, and d_m is the dimension of Hilbert space associated with m^{th} sub-system. To simplify, we denote $|i_1 i_2 ... i_N\rangle$ by $\left|\prod_{m=1}^{N} i_m\right\rangle$ and write an N qudit pure state as

$$\widehat{\rho} = \sum_{i_1 - i_N, j_1 - j_N} \left\langle \prod_{m=1}^{N} i_m \right| \widehat{\rho} \left| \prod_{m=1}^{N} j_m \right\rangle \left| \prod_{m=1}^{N} i_m \right\rangle \left\langle \prod_{m=1}^{N} j_m \right|. \quad (1)$$

The global partial transpose [1] of $\widehat{\rho}$ with respect to qudit p is defined as

$$\left\langle \prod_{m=1}^{N} i_m \right| \widehat{\rho}_G^{T_p} \left| \prod_{m=1}^{N} j_m \right\rangle = \left\langle j_p \prod_{m=1, m\neq p}^{N} i_m \right| \widehat{\rho} \left| i_p \prod_{m=1, m\neq p}^{N} j_m \right\rangle. \quad (2)$$

Recently, we proposed a characterization of N-partite states based on $K-$way negativities, where $2 \leq K \leq N$ [2]. The $K-$way partial transpose with respect to a subsystem of an N-partite composite system is constructed by partial transposition subject to specific constraints on transposed matrix elements for each value of K. The matrix elements of $K-$way partial transpose $\widehat{\rho}_K^{T_p}$ are given by

$$\left\langle \prod_{m=1}^{N} i_m \right| \widehat{\rho}_K^{T_p} \left| \prod_{m=1}^{N} j_m \right\rangle = \left\langle j_p \prod_{m=1, m\neq p}^{N} i_m \right| \widehat{\rho} \left| i_p \prod_{m=1, m\neq p}^{N} j_m \right\rangle, \quad if \quad \sum_m (1 - \delta_{i_m j_m}) = K, \quad (3)$$

$$\left\langle \prod_{m=1}^{N} i_m \right| \widehat{\rho}_K^{T_p} \left| \prod_{m=1}^{N} j_m \right\rangle = \left\langle \prod_{m=1}^{N} i_m \right| \widehat{\rho} \left| \prod_{m=1}^{N} j_m \right\rangle, \quad if \sum_m (1 - \delta_{i_m j_m}) \neq K, \quad (4)$$

where $\delta_{i_m, j_m} = 1$ for $i_m = j_m$, and $\delta_{i_m, j_m} = 0$ for $i_m \neq j_m$.

CP1110, *Quantum Communication, Measurement and Computing (QCMC), The Ninth International Conference,* edited by A. Lvovsky

GLOBAL AND PARTIAL $K-$WAY NEGATIVITY

Global Negativity [3, 4], defined as $N_G^p = \left(\left\|\rho_G^{T_p}\right\|_1 - 1\right) / (d_p - 1)$, measures the entanglement of subsystem p with its complement in a bipartite split of the composite system. Here $\|\rho\|_1$ is the trace norm of ρ. Global negativity vanishes on ppt-states and is equal to the entropy of entanglement on maximally entangled states. Partial $K-$way negativity is the contribution of $K-$way partial transpose to global negativity [2] of partially transposed state operator and measures $K-$partite coherence in an N-qudit pure state ($2 \leq K \leq N$). The partial $K-$way negativities provide measures of $K-$way entanglement for N-partite canonical states.

Global negativity with respect to a subsystem can be written as a sum of partial $K-$way negativities. Using the definition of trace norm and $Tr\left(\widehat{\rho}_G^{T_p}\right) = 1$, we get

$$N_G^p = -\frac{2}{d_p - 1}\sum_i \left\langle \Psi_i^{G-}\right| \widehat{\rho}_G^{T_p} \left|\Psi_i^{G-}\right\rangle = -\frac{2}{d_p - 1}\sum_i \lambda_i^{G-}, \tag{5}$$

where λ_i^{G-} and $\left|\Psi_i^{G-}\right\rangle$ are the negative eigenvalues and eigenvectors of $\widehat{\rho}_G^{T_p}$. The global transpose with respect to subsystem p, may also be written as

$$\widehat{\rho}_G^{T_p} = \sum_{K=2}^{N} \widehat{\rho}_K^{T_p} - (N-2)\widehat{\rho}. \tag{6}$$

Using Eq. (6) in Eq. (5), and defining partial $K-$way negativity E_K^p ($K = 2$ to N) as

$$E_K^p = -\frac{2}{d_p - 1}\sum_i \left\langle \Psi_i^{G-}\right| \widehat{\rho}_K^{T_p} \left|\Psi_i^{G-}\right\rangle, E_0^p = -\frac{2(N-2)}{d_p - 1}\sum_i \left\langle \Psi_i^{G-}\right| \widehat{\rho} \left|\Psi_i^{G-}\right\rangle \tag{7}$$

we may split the global negativity for qudit p as $N_G^p = \sum_{K=2}^{N} E_K^p - E_0^p$.

PARTIAL 3-WAY NEGATIVITIES AND THREE TANGLE

Partial $K-$way negativities of pure three qubit generic canonical state [5] have been found to be related [6] to three tangle [7] and tangle [8]. Writing the generic three qubit in canonical form with respect to qubit A as

$$\Psi = a|000\rangle + be^{i\varphi}|100\rangle + c|110\rangle + d|101\rangle + f|111\rangle, \tag{8}$$

we obtain $N_G^A = 2ag$, and the products $E_3^A N_G^A = 4a^2 f^2$, and $E_2^A N_G^A = 4a^2\left(c^2 + d^2\right)$, where $g = \sqrt{c^2 + d^2 + f^2}$. The tangle, defined to be a measure of entanglement of qubit A with subsystem BC [8], for the pure state of Eq. (8) is given by $\tau_{A(BC)} = 4a^2 g^2$. The three tangle [7] defined as $\tau_3 = \tau_{A(BC)} - \tau_{AB} - \tau_{AC}$, is known to measure the tripartite entanglement of the pure state. For the state Ψ, $\tau_3 = 4a^2 f^2 = E_3^A N_G^A$ and

$\tau_{AB} + \tau_{AC} = \tau_{A(BC)} - \tau_3 = E_2^A N_G^A$. For the special case of three qubit states, we obtain the inequalities

$$N_G^p \geq E_2^p, \qquad N_G^p \geq E_3^p, \tag{9}$$

the former being equivalent to the CKW inequality [7], $\tau_{A(BC)} \geq \tau_{AB} + \tau_{AC}$, on which the definition of three tangle is based.

FOUR QUBIT CANONICAL STATES

For qubit p in a four qubit state in normal form $\left(N_G^p\right)^2 = N_G^p E_4^p + N_G^p E_3^p + N_G^p E_2^p - N_G^p E_0^p$, that is there exist monogamy inequalities $\left(N_G^p\right)^2 \geq N_G^p E_K^p$, where $2 \leq K \leq 4$. It has been shown in ref. [9] that a pure state of four qubits can be transformed by Stochastic local operations and classical communication (SLOCC) to a state belonging to one of a set of nine families of states. We classify the nine families of states by using global and partial K-way negativities. These measures are polynomial functions of local invariants characterizing each state as such entanglement measures.

Class I - $E_4 \neq 0, E_3 = 0, E_2 \neq 0$

The families of states, G_{abcd}, L_{abc_2}, $\mathrm{L}_{a_2b_2}$, and $\mathrm{L}_{a_2 0_{3\oplus 1}}$ of ref. [9] lie in class I. A common feature of these families of states is a positive three way partial transpose, independent of the qubit with respect to which partial transpose is constructed. In the canonical form, the states do not have three qubit correlations. All the states have genuine 4−partite entanglement, which is destroyed when a measurement is made on the state of a single qubit. The states may have four-partite entanglement due to pairwise entanglement, as well. When $E_2 \neq 0$, reduced two qubit mixed states are entangled states.

Class II - $E_3 \neq 0$

The sets of states L_{ab_3} and L_{a_4}, having 4-partite, tripartite and bi-partite entanglement are grouped in class II, along with the states $\mathrm{L}_{0_{5\oplus 3}}$, $\mathrm{L}_{0_{7\oplus 1}}$, and $\mathrm{L}_{0_{3\oplus 1}0_{3\oplus 1}}$ having $E_3 \neq 0$ while $E_4 = 0$. For the set of normalized states L_{ab_3} in the general form

$$\begin{aligned} L_{ab_3} &= a\left(|0000\rangle + |1111\rangle\right) + \frac{a+b}{2}\left(|0101\rangle + |1010\rangle\right) \\ &\quad + \frac{a-b}{2}\left(|0110\rangle + |1001\rangle\right) \qquad (10) \\ &\quad + ic\left(|0001\rangle + |0010\rangle + |0111\rangle + |1011\rangle\right), \qquad (11) \end{aligned}$$

fig. (2) displays the contour plots of global negativity and partial two, three and four way negativities calculated numerically from state operator partially transposed with respect to qubit A.

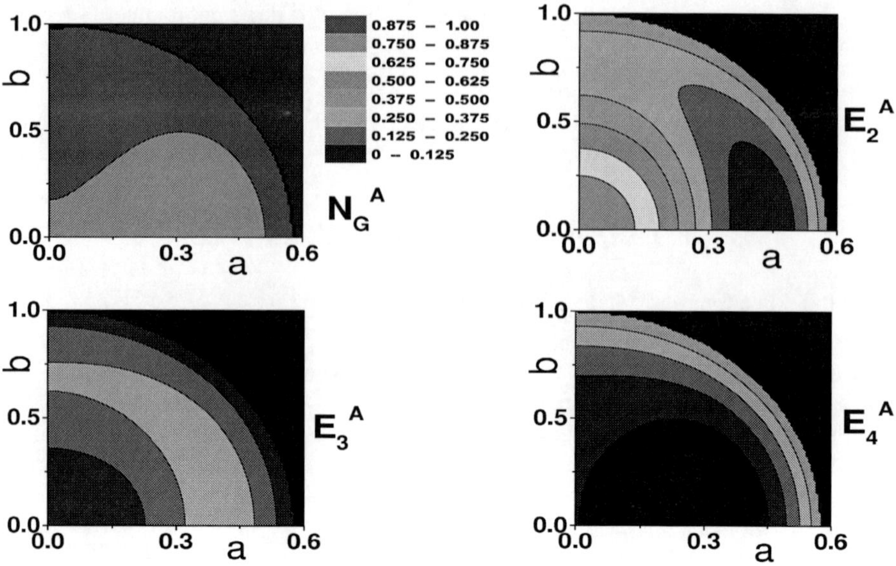

FIGURE 1. Contour plots of N_G^A, E_4^A, E_3^A and E_2^A for the states L_{ab_3} versus parameters a and b.

An N-qubit state is maximally entangled when for the state in canonical form $N_G^p = E_N^p = 1$, for $p = 1$ to N. Only, $N_G^p = 1$ for $p = 1$ to N, does not ensure maximal entanglement. Partial four way negativity of a pure state measures genuine 4-partite entanglement of the state. Three-way partial negativity determines the probabilistic entanglement that becomes available to the three parties after the fourth party measures the state of the qubit it holds. Two-way negativity measures the pairwise entanglement. Local unitary rotations may enhance a given set of partial K-way negativities at the cost of others.

Financial Support from Fundacao Araucaria and Faep UEL PR Brazil is acknowledged.

REFERENCES

1. A. Peres, Phys. Rev. Lett. 77, 1413 (1996).
2. S. Shelly Sharma and N. K. Sharma, Phys. Rev. A 76, 012326 (2007), Phys. Rev. A 77, 042117 (2008), arXiv: quant-ph/0608062.
3. K. Zyczkowski, P. Horodecki, A. Sanpera, and M. Lewenstein, Phys. Rev. A 58, 883 (1998).
4. G. Vidal and R. F. Werner, Phys. Rev. Vol. 65, 032314 (2002).
5. A. Acin, A. Andrianov, L. Costa, E. Jane, J.I. Latorre, R. Tarrach, Phys. Rev. Lett. 85, 1560 (2000).
6. S. Shelly Sharma and N. K. Sharma, Phys. Rev. A 78, 012113 (2008).
7. V. Coffman, J. Kundu, and W. K. Wootters, Phys. Rev. A 61, 052306 (2000).
8. W. K. Wootters, Phys. Rev. Lett. 80, 2245 (1998).
9. F. Versraete, J. Dehaene, B. De Moor, and H. Verschelde, Phys. Rev. A65, 052112 (2002).

The Consumption of Reference Resources

G. A. White J. A. Vaccaro H. M. Wiseman

Center for Quantum Computer Technology, Center for Quantum Dynamics, School of Science, Griffith University, Brisbane, 4111 Australia

Abstract. Under the operational restriction of the U(1)-superselection rule, states that contain coherences between eigenstates of particle number constitute a resource. Such resources can be used to facilitate operations upon systems that otherwise cannot be performed. However, the process of doing this consumes reference resources. We show this explicitly for an example of a unitary operation that is forbidden by the U(1)-superselection rule.

Keywords: Reference Frames, Superselection rules, Quantum information
PACS: 03.67.-a, 03.67.Mn, 03.65.Ud, 03.65.Ta

INTRODUCTION

Superselection rules (SSRs) are the constraint that one cannot observe coherences between eigenstates of a particular quantity [1]. They were originally discovered by Wight, Wickman and Wigner in the context of conservation laws [2]. However, it is now customary to think of them as arising from a practical restriction that one may need to consider due to the presence of a conservation law, or otherwise [1]. A SSR can in general be alleviated by an ancillary system that acts as a reference for the coherences [3]. Under a SSR there exists a group $\{T(g) : g \in G\}$ of unitary operators whose effect on the system is undetectable. The notation we adopt here is that $\{T(g) : g \in G\}$ is a representation of the abstract symmetry group $G = \{g\}$. Consider a system which has been prepared in the state ρ. As ρ is indistinguishable from $T(g)\rho T^{\dagger}(g)$ for any $g \in G$, the *effective* state system of the system is given by the "twirl" $\mathscr{G}_G[\rho]$ of ρ which is defined as [3]

$$\mathscr{G}_G[\rho] = \frac{1}{|G|}\sum_{g\in G} T(g)\rho T^{\dagger}(g) , \tag{1}$$

where $|G|$ is the order of G. This indistinguishability implies that the SSR restricts the allowed operations $\mathscr{O}$, including measurements and unitaries, that can be performed upon the system. The allowed operations are those that are G invariant in the sense that [4]

$$\mathscr{O}[\hat{T}(g)\rho\hat{T}^{\dagger}(g)] = \hat{T}(g)\mathscr{O}[\rho]\hat{T}^{\dagger}(g) \quad \forall g \in G . \tag{2}$$

The definition of the twirl in Eq. (1) is easily extended to composite systems. We restrict ourselves to composite systems where the component subsystems have the same symmetry described by the symmetry group, G. This means that the representation of the group G for the composite system is generated by the tensor product of the representations for the subsystems. Under this restriction the twirl of a composite system in state ρ is given by Eq. (1) where the composite-system operators $T(g)$ have a

CP1110, *Quantum Communication, Measurement and Computing (QCMC), The Ninth International Conference,* edited by A. Lvovsky

decomposition of the form

$$T(g) = T_1(g) \otimes T_2(g) \otimes \cdots \quad \forall g \in G\,. \tag{3}$$

Here $T_i(g)$ acts on the Hilbert space of subsystem labeled i.

EXPLICIT EXAMPLE OF A NON U(1)-INVARIANT UNITARY

As shown in the previous section, the effect of a SSR can be to reduce the purity of a particular state. Consider an arbitrary state, ρ, that may be mixed and has a von-Neumann entropy equal to $S(\rho)$. The von Neumann entropy of the effective state $\mathscr{G}\rho$ is greater that that of ρ, in general [4, 5]. The increase in the von Neumann entropy is a measure of the *asymmetry*, $A_G(\rho)$, of the state ρ with respect to the symmetry group G. The asymmetry $A_G(\rho)$ of ρ is defined as [6]

$$A_G(\rho) = S(\mathscr{G}_G[\rho]) - S(\rho) \tag{4}$$

where $S(\rho)$ is the von Neumann entropy of ρ. $A_G(\rho)$ has been shown to quantify the ability of an ancilla in state ρ to act as a reference and partially alleviate the effects of the G SSR [6].

In general the use of a reference ancilla to perform a non U(1)-invariant operation on the system may increase the asymmetry of the system with respect to the U(1) group of phase translations. Using a reference ancilla repeatedly in this way will reduce the reference ability of the ancilla. This effect has been studied explicitly for the case where a reference is used to facilitate a phase measurement [7]. Here we wish to examine this process for the case of a non U(1)-invariant unitary operation. Consider a system that is in the vacuum state

$$|\psi\rangle_S = |0\rangle_S\,. \tag{5}$$

The operation F_S on the system for which

$$F_S(|0\rangle_S) = \frac{1}{\sqrt{2}}(|0\rangle_S + e^{i\theta}|1\rangle_S) \tag{6}$$

is forbidden under the SSR. However, an analogous operation can be performed on the coherent subspaces of the combined reference ancilla and system as follows. We define an analogous operation on the basis $\{|n\rangle_R|0\rangle_S, |n-1\rangle_R|1\rangle_S\}$ of the coherent subspaces representing n particles by

$$F_{\mathrm{RS}}^{(n)}|n\rangle_R|0\rangle_S = \frac{1}{\sqrt{2}}(|n\rangle_R|0\rangle_S + |n-1\rangle_R|1\rangle_S) \tag{7}$$

for $n = 1, 2, \ldots, N$. One can perform this operation with an ancilla say, of the form $|n\rangle$ and produce the the right hand side of (7). However, upon performing the operation one can easily see that the partial trace over the reference would leave the system in a maximally mixed state. To minimize this effect use a phase state of the form [8]

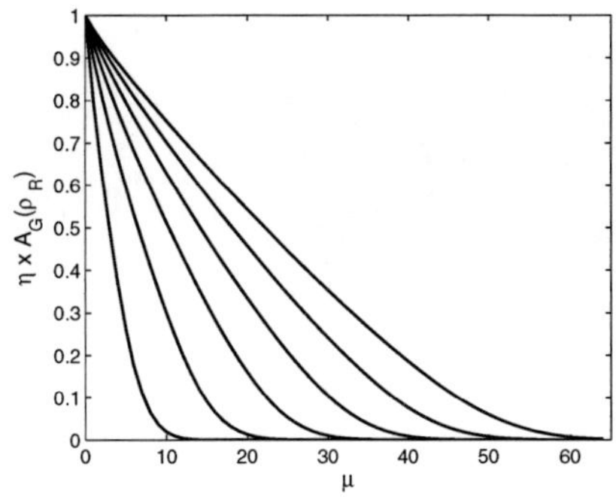

FIGURE 1. The consumption of the reference ancilla is demonstrated by the reduction of the normalized asymmetry for reference ancilla of $M = 5, 10, 15, 20, 25$ and 30 as a function of the number of times, μ, the reference is used. Here the normalization, η, is given by $\log_2[D_R]$.

$$|\psi\rangle_R = \frac{1}{\sqrt{N+1}} \sum_{n=0}^{N} e^{i\theta n} |n\rangle_R \,, \tag{8}$$

as the state of the reference ancilla. This state has the property that it maximizes the asymmetry of the reference state which has at most N particles. To further simplify matters we set $\theta = 0$. The forbidden operation, $\mathscr{F}_S$. can be mapped onto a globally G-invariant operation that acts on the combined state to give

$$\begin{aligned} |\Psi\rangle_{RS} \equiv F_{RS} |\psi\rangle_R |0\rangle_S &= \frac{1}{\sqrt{2(N+1)}} \sum_{n=1}^{N} [|n\rangle_R |0\rangle_S \\ &+ |n-1\rangle_R |1\rangle_S)] + \frac{1}{\sqrt{N+1}} |0\rangle_R |0\rangle_S \,. \end{aligned} \tag{9}$$

This state is entangled. Therefore the reduced state of the reference ancilla is mixed. Consider a system in the vacuum state $|0\rangle$. We begin with a reference ancilla in a phase state and perform the forbidden operation outlined above. The reduced state of the reference is then given by

$$\rho_R' = Tr_S[|\Psi\rangle\langle\Psi|] \,. \tag{10}$$

The reference ancilla is then used to facilitate the same operation on another system which is also prepared in the state $|0\rangle$. This process is repeated μ times and the resources, i.e. the asymmetry, of the reference ancilla is consumed as μ increases. The consumption of the reference ancilla for reference ancillae of various size is shown in figure 1 as a function of the number of times, μ, the reference is used to perform the U(1)-SSR forbidden operation, F_S.

Consider now the reduced state of the system. Due to the entanglement between the system and the reference ancilla, the reduced state of the system is mixed and therefore the fidelity between the system with the state on the right hand side of (6) is not perfect.

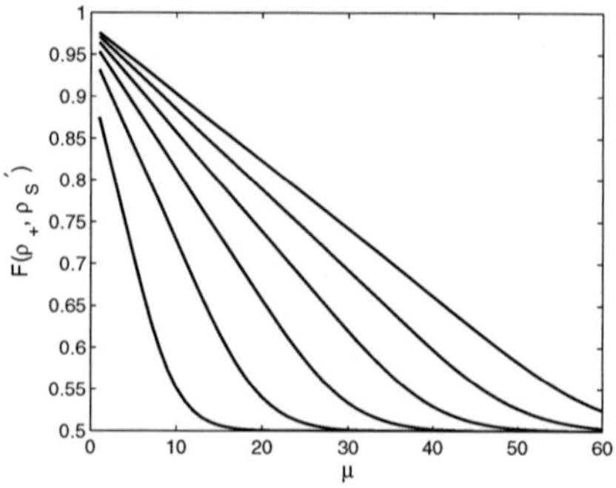

FIGURE 2. The degradation in fidelity of the system state with a $|+\rangle$ state for $M = 5, 10, 15, 20, 25$ and 30 where μ is the number of times the reference has been used.

To see this we first define the fidelity of the operation as

$$F(\rho_S', \rho_+) = \mathrm{Tr}[\rho_S' |+\rangle\langle+|] \tag{11}$$

where

$$\rho_S' = \mathrm{Tr}_R[F_S \rho_R |0\rangle_S\langle 0|] \text{ and } |+\rangle = \frac{1}{\sqrt{2}}(|0\rangle + |1\rangle) \,. \tag{12}$$

The fidelity the final state of the system shown in Figure 2. We see that the reference's ability to facilitate a non U(1)-invariant unitary reduces as the reference resources of the reference ancilla are consumed. The similarity of the trends in figures 1-2 show that as the reference show that as the reference is consumed its ability to facilitate non U(1)-invariant operations diminishes.

ACKNOWLEDGMENTS

This work was supported by the Australian Research Council

REFERENCES

1. S. D. Bartlett, T. Rudolph, and R. W. Spekkens **79**, 555 (2007).
2. G. C. Wick, A. S. Wightman, and E. P. Wigner **88**, 102 (1952).
3. Y. Aharanov, and L. Susskind **155**, 1428 (1967).
4. S. D. Bartlett, and H. M. Wiseman **91**, 097903 (2003).
5. H. M. Wiseman, and J. A. Vaccaro **91**, 097902 (2003).
6. J. A. Vaccaro, F. Anselmi, H. M. Wiseman, and K. Jacobs, *Phys. Rev. A* **77**, 032114 (2008).
7. S. D. Bartlett, T. Rudolph, R. W. Spekkens, and P. S. Turner., *New. J. Phys.* **8**, 58 (2006).
8. D. T. Pegg, and S. M. Barnett, *Europhys. Lett.* **6**, 483 (1988).

A Universal Operator Theoretic Framework for Quantum Fault Tolerance

Gerald Gilbert and Yaakov S. Weinstein

Quantum Information Science Group, MITRE, 260 Industrial Way West, Eatontown, NJ 07724, USA

Abstract. We extend standard fault tolerance theory and Kitaev's model for quantum computation so as to enable quantitative determination of design parameters for quantum computers that ensure that the overall computation yields a correct *final result* with some prescribed probability, as opposed to merely ensuring that the desired *final quantum state* is obtained. Our extension allows us to explicitly calculate the number of levels of error correction concatenation needed to achieve a correct final result with some prescribed success probability and demonstrate a clear connection between error correction and fault tolerance.

Keywords: fault tolerance, Kitaev model, Quantum Computer Condition, concatenation
PACS: 03.67.Lx, 03.67.Pp

The purpose of any computation is to compute the value of a function for specified values of the variables on which the function depends. In the case of a quantum computer, the final result is obtained from the outcome of a final measurement performed on the final quantum state produced by the quantum computation. In this paper, the term "quantum computation" refers solely to the dynamical evolution of the qubits from some initial quantum state to some final quantum state but does *not* include the final measurement from which the final result of the overall computation is obtained.

Since the final measurement produces, in general, a probabilistically distributed set of outcomes, the question arises as to whether or not reliable *final* results for the overall computation can be obtained. A partial answer to this question follows from the quantum computational model formalized by Kitaev ([1], §4.1). Kitaev's model identifies a bound, p, on the probability that the final result of the overall computation is incorrect due to the indeterminism of the final measurement. It follows that reliable overall computation can be achieved for sufficiently small p. However, Kitaev's model assumes that the quantum computation that produces the final quantum state is perfectly implemented, with no errors. In the present paper, we extend Kitaev's model so as to take into account the combined effects of errors that arise in the course of the quantum computation, which fall in the purview of fault-tolerance, the errors due to the indeterminacy of the final measurement, and the interplay between them.

The Kitaev model [1] can be depicted by the following diagram:

$$\begin{array}{ccc} \mathbf{T}(H_{\text{logical}}) & \xrightarrow{\;G\;} & \mathbf{T}(H_{\text{logical}}) \\ \mathbf{I_{dm}} \Big\uparrow & p & \Big\downarrow \mathbf{O_{dm}} \\ X & \xrightarrow[\;F\;]{} & Y \end{array} \quad . \tag{1}$$

CP1110, *Quantum Communication, Measurement and Computing (QCMC), The Ninth International Conference,* edited by A. Lvovsky

This diagram expresses the fact that the output of the quantum computation $G : \rho \mapsto U\rho U^\dagger$, where ρ is the initial density matrix of the quantum computer, is intended to compute the function F, where $F : X \to Y$ is an instance of the overall computational problem. $\mathbf{I_{dm}}$ is an initialization map, which maps the input space X into states of the logical Hilbert space H_{logical}, and $\mathbf{O_{dm}}$, given by a projection-valued measure $\{\mathrm{E}_y\}_{y\in Y}$, is the output map, which maps the output of the operation onto the output space Y. The symbol in the center of the diagram denotes the probabilistic inaccuracy, p, associated to the output of the overall computation. If $p = 0$, the computation always produces the correct result, and the diagram commutes. Thus, the quantity $1 - p$ furnishes a lower bound on the probability of the success of the overall computation. We note that p may be non-vanishing, even if the quantum algorithm is perfectly implemented, due to the fact that the measurement of the final quantum state is necessarily quantum probabilistically distributed. Given an input $x \in X$, the final result of the overall computation is distributed according the probability measure on Y as follows: the probability $\mathrm{Pr}_x(y)$ of a singleton $y \in Y$ is $\mathrm{tr}(\sqrt{\mathrm{E}_y} U \mathbf{I_{dm}}(x) U^\dagger \sqrt{\mathrm{E}_y})$. Then, the *near commutativity* of Diagram (2) means that for each $x \in X$, the probability measure $\mathrm{Pr}_x(y)$ is concentrated at $y = F(x)$, so that $F(x)$ is the final result of the overall computation with high probability. Thus, Kitaev's formulation [1] requires that the Diagram (2) be nearly commutative in a probabilistic sense, that is

$$\mathrm{tr}\left(\mathbf{O_{dm}}\left(G\left(\mathbf{I_{dm}}\left(x\right)\right)\right)\right) \equiv \mathrm{tr}\left(\sqrt{\mathrm{E}_{F(x)}}U\mathbf{I_{dm}}(x)U^\dagger\sqrt{\mathrm{E}_{F(x)}}\right) \geq 1 - p \tag{2}$$

In other words, the measurement of the final quantum state gives the correct final result for the overall computation with probability greater than $1 - p$.

We now extend Kitaev's model by allowing for residual errors in any realistic implementation of a quantum computation, even upon successful application of fault-tolerance techniques. We consider a function F, and an implementation of an overall computation that is intended to calculate F for some value of the input. The relationship between the *ideal*, error-free quantum computation, $G(\rho) = U\rho U^\dagger$, defined on the logical Hilbert space, H_{logical}, and the *actual* dynamical map, P, implemented by the physical quantum computer on the computational Hilbert space, H_{comp}, can be described by the following diagram:

$$\begin{array}{ccc} \mathbf{T}(H_{\text{comp}}) & \xrightarrow{P} & \mathbf{T}(H_{\text{comp}}) \\ \mathcal{M}_{\{\text{l}\to\text{c}\}}\Big\uparrow & \alpha & \Big\downarrow\mathcal{M}_{\{\text{c}\to\text{l}\}} \\ \mathbf{T}(H_{\text{logical}}) & \xrightarrow[G]{} & \mathbf{T}(H_{\text{logical}}) \end{array} , \tag{3}$$

where the superoperators $\mathcal{M}_{\{\text{l}\to\text{c}\}}$ and $\mathcal{M}_{\{\text{c}\to\text{l}\}}$ are *linking maps* that are mathematically required to relate states in the logical space H_{logical} to states in the computational space H_{comp}. We formally express the content of Diagram (3) *via*:

$$\|\mathcal{M}_{\{\text{c}\to\text{l}\}}(P \cdot (\mathcal{M}_{\{\text{l}\to\text{c}\}}(\rho))) - G(\rho)\|_1 \leq \alpha \, , \tag{4}$$

where $\|\cdot\|_1$ signifies the Schatten 1-norm. The quantity on the left-hand-side of this inequality, the implementation inaccuracy, furnishes a measure of the inaccuracy of the actual implementation of G by P. The quantity α is a prescribed bound on the implementation inaccuracy. Thus, the diagram states that P can implement the idealized, perfect quantum computation, G, with an inaccuracy no greater than α. We refer to the entire inequality as the *Quantum Computer Condition* (QCC).

To establish the utility of practical quantum computers, what is needed now is a relation that connects the actual, device-implemented quantum computation, P, to the intended calculation of F. This is established by noting:

$$\mathrm{tr}\Big(\mathbf{O_{dm}}(P^{\mathcal{M}}(\mathbf{I_{dm}}(x)))\Big) > -\alpha + 1 - p, \tag{5}$$

where we have used the properties of the Schatten 1-norm and the QCC. The physical meaning of this result can be transparently expressed by combining the content of Diagrams (1) and (3) to yield

$$\begin{array}{ccc} \mathbf{T}(H_{\mathrm{comp}}) & \xrightarrow{P} & \mathbf{T}(H_{\mathrm{comp}}) \\ \tilde{\mathbf{I}}_{\mathbf{dm}} \Big\uparrow & \alpha + p & \Big\downarrow \tilde{\mathbf{O}}_{\mathbf{dm}} \\ X & \xrightarrow[F]{} & Y \end{array} . \tag{6}$$

This diagram is nearly commutative in the sense that

$$\mathrm{tr}\left(\mathbf{O_{dm}}\left(P^{\mathcal{M}}\left(\mathbf{I_{dm}}\left(x\right)\right)\right)\right) \equiv \mathrm{tr}\Big(\tilde{\mathbf{O}}_{\mathbf{dm}}(P(\tilde{\mathbf{I}}_{\mathbf{dm}}(x)))\Big) > 1 - (p + \alpha)\,, \tag{7}$$

where $\tilde{\mathbf{I}}_{\mathbf{dm}} \equiv \mathcal{M}_{\{\mathrm{l}\to\mathrm{c}\}} \circ \mathbf{I_{dm}}$, $\tilde{\mathbf{O}}_{\mathbf{dm}} \equiv \mathbf{O_{dm}} \circ \mathcal{M}_{\{\mathrm{c}\to\mathrm{l}\}}$.

Thus the total probability of failure that the overall computation yields the correct *final* result is bounded by the sum of two terms: (1) an upper bound on the failure probability for the overall computation when the quantum computation is perfectly implemented without errors, and (2) an upper bound on the implementation inaccuracy of the actual quantum computation implemented by a practical device. It is not immediately obvious that these quantities should combine in such a simple way. The quantity p represents a bound on probabilities while α is a bound on the normed difference between the states resulting from idealized and actual quantum computations, respectively.

As an example suppose we wish to build a quantum computer which provides the correct solution to some problem with probability $1 - \hat{p}$ or better. From eq.(7) we infer that the failure probability of the overall computation is bounded by $p + \alpha$. We therefore require $p + \alpha = \hat{p}$. The quantity p is an intrinsic characteristic of the quantum algorithm, and can always, in principle, be determined. We therefore require $\alpha = \hat{p} - p$. From eq.(4), we see that this criterion is met if

$$\|\mathcal{M}_{\{\mathrm{c}\to\mathrm{l}\}}(P \cdot (\mathcal{M}_{\{\mathrm{l}\to\mathrm{c}\}}(\rho))) - G(\rho)\|_1 \leq \hat{p} - p \tag{8}$$

for all ρ.

Fault tolerance theory can be used to determine the probability ε_{QC} that the quantum computation fails to produce the desired final quantum state. We therefore write the final state of the *logical* qubits resulting from the actual quantum computation as

$$\mathcal{M}_{\{\mathrm{c}\to\mathrm{l}\}}(P \cdot (\mathcal{M}_{\{\mathrm{l}\to\mathrm{c}\}}(\rho))) = (1 - \varepsilon_{QC})G(\rho) + \varepsilon_{QC}\rho_{err}\,, \tag{9}$$

where ρ_{err} arises from errors. Eq. (8) then becomes

$$||\varepsilon_{QC}[\rho_{err} - G(\rho)]||_1 = \varepsilon_{QC}||\rho_{err} - G(\rho)||_1 \leq \hat{p} - p \ . \tag{10}$$

Since, $||\rho_{err} - G(\rho)||_1 \leq 2$, eqs. (10) and (8) will be satisfied provided

$$\varepsilon_{QC} \leq \frac{1}{2}(\hat{p} - p) \ . \tag{11}$$

This is a new, quite general result relating the fault tolerance theoretic parameter ε_{QC} to the constraints we have derived, which ensure that the overall computation yield the correct final result with some prescribed success probability.

In [3] we consider the fault tolerance approach described in [2] and explicitly determine the number of concatentation levels necessary to give the correct final result with probability $\hat{p}$. In addition, we note that Eq. 4 is an operator theoretic approach to quantum error correction which can be extended to an operator based formulation of quantum fault tolerance [4]. To see this recall that the general error correction condition can be written as $V_{dec}R\varepsilon V_{enc}\rho = \rho$, where V_{dec} and V_{enc} are the decoding and encoding operations of suitable error correction codes and/or error avoidance techniques, R is the recovery operation and ε is the error operator. The QCC includes error correction as a special case when the superoperator P includes all dynamics associated with encoding, decoding, error and recovery operations, U is the identity operation and $\alpha = 0$. However, the QCC extends quantum error correction to quantum fault tolerance by including the full (*i.e.*, non-identity) dynamical evolution of the quantum computer and incorporating a finite α, based on the realization that no error correction scheme will ever be perfect. As an example of its application to fault tolerance the QCC allows us to formulate the following criterion for the success of a given level of concatenation:

$$\frac{\sup_\rho ||\tilde{P}^{(i+1)}\rho - U\rho U^\dagger||_1}{\sup_\rho ||\tilde{P}^{(i)}\rho - U\rho U^\dagger||_1} < 1 \ . \tag{12}$$

where the superscripts refer to the concatentation level. This criterion incoporates not just the probability of errors but also the magnitude of the errors.

In summary, in this paper we have introduced a methodology for determining design parameters for a quantum computer which ensures that an overall computation of interest yields a correct *final result* with some prescribed probability of success.

This research was supported under MITRE Technology Program Grant 07MSR205.

REFERENCES

1. A. Kitaev, *Russian Mathematical Surveys* **52**, 1191-1249 (1997).
2. J. Preskill, *Proc. Roy. Soc. Lond. A* **454**, 385-410 (1998).
3. G. Gilbert, M. Hamrick, and Y.S. Weinstein, to be published *Quant. Inf. Proc.*
4. G. Gilbert, M. Hamrick, Y.S. Weinstein, V. Agarwal, A.R. Calderbank, *arXiv:0709.0128*.

The Bloch Representation of quantum states

Renan Fargetton and Pablo Arrighi

Laboratoire d'informatique de Grenoble, BP 53 - 38041 Grenoble, France

Abstract. The Bloch Sphere is widely used as a graphical representation of a qubit. The purpose of this paper is to highlight some new properties of the canonical generalisation of the Bloch Sphere: the Bloch Representation (BR). We give several properties of this representation that makes it a useful representation. We then show the relation between this representation and the probabilities for canonical measurements of the quantum system.

Keywords: Bloch,probabilities,quantum information,quantum computation, Bloch representation
PACS: 03.67.Mn

GENERALITIES ABOUT THE BLOCH REPRESENTATION (*BR*)

The Bloch Sphere [1] is based upon the Pauli operators[2] which form an orthonormal basis for the real vector space of hermitian matrices of dimension 2×2 endowed with the scalar product $(A,B) = \frac{1}{2}Tr(AB)$. The *BR* uses a similar basis (Generalized Pauli operators) for the $2^n \times 2^n$ hermitian matrices with the scalar product $(A,B) = \frac{1}{2^n}Tr(AB)$: $\{\sigma_{i_1..i_n} = \sigma_{i_1} \otimes \sigma_{i_2} \otimes ... \otimes \sigma_{i_n} / i_1, i_2, .., i_n \in [0,3]\}$. Because density matrices are positive operators (hence hermitian) with trace one, a density matrix for *n* qubits can be decomposed on this basis:

Definition 1 (Bloch Representation (*BR*)) [3] *Let ρ be a density matrix for a n qubits system. It decomposes as:*

$$\rho = \frac{1}{2^n} \sum_{i_1,..,i_n=0}^{3} r_{i_1..i_n} \text{ with } r_{i_1..i_n} = Tr(\rho\, \sigma_{i_1..i_n}) \quad (1)$$

The BR of ρ is the n-dimensional tensor $\left(r_{i_1..i_n}\right)_{i_1,..,i_n \in [0,3]}$ *(with* $r_{0..0} = Tr(\rho) = 1$*).*

BR properties

Trace-out and block structure. The first reason to study the *BR* is its rich structure. It makes more obvious some properties of the given quantum state. Let consider the case

---

[1] F. Bloch, Phys. Rev. 70, 460 (1946).

[2] $\sigma_0 = \begin{pmatrix} 1 & 0 \\ 0 & 1 \end{pmatrix}$ $\sigma_1 = \begin{pmatrix} 1 & 0 \\ 0 & -1 \end{pmatrix}$ $\sigma_2 = \begin{pmatrix} 0 & -i \\ i & 0 \end{pmatrix}$ $\sigma_3 = \begin{pmatrix} 0 & 1 \\ 1 & 0 \end{pmatrix}$

[3] Hassan & Joag, *An Experimentally accessible geometric measure for entanglement in N-qubit pure states*, quant-ph/0804.2062v3

CP1110, *Quantum Communication, Measurement and Computing (QCMC), The Ninth International Conference,* edited by A. Lvovsky

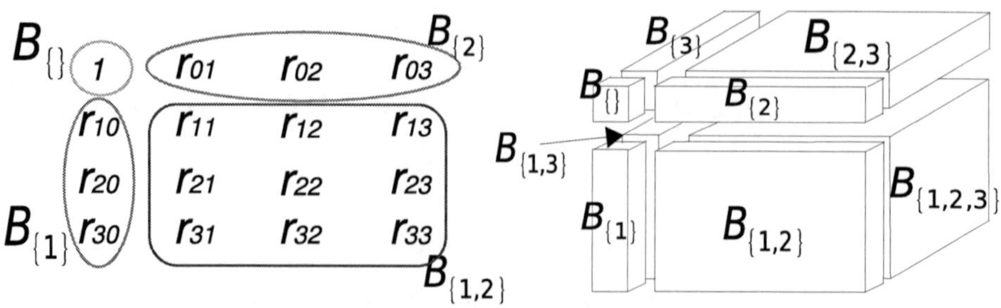

FIGURE 1. *Bloch representation for two qubits (left) and three qubits (right). The blocks $B_{\{1\}}$,$B_{\{2\}}$ and $B_{\{3\}}$ are the Bloch vectors for each qubits. The blocks labeled $B_{\{1,2\}}$, $B_{\{2,3\}}$ and $B_{1,3}$ are the second order blocks : they encode the joint information between the corresponding qubits. Finally the block $B_{\{1,2,3\}}$ is the third order block : it encodes for the joint information between the three qubits.*

of a system of 2 qubits A and B with density matrix $\rho_{AB} = 1/4\sum_{ij} r_{ij}\,\sigma_{ij}$. The partial density matrix for the first qubit is $\rho_A = Tr_B(\frac{1}{4}\sum_{i,j=0}^{3} r_{ij}\,\sigma_i \otimes \sigma_j) = \frac{1}{4}\sum_{i,j=0}^{3} r_{ij}\,\sigma_i Tr(\sigma_j) = \frac{1}{2}\sum_{i=0}^{3} r_{i0}\,\sigma_i$. Translated in the language of the *BR* and generalized to n qubits we obtain:

Property 1 (Trace-out in the *BR*) *For a BR* $\left(r_{i_1..i_n}\right)_{i_1,..,i_n\in[0,3]^n}$ *of n qubits and A a subset of $\{1,..,n\}$. The partial BR for the qubits in the subset A is* $\left(r_{i_1..i_n}\right)_{k\in A,\, i_k\in[0,3]}^{k\notin A,\, i_k=0}$

Hence the information about the partial state of a subsystem is encoded in a sub-tensor of the *BR* . This is a remarkable property of the *BR* compared with density matrices as it provides a way in which to speak of qubit 1, 2 or their correlations in a separate manner. This is illustrate by figure 1 where we introduce the notations $B_{\{k\}}$, $B_{\{k,l\}}$, etc that correspond to the Block Structure of the *BR* .

Product states. By definition the product states can be defined by the partial density matrix for each qubit. The translation of this property to the Bloch representation is that the first order blocks $B_{\{i\}}$ are sufficient to describe the product states. The second to n^{th} order blocks contain no additional information for those states:

Property 2 (Product state) *For a product state with BR* $(r_{i_1,\dots,i_n})_{i_1,\dots,i_n\in[0,3]^n}$*, we have:*

$$\forall i_1,\dots,i_n \in [0,3]^n,\ r_{i_1,\dots,i_n} = r_{i_1,0,\dots,0}\, r_{0,i_2,0,\dots,0} \cdots r_{0,\dots,0,i_n}$$

Local unitary transformations. The effect of a unitary on the qubit k is to apply the same rotation as to the Bloch vector in the Bloch Sphere to all the columns in the *BR* :

Property 3 (Local unitary) *Let* $(r_{i_1,..,i_n})_{i_1,..,i_n\in[0,3]^n}$ *be a BR and U a unitary transformation on one qubit. This unitary applied on qubit k results in the BR given by :*

$$\forall l \in [1,n], l \neq k, i_l \in [0,3], \left\{ \begin{array}{l} r'_{i_1\dots i_{k-1}0 i_{k+1}\dots i_n} = r_{i_1\dots i_{k-1}0 i_{k+1}\dots i_n} \\ \forall i_k \in [1,3], r'_{i_1\dots i_n} = \sum_{j_k\in[0,3]} O_{i_k j_k}\, r_{i_1\dots i_{k-1} j_k i_{k+1}\dots i_n} \end{array} \right.$$

With $(O_{i_kj_k})_{i_kj_k}$ a rotation matrix in $SO(3)$ defined by $O_{i_kj_k} = Tr(U^\dagger \sigma_{i_k} U \sigma_{j_k})$.

We notice that this transformation leaves $(r_{i_1...i_{k-1}0i_{k+1}...i_n})_{i_1...i_{k-1}0i_{k+1}...i_n}$ unchanged. This is not surprising as this subtensor corresponds to the partial density matrix for qubits other than qubit k.

BR for Pure states. We aim at an intrinsic definition of a quantum state in the Bloch Representation (*i.e.* without reference to its density matrix). The main difficulty is that every hermitian operator can be decomposed using equation (1), and not just positive operators. But we has been able to characterize the set of pure quantum states:

Theorem 1 (*BR* for Pure states) *A tensor $\left(r_{i_1...i_n}\right)_{i_1,...,i_n}$ is a Bloch Representation of a pure state if and only if*[4]*:*

$$\forall K, r_K = \sum_{I,J \in [0...3]^n} \frac{r_I r_J Tr(\sigma_I \sigma_J \sigma_K)}{2} \text{ and } r_{0...0} = 1$$

PROBABILITIES IN THE BLOCH REPRESENTATION

In this section we relate the *BR* to its probabilistic interpretation. We gain an understanding of the r_{ij} values in the *BR* as the expectation values of the σ_{ij} observables.

Bloch random variables. The vector $(r_{0..0i_k0..0})_{i_k \in [1..3]}$ is the Bloch Vector for qubit k. For each of the $r_{0..0i_k0..0}$ coefficients we associate a random variable $X_i^{(k)}$ that describes the resulting statistics for the measurement of qubit k by the observable σ_i :

Definition 2 (Bloch Random variables) *Let $(r_{i_1..i_n})_{i_1..i_n}$ be a BR. The Bloch random variables $(X_i^{(k)})_{i \in \{1,2,3\}, k \in [1..n]}$ are random variables with outcomes $b \in \{-1, 1\}$ and associate probabilities $P(X_i^{(k)} = b) = \frac{1}{2}(1 + b\, r_{0..0i_k0..0})$.*

Notice that $E(X_i^{(k)}) = P(X_i^{(k)} = 1) - P(X_i^{(k)} = -1) = r_{0..0i_k0..0}$. Hence the coefficients in the Bloch vectors are the expectations values of the associated Bloch random variables.

Measurements. We associate a Bloch random variable to a projective measurement which may modify the state of the *BR* in the following way:

Property 4 (Measurements) *Let $(r_{i_1,..,i_n})_{i_1,...,i_n \in [0,3]^n}$ be a BR. The measurement along the σ_j observable ($j \in [1,3]$) of the qubit k with result $b \in \{-1, 1\}$ projects the BR in the state $(r'_{i_1,..,i_n})_{i_1,...,i_n \in [0,3]^n}$ given by :*

$$\forall i_1, .., i_n \in [0,3]^n, r'_{i_1...i_n} = (\delta_{i_k 0} + \delta_{i_k j}) \frac{r_{i_1...i_n} + b r_{i_1...i_{k-1}(i_k - j)i_{k+1}...i_n}}{2P(X_j^{(k)} = b)}$$

[4] $K = k_1 \ldots k_n \in [0 \ldots 3]^n$, $\sigma_I = \sigma_{i_1} \otimes \ldots \otimes \sigma_{i_n}$(resp.$J, K$)

Joint Probabilities. If we measure qubit 1 with a σ_i observable and qubit 2 with a σ_j observable ($i,j \in [1..3]$), the probability to obtain the outcome $\{b_1,b_2\}$ with $b_1,b_2 \in [-1,1]$ is :

$$P(X_i^{(1)} = b_1 \cap X_j^{(2)} = b_2) = P(X_j^{(2)} = b_2 | X_i^{(1)} = b_1)\, P(X_i^{(1)} = b_1)$$
$$= \frac{1+b_2 r'_{0j}}{2} P(X_i^{(1)} = b_1) = \frac{P(X_i^{(1)} = b_1) + b_2 \frac{r_{0j}+b_1 r_{ij}}{2}}{2} = \frac{1+b_1 r_{i0} + b_2 r_{0j} + b_1 b_2 r_{ij}}{4}$$

As this formula is symetric for the two qubits it proves that the order of measurements does not matter [5]. Furthemore the formula generalizes nicely :

Theorem 2 (joint probabilities) *Let $(r_{i_1,..,i_n})_{i_1,..,i_n \in [0,3]^n}$ be a BR. The probability to obtain outcomes $(b_k)_{k\in[1,n]}$ for measurements along σ_{i_k} observables [6] is given by :*

$$P(\cap_{k\in[1,n]} X_{i_k}^{(k)} = b_k) = \frac{1}{2^n} \sum_{p_1 \dots p_n \in \{0,1\}^n} \Big(\prod_{k\in[1,n]} b_{i_k}{}^{p_k} \Big)\, r_{(p_1 i_1)\dots(p_n i_n)}$$

BR coefficients as expectation values of parity random variables. We now define parity random variables and connect this to the *BR* , and get to our main result.

Definition 3 (Parity random variables) *The parity random variables $X_{i_1\dots i_n}^{(1,..,n)}$ for a set of qubits $[1\dots n]$ are defined by $P(X_{i_1\dots i_n}^{(1,..,n)} = b) = P(X_{i_1}^{(1)} \dots X_{i_n}^{(n)} = b)$ with $b \in \{-1,1\}$.*

Property 5 (Parity random variables in the *BR*) *Let $(r_{i_1,..,i_n})_{i_1,..,i_n \in [0,3]^n}$ be a BR. The probability of outcomes $b \in \{-1,1\}$ for parity random variables is $P(X_{i_1\dots i_n}^{(1,..,n)} = b) = \frac{1}{2}(1 + b\, r_{i_1\dots i_n})$.*

As a consequence the expectation value for parity random variable is $r_{i_1\dots i_n}$:

Theorem 3 (*BR* coefficients as expectation values of parity random variables) *Let $(r_{i_1\dots i_n})_{i_1\dots i_n}$ be a BR. $\forall i_1 \dots i_n \in [0,3]$, then $r_{i_1\dots i_n} = E(X_{i_1,\dots,i_n}^{(1,\dots,n)})$*

CONCLUSION

We have described the Bloch Representation and some of its properties. We have related it to its probabilistic interpretation : the *BR* coefficients have a simple interpretation as expectation values of the parity random variables (Theorem 2). The *BR* is a representation of the complete statistics of a quantum system. A remaining important question for further study is how to give an instrinsic definition of the *BR* (without reference to the density matrix). We made a step in this direction for pure states (Theorem 1).

---

[5] if the measurements are made on two differents qubits. Otherwise the joint probability can not be well defined since two measurements on the same qubits are not commutative.
[6] $\forall k$, $b_k \in \{-1,1\}$ and $i_k \in [1,3]$

Control of subspaces of quantum states

Ilya G. Kovalenko and Dmitry V. Dodin

Physics Department, Volgograd State University, Universitetskij prospekt 100, Volgograd 400062, Russia; igk@vlink.ru, Dodin_DV@mail.ru

Abstract. Not only preparation of a state but also control of a state via unitary evolution or measurement are important in quantum informatics. Such a control is usually performed by unitary evolution (measurement) of a state as a whole or by evolution of the subsystems. We define a generalized evolution of quantum states based on selection of subspaces of states of a quantum system which evolve, for some reason, as an aggregate. Such a procedure can be considered as control of 'subsystems' both in composite and non-composite quantum systems. The concepts of partial reduction, partial unitary evolution and partial diagonalization emerge naturally in this approach. We introduce partial eigenvalue problem for an operator. At the end we consider an example of a 5-dimensional system in which control of 2.5-dimensional 'subsystem' is being maintained.

Keywords: Unsharp measurements, quantum control
PACS: 00.03.65.Ca; 00.03.65.Yz

MOTIVATION

The state of a compound system is described by a vector in a f-dimensional space. As a result the density matrix has the dimension $f \times f$: $\widehat{\rho}^{(f)}$.

Reduction to a r-dimensional space occurs after measurement of the subsystem (distant partner). The choice of subspace is determined by what measurement over subsystem is carried out. The subspace itself contains information (record) about the state of the resulting subsystem

$$\widehat{\rho}^{(f)} = \widehat{\rho}^{(p)} \otimes \widehat{\rho}^{(r)} = \left(\begin{array}{c|c|c} \rho_{11}\widehat{\rho}^{(r)} & \dots & \rho_{1p}\widehat{\rho}^{(r)} \\ \hline \dots & \dots & \dots \\ \hline \rho_{p1}\widehat{\rho}^{(r)} & \dots & \rho_{pp}\widehat{\rho}^{(r)} \end{array} \right) \xrightarrow[reduction]{} \rho_{ii}\widehat{\rho}^{(r)} . \tag{1}$$

In this example all blocks of $\widehat{\rho}^{(f)}$ are exactly equal up to a multiplier; hence the results of measurement turn out to be always identical to a state $\widehat{\rho}^{(r)}$.

If the system is entangled, the subsystem states $\widehat{\rho}_{ii}^{(r)}$ are different and the multipliers ρ_{ii} describe a probability to find the reduced subsystem in the given state. Mathematically, each block selects an individual subspace; thus, different subspaces contain records about different states of the subsystem.

In such a problem one implies an algebraic constraint between the dimensions of the system and the subsystems: $f = r \cdot p$. In the case of finite-dimensional Hilbert space f is a composite integer number. Following this approach, if f is a prime integer, the system cannot be a composite one as a matter of principle.

CP1110, *Quantum Communication, Measurement and Computing (QCMC), The Ninth International Conference,* edited by A. Lvovsky

Is it possible to define the conception of subsystems in a general case? We show that it is indeed possible by cancelling the requirement of the integer-dimensionality of Hilbert space for the subsystems.

DIFFERENT-DIMENSIONAL SUBSYSTEMS

For demo purposes we apply the developing formalism to a 5-dimensional system.

There are two ways to divide a 5-dimensional system into two subsystems, $5 = 2+3 = 1+4$:

$$\left(\begin{array}{c|c} 3\times 3 & \\ \hline & 2\times 2 \end{array}\right) \quad \text{or} \quad \left(\begin{array}{c|c} 4\times 4 & \\ \hline & 1\times 1 \end{array}\right). \tag{2}$$

At this rate it is natural to consider the density matrix a table of blocks rather than numbers. We generalize the algebraic operations over matrices writing them down as operations with blocks. A more detailed description of this formalism is given in [1].

The specific case of different-dimensional subspaces does not allow us to arrange an algebraical sum of blocks with different sizes therefore we are forced to use a representation where the reduced system is not described by the matrix of lesser dimension but by the block in the matrix of initial dimension n:

$$\left(\begin{array}{c|c} \cdots & \cdots \\ \hline \cdots & \cdots \end{array}\right) \xrightarrow[reduction]{} \left(\begin{array}{c|c} \cdots & 0 \\ \hline 0 & 0 \end{array}\right) \equiv \widehat{S}. \tag{3}$$

This formula gives an example of selective measurement.

Non-selective measurement (mixture of outcomes of several measurements) is represented by the block-diagonal matrix

$$\widehat{N} = \sum_i \widehat{S}_i = \left(\begin{array}{c|c|c} s_1 & & \\ \hline & s_2 & \\ \hline & & \cdots \end{array}\right) = \bigoplus_i s_i, \tag{4}$$

where s_i are the blocks with dimensions $\dim s_i = r_i$, $\sum_i r_i = n$ and $n = \dim \widehat{N}$.

The number of blocks in the density matrix $\widehat{\rho}$ specifies the dimension of the reducing subsystem ("p"). If the blocks are of unequal dimensions, the effective (mean) dimension of the reduced subsystem ("r") can be a non-integer rational number:

$$r = \frac{\sum_i r_i}{p} = \frac{f}{p}. \tag{5}$$

In the example considered $f = 5$, $r_1 = 3$, $r_2 = 2$, $r = 2.5$.

The incomplete splitting of beams in the SG-analyzer can be taken as an elementary example. For the spin $s = 2$ particles the Hilbert space is 5-dimensional. A two-dimensional subsystem can be defined as one composed of two beams with $s_z = 2, 1, 0$ and $s_z = -1, -2$.

PARTIAL EIGENVALUE PROBLEM

The procedure of block diagonalization is inherently partial diagonalization. In what follows we name all procedures operating with blocks as partial ones.

The problem of block diagonalization is tightly bound with the partial eigenvalue problem. We designate partial eigenvalues $\widehat{\Lambda}_i$ as matrices with dimensions r_i. Mathematical expression of this problem should be

$$\widehat{\rho}|X) = |X)\widehat{\Lambda}. \tag{6}$$

Here

$$|X) = \langle |x_1\rangle, |x_2\rangle, ..., |x_r\rangle | \tag{7}$$

is the generalization of the ket-vector which we call *joy*-vector. Hermitian conjugate construction

$$(X| = |\langle x_1|, \langle x_2|, ..., \langle x_r|\rangle \tag{8}$$

will be referred to as *sad*-vector.

As in the case of traditional eigenvalue problem one can obtain a characteristic equation which generalizes to

$$\det[I^{\dim\Lambda} \otimes \widehat{\rho} - \widehat{\Lambda} \otimes I^{\dim\rho}] = 0. \tag{9}$$

Physically, partial eigenvalues provide the outcome of unsharp measurements, that is, measurements realized by a device with poor accuracy.

The developed methodology provides a new vision of the concept of a subsystem. In a common sense the composite system is combined through separate objects called subsystems (*multiplicity principle*). In our approach any object even if it is non-composite in a common sense can be decomposed into component parts considered as subsystems (*division principle*).

PARTIAL TRANSFORMATIONS AND CONTROL OF SUBSYSTEMS OF NON-COMPOSITE QUANTUM STATES

The logic of reasoning allows us to introduce a class of unitary transformations that control subsystems in non-composite systems. The operator of unitary transformation can be written as

$$\widehat{U} = \sum_i |X_i)\widehat{T}_i(X_i|, \tag{10}$$

where $\widehat{T}_i$ are transformation matrices with dimensions $r_i \times r_i$ acting within corresponding subspaces. We may thus treat these partial unitary evolutions as a *control of fractional-dimensional subsystems in the non-composite quantum systems*.

SUMMARY

Let us enumerate the principal results of the present work.

1. A concept of subsystem for a quantum system with an arbitrary dimension f is introduced.
2. If f is a prime integer number, the subsystem has a fractional dimension.
3. Extraction of subsystems is closely related to the procedure of matrix block diagonalization.
4. The partial eigenvalue problem is posed. The corresponding characteristic equation is written down.
5. The partial eigenvalues correspond to a set of unsharp measurements.
6. The class of untitary transformations specifying subsystems control in the non-composite quantum systems.

REFERENCES

1. D.V. Dodin and I.G.Kovalenko. *Decomposition of non-composite systems.* QCMC-2008, this volume.

Analysis of a tripartite Bell inequality for 3-qubit states

S. Ghose*, N. Sinclair†,*, S. Debnath**,* and A. Kabra**,*

*Department of Physics and Computer Science, Wilfrid Laurier University, Waterloo, Canada
†Institute for Quantum Information Science, University of Calgary, Canada
**Indian Institute of Technology, Mumbai, India

Abstract. We consider 3-qubit entangled pure states of the form $\alpha|000\rangle + \beta|111\rangle$ and explore nonlocal correlations between the qubits by analyzing a Bell inequality belonging to the Mermin-Klyshko (MK) set of N-qubit inequalities. We find an expression for the maximum expectation value of the MK Bell operator and show that it does not increase smoothly with the amount of entanglement between the qubits. A subset of these tripartite entangled states does not violate the bound of 2 imposed by local hidden variable theories.

Keywords: Bell inequalities, multiqubit entanglement, nonlocality
PACS: 03.65.Ud, 03.67.Mn, 03.67.-a

INTRODUCTION

Entanglement is one of the most remarkable features of quantum mechanics and a crucial ingredient for various quantum information processing applications [1]. The nonclassical correlations inherent in entangled states can result in violations of Bell-type inequalities [2] that place upper bounds on correlations between spatially separated systems allowed by *local realistic* theories. Whereas 2-qubit systems have been widely studied, the relationship between entanglement and Bell inequality violations in systems of three or more qubits is more challenging to explore because of the existence of bipartite as well as tripartite and/or multipartite correlations between the qubits. A detailed analysis of the relationship between entanglement and nonlocality in multiqubit systems is still lacking and is an important step in the development of a complete theory of multipartite entanglement.

In this paper, we analyze a tripartite Bell inequality with respect to a family of 3-qubit states having the form

$$|\psi\rangle = \alpha|000\rangle + \beta e^{i\gamma}|111\rangle. \tag{1}$$

These states belong to the GHZ class [3] and include the GHZ state corresponding to $\alpha = \beta = 1/\sqrt{2}, \gamma = 0$. The GHZ state has been the focus of several experimental and theoretical studies and has been shown to be a useful resource for quantum computation and communication tasks [4, 5, 6, 7]. Our goal is to investigate nonlocality in the states described by Eq. (1). We consider the Mermin-Klyshko (MK) set of N-qubit inequalities [8, 9] and focus on the 3-qubit case $N = 3$, which yields the inequality $|\langle M\rangle| \leq 2$ [8]. M is a Bell-type operator describing a combination of local spin measurements performed on each qubit [8]. Violation of this inequality indicates that no choice of local

CP1110, *Quantum Communication, Measurement and Computing (QCMC), The Ninth International Conference*, edited by A. Lvovsky

hidden variables can explain the correlations between certain measurement outcomes, thereby implying the existence of nonlocal correlations between the qubits.

We find an expression for the maximum expectation value of the 3-qubit MK Bell-type operator M for the states described in Eq. (1). We show that when $\alpha\beta \leq 1/4$, the bound of 2 set by local realism is not violated although the states have genuine tripartite entanglement. Previous studies [10] had numerically found approximate estimates of the maximum expectation value of M. Our method yields an exact expression that confirms these numerical estimates, and also shows that there is a sharp change in the maximum value of $|\langle M\rangle|$at $\alpha\beta = 1/\sqrt{20}$ due to a sudden switch in the optimal measurement angles.

3-QUBIT ENTANGLEMENT AND NONLOCALITY

Bipartite entanglement between two qubits can be explicitly quantified in terms of the concurrence $\mathscr{C}$ of a given state [11], with $\mathscr{C} = 0$ for a separable state and $\mathscr{C} = 1$ for a maximally entangled Bell state. Coffman, Kundu and Wootters formulated a measure of 3-qubit entanglement known as the 3-tangle τ_{ABC} [12], which is defined in terms of the concurrence of the constituent bipartite subsystems as

$$\tau_{ABC} = \mathscr{C}^2_{A(BC)} - \mathscr{C}^2_{AB} - \mathscr{C}^2_{AC} \tag{2}$$

for qubits A, B and C. The concurrence $\mathscr{C}_{AB}$ quantifies the entanglement of qubit A with qubit B after tracing out qubit C. Similarly $\mathscr{C}_{AC}$ measures the entanglement between qubits A and C when one traces out qubit B. The concurrence $\mathscr{C}_{A(BC)}$ is a measure of the entanglement between qubit A and the joint state BC. The 3-tangle is invariant under permutation of these indices and is bounded as $0 \leq \tau_{ABC} \leq 1$. $\tau_{ABC} = 0$ if one of the qubits is separable from the other two, and $\tau_{ABC} = 1$ for the maximally entangled GHZ state. For the states of Eq. 1, $\mathscr{C}_{AC} = 0$, $\mathscr{C}_{AB} = 0$ and $\mathscr{C}^2_{A(BC)} = 4\alpha^2\beta^2$. Therefore,

$$\tau_{ABC} = 4\alpha^2\beta^2. \tag{3}$$

The 3-qubit MK Bell-type operator is defined as [8, 9]

$$M \quad = \quad ABC' + AB'C + A'BC - A'B'C' \tag{4}$$

Here A or A' is a local spin measurement in a randomly chosen basis performed on the first qubit. Similarly B or B' is performed on the second qubit, and C or C' is performed on the third qubit. Any theory based on local realism predicts that the maximum expectation value of M is 2 [8, 9]. Quantum entangled states can violate this bound up to a maximum value of $|\langle M\rangle| = 4$, which is achieved by the maximally entangled GHZ state [8]. In the following section we calculate the maximum expectation value of M for the states in Eq. (1) and relate it to the amount of tripartite entanglement τ_{ABC} in the state.

MAXIMUM EXPECTATION VALUE OF THE MK OPERATOR

The operators ABC in Eq. (4) can be explicitly expressed in terms of the Pauli matrices $\vec{\sigma}$ and unit vectors $\vec{a}, \vec{b}$ and $\vec{c}$ as $ABC = \vec{a} \cdot \vec{\sigma}_A \otimes \vec{b} \cdot \vec{\sigma}_B \otimes \vec{c} \cdot \vec{\sigma}_C$. Writing each unit vector in terms of the spherical angles, θ_i and ϕ_i, corresponding to the i-th qubit, and using Eq. (3), the expectation value of the operator M in Eq. (4) can be written as

$$\langle M \rangle = \langle \psi | M | \psi \rangle = c_1 \sqrt{1 - \tau_{ABC}} + c_2 \sqrt{\tau_{ABC}}, \tag{5}$$

where $|\psi\rangle$ is given by Eq. (1), and c_1 and c_2 are given by

$$\begin{aligned} c_1 &= \cos\theta_a(\cos\theta_b\cos\theta_{c'} + \cos\theta_{b'}\cos\theta_c) + \cos\theta_{a'}(\cos\theta_b\cos\theta_c - \cos\theta_{b'}\cos\theta_{c'}) \\ c_2 &= \sin\theta_a(\sin\theta_b\sin\theta_{c'}\cos\phi_{abc'} + \sin\theta_{b'}\sin\theta_c\cos\phi_{ab'c}) \\ &\quad + \sin\theta_{a'}(\sin\theta_b\sin\theta_c\cos\phi_{a'bc} - \sin\theta_{b'}\sin\theta_{c'}\cos\phi_{a'b'c'}) \end{aligned} \tag{6}$$

with $\cos\phi_{abc} = \cos(\phi_a + \phi_b + \phi_c - \gamma)$. We find that c_1 and c_2 obey the relationship

$$|c_2| \leq 4 - 2|c_1|, \tag{7}$$

where $|c_1| \leq 2$ and $|c_2| \leq 4$. Using this inequality to substitute for c_2 in Eq. (5),

$$\langle M \rangle \leq 4\sqrt{\tau_{ABC}} + (\sqrt{1 - \tau_{ABC}} - 2\sqrt{\tau_{ABC}})|c_1|. \tag{8}$$

It is clear that when $\tau_{ABC} \leq 1/5$, the coefficient of $|c_1|$ in Eq. (8) is positive and hence the above bound is a maximum of $2\sqrt{1 - \tau_{ABC}}$ if $|c_1|$ is equal to its maximum value of 2. This upper bound can be achieved by picking the unit vectors to be pointing along the positive or negative $z-$ axis (Eq. (6)). In this case, $c_2 = 0$ (Eq. (6)). Conversely, when $\tau_{ABC} \geq 1/5$, the coefficient of $|c_1|$ in Eq. (8) is negative and the bound in Eq. (8) is a maximum of $4\sqrt{\tau_{ABC}}$ if $c_1 = 0$. This maximum value can be achieved by picking unit vectors in the x-y plane that yield $|c_2| = 4$ (Eq. (6)). Thus the upper bound given in Eq. (8) can always be achieved by an appropriate choice of unit vectors. Hence we have obtained the following expression relating the maximum value of $\langle M \rangle$ to τ_{ABC},

$$\max(\langle M \rangle) = \max(2\sqrt{1 - \tau_{ABC}}, 4\sqrt{\tau_{ABC}}). \tag{9}$$

We have confirmed this expression by numerically maximizing $\langle M \rangle$ as a function of τ_{ABC}. Our exact expression also confirms previous numerical estimates [10]. Our analysis shows that the two regimes in Eq. (9) correspond to the switching of the optimal measurement unit vectors between lying in the x-y plane and pointing along the z-axis.

The above relationship is quite different from the relationship between bipartite entanglement and violation of the Bell-CHSH inequality [13] for 2-qubit states of the same form [14, 15]. Unlike the 2-qubit case, the maximum value of $\langle M \rangle$ first decreases with increasing $\tau > 0$ and only increases once $\tau > 1/5$. It is clear that the bound of 2 satisfied by local hidden variable theories is only violated when $\tau > 1/4$. This is also in contrast to the 2-qubit case where the the CHSH inequality is violated by all pure entangled 2-qubit states.

CONCLUSIONS

In conclusion, we have found the maximum expectation value of the well-known 3-qubit MK Bell operator for the set of states described by Eq. (1). Our results show that some entangled states do not violate the MK inequality. On the other hand, we note that violation of the MK inequality does not necessarily indicate the existence of nonlocal correlations between all three qubits, since biseparable states can also violate this inequality [16, 17]. We plan to analyze a more suitable inequality [18] for identifying genuine tripartite nonlocality in future work.

ACKNOWLEDGMENTS

We thank R. Stock and S. Bandyopadhyay for useful discussions. SG was supported by an NSERC Discovery grant.

REFERENCES

1. M. A. Nielsen and I. L. Chuang *Quantum Computation and Quantum Information,* Cambridge University Press, Cambridge (2000).
2. J. S. Bell, Physics, **1**, 195 (1964).
3. W. Dur, G. Vidal, and J. I. Cirac, Phys. Rev. A **62**, 062314 (2000).
4. R. Laflamme, E. Knill, W. H. Zurek, P. Catasti, and S. V. S. Mariappan, Phil. Trans. Roy. Soc. Lond. A **356**, 1941 (1998); D. Bouwmeester, J.-W. Pan, M. Daniell, H. Weinfurter, and A. Zeilinger, Phys. Rev. Lett. **82**, 1345 (1999); C. F. Roos, M. Riebe, H. Häffner, W. Hänsel, J. Benhelm, G. P. T. Lancaster, C. Becher, F. Schmidt-Kaler and R. Blatt, Science **304**, 1478 (2004); R. J. Nelson, D. G. Cory, and S. Lloyd, Phys. Rev. A **61**, 022106 (2000).
5. D. Gottesman and I. L. Chuang, Nature **402**, 390 (1999); D. E. Browne, and T. Rudolph, Phys. Rev. Lett. **95**, 010501 (2005); R. Raussendorf, D. E. Browne, and H. J. Briegel, Phys. Rev. A **68**, 022312 (2003); F. M. Spedalieri, H. Lee, and J. P. Dowling, Phys. Rev. A **73**, 012334 (2006).
6. A. Karlsson and M. Bourennane, Phys. Rev. A **58**, 4394 (1998).
7. M. Hillery, V. Buđek, and A. Berthiaume, Phys. Rev. A **59**, 1829 (1999); W. Tittel, H. Zbinden, and N. Gisin, Phys. Rev. A **63**, 042301 (2001). H. Lee, J. Lim, and H. Yang, Phys. Rev. A **73**, 042305 (2003); X.-R. Jin, X. Jia, Y.-Q. Zhang, S. Zhang, S.-K. Hong, K.-H. Yeon and C.-I. Um, Phys. Lett. A **354**, 67 (2006);
8. N. D. Mermin, Phys. Rev. Lett. **65**, 1838 (1990).
9. A. V. Belinskii and D. N. Klyshko Phys. Usp. **36**, 653 (1993).
10. V. Scarani and N. Gisin, J. Phys. A **34**, 6043 (2001).
11. W. K. Wootters, Phys. Rev. Lett., **80**, 2245 (1998).
12. V. Coffman, J. Kundu, and W. K. Wootters, Phys. Rev. A **61**, 052306 (2000).
13. J. F. Clauser, M. A. Horne, A. Shimony, and R. A. Holt, Phys. Rev. Lett., **23**, 880 (1969).
14. N. Gisin, Phys. Lett. A **154**, 201 (1991).
15. S. Popescu and D. Rohrlich, Phys. Lett. A **166**, 293 (1992).
16. D. Collins, N. Gisin, S. Popescu, D. Roberts, and V. Scarani, Phys. Rev. Lett. **88**, 170405 (2002).
17. J. L. Cereceda, Phys. Rev. A **66**, 024102 (2002).
18. G. Svetlichny, Phys. Rev. D **35**, 3066 (1987).

Decomposition of non-composite quantum systems

Dmitry V. Dodin and Ilya G. Kovalenko

Physics Department, Volgograd State University, Universitetskij prospekt 100, Volgograd 400062, Russia; Dodin_DV@mail.ru, igk@vlink.ru

Abstract. A composite quantum system is a fundamental object in quantum informatics. A composite quantum system is considered a system whose dimension is a composite integer number whereas the prime integers may only correspond to non-composite systems. Entanglement and the procedure of reduction to a subsystem via averaging over environment are only formulated for composite systems. In the present report we introduce a procedure of decomposition into 'subsystems' for a system whose dimension is an arbitrary number including a prime integer. This procedure generalizes the standard procedure of reduction. In the particular case of a non-composite system the space of states of the decomposed system splits into direct sum of subspaces of states and the average dimension of subspaces is a fractional number. We compare the properties of the introduced procedure of decomposition and the standard separation into subsystems.

Keywords: Unsharp measurements
PACS: 00.03.65Ca; 00.03.65.Yz

BASIS

As is known, projective measurements in quantum mechanics are described with the aid of projectors $P_i = |i\rangle\langle i|$ with the properties

$$P_i P_j = 0 \quad (j \neq i), \qquad P_i P_i = P_i, \qquad P_i = P_i^{\dagger}, \tag{1}$$

where $|i\rangle$ are the basic vectors. The set of projectors has a property of completeness

$$\sum_i P_i = \sum_i P_i P_i = \sum_i P_i P_i^{\dagger} = I^{(f)}. \tag{2}$$

Here the right upper index refers to the dimension of the system.

Reduction of the system with the dimension $f = rp$ in this scheme is considered by us an operator sum of the form [1]

$$\hat{\rho}^{(r)} = \sum_i \langle i| \otimes I^{(p)} \cdot \hat{\rho}^{(f)} \cdot |i\rangle \otimes I^{(p)}. \tag{3}$$

Reduction in the experiment corresponds to total neglect of information about the state of subsystem over which reduction is fulfilled. Mathematically, it is expressed in averaging over subspaces of the states of large system. Meanwhile, the dimension of the subspace over which averaging p is fulfilled is a divisor of the dimension of space of states of the complete system f and specifies in this approach the dimension of reduced system as a ratio $r = f/p$.

CP1110, *Quantum Communication, Measurement and Computing (QCMC), The Ninth International Conference,* edited by A. Lvovsky

We raise the problem of generalizing the procedure of reduction expressed as in (3) in case when the space of complete system is divided into subspaces of different dimensions. In the general case the dimension of the complete space is represented by the sum of dimensions of subspaces $f = p_1 + p_2 + ... + p_r$ and only in the case when all dimensions of all subspaces are identical $p_1 = p_2 = ... = p_r = p$ we get $f = pr$.

A solution to this problem consists in assuming the equality (2) as a primary one and abandoning the property of self-conjugacy of the operators P_i. We considered as such the operators

$$M_k = \langle |x_1\rangle, |x_2\rangle, ..., |x_r\rangle| = |X). \tag{4}$$

On the pictographic analogy of such a notation $|..)$ with known internet-meme we call them *joy*-vectors. Hermitian conjugate construction

$$M_k^\dagger = |\langle x_1|, \langle x_2|, ..., \langle x_r|\rangle = (X| \tag{5}$$

will be referred to as a *sad*-vector.

Now

$$M_k \neq M_k^\dagger, \tag{6}$$

but at the same time an equality

$$\sum_k M_k M_k^\dagger = I^{(f)} \tag{7}$$

holds.

The subsequent explanation will be evident from an example.

$$I^{(5)} = \begin{pmatrix} 1 & 0 & 0 & 0 & 0 \\ 0 & 1 & 0 & 0 & 0 \\ 0 & 0 & 1 & 0 & 0 \\ 0 & 0 & 0 & 1 & 0 \\ 0 & 0 & 0 & 0 & 1 \end{pmatrix} = M_1 M_1^\dagger + M_2 M_2^\dagger, \tag{8}$$

where

$$M_1 = \begin{pmatrix} 1 & 0 \\ 0 & 1 \\ 0 & 0 \\ 0 & 0 \\ 0 & 0 \end{pmatrix} \equiv |1), \quad M_2 = \begin{pmatrix} 0 & 0 & 0 \\ 0 & 0 & 0 \\ 1 & 0 & 0 \\ 0 & 1 & 0 \\ 0 & 0 & 1 \end{pmatrix} \equiv |2). \tag{9}$$

The set of such joy-vectors allows to decompound initial density matrix into blocks

$$\hat{\rho}_{kn} = M_k^\dagger \hat{\rho}^{(5)} M_n \equiv (k|\hat{\rho}^{(5)}|n), \tag{10}$$

that have the sense of "coarse matrix elements".

Joy-vectors M_k execute projection of the system state onto corresponding subspace of states

$$\hat{\rho}_{kk} = M_k^\dagger \hat{\rho}^{(5)} M_k. \tag{11}$$

For instance, the block $\hat{\rho}_{11} = M_1^\dagger \hat{\rho}^{(5)} M_1 = \hat{\rho}^{(2)} Sp(\hat{\rho}_{11})$ is the product of the density matrix of corresponding 2-dimensional subspace and probability of that projection onto this subspace equal to $Sp(\hat{\rho}_{11})$ will be held in process of measurement.

It is useful to define operators further

$$P_{11} = M_1 M_1^\dagger = \begin{pmatrix} 1 & 0 & 0 & 0 & 0 \\ 0 & 1 & 0 & 0 & 0 \\ 0 & 0 & 0 & 0 & 0 \\ 0 & 0 & 0 & 0 & 0 \\ 0 & 0 & 0 & 0 & 0 \end{pmatrix}, \quad P_{22} = M_2 M_2^\dagger = \begin{pmatrix} 0 & 0 & 0 & 0 & 0 \\ 0 & 0 & 0 & 0 & 0 \\ 0 & 0 & 1 & 0 & 0 \\ 0 & 0 & 0 & 1 & 0 \\ 0 & 0 & 0 & 0 & 1 \end{pmatrix}. \quad (12)$$

This makes it possible to expand initial matrix as an algebraic sum of the block matrices

$$\hat{\rho}^{(5)} = P_{11}\hat{\rho}^{(5)}P_{11} + P_{22}\hat{\rho}^{(5)}P_{22} + P_{11}\hat{\rho}^{(5)}P_{22} + P_{22}\hat{\rho}^{(5)}P_{11} = \sum_{nk} |n)(n|\hat{\rho}^{(5)}|k)(k|. \quad (13)$$

This is an analogy of ket-bra expansion which we call *joy-sad* expansion. Thus, if the set of operators $\{M_k\}$ is given, it would be sufficient to build the system of operators $\{P_{kk} = M_k M_k^\dagger\}$ for matrix block decomposition.

GENERALIZED REDUCTION

Given the set of operators $\{M_k\}$.

As appears from the relations (11) "matrix elements" calculated in this basis are generally not matrices of equal size. Therefore we cannot apply algebraic sum to them as in the case of standard reduction. Instead, we determine the procedure of *pre-reduction* with an expression

$$\tilde{\rho}^{(f)} = \bigoplus_k M_k^\dagger \cdot \hat{\rho}^{(f)} \cdot M_k, \quad (14)$$

or, which is equivalent, with an operator sum

$$\tilde{\rho}^{(f)} = \sum_k P_{kk}\hat{\rho}^{(f)}P_{kk}. \quad (15)$$

The *matrix of pre-reduction* obtained is treated as the density matrix of the system after averaging over the outcomes of measurement in the basis with the dimension that is not a divisor of the system dimension. Notice that the *mean effective dimension* of the block defined as ratio of dimensions of blocks to their number and equal to f/p is generally a *fractional* number.

The physical interpretation of such a procedure is as follows. One uses the device whose resolution is not sufficient for selection of a concrete vector of state in measurement but relates the outcome completely to the subspace of possible states. Thus, the formalism suggested is ideal for description of unsharp measurements.

For an example of 5-dimensional system considered

$$\rho^f = \begin{pmatrix} \rho_{11} & \rho_{12} & \rho_{13} & \rho_{14} & \rho_{15} \\ \rho_{21} & \rho_{22} & \rho_{23} & \rho_{24} & \rho_{25} \\ \rho_{31} & \rho_{32} & \rho_{33} & \rho_{34} & \rho_{35} \\ \rho_{41} & \rho_{42} & \rho_{43} & \rho_{44} & \rho_{45} \\ \rho_{51} & \rho_{52} & \rho_{53} & \rho_{54} & \rho_{55} \end{pmatrix} \tag{16}$$

and basis

$$M_1 = \begin{pmatrix} 1 & 0 \\ 0 & 1 \\ 0 & 0 \\ 0 & 0 \\ 0 & 0 \end{pmatrix}, \qquad M_2 = \begin{pmatrix} 0 & 0 & 0 \\ 0 & 0 & 0 \\ 1 & 0 & 0 \\ 0 & 1 & 0 \\ 0 & 0 & 1 \end{pmatrix} \tag{17}$$

we get after pre-reduction

$$\tilde{\rho}^{(f)} = M_1^{\dagger} \cdot \hat{\rho}^{(f)} \cdot M_1 \oplus M_2^{\dagger} \cdot \hat{\rho}^{(f)} \cdot M_2 = \begin{pmatrix} \rho_{11} & \rho_{12} & 0 & 0 & 0 \\ \rho_{21} & \rho_{22} & 0 & 0 & 0 \\ 0 & 0 & \rho_{33} & \rho_{34} & \rho_{35} \\ 0 & 0 & \rho_{43} & \rho_{44} & \rho_{45} \\ 0 & 0 & \rho_{53} & \rho_{54} & \rho_{55} \end{pmatrix} = \begin{pmatrix} r_1\hat{\rho}^{(2)} & 0 \\ 0 & r_2\hat{\rho}^{(3)} \end{pmatrix}, \tag{18}$$

$$r_1 = \rho_{11} + \rho_{22}, \qquad r_2 = \rho_{33} + \rho_{44} + \rho_{55}. \tag{19}$$

Here r_1 is the probability to find the system in a 2-dimensional subspace after measurement whereas r_2 is the probability to find it in a 3-dimensional one. The density matrices $\hat{\rho}^{(2)}$ and $\hat{\rho}^{(3)}$ describe the states in subspaces of corresponding dimensions.

SUMMARY

Let us enumerate the principal results of the present work.

1. A suitable mathematical formalism for description of unsharp measurements is suggested.
2. A generalization of the procedure of reduction to the case when the dimension of reducing subsystem is not a divisor of the dimension of complete system. An effective dimension of reduced system appears to be fractional.
3. An idea of pre-reduction matrix is introduced. This matrix describes the distribution of probabilities to find the system in a state with the prescribed dimension as the outcome of unsharp measurement.

REFERENCES

1. D. V. Dodin, I. G. Kovalenko. *Proc. SPIE*, **7023**, 702306 (2008).

Quantum Primitives

Joseph Spring

Quantum Information & Probability Group,
School of Computer Science, University of Hertfordshire,
Hatfield, Herts AL10 9AB, UK
j.spring@herts.ac.uk

Abstract. We explore possible characterisations of entanglement classes which may be interpreted as gates acting on globally distributed systems. The cyclic nature of a selection of *entanglement gates*, primitives, is explored commencing with gates for generating Bell, GHZ and W states.

Keywords: Entanglement, Irreducibility, Prime, Separability
PACS: 02.20-a, 03.65.Fd, 03.67.Ac, 03.67.Bg, 03.67.Dd

INTRODUCTION

Cyclic Groups and Finite Fields. [1, 2, 3]. Central to classical error detection schemes such as the cyclic redundancy check, found at the link layer in network protocols such as BISYNC, HDLC, DDCMP, CSMA and token ring, is the concept of a cyclic structure operating within a finite field. These may take various forms such as $\mathbb{Z}_p$ with p prime, $p \in \mathbb{N}^+$, $\mathbb{Z}_p[x]\backslash f$ in which f is an irreducible divisor, or involve geometric constructions such as $E_p(a,b)$, in for example, elliptic and hyperelliptic settings. Related algebraic and geometric structures appear within a quantum setting in for example [4, 5, 6, 7, 8, 9, 10], whilst within a classical setting they are to be found in symmetric and asymmetric cryptography, where they play a major role. Motivated by alternative applications for cyclic structures, the role of primes and irreducibilty in both classical and quantum settings, and the similarity between entanglement and irreducibilty from an algebraic perspective, we consider properties of gates used to generate such states.

Ambient Space, Primes and Entanglement. [11]. Entanglement and primes share a common property in that their status is dependent upon the space in which they are perceived to belong. 17 = (4 + i)(4 - i) for example is prime over $\mathbb{Z}$ but not over $\mathbb{Z}(i)$, likewise $X^2 + 1$ is irreducible for $\mathbb{R}[x]$ when the roots are restricted to $\mathbb{R}$ but not over $\mathbb{C}$. Likewise, Bell states are seen to be entangled provided they are restricted to the action of operators from $\mathscr{B}(\mathscr{H}) \otimes \mathscr{B}(\mathscr{H})$ with $\mathscr{H}$ a one particle, two dimensional Hlbert space. However if we extend the operator space to $\mathscr{B}(\mathscr{H} \otimes \mathscr{H})$ then the Bell states are seen to be separable. It is the potential for global operators that characterises a state as either entangled or separable. Here we consider 'Bell gates' from $\mathscr{B}(\mathscr{H} \otimes \mathscr{H})$, $\mathscr{B}(\mathscr{H} \otimes \mathscr{H} \otimes \mathscr{H})$ and related spaces.

Bell States. Bell states are entangled states over $\mathscr{B}(\mathscr{H}) \otimes \mathscr{B}(\mathscr{H})$ and may be generated by a combination of local unitary Hadamard and Identity gates in $\mathscr{B}(\mathscr{H})$ followed by a global unitary CNOT gate in $\mathscr{B}(\mathscr{H} \otimes \mathscr{H})$.

CP1110, *Quantum Communication, Measurement and Computing (QCMC), The Ninth International Conference,* edited by A. Lvovsky

GATES

Bell Gates. Consider the gate

$$B = CNOT.H \otimes I = \frac{1}{\sqrt{2}} \begin{bmatrix} I & I \\ X & -X \end{bmatrix} \text{ with } X \text{ and } I \text{ Pauli Operators acting on } \mathscr{H}$$

The action of the Bell gate on $x \in \{|00\rangle, |01\rangle, |10\rangle, |11\rangle\}$ results in each of the Bell states $\frac{1}{\sqrt{2}}(|00\rangle + |11\rangle), \frac{1}{\sqrt{2}}(|01\rangle + |10\rangle), \frac{1}{\sqrt{2}}(|00\rangle - |11\rangle), \frac{1}{\sqrt{2}}(|01\rangle - |10\rangle)$ in $\mathscr{H} \otimes \mathscr{H}$.

The gate B generates a cyclic group of order 8, with orbit:

$$\frac{1}{\sqrt{2}} \begin{bmatrix} I & I \\ X & -X \end{bmatrix}, \quad \frac{1}{2} \begin{bmatrix} I+X & I-X \\ -(I-X) & I+X \end{bmatrix}, \quad \frac{1}{\sqrt{2}} \begin{bmatrix} X & I \\ X & -I \end{bmatrix}, \quad \frac{1}{2} \begin{bmatrix} X & 0 \\ 0 & X \end{bmatrix},$$

$$\frac{1}{\sqrt{2}} \begin{bmatrix} X & X \\ I & -I \end{bmatrix}, \quad \frac{1}{2} \begin{bmatrix} I+X & -(I-X) \\ I-X & I+X \end{bmatrix}, \quad \frac{1}{\sqrt{2}} \begin{bmatrix} I & X \\ I & -X \end{bmatrix}, \quad \begin{bmatrix} I & 0 \\ 0 & I \end{bmatrix}$$

Assigning each of the given Bell states in different orders to the basis states $\{|00\rangle, |01\rangle, |10\rangle, |11\rangle\}$ generates 4! different possible gates from B. Different orders result for the orbits of the cyclic groups generated by these gates. A self adjoint version of B, for example, such as $\frac{1}{\sqrt{2}} \begin{bmatrix} X & I \\ I & -X \end{bmatrix}$ generates a cyclic group of order 2 whilst one such as $\frac{1}{\sqrt{2}} \begin{bmatrix} I & I \\ iY & -iY \end{bmatrix}$ generates a cyclic group of order 24.

It may be shown that of the 4! possible gates suggested, 4 generate cyclic groups of order 24, 4 generate cyclic groups of order 12, 10 generate cyclic groups of order 8, 4 of order 4 and 2 of order 2. The orbits involve operators, each of whose action on $\{|00\rangle, |01\rangle, |10\rangle, |11\rangle\}$, either result in 4 entangled states or 4 separable states. The cyclic groups generated by each of the 4! gates are likewise either gates generating separable states or gates generating entangled states. In any cyclic group generated there is found to be either 1, 2 or 4 gates whose action leads to entangled states. B for example has just 2 gates leading to entangled states, B and B^5.

Let $B_1 = \frac{1}{\sqrt{2}} \begin{bmatrix} I & I \\ -X & X \end{bmatrix}$ and $B_2 = \frac{1}{\sqrt{2}} \begin{bmatrix} X & X \\ -I & I \end{bmatrix}$ be two permutations of B. Then $B_1 = -B_2^5$.

Likewise for $B_3 = \frac{1}{\sqrt{2}} \begin{bmatrix} X & X \\ I & -I \end{bmatrix}$ we find that $B = B_3^5$. It follow that the cyclic group <B> generated by gate B is identical to that generated by gate B_3, so $< B >=< B_3 >$ and $< B_1 >=< -B_2^5 >$.

Theorem The general form for the operator B generating GHZ states is

$$\frac{1}{\sqrt{2}} \begin{bmatrix} \otimes I & \otimes I \\ \otimes X & -\otimes X \end{bmatrix}$$

Proof. This follows by construction. We commence with the sum of the least binary value $|000\ldots0\rangle$ and the greatest binary value $|111\ldots1\rangle$. This gives the first entry in the

I Pauli operator and the first entry in the *X* Pauli operator. We then add one to the least binary value and subtract one from the greatest binary value to obtain the second column in the *I* operator and the second column in the *X* operator. Continuing in this way to the half way point in our construction generates $\otimes I$ for the upper half of the gate and $\otimes X$ for the lower half of the gate. Following this we take the 'conjugate' of the second part of the sum from the first column, then the second column, then ... column generating $\otimes I$ for the remaining upper half of the gate and -$\otimes X$ for the remaining lower half of the gate. The scalar follows by normalisation. □

The orbit for the given GHZ gate is found to be 8.

W Gates. The smallest *W* gate is $2^3 \times 2^3$ leading to $2^3!$ different gates that one may use to generate *W* states from the 2^3 basis states $\{|ijk\rangle\}$ in which $i, j, k \in \{0,1\}$.

Definition We define a starting global gate *W* for describing the generation of W-states in $\mathscr{H} \otimes \mathscr{H} \otimes \mathscr{H}$ to be:

$$W = \frac{1}{\sqrt{3}} \begin{bmatrix} I & e_3 & X & e_1 \\ -e_3 & -I & e_1 & X \\ -X & e_4 & I & -e_2 \\ e_4 & -X & e_2 & -I \end{bmatrix}$$

Proposition The gate $W \in \mathscr{B}(\mathscr{H} \otimes \mathscr{H} \otimes \mathscr{H})$ is a cyclic operator of order 8.

Proof. $W^2 = \frac{1}{\sqrt{3}} \begin{bmatrix} 0 & 0 & X & 0 \\ 0 & 0 & 0 & -X \\ -X & 0 & 0 & 0 \\ 0 & X & 0 & 0 \end{bmatrix}$,

$W^4 = -I$, and $W^8 = I$.

□

A greater degree of complexity emerges in the cyclic groups generated by *W* gates. These range from gates generating cyclic groups of order 2, for example:

$$W_1 = \frac{1}{\sqrt{3}} \begin{bmatrix} I & |1\rangle\langle 0| & X & |0\rangle\langle 0| \\ -|1\rangle\langle 0| & -I & |0\rangle\langle 0| & X \\ -X & -|1\rangle\langle 1| & I & -|0\rangle\langle 1| \\ |1\rangle\langle 1| & -X & |0\rangle\langle 1| & -I \end{bmatrix}$$

to gates such as

$$W_2 = \frac{1}{\sqrt{3}} \begin{bmatrix} I & |1\rangle\langle 0| & X & |0\rangle\langle 1| \\ -|1\rangle\langle 0| & -I & |0\rangle\langle 0| & I \\ -X & |1\rangle\langle 1| & I & -|0\rangle\langle 0| \\ |1\rangle\langle 1| & -X & |0\rangle\langle 1| & -X \end{bmatrix}$$

.

EL-GAMAL

The generation of cyclic elements lends itself to encryption protocols such as El-Gamal, illustrating in principle, the possibility for masking the choice of gate employed at different stages within an application. Further research in this, the above and related areas has been carried out. The details will appear elsewhere.

ACKNOWLEDGMENTS

Many thanks to all those at QCMC08 for their warm hospitality and stimulating discussions.

REFERENCES

1. L. L. Peterson and B. S. Davie, *Computer Networks: A Systems Approach*, Morgan Kaufmann, Elsevier,2007.
2. Henri Cohen and Gerhard Frey (et. al.), *Handbook of Elliptic and Hyperelliptic Curve Cryptography*, Discrete Mathematics and its Application, Chapman & Hall/CRC, 2006.
3. Joan Daemen and Vincent Rijmen, *The Design of Rijndael, AES - The Advanced Encryption Algorithm*, Springer - Verlag, 2002.
4. Stephane Beauregard, Gilles Brassard and Jose Manuel Fernandez, *Quantum Arithmetic on Galois Fields*, quant-ph 0301163, pp 1-29, 2003.
5. Kathleen S. Gibbons, Mahew J. Hoffman and William K. Wooters, *Discrete Phase Space based on Finite Fields*, quant-ph/0401155, 2004.
6. Haret Rosu, Michael Planat and Metod Saniga, *From Finite Projective Geometry to Quantum Phase Enciphering*, Quantum Communication, Measurement and Computing, edited by S. M. Barnett *et al.*, American Institute of Physics, New York, 2004.
7. Michael Planat, Haret Rosu,Serge Perrine and Metod Saniga, *Finite Algebraic geometric Structures Underlying Mutually Unbiased Quantum Measurements*, R. Buchanan et al. (es.); Time, Quantum and the Subjective;quant-ph/0503159, World Scientific Publishing Co., pp409-426, 2004.
8. Michael Planat, *Huyghens, Bohr, Riemann and Galois Phase Locking*, International Journal of Modern Physics B;quant-ph/0510044, World Scientific Publishing Co., pp 1-18, 2005.
9. Hashang Heydari, *Geometical Structure of Entangled States and Secant Variety*, quant-ph/0611144, pp 1-7, 2006.
10. J. A. Vaccaro, J. Spring and A. Chefles *Quantum Protocols for Anonymous Voting and Surveying*, Physical Review A,75, 012333, 2007.
11. Joseph Spring, *Entanglement and Irreducibilty*, Quantum Communication, Measurement and Computing 8, AIP, 2006.

CHAPTER 3

POPESCU-ROHRLICH BOXES

Towards a Graph-Theoretic Characterization of Popescu-Rohrlich Box Simulation Strategies

Samuel C. Fletcher

Dept. of Logic and Philosophy of Science, University of California, Irvine, Irvine, CA 92617, USA

Abstract. The propensity for certain quantum system to consistently violate Bell inequalities—dubbed quantum nonlocality—has recently been found to be an information theoretic resource in quantum communication that is distinct from entanglement. Accordingly, a useful measurement of "how much" of this resource is present in a given system is needed. One candidate, the PR-box (sometimes called a nonlocal machine), has been suggested as a possible benchmark, but one of the difficulties in using it has been to determine exactly what kinds of quantum systems are simulable by it. The present work is a first step towards a graph-theoretic understanding of PR-boxes that may prove useful in this endeavor.

Keywords: nonlocality, PR-boxes, simulation of quantum states
PACS: 02.10.Ox, 03.65.Ud, 03.67.-a

INTRODUCTION

Quantum nonlocality arises from measurements on spatially separated entangled quantum mechanical systems. These measurements are correlated to an extent that cannot be explained by any local hidden variable theory, *i.e.* they violate some Bell inequality. Explicitly, this obtains from the conditional probabilities of measuring outputs r_A and r_B, given input measurements A_i and B_j, not factoring as they would with "independent" events:

$$P(r_A, r_B|A_i, B_j) \equiv P_{ij}(r_A, r_B) \neq P_i(r_A)P_j(r_B).$$

While these correlations cannot be used for signaling, they can nevertheless reduce the number of signaling bits necessary for certain communication complexity problems [1]. Thus nonlocality can be used as an information-theoretic resource [2], which has led to a search for quantifying measures of nonlocality, just as such measures have been sought for entanglement [3]. One of the first suggestions, by Popescu and Rohrlich [4], was to consider the theoretical construct of a maximally nonlocal machine, that is, a hypothetical EPR-like device whose correlations are constrained not by quantum mechanics but only by the bare requirements that they be nonsignaling. This condition ensures that these "PR-boxes" maximally violate their corresponding Bell-type inequalities by yielding the latters' algebraic maxima, in contrast to a smaller bound for local hidden variable theories and quantum mechanics. In particular, a PR-box always produces perfect correlation or anti-correlation according to the governing equation

$$xy = a + b \pmod 2,$$

where x and y are the input bit values for Alice and Bob, respectively, and a and b are similarly their output bits (see Fig. (1)). Using such PR-boxes and local operations, can

CP1110, *Quantum Communication, Measurement and Computing (QCMC), The Ninth International Conference,* edited by A. Lvovsky

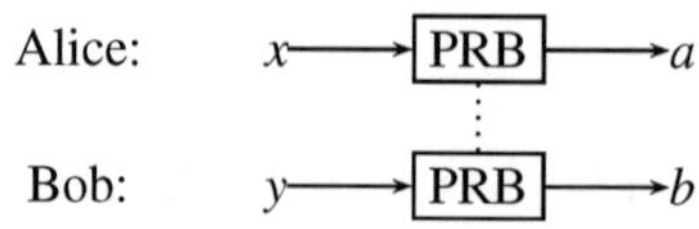

FIGURE 1. Each party passes a bit into his or her part of the PR-box, which in turn outputs a bit. The dotted line represents the fact that the results are correlated despite the distance between Alice and Bob.

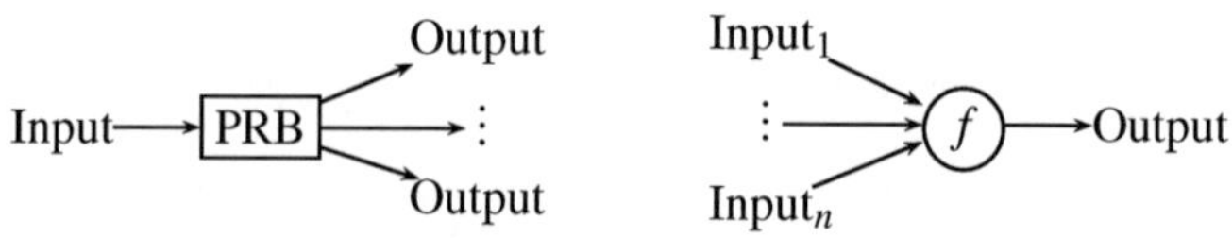

FIGURE 2. Pictured on the left is a PRB, which takes one input; note that each of the outputs must be the same. On the right we have a Boolean function, the only other "building block" for the case at hand.

one then simulate the correlations of general quantum systems [5]? Part of the difficulty in answering this question arises from the fact that it is *prima facie* unclear how one could exhaust all the possible simulation strategies for a given number of PR-boxes. The present investigation suggests that this might be achieving by taking the "box" analogy seriously and treating PR-boxes like elements of a circuit – or more generally, as nodes in a directed graph.

APPLICATION OF GRAPH THEORY

In Fig. (2) are the representations of the two types of atomic operations a party can perform. We see that the PR-box "node" takes one input and can output the same bit to many other nodes, while general local operations can always be written as a set of Boolean functions. In addition to PR-boxes and these Boolean functions, which have in-degree and out-degree 1, resepectively, each party must have an input node—*i.e.* they must choose at least one "measurement setting"—that is a *source* and an output node of in-degree 1 that is a *sink*, since each party must eventually output some bit value that will later be used to check for correlation. Note that for any given graph the particular bit values that the input node outputs, as well as the particular Boolean function represented by the corresponding Boolean node, are left unspecified unless otherwise noted. This allows for full generality in the foregoing analysis, and also curtails concerns over simulation strategies that are probabilistic. For while there is nothing physically problematic with such strategies, they are formed by some probability distribution over the deterministic strategies; hence their characterization is entailed by the characterization of the deterministic strategies.

There are a few more facts to consider about these simulation graphs that will help in our analysis. First, a graph G is usually defined as the union of a set of vertices or nodes V with a set of edges E that, for directed graphs, are ordered pairs of nodes. But it will also be useful to use *in-out* notation, in which we write of a particular node N

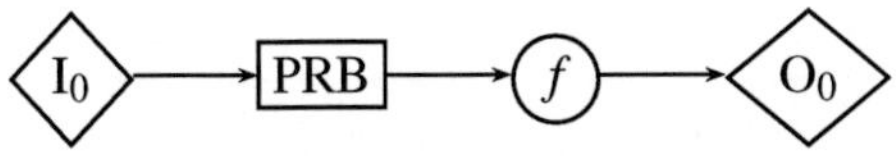

FIGURE 3. Visual representation of the smallest 1-general simulation graph.

that $N \sim (x_1,\ldots,x_i;y_1,\ldots,y_j)$, where the list before the semicolon is the set of nodes from which there is an edge to N and the list after is set of nodes to which there is an edge from N. Second, each of the graphs we are considering must be acylic to avoid causality problems. Third, there must be a path from the input to the output node. If all these conditions are met, then we have a *simulation graph.* Finally, we should consider rules for *simulation isomorphism*, which holds for two simulation graphs when they can simulate the same states. At this point a specific extensional definition is not necessary, but some intentional cases are:

1. Simulation graphs with edges of the form (I_0, f), where f is a Boolean node, are simulation isomorphic to graphs with those edges removed that are otherwise identical.
2. Simulation graphs with edges (f,g) between Boolean nodes are simulation isomorphic to graphs that are identical except for having $f \sim (x_1,\ldots,x_i;f(x))$ and $g \sim (y_1,\ldots,y_j;g(y))$ replaced by $h \sim (\{x_k\}_{k=1}^{i} \cup \{y_k\}_{k=1}^{j};z)$.
3. Simulation graphs with sources or sinks that are not I_0 and O_0, respectively, are simulation isomorphic to graphs with those nodes removed.

Definition. A simulation graph G_n is called *n-general* when it contains at least n PRB nodes and when graphs obtained by adding Boolean nodes or edges to G_n are simulation isomorphic to it.

Definition. An n-general simulation graph G_n is called the *smallest* when any proper n-general simulation subgraph of G_n is not general. Equivalently, it is an n-general subgraph of each n-general graph.

Theorem. *The smallest 1-general simulation graph, which is shown in Fig. (3), is specified by*

$$\text{PRB} \sim (I_0; f), \qquad f \sim (\text{PRB}; O_0). \tag{1}$$

Proof. Let $f \sim (x_1,\ldots,x_n;y)$ be a Boolean node in the graph G_1. For the simplest general graph, none of the entering edges of f can be Boolean nodes or the input node, and the output node has no exiting edges. Thus the only possible entering edge is from a PRB. Similarly, the exiting edge of f can only be to a PRB or the output node. But if there is only one PRB, then any Boolean node must have the form $f \sim (PRB; O_0)$. Since each of PRB and O_0 are unique in the graph, there must be only one Boolean node.

Now consider the entering edge for the sole PRB. It cannot be from a Boolean node for we have just shown that any such node's exiting edge must be to O_0. This leaves the input I_0 node as the only option, since there are no other PRBs and the output node has no exiting edges. As for its outputs, it cannot be to another PRB, nor to the input node. Since the output node only has one entering edge that is already taken by f, this forces the form $\text{PRB} \sim (I_0; f)$. □

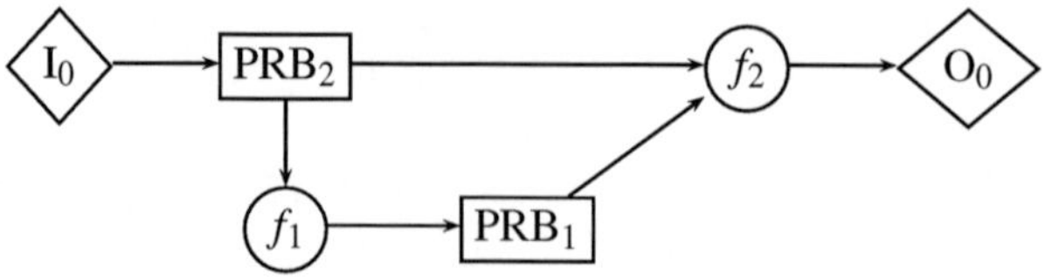

FIGURE 4. Visualization of the smallest 2-general simulation graph.

Theorem. *The smallest 2-general simulation graph G_2, shown in Fig. (4), is of the form*

$$\begin{aligned} \mathrm{PRB}_1 &\sim (f_1; f_2) & f_1 &\sim (\mathrm{PRB}_2; \mathrm{PRB}_1) \\ \mathrm{PRB}_2 &\sim (I_0; f_1, f_2) & f_2 &\sim (\mathrm{PRB}_1, \mathrm{PRB}_1; O_0). \end{aligned} \tag{2}$$

Proof. First we must show that G_2 is 2-general by showing that adding neither Boolean nodes and edges nor solely edges yields a graph that is simulation isomorphic to G_2. If we add only edges, they must emanate from I_0, PRB_1, or PRB_2 and end at f_1 or f_2. This gives six possibilities. Both (I_0, f_1) and (I_0, f_2) fail because the resulting graph is isomorphic to the one at hand; (PRB_1, f_1) fails because it would make the graph cyclic; and the remaining three are already edges of the graph. If we add both Boolean nodes and edges, like before the new edges must end at some Boolean node, but any such examples are isomorphic to the case in which no new Boolean nodes have been added.

To show that it is smallest as well, we must try to remove some edges or nodes and show that the resulting graph is still 2-general. With regards to just edges, the only possible choice that doesn't as well eliminate a node is (PRB_2, f_2), but the graph resulting from such a move can be entirely simulated by G_2 by picking f_2 to be insensitive to the input from PRB_2. If we try to eliminate a node as well, the only possible choices are f_1 and PRB_1, but either of these makes the graph no longer 2-general. □

ACKNOWLEDGMENTS

My deepest thanks go to Hans Halvorson of Princeton University who supervised the thesis from which this work came. I am also grateful to the QCMC committee for granting me the funding that allowed me to attend QCMC2008.

REFERENCES

1. R. Cleve and H. Buhrman. "Substituting quantum entanglement for communication." *Phys. Rev. A* **56** 1201 (1997).
2. J. Barrett et al. "Nonlocal correlations as an information-theoretic resource." *Phys. Rev. A* **71** 022101 (2005).
3. J. Barrett and S. Pironio. "Popescu-Rohrlich Correlations as a Unit of Nonlocality." *Phys. Rev. Lett.* **95** 140401 (2005).
4. S. Popescu and D. Rohrlich. "Quantum Nonlocality as an Axiom." *Found. of Phys.* **24** 379–385 (1994).
5. N. Brunner, N. Gisin, and V. Scarani. "Entanglement and non-locality are different resources." *New J. Phys.* **7** 88 (2005).

How Non-Local are n Noisy Popescu-Rohrlich Machines?

Matthias Fitzi*, Esther Hänggi*, Valerio Scarani† and Stefan Wolf*

*Computer Science Department, ETH Zürich, Switzerland [1]
†Centre for Quantum Technologies and Department of Physics, National University of Singapore, Singapore [2]

Abstract. We show that the local part of n symmetric ε-PRMs is of order $\Theta(\varepsilon^{\lceil n/2 \rceil})$ and that the local part of n maximally biased (asymmetric) δ-PRMs is exactly $(3\delta)^n$.

Keywords: Local part, non-local box, non-signaling
PACS: 03.65.Ud,03.67.-a,87.19.lo

INTRODUCTION

The behavior of a bipartite input/output system $P_{XY|UV}$ is *non-local* if it cannot be explained by pre-shared information. For example, the measurement choice/outcome behavior of certain *entangled* quantum states is non-local. As an application, non-locality can imply device-independent unconditional secrecy in quantum cryptography [1]: hidden parameters that do not exist cannot be known by the adversary; and the stronger the non-locality the more secret is the respective information. Non-local correlations can also be seen as a resource to fulfill distributed tasks [2].

Non-locality of a binary input/output system is typically characterized by the Popescu-Rohrlich Machine (PRM) [3] that, on inputs X and Y, produces random outputs U and V such that $X \oplus Y = U \cdot V$. Quantum mechanically, PRM behavior can only be simulated with an accuracy of roughly 85% [4] whereas the classical limit is 75% [5].

The question of how much non-locality there is in a given system's behavior — where non-locality is quantified by partitioning the behavior into a local part of maximal weight and the remaining non-local part — has first been asked in [6] (see also [7]). We study here the local part of (a number of) imperfect PRMs, e.g., the local part of a perfect PRM is zero. Our main result is that the local part of n symmetric ε-PRMs is of order $\Theta(\varepsilon^{\lceil n/2 \rceil})$ and that the local part of n maximally biased δ-PRMs is exactly $(3\delta)^n$ (see also [8]).

DEFINITIONS

Note that we restrict ourselves to bipartite systems although generalizations to more parties are possible. These bipartite systems take an input and yield an output from a

---

[1] Supported by the Swiss National Science Foundation.
[2] Supported by the National Research Foundation and Ministry of Education, Singapore.

CP1110, *Quantum Communication, Measurement and Computing (QCMC), The Ninth International Conference,* edited by A. Lvovsky

well-defined alphabet on each side (i.e., to each party) and can be characterized by a conditional probability distribution $P_{XY|UV}(x,y,u,v)$ where U and V are the inputs, and X and Y are the outputs on the respective sides.

DEFINITION 1 *A bipartite conditional probability distribution $P_{XY|UV}(x,y,u,v)$ is called* non-signaling *if the two parties cannot use it to transmit information, i.e.,*

$$\sum_x P_{XY|UV}(x,y,u,v) = \sum_x P_{XY|UV}(x,y,u',v) \; \forall y,v$$

and similar when the role of the two parties are reversed. It is called local deterministic *if it can be written as*

$$P_{XY|UV} = \delta_{x,f(u)} \cdot \delta_{y,g(v)} \, ,$$

where $f: U \to X$ and $g: V \to Y$ and δ is the Kronecker symbol; and it is local *if it is a convex combination of local deterministic probability distributions.*

We will only consider non-signaling probability distributions in this paper. Note that the space of all non-signaling probability distributions over a certain input/output alphabet is convex. All local probability distributions can be simulated by two distant parties using a pre-agreed strategy and shared randomness — the shared randomness determines which local deterministic probability distribution to use, and respective output is then a deterministic function of the input (on the same side).

DEFINITION 2 *Given a bipartite non-signaling probability distribution $P_{XY|UV}$, the maximum p, $0 \leq p \leq 1$, such that $P_{XY|UV}$ can be written as the convex combination of a local and a non-signaling probability distribution is called its* local part*:*

$$P_{XY|UV} = p \cdot P_{local} + (1-p) \cdot P_{ns} \, .$$

A probability distribution is local if and only if its local part is equal to one. However, in the special case of probability distributions taking binary input and giving binary output, there is a simple inequality which can be used to determine if a probability distribution is local.

PROPOSITION 1 (Bell [5]) *A bipartite probability distribution $P_{XY|UV}$ taking binary input and giving binary output is non-local if, for uniform inputs,*

$$P(X \oplus Y = U \cdot V) > 0.75 \, .$$

Note that, up to relabelling of the inputs and outputs, the above condition is in fact equivalent to non-locality. After [9], we denote the condition $X \oplus Y = U \cdot V$ by *CHSH-condition*.

For more than two inputs and outputs, the following Lemma 1 will be of use.

LEMMA 1 *Consider two non-signaling probability distributions $P_{XY|UV}$ and $P_{ns,1}$. The former one can be written as $P_{XY|UV} = p \cdot P_{ns,1} + (1-p) \cdot P_{ns,2}$ (where $P_{ns,2}$ is a second non-signaling probability distribution) if and only if*

$$p \cdot P_{ns,1}(x,y,u,v) \leq P_{XY|UV}(x,y,u,v) \; \forall x,y,u,v \, .$$

Proof. FORWARD DIRECTION: Since both $P_{XY|UV}$ and $P_{ns,1}$, are normalized and non-signaling, $P_{ns,2}$ is also normalized and non-signaling (both properties are linear). Now, since $P_{ns,2}(x,y,u,v) = (1/(1-p))(P_{XY|UV}(x,y,u,v) - p \cdot P_{ns,1}(x,y,u,v))$, which is larger than zero by assumption, the forward direction follows. REVERSE DIRECTION: Assume that $p \cdot P_{ns,1}(x,y,u,v) > P_{XY|UV}(x,y,u,v)$ for some x,y,u,v. Then $P_{ns,2}(x,y,u,v) < 0$ and thus $P_{ns,2}$ is not a probability distribution. □

We will study in this paper two particular non-signaling probability distributions:

DEFINITION 3 *A* symmetric ε-PRM *(denoted by* $P^{1,\varepsilon}_{XY|UV}$ *for one* ε*-PRM) and a* maximally biased δ-PRM *are bipartite conditional probability distribution given by the probability tables below.*

ε-PRM=

V	Y \ X	U=0: 0	U=0: 1	U=1: 0	U=1: 1
0	0	$\frac{1}{2}-\frac{\varepsilon}{2}$	$\frac{\varepsilon}{2}$	$\frac{1}{2}-\frac{\varepsilon}{2}$	$\frac{\varepsilon}{2}$
	1	$\frac{\varepsilon}{2}$	$\frac{1}{2}-\frac{\varepsilon}{2}$	$\frac{\varepsilon}{2}$	$\frac{1}{2}-\frac{\varepsilon}{2}$
1	0	$\frac{1}{2}-\frac{\varepsilon}{2}$	$\frac{\varepsilon}{2}$	$\frac{\varepsilon}{2}$	$\frac{1}{2}-\frac{\varepsilon}{2}$
	1	$\frac{\varepsilon}{2}$	$\frac{1}{2}-\frac{\varepsilon}{2}$	$\frac{1}{2}-\frac{\varepsilon}{2}$	$\frac{\varepsilon}{2}$

δ-PRM=

V	Y \ X	U=0: 0	U=0: 1	U=1: 0	U=1: 1
0	0	$\frac{1}{2}-\frac{\delta}{2}$	0	$\frac{1}{2}-\frac{\delta}{2}$	0
	1	δ	$\frac{1}{2}-\frac{\delta}{2}$	δ	$\frac{1}{2}-\frac{\delta}{2}$
1	0	$\frac{1}{2}-\frac{\delta}{2}$	0	0	$\frac{1}{2}-\frac{\delta}{2}$
	1	δ	$\frac{1}{2}-\frac{\delta}{2}$	$\frac{1}{2}+\frac{\delta}{2}$	0

SYMMETRIC ε-PRMS

We now study the case of symmetric ε-PRMs ($\varepsilon \in [0, 0.25]$), i.e., PRMs that fulfill the CHSH-condition with probability $1-\varepsilon$ for all inputs and unbiased output bits.

By Lemma 1, we can write any non-signaling probability distribution as

$$P_{XY|UV} = p_i \cdot P_{ld,i} + (1 - \sum_i p_i) \cdot P_{ns}$$

where $P_{ld,i}$ are the different local deterministic strategies fixed by the input and output size. Together with the definition of the local part this implies the following two lemmas.

LEMMA 2 *The local part is the optimal value of the linear program:*

$$max\colon \sum_i p_i \quad s.t. \quad \sum_i p_i \cdot P_{l-d,i}(x,y,u,v) \leq P_{XY|UV}(x,y,u,v) \quad and \quad p_i \geq 0\,.$$

LEMMA 3 *The local part of* $P^{1,\varepsilon}_{XY|UV}$ *is* 4ε.

Now, consider the case of two independent symmetric ε-PRMs. We can write these two machines as one single machine taking 2 input bits and giving 2 output bits on each side:

$$\begin{aligned} P^{2,\varepsilon}_{XY|UV}(x,y,u,v) &= P^{2,\varepsilon}_{XY|UV}((x_1x_2),(y_1y_2),(u_1u_2),(v_1v_2)) \\ &= P^{1,\varepsilon}_{XY|UV}(x_1,y_1,u_1,v_1) \cdot P^{1,\varepsilon}_{XY|UV}(x_2,y_2,u_2,v_2)\,. \end{aligned}$$

Obviously it is always possible to write each of the two machines separately as a combination of one local and one non-local machine. This would give a local weight of $(4\varepsilon)^2$. However, the local part might be larger and, indeed, Lemma 2 and 3 show that the local part of two symmetric ε-PRMs is the same as the local part of one single symmetric ε-PRM.

LEMMA 4 $P^{2,\varepsilon}_{XY|UV} = (4\varepsilon) \cdot P^{2,local}_{XY|UV} + (1-4\varepsilon) \cdot P^{2,0}_{XY|UV}$.

This shows that it is neither possible to use two symmetric ε-PRMs in parallel to create a better ε-PRM, nor to create a more secure bit from the outputs of two ε-PRMsby applying a function (where we assume that the inputs are public).

We now consider the case of any number n of independent symmetric ε-PRMs.

LEMMA 5 *For every local deterministic strategy for n PRMs, there always exist inputs u and v such that $x_i \oplus y_i \neq u_i \cdot v_i$ for at least $n/2$ of the indices i.*

Proof. Assume, wlog, that $x(\vec{0}) = \vec{0}$. Consider the case $u = \vec{0}$. For at most k out of the n instances to fail, $y(v)$ must have Hamming weight at most k (independently of v). Now, consider $y(\bar{x}(\vec{1}))$: For all n instances to be correct for Input $u = \vec{1}$, $y(\bar{x}(\vec{1}))$ must be equal to $x(\vec{1})$ exactly at the positions where $x(\vec{1})$ is '1,' i.e., $y(\bar{x}(\vec{1})) = \vec{1}$. Thus, for at most k instances to fail, $y(\bar{x}(\vec{1}))$ must have Hamming weight at least $n-k$. Since $k < n/2$, this contradicts the fact that $y(v)$ must have Hamming weight at most k. □

THEOREM 1 *The local part of n symmetric ε-PRMs is of order $\Theta(\varepsilon^{\lceil \frac{n}{2} \rceil})$.*

Proof. It is easy to see that this order can be reached as a local part of $(4\varepsilon)^{\lceil \frac{n}{2} \rceil}$ can be achieved by combining the ε-PRMs in pairs. On the other hand, Lemma 5 implies that it is of order $O(\varepsilon^{\lceil \frac{n}{2} \rceil})$. □

MAXIMALLY BIASED δ-PRMS

Consider a PRM which fullfills the CHSH-condition in three out of the four input-cases with proability $1-\delta$ and in the fourth case perfectly, and where the output bit X is maximally biased towards zero.

A simple maximization shows that the local part of one maximally biased δ-PRM is 3δ. More generally, it is possible to show

THEOREM 2 *The local part of n maximally biased δ-PRMs is $(3\delta)^n$.*

REFERENCES

1. J. Barrett, L. Hardy, and A. Kent, *Physical Review Letters* **95**, 010503 (2005).
2. W. van Dam (2005), quant-ph/0501159.
3. S. Popescu, and D. Rohrlich, *Foundations of Physics* **24**, 379–385 (1994).
4. B. S. Cirel'son, *Letters in Mathematical Physics* **4**, 93–100 (1980).
5. J. S. Bell, *Physics* **1**, 195–200 (1964).
6. A. C. Elitzur, S. Popescu, and D. Rohrlich, *Physics Letters A* **162**, 25–28 (1992).
7. V. Scarani, *Physical Review A* **77**, 042112 (2008).
8. M. Fitzi, E. Hänggi, V. Scarani, and S. Wolf (2008), quant-ph/0811.1649.
9. J. F. Clauser, M. A. Horne, A. Shimony, and R. A. Holt, *Physical Review Letters* **23**, 880–884 (1969).

The Universality of Non-Local Boxes

Manuel Forster and Stefan Wolf

Computer Science Department, ETH Zürich, ETH Zentrum, CH-8092 Zürich, Switzerland.[1]

Abstract. One of the most fascinating consequences of quantum theory is *non-locality*, i.e., the fact that the measurement outcomes of (spatially separated) parts of a system can have a correlation unexplainable by shared classical information. Note that at the same time, these correlations are *non-signaling* and do not allow for message transmission. Popescu and Rohrlich have defined a *non-local box* as a "basic building block of non-locality" and initiated a systematic study of non-local correlations and their applications. However, they left open the question of whether *any* non-signaling correlation can be simulated by such non-local boxes. We show that the answer is *yes* with respect to arbitrarily accurate approximations.

Keywords: non-signaling, non-locality, non-local box
PACS: 03.67.-a,03.65.Ud

INTRODUCTION

In probability theory, the term correlation is often used to indicate a departure from independence. When two separated parts of a quantum state are measured, then the outcomes can be correlated. While this may be surprising from a physical point of view, it is much less from an information-theoretical perspective since such correlations could have arisen by randomness shared when the two particles were generated. In this note we address correlations of a stronger kind - those which are unexplainable by shared randomness. More precisely, we study correlations between the outcomes of the two ends of a bipartite input-output *system*, characterized by a conditional probability distribution $P(ab|xy)$. Let x and a stand for the input and output on the left-hand side of the system, and y and b for the corresponding values on the right-hand side.

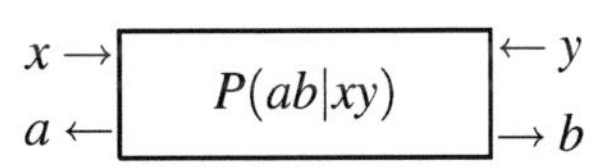

We call such a behavior *local* if it is explainable by shared classical information and *signaling* if it allows for message transmission in either direction.

Local correlations satisfy certain linear inequalities known as *Bell inequalities* [1]; in other words, violation of such inequalities indicates non-locality. John Bell was the first to realize that entangled quantum states can have a non-local behavior under measurements if x and y are the choice of the measurements to be carried out, and a and b are the measurement results. The non-local behavior of quantum systems was highlighted by

[1] Supported by the Swiss National Science Foundation (SNSF).

CP1110, *Quantum Communication, Measurement and Computing (QCMC), The Ninth International Conference,* edited by A. Lvovsky

Einstein, Podolsky and Rosen [2], and what was originally (erroneously [1]) considered a witness for the incompleteness of quantum mechanics, has proven to be a very useful information theoretic resource [3].

Non-Signaling Systems

In accordance with relativity theory, the joint behavior of quantum systems must be *non-signaling*: otherwise superluminal message transmission would be possible. A binary input-output system characterized by a conditional probability distribution $P(ab|xy)$ is *non-signaling* if one cannot signal from one side to the other by the choice of the input. This means that the marginal probabilities $P(a|x)$ and $P(b|y)$ are independent of y and x, respectively, i.e.,

$$\sum_b P(ab|xy) = \sum_b P(ab|xy') \equiv P(a|x) \ \forall a,x,y,y',$$
$$\sum_a P(ab|xy) = \sum_a P(ab|x'y) \equiv P(b|y) \ \forall b,x,x',y.$$

Using a non-signaling system a party receives its output immediately after giving its input, independently of whether the other has given its input already. Note that this is possible since the system is non-signaling.

The Non-Local Box

We are interested in systems that are *neither signaling nor local*. It may not be obvious that such correlations can be defined, which are impossible to be simulated by quantum mechanics. Surprisingly, they can, and an example is the *non-local box* (NLB for short) or *Popescu-Rohrlich (PR) machine* [4]. The NLB is a system violating the Clauser-Horne-Shimony-Holt (CHSH) inequality [5] to the algebraic maximum. Its behavior is as follows: We have $x,y,a,b \in \{0,1\}$; a and b are uniform and independent of (x,y), but $a+b \equiv xy \bmod 2$ always holds. Its probability distribution is, thus,

$$P_{\mathrm{NLB}}(ab|xy) = \begin{cases} 1/2 & \text{if } xy = a+b \bmod 2, \\ 0 & \text{otherwise.} \end{cases}$$

The PR machine cannot occur as the behavior of a quantum state [6], but it can be approximated with an accuracy of roughly 85%, whereas 75% is the local limit.

MOTIVATION

With the definition of the non-local box, Popescu and Rohrlich raised the question why quantum mechanics is not maximally non-local, that is, why quantum non-locality is not solely constrained by the non-signaling conditions. It is hence possible to study non-locality without reference to quantum mechanics, but simply as a property of joint probability distributions. The non-local box has since been considered as a possible candidate for an “atom” or “basic building block” for non-locality. Clearly, this notion makes more sense if, actually, non-local boxes allow the realization of *any* non-signaling system, that is, if two parties can non-interactively simulate any non-signaling system by local operations on shared randomness and a finite quantity of non-local boxes.

Previous Work

In [7], and independently in [8], it is proven that every non-signaling system with *binary outputs* $a,b \in \{0,1\}$ can be simulated by local operations on a finite number of NLBs. In [9], the alternative case of *binary inputs* $x,y \in \{0,1\}$ is solved. In the latter case, the simulation can be made arbitrarily close, but not perfect. In certain cases perfect simulation is known to be impossible [10].

OUR CONTRIBUTION

We show that, on the other hand, if one is willing to accept an arbitrarily small error, then non-local boxes *are* universal.

Main Result. *Any bipartite non-signaling system can be approximated arbitrarily closely by shared randomness and non-local boxes.*

The proof consists of two parts. The first part defines a proper subset $\mathcal{D}$ of all non-signaling systems $\mathcal{P}$ with the following properties: Let $\mathcal{D}$ include all non-signaling systems D_d with output alphabet $a,b \in S = \{0,1,...,d-1\}$, arbitrary input sets $x \in X$, $y \in Y$ and the joint probability distribution

$$P_{D_d}(ab|xy) = \begin{cases} 1/d & \text{if } f_{xy}(a) = b, \\ 0 & \text{otherwise,} \end{cases}$$

where $f_{xy} : S \rightarrow S$ is a permutation on the output set depending on x,y.

We define a simulation protocol as a number of classical operations on *resource* systems and (if needed) shared randomness. The operations are subsequently and non-interactively executed by two parties (usually called Alice and Bob) which finally simulate a certain *target* system. To complete the first part we give the construction of such a simulation protocol leading to the following statement:

$\mathcal{D}$ is universal. *For any system $P \in \mathcal{P}$ we can find $D_d[P] \in \mathcal{D}$ such that Alice and Bob approximate P arbitrarily well by a simulation protocol on $D_d[P]$.*

Henceforth, we only have to care about the well-characterized proper subset $\mathcal{D}$ of non-signaling systems.

The second part of the proof is inductive on the output alphabet size d, i.e., it reduces the problem of simulating any D_d to simulating a finite number of systems D_{d-1}. Again, a simulation protocol is given. That is, we constructively show the following induction step:

Induction on d**.** *For any alphabet size* $d > 2$ *any system* $D_d \in \mathscr{D}$ *can be simulated, within any error probability* $\delta_d > 0$*, by a simulation protocol using shared randomness and a finite number of approximations to a system* $D_{d-1} \in \mathscr{D}$*.*

For the final step of the argument, we use the fact that any $D_2 \in \mathscr{D}$, i.e., any distributed Boolean function, can be simulated by local operations on a finite number of non-local boxes [11].

Detailed proofs of the claimed results can be found in [12].

CONCLUDING REMARKS

We have shown that any non-signaling bipartite system can be approximated arbitrarily well by non-local boxes. This complements an earlier result [10] which states that a *perfect* simulation is not always possible.

Our result implies that the non-local box can be considered a basic unit of bipartite non-locality. Another implication is that every inner correlation, that is, any non-signaling correlation which is not element of the convex hull of the non-signaling polytope, can be simulated perfectly by non-local boxes.

It is a challenging open problem whether the same holds for *multi*-partite non-signaling systems. If this fails to be true, a natural question would be whether there still exists a single system, or a finite number of systems, which is universal in the same sense as the NLB in the bipartite case.

REFERENCES

1. J. Bell, *Physics* **1**, 195–200 (1964).
2. A. Einstein, N. Rosen, and B. Podolsky, *Phys. Rev.* **47**, 777 (1935).
3. M. A. Nielsen, and I. L. Chuang, *Quantum Computation and Quantum Information*, Cambridge University Press, 2000.
4. S. Popescu, and D. Rohrlich, *Foundations of Physics* **24**, 379–385 (1994).
5. J. Clauser, M. Horne, A. Shimony, and R. Holt, *Physical Review Letters* **23**, 880–884 (1969).
6. B. S. Tsirelson, *Letters in Mathematical Physics* **4**, 93–100 (1980).
7. N. S. Jones, and L. Masanes, *Physical Review* **72**, 052312 (2005).
8. J. Barrett, and S. Pironio, *Physical Review Letters* **95**, 140401 (2005).
9. J. Barrett, N. Linden, S. Massar, S. Pironio, S. Popescu, and D. Roberts, *Physical Review* **71**, 022101 (2005).
10. F. Dupuis, N. Gisin, A. Hassidim, A. A. Methot, and H. Pilpel, *arXiv:quant-ph/0701142* (2007).
11. W. van Dam, *arXiv:quant-ph/0501159* (2005).
12. M. Forster, and S. Wolf, *arXiv:quant-ph/0808.0651v1* (2008).

CHAPTER 4

QUANTUM CHANNELS

A Lossy Bosonic Quantum Channel with Non-Markovian Memory

Oleg V. Pilyavets[*,†] and Vadim G. Zborovskii and Stefano Mancini[†]

[*]*P. N. Lebedev Physical Institute, Leninskii Prospect 53, Moscow 119991, Russia*
[†]*Dipartimento di Fisica, Università di Camerino, I-62032 Camerino, Italy*

Abstract. We provide a simple model to study memory effects in a lossy bosonic quantum channel over arbitrary number of uses. The noise correlation among different uses is introduced by contiguous modes interactions which results in an exponential decay of the correlations over channel uses (modes). This model allows us to characterize the asymptotic behavior of the channel for classical information transmission. This submission is based on the work [1].

Keywords: Gaussian quantum channel, entanglement, transmission rates
PACS: 03.67.Hk, 03.65.Yz, 42.50.Dv, 03.67.Mn

LOSSY BOSONIC CHANNELS

The model for a lossy bosonic channel $\mathscr{L}$ is depicted in Fig.1. The covariance matrices for input state, environment state and output state are labeled by V_{in}, V_{env} and V_{out}, respectively. V_{cl} is covariance matrix for classical information encoding in input state using random displacement. An output of the channel avaraged over encoding V_{cl} is labeled by $\overline{V}_{\text{out}}$.

To introduce the memory effect among different channel uses, we consider the environment modes initially in a correlated state, actually a multimode squeezed vacuum. We are going to use multimode squeezed states depending on a single parameter, so that we choose the covariance matrix V_{env} as

$$V_{\text{env}} = \frac{1}{2}\begin{bmatrix} \exp(s\Omega) & 0 \\ 0 & \exp(-s\Omega) \end{bmatrix}. \tag{1}$$

Here, the $n \times n$ matrix Ω is taken to be

$$\Omega = \begin{pmatrix} 0 & 1 & \cdots\cdots\cdots & & & 0 \\ 1 & 0 & 1 & \cdots\cdots & & 0 \\ \vdots & 1 & 0 & 1 & .. & 0 \\ \vdots & \vdots & \ddots & \ddots & \ddots & \\ \vdots & \vdots & & \ddots & \ddots & 1 \\ 0 & 0 & \cdots\cdots & & 1 & 0 \end{pmatrix}, \tag{2}$$

and the parameter $s \in \mathbb{R}$ represents the memory strength. Notice that the memory effect is symmetric among all modes and decays over the number of uses. Actually,

CP1110, *Quantum Communication, Measurement and Computing (QCMC), The Ninth International Conference,* edited by A. Lvovsky

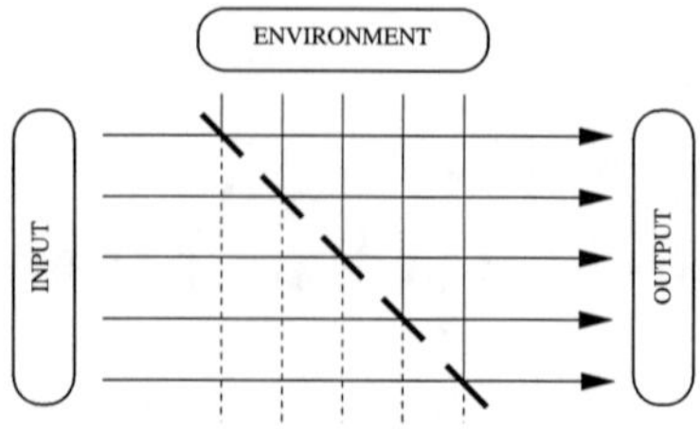

FIGURE 1. The model for a lossy bosonic channel $\mathscr{L}$. Each input mode (left-right line), representing one use of the channel, interacts with the corresponding environment mode (top-bottom line) through a beam-splitter with transmittivity η. To introduce memory effects, environment modes are initially considered in a correlated state.

if we consider the set of V_{env} matrix elements belonging to a fixed row as a discrete function, this is well fitted by a Gaussian having width $2\sqrt{|s|}$. Such exponential decay of correlations makes the noise in the channel non-Markovian and the channel not forgetful.

Since we would explore the usefulness of entangled inputs in the presence of memory, we then consider input state having a covariance matrix similar to that of Eq.(1), that is,

$$V_{\text{in}} = \frac{1}{2}\begin{bmatrix} \exp(r\Omega) & 0 \\ 0 & \exp(-r\Omega) \end{bmatrix}, \tag{3}$$

where $r \in \mathbb{R}$ is the entanglement parameter. To respect the input energy constraint, the added number of photons per mode by entanglement

$$\frac{1}{2n}\text{Tr}(V_{\text{in}}) - \frac{1}{2},$$

must necessarily be a fraction of N, say $\theta_n N$ with $0 \leq \theta_n \leq 1$. This limits the range of r to a finite interval of $\mathbb{R}$.

Besides entanglement in input, we could introduce classical correlations as well through V_{cl}. Since entanglement adds a fraction $\theta_n N$ of photons per mode, the diagonal of V_{cl} must contain a fraction $(1-\theta_n)N$ of photons, while off diagonal could introduce correlations. We choose the covariance matrix V_{cl} describing classical noise to be of the same form of V_{env} and V_{in}, that is,

$$V_{\text{cl}} := \frac{2nN(1-\theta_n)}{\text{Tr}(Y)} Y, \tag{4}$$

where

$$Y = \frac{1}{2}\begin{bmatrix} \exp(y\Omega) & 0 \\ 0 & \exp(-y\Omega) \end{bmatrix},$$

with the parameter $y \in \mathbb{R}$ describing classical correlations.

CLASSICAL CAPACITY AND RATES

We assume that maximization of Holevo-χ overall possible input states and encodings V_{cl} taking them to be Gaussian allows us to reach capacity. By applying the model of

Sec. we get the following result for the classical capacity of asymptotic behavior of the channel:

$$C := \lim_{n\to\infty} C_n = \max_{r,y\in\mathbb{R}} \frac{1}{\pi}\int_0^{\pi}\left[g\left(\left|\nu^{(\overline{V}^{\text{out}})}\right| - \frac{1}{2}\right) - g\left(\left|\nu^{(V^{\text{out}})}\right| - \frac{1}{2}\right)\right] d\xi, \tag{5}$$

where $\nu^{(V^{\text{out}})}$ and $\nu^{(\overline{V}^{\text{out}})}$ are given by equations:

$$\nu_k^{(V^{\text{out}})} = \frac{1}{2}\left[\eta^2 + (1-\eta)^2 + 2\eta\,(1-\eta)\cosh\varphi_{sr}\right]^{1/2}, \tag{6}$$

$$\begin{aligned}\nu_k^{(\overline{V}^{\text{out}})} &= \frac{1}{2}\left[\eta^2 + (1-\eta)^2 + \eta^2 K^2 + 2\eta\,(1-\eta)\cosh\varphi_{sr}\right. \\ &+ \left. 2\eta^2 K\cosh\varphi_{ry} + 2\eta(1-\eta)K\cosh\varphi_{sy}\right]^{1/2},\end{aligned} \tag{7}$$

where

$$\varphi_{lm} := 2\,(l-m)\cos\xi,\;\; l,m = r,y,s,$$

$$K = \frac{2(N+1/2) - I_0(2r)}{I_0(2y)}. \tag{8}$$

I_0 denotes Bessel function of the first kind and zero order. The function $g(x)$ is defined as

$$g(x) = (x+1)\log_2(x+1) - x\log_2 x.$$

We are now going to consider information transmission rates by conventional decoding procedures such as heterodyne and homodyne measurements. Mutual information for heterodyne case is given by relation

$$I[Z:A] = \frac{1}{2}\log_2\left|\left(\overline{V}_{\text{out}} + \frac{1}{2}\right)\left(V_{\text{out}} + \frac{1}{2}\right)^{-1}\right|. \tag{9}$$

and for homodyne case is given by relation

$$I[\Re Z:\Re A] = \frac{1}{2}\log_2\left|\left(\overline{V}_{\text{out}}^{(11)}\right)\left(V_{\text{out}}^{(11)}\right)^{-1}\right|. \tag{10}$$

By using the model of Sec. we explicitly get the rate for asymptotic behavior of the channel:

$$F := \lim_{n\to\infty} F_n = \max_{r,y\in\mathbb{R}} \frac{\flat}{\pi}\int_0^{\pi}\log_2\left[1 + \flat^{-1}K\frac{\eta\, e^{2y\cos\xi}}{\eta\, e^{2r\cos\xi} + (1-\eta)\, e^{2s\cos\xi} + (2\flat - 1)}\right] d\xi.$$

Here, $\flat = 1$ and $\flat = 1/2$ give heterodyne and homodyne rates, respectively.

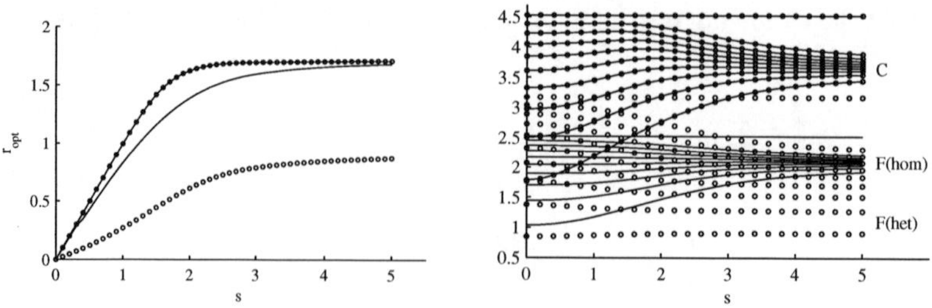

FIGURE 2. On the left, the optimal r value for the quantities C (solid line with circles), F ($\flat = 1$ line with circles only), and F ($\flat = 1/2$ pure solid line) is shown versus s. The value of the other parameters are $N = 8, \eta = 0.7$. On the right, the quantity C (solid lines with circles), F ($\flat = 1$ lines with circles only) and F ($\flat = 1/2$ pure solid lines) are plotted versus s for values of η going from 0.1 (bottom curve) to 1 (top curve) with step 0.1. The value of the other parameter is $N = 8$.

CONCLUSION

We have shown the usefulness of entangled inputs in the presence of memory. In particular, we have shown that they can enhance the classical capacity with respect to the memoryless case. This has been done by maximizing the Holevo-χ quantity over a class of Gaussian states. We have also shown the utility of entangled inputs for rates achievable by conventional decoding procedures (heterodyne and homodyne measurements), although for most interesting from physical point of view cases these do not exceed the classical capacity of the memoryless channel.

The usefulness of entangled inputs is summarized in Fig. 2 (left) where we can see that the optimal entanglement degree smoothly varies with the degree of memory. Still, in Fig. 2 (right) we can see the enhancement of the classical capacity in the presence of memory for channels characterized by different values of transmittivity η. We also notice that for increasing values of s, curves corresponding to different values of η (except $\eta = 1$) flow together towards a single value. This is presumably due to the fact of saturation of the optimal r value caused by energy restriction at the input of channel.

Also, in Fig.2-right we can see the rates achievable by heterodyne and homodyne measurements with their behavior dependent of choosed values of η. These types of measurements reflects the capacity behavior each tending to the same own limit rate value for all values of η taking big values of s. One can see the possibility to overcome memoryless channel capacity using entangled inputs for rates if transmittivity η is very small and values of s are big (e.g., a channel with $s = 5$ and $N = 8$ allows to overcome its memoryless capacity using heterodyne measurements if its $\eta \lesssim 0.02$ and using homodyne measurements if its $\eta \lesssim 0.1$). In addition to one can see that effect of memory in the channel improves heterodyne rates if $\eta \lesssim 0.1$ and homodyne rates if $\eta \lesssim 0.5$.

REFERENCES

1. O. V. Pilyavets, V. G. Zborovskii, and S. Mancini, *Phys. Rev.* A **77**, 052324 (2008).

The Algebraic Measure of a Hidden Markov Quantum Memory Channel

I. Akhalwaya*, J. Wouters†, M. Fannes† and F. Petruccione*,**

*Quantum Research Group, School of Physics, Westville, University of KwaZulu-Natal, Private Bag X54001, Durban, 4000, South Africa
†Instituut voor Theoretische Fysica, Katholieke Universiteit Leuven, B-3001 Heverlee, Belgium
**National Institute for Theoretical Physics, UKZN, South Africa

Abstract. The classical product state capacity of a noisy quantum channel with memory is investigated. A forgetful noise-memory channel is constructed by Markov switching between two depolarizing channels which introduces non-Markovian noise correlations between successive channel uses. This function of a Markov process can be reformulated as an algebraic measure. This framework provides an expression for the asymptotic entropy rate and thus enables the calculation of the classical capacity. The effects of the hidden-Markovian memory on the capacity are explored. An increase in noise-correlations is found to increase the capacity.

Keywords: Quantum Information, Forgetful Channels, Classical Capacity, Noise Memory, Markov
PACS: 03.67.Hk, 89.70.Kn

INTRODUCTION

Great strides have been made in understanding the capacity of quantum channels. For example, the celebrated Holevo-Schumacher-Westmoreland (HSW) theorem [1] gives an expression for the classical capacity of a noisy quantum channel with product state inputs. The memoryless channel restriction has since been extended to, so called, *forgetful* memory channels [2]. The inclusion of memory is the next step in the attempt of accurately modelling the complicated noise-correlated real world.

We construct a forgetful channel and incorporate memory effects by Markov switching between two sub-channels. We investigate the classical product state capacity of this channel. This involves the entropy of the classical output of the channel. To manage this complicated conditional dependence, we use the hidden Markov nature of the process to reformulate the problem using the algebraic measure construction [3]. The algebraic measure approach allows us to derive an expression for the asymptotic entropy rate. We then explore the effects that our non-Markovian memory has on the classical product state capacity.

CONSTRUCTION OF THE CHANNEL

The forgetful channel is constructed by combining two memoryless single qubit depolarizing channels ($\mathscr{E}_0$ and $\mathscr{E}_1$), switching between them using a two-state Markov chain ($Q = (q_{ij}), \quad i, j \in \{0, 1\}$). Thus, Q is the 2×2 Markovian channel selection matrix with q_{ij} being the probability of switching from channel i to channel j.

CP1110, *Quantum Communication, Measurement and Computing (QCMC), The Ninth International Conference,* edited by A. Lvovsky

The depolarizing channels can be written as: $\mathscr{E}_i(\rho) = x_i^0\rho + x_i^1(\mathbf{1}-\rho)$. These single qubit channels can be thought of as probabilistically mixing the identity channel (with probability x_i^0) and 'flip' channel (with probability $x_i^1 = 1 - x_i^0$) acting on a single qubit density operator ρ. However this rewriting is only completely positive for $1/3 \leq x_i^0 \leq 1$.

The built-up channel corresponding to n successive uses of the single qubit sub-channels, Λ_n, is constructed as follows:

$$\Lambda_n = \rho_1 \otimes \ldots \otimes \rho_n \mapsto \sum_{i_1,\ldots,i_n} \gamma_{i_1} q_{i_1 i_2} \cdots q_{i_{n-1} i_n} \mathscr{E}_{i_1}(\rho_1) \otimes \ldots \otimes \mathscr{E}_{i_n}(\rho_n) .$$

The sum is over all possible Markov paths $(i_1,\ldots,i_n) \in \{0,1\}^n$ and each term is a tensor product of the selected sub-channels weighted by the probability of occurrence (γ_i is the initial probability of selection). We calculate the capacity with this n-use form of the channel and then regularise to take the limit as n goes to infinity. The HSW theorem [1] built on by the forgetful channel extension [2] gives us an expression for the capacity $C^1_{\text{Classical}}$ (classical product state):

$$C^1_{\text{Classical}}(\Lambda_n) = \chi^*(\Lambda_n) ,$$

where χ^* is the maximisation over all possible input ensembles (effectively only pure states) of the Holevo χ quantity.

For this channel, the maximum is obtained using the uniformly distributed computational basis states [5]. By taking the asymptotic average of the n-product classical capacity, we arrive at the capacity of the memory channel [2]:

$$C^* = \lim_{n\to\infty} \frac{1}{n} C^1_{\text{Classical}}(\Lambda_n)) = 1 - \lim_{n\to\infty} \frac{1}{n} S(\Lambda_n(\rho)) . \tag{1}$$

ALGEBRAIC MEASURE OF THE CHANNEL

An algebraic measure, μ, is a translational-invariant measure on a set $\{0,\ldots,q-1\}^{\mathbb{Z}}$, with probabilities determined by matrices E_a with positive entries, one for each of the q states. The probability of a sequence is obtained by applying a positive linear functional σ to a matrix product of the corresponding matrices of the states of the sequence: $\mu(i_1,\ldots,i_n) = \sigma(E_{i_1}\ldots E_{i_n})$. This matrix algebraic construction is the reason for the name *Algebraic Measure*, studied in detail in Ref. [3]. Every measure that is a function of a Markov process (also called hidden Markov) is an algebraic measure and remarkably, the converse holds too. The relationship between the hidden Markov measure, say μ' on $K^{\mathbb{Z}}$, and the underlying Markov measure ν with the Markov property on $L^{\mathbb{Z}}$ is through a 'tracing' function $\Phi : L \to K$, as follows (L and K are finite sets of states):

$$\mu'((\omega_m,\ldots,\omega_n)) = \sum_{\substack{\varepsilon_m,\ldots,\varepsilon_n \\ \Phi(\varepsilon_m)=\omega_m \ldots \Phi(\varepsilon_n)=\omega_n}} \nu((\varepsilon_m,\ldots,\varepsilon_n)) ,$$

where $\omega_m,\ldots,\omega_n \in K$ and $\varepsilon_m,\ldots,\varepsilon_n \in L$.

The underlying Markov process for the overall quantum channel has a four state configuration space corresponding to channel selection and error occurrence: $K =$

$\{(00),(01),(10),(11)\}$. The first index indicates which depolarizing channel has been chosen and the second indicates whether a bit flip occurred. The transition matrix, E, for this process is then given by $E_{(ij)(i'j')} = q_{ii'}x_{i'}^{j'}$: the probability of going from (ij) to $(i'j')$ is given by the switching probability from channel i to i', $q_{ii'}$, multiplied by the probability that channel i' produces the error-configuration j', $x_{i'}^{j'}$.

The function that produces the correct hidden Markov process is then given by

$$\Phi((i,j)) = j\,.$$

This function reflects the fact that we are unaware of the choice of channel that has been made. The only effect that is visible from the outside is whether or not an input qubit has been flipped. Thus, Φ has to 'trace out' the choice of channel. Φ maps into the two-state error configuration space containing 'no flip' and 'flip': $L = \{0,1\}$.

With the successful construction of the algebraic measure in terms of the traced out matrices, we try to calculate the asymptotic entropy rate. In [3, 4], an expression for this rate was derived:

$$s(\mu) = \sum_{a\in L}\int_{\mathscr{B}} \phi(d\nu)h_a(\nu)\,. \tag{2}$$

where $\mathscr{B}$ is the set of probability measures on the set of q states and h_a is an entropy-like function. The measure ϕ is the fixed point of a transformation T_μ determined by the matrices of the algebraic measure E_a. We have used this transformation to generate ϕ iteratively. The iterative procedure can be simulated numerically allowing us to approximate the entropy and thus the capacity.

RESULTS

In order to simplify interpretation, we symmetrise the Markov switching process by the substitution $q_{00} \to (1+s)/2$, $q_{10} \to (1-s)/2$. We parametrize the error probabilities by their average and difference: $x_0^0 \to a+d$, $x_1^0 \to a-d$. By defining a as the average probability of no-error of the two sub-channels and studying the difference, d, we are pre-empting the result that it is the average and difference of the two sub-channels that are important to the capacity and not the absolute probabilities.

The main result is that the capacity increases with stronger noise-correlations which manifests itself in two ways. Firstly, if we make the switching more correlated (s away from 0) the capacity increases and secondly, if we increase the difference between the two sub-channels the capacity also increases.

In Figure 1, d is set to the maximum possible value while keeping an average of a ($d = \min[a-1/3, 1-a]$). The capacity is plotted against varying a and s. We can see that the capacity increases as the noise-correlation (s) gets stronger. When $a = 2/3$, d attains its maximum $(1/3)$ and the effect of increasing s on the capacity is greatest. Another interesting observation is the case when the two sub-channels average to the maximally mixing channel ($a = 1/2$, ordinarily with zero capacity), taking into account memory effects there is a non-zero capacity.

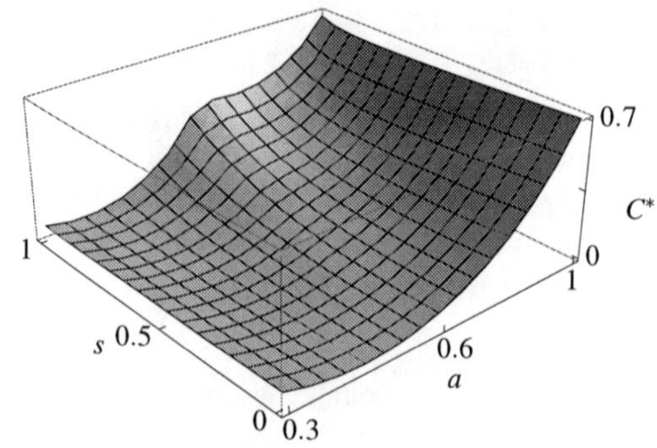

FIGURE 1. Noisy-on-average but Distinct allows increase in Capacity

CONCLUSION

We have constructed a simple forgetful noise-memory quantum channel. The noise-correlation is a function of the underlying hidden Markov process which allowed us to construct the algebraic measure. We used the measure in an algebraic asymptotic entropy expression. Without this, the entropy would be very difficult to compute, involving exponentially many Markov paths.

We studied the effects that the noise correlations had on the classical capacity and discovered that the capacity increases with stronger correlations. This is sensible because the correlations can be used to combat the noise when coding information.

ACKNOWLEDGMENTS

We would like to acknowledge N. Datta and T. Dorlas for the idea of the channel construction and valuable assitance. This work is based upon research supported by the South African Research Chair Initiative of the Department of Science and Technology and National Reseach Foundation.

REFERENCES

1. A.S. Holevo, *The Capacity of the Quantum Channel with General Signal States*, IEEE Trans. Inform. Theory 44, 269; 1998; B. Schumacher and M.D. Westmoreland, *Sending classical information via noisy quantum channels*, Phys. Rev. A, 56, 131; 1997.
2. D. Kretschmann and R.F. Werner, *Quantum Channels with Memory*, Phys. Rev. A, 72(6):62323, 2005.
3. M. Fannes, B. Nachtergaele, and L. Slegers, *Functions of Markov processes and algebraic measures*, Rev. Math. Phys. 4, 39; 1992.
4. D. Blackwell, *The Entropy of Functions of Finite State Markov Chains*, Trans. First Prague Conference on Information Theory, Decision Functions, and Random Processes, Prague, 1957, 13-20.
5. C. King, *The Capacity of the Quantum Depolarizing Channel*, IEEE Transactions on Information Theory, 49(1), 2003.

Fidelity criterion for quantum-domain continuous variable channels

R. Namiki*, M. Koashi† and N. Imoto†

*Department of Physics, Graduate School of Science, Kyoto University, Kyoto 606-8502, Japan
†CREST Research Team for Photonic Quantum Information, Division of Materials Physics, Department of Materials Engineering Science, Graduate school of Engineering Science, Osaka University, Toyonaka, Osaka 560-8531, Japan

Abstract. We provide a general quantum benchmark for transmission/storage of coherent states in terms of the average fidelity. It leads to another benchmark described in terms of an average quadrature noise. While the fidelity criterion provides a sufficient condition for quantum domain of any continuous variable channel, for Gaussian channels, it is turn out to be necessary and sufficient.

Keywords: Quantum teleportation/memory, Quantum benchmark, Entanglement breaking channel
PACS: 03.67.Hk, 03.65.Ud, 03.67.Mn, 42.50.Xa

In quantum information science, the manipulation of quantum system is considered to be a channel that transforms a set of quantum states to another set of quantum states. A fundamental distinction is posed on the channel whether or not it can be simulated by a *measure-and-prepare* (M&P) scheme. The M&P scheme implies that the output of the channel is produced merely based on the classical data processing from the measurement outcomes and that the channel action breaks quantum entanglement shared between the input of the channel and other systems [1]. Therefore, a natural benchmark for the *quantum-domain* (QD) operation of a given experimental quantum manipulation is that the channel is outperforming any M&P scheme. Mathematically, a QD channel E implies that the unnormalized state given by Jamiolkowski isomorphism, $\hat{J}_{AB} \equiv 1_A \otimes E_B |\Psi\rangle\langle\Psi|$, is inseparable. Here, $|\Psi\rangle \equiv \sum_n |n\rangle_A |n\rangle_B$, and $\{|n\rangle\}$ is an orthonormal basis that spans the support of E. Physically, a QD channel implies that it can transmit a quantum correlation. This means storage of a quantum correlation when we consider the channel acting in the time direction such as quantum memory. If the state $\hat{J}_{AB}$ is separable, say $\hat{J}_{AB} = \sum_i |\phi_i\rangle\langle\phi_i| \otimes |\psi_i\rangle\langle\psi_i|$, we can write E as a M&P scheme explicitly, $E(\hat{\rho}) = \sum_i \langle\phi_i^*|\hat{\rho}|\phi_i^*\rangle |\psi_i\rangle\langle\psi_i|$ where $|\phi_i^*\rangle = \sum_n \langle\phi_i|n\rangle|n\rangle$. The M&P scheme is also referred to as the *entanglement breaking channel.*

Hammerer *et al.* [2] have established the criterion for the QD operation of continuous-variable (CV) channels by proving a limit of the average fidelity achievable by the M&P schemes, F_c, assuming an input set of coherent states: Surpassing F_c ensures the QD operation for transmission and storage of coherent states. This criterion gave a proof for the long-standing conjecture of CV quantum teleportation about F_c [3, 4] and provided a firm foundation for the central claims of experimental CV quantum teleportation [4, 5, 6] and quantum memory [7]. However, the application of this criterion is limited to the *unit-gain* (UG) channels where the coherent-state amplitudes of the input state and the output state are expected to be the same [6, 8]. Recently, this fidelity criterion was generalized

CP1110, *Quantum Communication, Measurement and Computing (QCMC), The Ninth International Conference,* edited by A. Lvovsky

to be suitable for a non-unit-gain condition by considering attenuation/amplification of the coherent-state amplitude, and it was also shown that the framework to measure the average fidelity is capable of detecting all Gaussian channel in QD [9]. In this report we rewrite the criterion for quantum-domain channels given in Ref. [9] rather in the form of mathematical theorem on quantum channels. A criterion for quantum-domain channels is simultaneously a inseparable criterion on their isomorphism states $\hat{J}_{AB}$.

We consider a transformation task that modulates the amplitude of the coherent states by the factor of $\eta > 0$, $\{|\alpha\rangle\} \to \{|\sqrt{\eta}\alpha\rangle\}$. We assume that the prior distribution of coherent states is a symmetric Gaussian function

$$p_\lambda(\alpha) = \frac{\lambda}{\pi}\exp(-\lambda|\alpha|^2). \tag{1}$$

The average fidelity of a channel E with respect to the task is defined as

$$F(E,\eta,\lambda) \equiv \int d^2\alpha p_\lambda(\alpha)\langle\sqrt{\eta}\alpha|E\left(|\alpha\rangle\langle\alpha|\right)|\sqrt{\eta}\alpha\rangle. \tag{2}$$

The upperbound of the average fidelity achievable by the M&P scheme is [9]

$$F_c(\eta,\lambda) \equiv \sup_{\{\hat{M}_k\},\{\hat{\rho}_k\}} \int d^2\alpha p_\lambda(\alpha)\sum_k\langle\alpha|\hat{M}_k|\alpha\rangle\langle\sqrt{\eta}\alpha|\hat{\rho}_k|\sqrt{\eta}\alpha\rangle = \frac{1+\lambda}{1+\eta+\lambda} \tag{3}$$

where both of the measurement $\{\hat{M}_k\}$ and state preparation $\{\hat{\rho}_k\}$ are optimized. Hence, if $F(E,\eta,\lambda)$ surpasses $F_c(\eta,\lambda)$ one can verify the given channel E is in QD from the set of measured fidelity $\langle\sqrt{\eta}\alpha|E\left(|\alpha\rangle\langle\alpha|\right)|\sqrt{\eta}\alpha\rangle$. Note that the classical bound on the fidelity for a more general class of tasks $\{\hat{U}|\alpha\rangle\} \to \{\hat{V}|\sqrt{\eta}\alpha\rangle\}$ is also $F_c(\eta,\lambda)$ where $\hat{U}$ and $\hat{V}$ are any unitary operators. Thus, we can select $\hat{U}$ and $\hat{V}$ as well as η to ensure the QD operation. Note also that the criterion reduces to the UG case when $\eta = 1$ and we find the well-known classical limit-fidelity of $F_c(1,0) = 1/2$ with the uniform distribution $\lambda \to 0$. The UG condition $\eta = 1$ means that the channel is expected to be identical transformation or unitary process because the sets of states $\{\hat{U}|\alpha\rangle\},\{\hat{V}|\alpha\rangle\}$ are connected by an identity operation or unitary process as $\hat{U}|\alpha\rangle = (\hat{U}\hat{U}^\dagger\hat{V}^\dagger)\hat{V}|\alpha\rangle$ for any $|\alpha\rangle$. By introducing a parameter $\eta \neq 1$ we find a more direct quantum benchmark related to practical channels that includes essentially non-unitary processes. An extension for two-mode channels can be found in [11]. Assuming a transformation task for two non-orthogonal states, similar approach is given in [10].

Theorem 1.— Sufficient criterion for quantum domain:
For any entanglement breaking channel E, the average fidelity of E satisfies

$$F(E,\eta,\lambda) \leq \frac{1+\lambda}{1+\eta+\lambda} \tag{4}$$

where $\eta > 0$ and $\lambda > 0$. The equality is achieved by a double homodyne measurement and state preparation, which can be represented by the measurement operator $\hat{A}_\alpha \equiv |\frac{\sqrt{\eta}}{1+\lambda}\alpha\rangle\langle\alpha|/\sqrt{\pi}$. It gives the POVM element $\hat{M}_\alpha \equiv \hat{A}_\alpha^\dagger\hat{A}_\alpha = |\alpha\rangle\langle\alpha|/\pi$ and the corresponding state preparation $\hat{\rho}_\alpha \equiv \hat{A}_\alpha\hat{A}_\alpha^\dagger/\mathrm{Tr}(\hat{A}_\alpha\hat{A}_\alpha^\dagger) = |\frac{\sqrt{\eta}}{1+\lambda}\alpha\rangle\langle\frac{\sqrt{\eta}}{1+\lambda}\alpha|$.

Proof.— See Ref. [9] (This is the direct result of Eq. (3), and the equality can be confirmed by a direct substitution.). ■

In experiments, fidelity to a coherent state $|\alpha\rangle$ is directly determined by measuring the probability of photon detection after a displacement operation [9, 10]. In CV systems, homodyne measurement is commonly used and it might be useful to provide similar criterion in terms of the quadrature variances. Let us define the quadrature operators $\hat{x}_+ \equiv (\hat{a}+\hat{a}^\dagger)/\sqrt{2}$ and $\hat{x}_- \equiv (\hat{a}-\hat{a}^\dagger)/\sqrt{2}i$. Then we have the following criterion.

Corollary.— Sufficient for quantum domain with the average quadrature deviation: For any entanglement breaking channel E,

$$\bar{\delta} \geq 2(1-F_c(\eta,\lambda)) = \frac{2\eta}{1+\lambda+\eta} \tag{5}$$

holds where we defined the average quadrature mean-square deviation

$$\bar{\delta} = \int p_\lambda(\alpha)\mathrm{Tr}\left\{E\left(|\alpha\rangle\langle\alpha|\right)\left[(\hat{x}_+ - \sqrt{2\eta}\mathrm{Re}[\alpha])^2 + (\hat{x}_- - \sqrt{2\eta}\mathrm{Im}[\alpha])^2 - 1\right]\right\}d^2\alpha. \tag{6}$$

Proof.— Let us define the quadrature mean-square deviation from $\sqrt{\eta}\alpha$ by $V_\pm \equiv \langle\Delta\hat{x}_\pm^2\rangle + (\langle\hat{x}_\pm\rangle - \sqrt{\eta}\alpha_\pm)^2$. associated with the input $|\alpha\rangle$ and output $\hat{\rho}_\alpha \equiv E(|\alpha\rangle\langle\alpha|)$, where $\langle\cdot\rangle \equiv \mathrm{Tr}(\cdot\hat{\rho}_\alpha)$ and $\alpha_\pm \equiv \langle\alpha|\hat{x}_\pm|\alpha\rangle$. Since the sum of $V_\pm$ is $v \equiv V_+ + V_- = \mathrm{Tr}[(2\hat{a}^\dagger\hat{a}+1)\hat{D}(-\sqrt{\eta}\alpha)\hat{\rho}_\alpha\hat{D}^\dagger(-\sqrt{\eta}\alpha)]$ and $\mathrm{Tr}(\hat{a}^\dagger\hat{a}\hat{\rho}) \geq 1 - \mathrm{Tr}(|0\rangle\langle 0|\hat{\rho})$ for any state $\hat{\rho}$, we can bound $v \geq 3 - 2\langle\sqrt{\eta}\alpha|\hat{\rho}_\alpha|\sqrt{\eta}\alpha\rangle$. Averaging both sides over α with $p(\alpha)$, we find $\bar{\delta} = \bar{v} - 1 \geq 2(1-F(E,\eta,\lambda))$. Since $F(E,\eta,\lambda) \leq F_c(\eta,\lambda)$ for any entanglement breaking channel E from Theorem 1, we can see that $\bar{\delta} \geq 2(1-F_c(\eta,\lambda)) = 2\eta/(1+\lambda+\eta)$ holds if E is an entanglement breaking channel. ■

The quadrature operators are also important in the description of Gaussian states and Gaussian operations [12, 13, 14]. For the class of Gaussian channels, Theorem 1 becomes necessary and sufficient. In order to see this we introduce a Gaussian description:

Let us write the quadrature operators as a column vector $\hat{R} \equiv (\hat{x}_+, \hat{x}_-)^t$. The characteristic function of a density operator $\hat{\rho}$ is defined by $\phi(z) \equiv \mathrm{Tr}[\hat{\rho}\exp(i\hat{R}^t z)]$, where z is a real column vector. Then, $\hat{\rho}$ is a Gaussian state when it has the form $\phi(z) = \exp(id^t z - \frac{1}{2}z^t\gamma z)$. Here $d = \mathrm{Tr}(\hat{\rho}\hat{R})$ is the mean amplitude, and $\gamma = \mathrm{Tr}\hat{\rho}\,[\hat{R}\hat{R}^t + (\hat{R}\hat{R}^t)^t]/2 - dd^t$ is the correlation matrix [13, 14]. The physical requirement of the uncertainty relation is given by the condition $\gamma \geq (i/2)\Delta$ with $\Delta \equiv \begin{pmatrix} 0 & -1 \\ 1 & 0 \end{pmatrix}$. The Gaussian channel is a CPTP map that transforms the Gaussian states into the Gaussian states [13, 14]. Except for the freedom of the uniform displacements, any one-mode Gaussian channel can be described [14] by a pair of 2×2 matrices (K,M) that transforms γ and d as $\gamma' = K^t\gamma K + M$ and $d' = Kd$, respectively. In order that γ' is physical for any physical γ, $M \geq (i/2)(\Delta - K^t\Delta K)$ must be satisfied. We write the channel with (K,M) as $E_{K,M}$.

Theorem 2. — Necessary and Sufficient quantum domain criterion for Gaussian channel: A one-mode Gaussian channel E is in quantum domain iff there exist $\eta > 0$, $\lambda > 0$ and local unitary $\hat{U}$ and $\hat{V}$ such that

$$F(E',\eta,\lambda) \;>\; \frac{1+\lambda}{1+\eta+\lambda}. \tag{7}$$

where $E'(\hat{\rho}) \equiv \hat{V} E\left(\hat{U}\hat{\rho}\hat{U}^{\dagger}\right)\hat{V}^{\dagger}$ is a unitary equivalent channel of E.

Proof.— The sufficiency comes from Theorem 1. The necessity is proven by checking the inequality for the unitary-equivalent standard forms of the Gaussian channels [15]. With an appropriate choice of $\hat{U}$ and $\hat{V}$, the pair (K,M) of any one-mode QD Gaussian channel becomes either (i) $K_1 = I_2$, $M_1 = \begin{pmatrix} 1/2 & 0 \\ 0 & 0 \end{pmatrix}$ or (ii) $K_2 = \sqrt{\eta} I_2$, $M_2 = (\tilde{n} + |1-\eta|/2) I_2$ where I_2 is the identity matrix and $0 \le \tilde{n} < \min\{1,\eta\}$. By choosing the task $\{|\alpha\rangle\} \to \{|\alpha\rangle\}$, the fidelity of the channel (i) is $F(E_{K_1,M_1},1,\lambda) = [\det(\gamma+\gamma')]^{-1/2} = (2/3)^{1/2}$ where $\gamma = I_2/2$ is the correlation matrix of the coherent-state input. The fidelity is strictly higher than the classical boundary of $F_c(1,\lambda)$ if $\lambda < (2F(E_{K_1,M_1},1,\lambda) - 1)/(1-F(E_{K_1,M_1},1,\lambda)) \sim 3.45$. By choosing the task $\{|\alpha\rangle\} \to \{|\sqrt{\eta}\alpha\rangle\}$, the fidelity of the channel (ii) is $F(E_{K_2,M_2},\eta,\lambda) = [\det(\gamma+\gamma')]^{-1/2} = 2/(1+\eta+|1-\eta|+2\tilde{n})$, and $F(E_{K_2,M_2},\eta,\lambda) > F_c(\eta,\lambda)$ holds if $\tilde{n} < \min\{1,\eta\}$ and $\lambda < 2(\min\{1,\eta\} - \tilde{n})/(2\tilde{n} + |1-\eta| + \eta - 1)$. ■

In conclusion, we have provided a quantum benchmark for transmission/storage of coherent states described in terms of average fidelity (Theorem 1). It have lead to another benchmark described in terms of quadrature noise (Corollary). While the fidelity criterion provides a sufficient condition for quantum domain of any continuous variable channel, for Gaussian channels, it is turn out to be necessary and sufficient (Theorem 2). These results are directly applicable to the experimental success criterion for continuous-variable quantum devices with dissipation/amplification, such as storage of light for a quantum memory.

This work was supported by the Grant-in-Aid for the Global COE Program "The Next Generation of Physics, Spun from Universality and Emergence" from the Ministry of Education, Culture, Sports, Science and Technology (MEXT) of Japan. R.N. is supported by JSPS Research Fellowships for Young Scientists.

REFERENCES

1. M. Horodecki, P. W. Shor, and M. B. Ruskai, quant-ph/0301031, Rev. Math. Phys. **15**, 629-641 (2003).
2. K. Hammerer, M.M. Wolf, E.S. Polzik, and J.I. Cirac, Phys. Rev. Lett. **94,** 150503 (2005).
3. S. L. Braunstein, C.A. Fuchs, and J. Kimble, J. Mod. Opt **47,** 267 (2000).
4. A. Furusawa *et al.*, Science **282,** 706 (1998).
5. W.P. Bowen *et al.,* Phys. Rev. A **67,** 032302 (2003).
6. T. C. Zhang *et al.,* Phys. Rev. A **67,** 033802 (2003).
7. B. Julsgaard *et al.,* Nature 432, 482 (2004).
8. W.P. Bowen *et al.,* IEEE J. Sel. Top. Quantum Electron. **9**, 1519, (2003).
9. R. Namiki, M. Koashi, and N. Imoto, Phys. Rev. Lett. **101**, 100502 (2008).
10. R. Namiki, Phys. Rev. A 78, 032333 (2008).
11. T. Takano, M. Fuyama, R. Namiki, and Y. Takahashi, Phys. Rev. A **78**, 010307(R) (2008).
12. S. L. Braunstein, and P. van Loock, Rev. Mod. Phys. **77,** 513 (2005).
13. A.S. Holevo and R.F. Werner, Phys. Rev. A **63,** 032312 (2001); J. Eisert and M.B. Plenio, Int. J. Quant. Inf. **1,** 479 (2003).
14. A.S. Holevo, Prob. Info. Trans. **43** 1-11 (2007), quant-ph/0607051; F. Caruso, V. Giovannetti, and A.S. Holevo, New J. Phys. **8,** 310 (2006).
15. A.S. Holevo, ArXiv:0802.0235 [quant-ph].

CHAPTER 5

CLUSTER STATES

Generation of Four-Mode Continuous-Variable Cluster States

R. Ukai[*,†], M. Yukawa[*,†], S.C. Armstrong[*,**], J. Yoshikawa[*,†], P. van Loock[‡] and A. Furusawa[*,†]

[*]*Department of Applied Physics, School of Engineering, The University of Tokyo, 7-3-1 Hongo, Bunkyo-ku, Tokyo 113-8656, Japan*
[†]*CREST, Japan Science and Technology (JST) Agency, 5 Sanbancho, Chiyoda-ku, Tokyo 102-0075, Japan*
[**]*Laser Physics Centre, Research School of Physical Sciences and Engineering, Australian National University, Canberra, ACT 0200, Australia*
[‡]*Optical Quantum Information Theory Group, Institute of Theoretical Physics I and Max-Planck Research Group, Institute of Optics, Information and Photonics, Universität Erlangen-Nürnberg, Staudtstr. 7/B2, 91058 Erlangen, Germany*

Abstract.
Cluster states are sufficient resources for realizing quantum computation. Their implementations can be achieved via either discrete-variable systems (especially qubit systems) or continuous-variable systems. Here we report on the experimental generation of an important example of a continuous-variable cluster state, a four-mode linear cluster state.

Keywords: entanglements, cluster states
PACS: 03.67.Lx, 03.65.Ud

INTRODUCTION

The one-way quantum computation model [1, 2] is a conceptually interesting alternative to the standard circuit model. In this model, a special type of entanglement in form of a cluster state is used. Once a suitable cluster state is prepared, one can realize arbitrary unitary operations by only using single-mode measurements and feed-forward. While cluster states and cluster computations were first proposed for discrete-variable (DV) systems, the cluster-state approach was then also applied to continuous-variable (CV) systems through CV cluster computation [3, 4], relying upon CV cluster states [5]. There are already some experimental reports on small-scale DV cluster computations [6] and preparations of CV cluster states [7, 8]. In this paper, we shall first describe the link between DV and CV cluster states; then we will discuss our experimental preparation of CV four-mode cluster states.

DV CLUSTER STATES

Cluster states in DV systems are multi-qubit entangled states which can be expressed by using graphs. FIG. 1 shows several simple examples of cluster states; (a), (b), and (c) are called a three-qubit linear cluster state, a four-qubit linear cluster state, and a four-qubit

CP1110, *Quantum Communication, Measurement and Computing (QCMC), The Ninth International Conference,* edited by A. Lvovsky

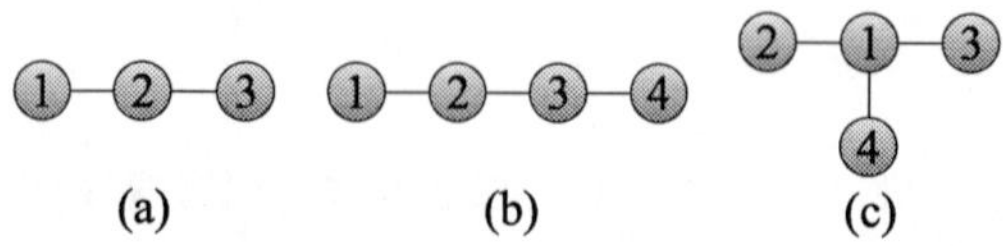

FIGURE 1. Examples of cluster states. In DV systems, (a), (b), and (c) are called a three-qubit linear cluster state, a four-qubit linear cluster state, and a four-qubit T-shape cluster state, respectively; in CV systems, these are referred to as a three-mode linear cluster state, a four-mode linear cluster state, and a four-mode T-shape cluster state, respectively.

T-shape cluster state, respectively. In this representation, each vertex corresponds a qubit and each edge represents an interaction between two qubits. Theoretically, these cluster states can be constructed in the following way[9]:

1. Prepare $|+\rangle^{\otimes n}$ states for an n-qubit cluster state.
2. Apply Controlled-Phase gate $\hat{C}_{Zij}$ between qubit-i and qubit-j if these two qubits are connected on its graph,

where $\hat{C}_{Zij}|x\ y\rangle = (-1)^{xy}|x\ y\rangle$ and $|\pm\rangle = \frac{1}{\sqrt{2}}(|0\rangle \pm |1\rangle)$ are the eigenstates of Pauli$-X$ operator $\hat{\sigma}_x$. For example, a three-qubit linear cluster state $|\psi_{3L}\rangle$ is

$$|\psi_{3L}\rangle = \hat{C}_{Z12}\hat{C}_{Z23}|+\rangle|+\rangle|+\rangle = \frac{1}{\sqrt{2}}(|+0+\rangle + |-1-\rangle) \tag{1}$$

Note that this state can be obtained by applying Hadamard gates on qubit-1 and qubit-3 of the GHZ state $|GHZ\rangle = \frac{1}{\sqrt{2}}(|000\rangle + |111\rangle)$.

Equivalently, cluster states may be represented via stabilizers. In general, a cluster state $|\Phi_C\rangle$ can be expressed as [1, 2]

$$\hat{\sigma}_x^{(a)} \bigotimes_{a' \in N(a)} \hat{\sigma}_z^{(a')} |\Phi_C\rangle = |\Phi_C\rangle \tag{2}$$

where $\hat{\sigma}_z$ is the Pauli$-Z$ operator and $N(a)$ is the set of nearest neighbor qubits of qubit a.

CV CLUSTER STATES

Similar to the DV case, cluster states in CV systems are multi-mode entangled states; they can also be represented using graphs. For example, FIG. 1 (a), (b), and (c) are refereed to as a three-mode linear cluster state, a four-mode linear cluster state, and a four-mode T-shape cluster state, respectively. In this case, each vertex corresponds to a mode and each edge represents an interaction between two modes. These CV cluster states can be built in the following way [10]:

1. Prepare $|p=0\rangle^{\otimes n}$ states for an n-mode cluster state.
2. Apply Controlled-Phase gate $\hat{C}_{Zij} = e^{2i\hat{x}_i\hat{x}_j}$ between mode-i and mode-j if these two modes are connected on its graph,

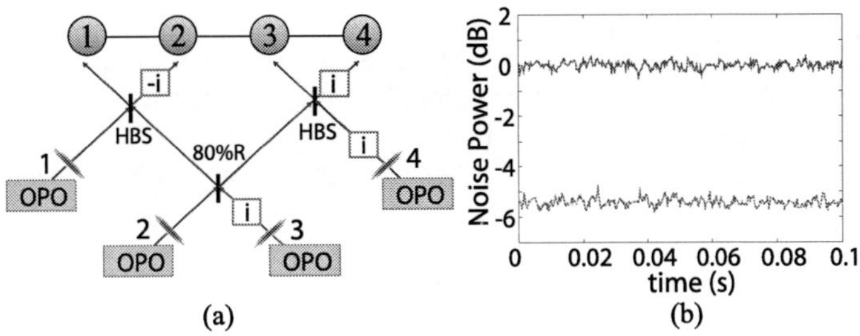

FIGURE 2. (a) is the schematic of generating a four-mode linear cluster state. Boxes including i and $-i$ are 90° and -90° rotations respectively. (b) is the measurement result of $\langle(\Delta(\hat{p}_1 - \hat{x}_2))^2\rangle$. The upper and lower lines are without squeezing and with squeezing, respectively. The measurement frequency is 1 MHz, resolution bandwidth is 30 kHz, and video bandwidth is 300Hz. All results are averaged 20 times.

where $|p=0\rangle$ is the eigenstate of the momentum operator $\hat{p}$ with eigenvalue zero. For example, the three-mode linear cluster state $|\psi_{3L}\rangle$ is

$$|\psi_{3L}\rangle = \hat{C}_{Z12}\hat{C}_{Z23}|p_1=0\rangle|p_2=0\rangle|p_3=0\rangle = \int_{-\infty}^{\infty}|p_1=a\rangle|x_2=a\rangle|p_3=a\rangle \mathrm{d}a \quad (3)$$

Note that this state can be obtained by applying Fourier gates on mode-1 and mode-3 of the GHZ state $|GHZ\rangle = \int_{-\infty}^{\infty}|x_1=a\rangle|x_2=a\rangle|x_3=a\rangle \mathrm{d}a$.

In the CV case, we may also use stabilizer operators. In general, a cluster state $|\Phi_C\rangle$ can be written as [10]

$$\hat{X}_a(s_a)\prod_{a'\in N(a)}\hat{Z}_{a'}(s_a)|\Phi_C\rangle = e^{-2is_a(\hat{p}_a - \Sigma_{a'\in N_a}\hat{x}_{a'})}|\Phi_C\rangle = |\Phi_C\rangle \quad (4)$$

where $\hat{X}_a(b)$ $\hat{Z}_a(b)$ are the position- and momentum-translation operators, respectively, and $N(a)$ is the set of nearest-neighbor modes of mode a.

For CV systems, it is convenient to describe the cluster states using the Heisenberg picture:

$$\hat{p}_a - \sum_{a'\in N_a}\hat{x}_{a'} = 0 \quad (5)$$

For finitely squeezed cluster states, the combinations of Eq.(5) have non-zero variances which cause errors when the states are used for cluster computation.

GENERATION OF CV FOUR-MODE CLUSTER STATES

Several ways of generating CV cluster states are known [10, 11]. We use the method of combining off-line squeezed light at beam splitters [10]. Experimentally, it is important to eliminate the antisqueezing components in the cluster states, as this reduces errors when the states are employed for quantum computation. This method of cluster generation can be used to create any types of multi-mode cluster states.

As examples, we generate three types of four-mode cluster states. FIG. 2(a) shows the schematic of generating a four-mode linear cluster state. A continuous-wave Ti:sapphire laser ($\lambda = 860$nm) is used as a laser resource. In order to generate four squeezed states, four optical parametric oscillators (OPOs) are pumped by second harmonic beams, which are generated in a cavity with a potassium niobate(KN) crystal. Each OPO contains a periodically poled $KTiOPO_4$(PPKTP) crystal. KN and PPKTP crystals are used as nonlinear media.

Four squeezed beams are combined at a 1:4 beam splitter and two half beam splitters. For this purpose, the relative phases of the two beams at the beam splitters must be properly adjusted.

The resulting cluster state is measured using homodyne detectors. The measurement results are summed up electronically and monitored by a spectrum analyzer.

FIG. 2(b) shows one of the measurement results and we obtain $\langle(\Delta(\hat{p}_1 - \hat{x}_2))^2\rangle = -5\ 4 \pm 0\ 2$ dB below the no-squeezing case. In the same way, we get $\langle(\Delta(\hat{p}_2 - \hat{x}_1 - \hat{x}_3))^2\rangle = -5\ 8$ dB, $\langle(\Delta(\hat{p}_3 - \hat{x}_2 - \hat{x}_4))^2\rangle = -5\ 3$ dB, and $\langle(\Delta(\hat{p}_4 - \hat{x}_3))^2\rangle = -5\ 8 \pm 0\ 2$. A sufficient condition for full inseparability [12] is

$$\left\{ \begin{array}{rcl} \langle(\Delta(\hat{p}_1 - \hat{x}_2))^2\rangle + \langle(\Delta(\hat{p}_2 - \hat{x}_1 - \hat{x}_3))^2\rangle & < & 1 \\ \langle(\Delta(\hat{p}_2 - \hat{x}_1 - \hat{x}_3))^2\rangle + \langle(\Delta(\hat{p}_3 - \hat{x}_2 - \hat{x}_4))^2\rangle & < & 1 \\ \langle(\Delta(\hat{p}_3 - \hat{x}_2 - \hat{x}_4))^2\rangle + \langle(\Delta(\hat{p}_4 - \hat{x}_3))^2\rangle & < & 1 \end{array} \right. \tag{6}$$

In our experiment, the values of the left-hand-sides are 0.34±0.02, 0.432±0.02, and 0.35±0.02, verifying that the generated state is fully inseparable. By changing the beam splitter transmittances and the relative phases of the beams, we also succeeded in creating a T-shape cluster state and a square-shape cluster state.

CONCLUSION

We briefly reviewed the definitions of DV and CV cluster states and their similarities. Then, the experimental generation of three types of four-mode cluster states was reported. Our method of creating Gaussian cluster states is universal and can be easily applied to the generation of more complicated and larger multi-mode cluster states.

REFERENCES

1. H. J. Briegel and R. Raussendorf, Phys. Rev. Lett. **86**, 910 (2001).
2. R. Raussendorf and H. J. Briegel, Phys. Rev. Lett. **86**, 5188 (2001).
3. N. C. Menicucci *et al.*, Phys. Rev. Lett. **97**, 110501 (2006).
4. P. van Loock, J. Opt. Soc. Am. B **24**, 340 (2007).
5. J. Zhang and S. L. Braunstein, Phys. Rev. A **73**, 032318 (2006).
6. P. Walther *et al.*, Nature (London) **434**, 169 (2005).
7. X. Su, A. Tan, X. Jia, J. Zhang, C. Xie, and K. Peng, Phys. Rev. Lett. **98**, 070502 (2007).
8. M. Yukawa, R. Ukai, P. van Loock, and A. Furusawa, Phys. Rev. A **78**, 012301 (2008).
9. M. A. Nielsen, Reports on Mathematical Physics **57**, 147 (2006).
10. P. van Loock, C. Weedbrook, and M. Gu, Phys. Rev. A **76**, 032321 (2007).
11. N. C. Menicucci, S. T. Flammia, H. Zaidi, and O. Pfister, Phys. Rev. A **76**, 010302(R) (2007).
12. P. van Loock and A. Furusawa, Phys. Rev. A **67**, 052315 (2003).

Fault-tolerant quantum computation in concatenation of verified cluster states

Keisuke Fujii and Katsuji Yamamoto

Department of Nuclear Engineering, Kyoto University, Kyoto 606-8501, Japan

Abstract. A novel scheme is presented for fault-tolerant quantum computation based on the cluster model. Some relevant logical cluster states are constructed in concatenation by post-selection through verification, without necessity of recovery operation, where a suitable code such as the Steane's 7-qubit code is adopted for transversal operations. This simple concatenated construction of verified cluster states achieves a high noise threshold $\sim 4\%$, and restrains the divergence of resources.

Keywords: fault-tolerant computation, one-way computation
PACS: 03.67.Lx, 03.67.Pp, 03.67.-a

In order to implement reliable computation in physical systems, the problem of noise should be overcome. Then, fault-tolerant quantum computation with error correction has been investigated [1, 2] (and references therein). A main motivation for QEC comes from the fact that in the circuit model the original qubits should be used throughout computation even if errors occur on them.

On the other hand, more robust computation may be performed in measurement-based quantum computers. An interesting computation model with error-correcting teleportation is proposed and provides a high noise threshold $\sim 3\%$ [2]. The cluster model or one-way computer [3] should also be considered for fault-tolerant computation [4].

In this work, we present a novel scheme of fault-tolerant quantum computation by making a better use of the unique feature of the cluster model [5]. Specifically, the fault-tolerant computation is implemented by *concatenated construction and verification of logical cluster states via one-way computation.* Although the high-fidelity preparation of ancilla states is adopted for the error-correcting teleportation [2], in the present scheme, even gate operations as cluster states are prepared and verified by post-selected computing. This can be viewed as correcting errors before the computation starts, say *error pre-correction.* This is quite distinct from the standard QEC, where errors are corrected after noisy operations, even via teleportation.

While high-fidelity state preparation is achieved by post-selection, huge resources are generally required due to the exponentially diminishing net success probability, which is a serious obstacle for scalability [2, 4]. We here succeed to overcome this dilemma in post-selection by presenting a systematic method of concatenation to construct logical cluster states through verification. As described in the following, the necessary post-selections are minimized and localized. Then, a high noise threshold $\sim 4\%$ is achieved by post-selection, while the resources usage is moderate, being comparable with or even less than the circuit-based QEC schemes. This concatenated cluster construction is implemented suitably by adopting a class of stabilizer codes of Calderbank-Shor-

CP1110, *Quantum Communication, Measurement and Computing (QCMC), The Ninth International Conference,* edited by A. Lvovsky

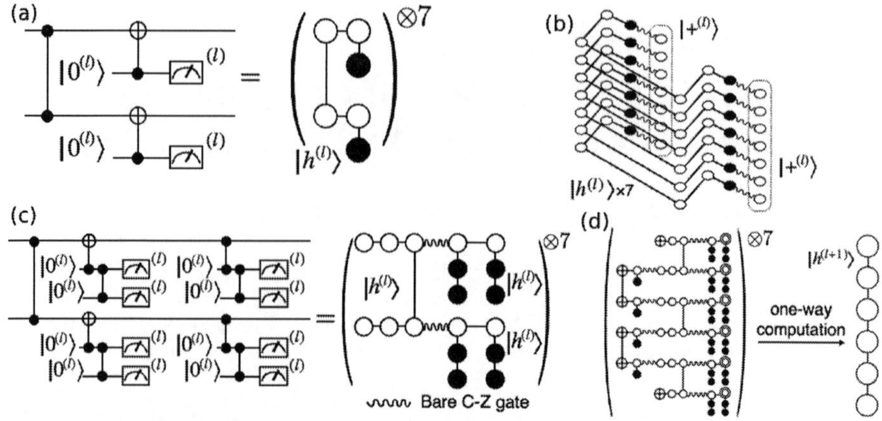

FIGURE 1. (a)C-Z gate with single verification at the level l. (b) Transfer of the ancilla level-l qubits, where $|+^{(l)}\rangle$ is connected with bare C-Z gates and transferred as $|0^{(l)}\rangle$ to • qubits in $|h^{(l)}\rangle$'s through one-bit teleportation. (c) C-Z gate with double verification at the level l. (d) The level-$(l+1)$ hexa-cluster construction with verifications through one-way computation.

Steane, e.g., the Steane's 7-qubit code. The logical measurements of Pauli operators as well as the Clifford gates, H, S and C-Z, are implemented transversally on such a quantum code. The non-Clifford π 8 gate is even operated for universal computation by preparing a specific qubit and making a transversal measurement [4].

(i) *Fundamental clusters:* A set of gate operations in one-way computation may be decomposed into fundamental clusters $|h^{(l)}\rangle$, $|0^{(l)}\rangle$, $|+^{(l)}\rangle$, which are composed of level-(l 1) qubits, to avoid divergence of resources in concatenation. The level-l qubits $|0^{(l)}\rangle$ and $|+^{(l)}\rangle$ are the l-th concatenated 7-qubit code states. The hexa-cluster $|h^{(l)}\rangle$, as a linear cluster of 6 level-(l 1)qubits, represents an elementary unit of gate operations. These fundamental clusters are combined by *bare* C-Z gates (transversal concatenation of physical C-Z gates without verification) to implement one-way computations for construction of the fundamental clusters of the next level.

(ii) *Verified C-Z gates:* A C-Z gate with single verification at the level-l is implemented by combining $7|h^{(l)}\rangle$'s and $2|+^{(l)}\rangle$'s, as shown in the right cluster diagram of Fig. 1 (a). This combination of clusters is schematically denoted by the symbol "⊗7" (henceforth used conveniently), which does not simply imply the tensor product but also includes the transfer of ancilla level-l qubits (marked with •) as shown in Fig 1 (b). As seen from the circuit equivalent in Fig. 1 (a), the error syndromes of the 2 level-l qubits through the C-Z gate are extracted for verification [1]. A C-Z gate with double verification is also implemented by combining 7×3 $|h^{(l)}\rangle$'s and $8|+^{(l)}\rangle$'s in Fig. 1 (c). Here, the errors in the ancilla $|+^{(l)}\rangle$'s ($|0^{(l)}\rangle$'s via teleportation) are even detected for higher fidelity. It should be remarked that at the beginning of concatenation the verified level-1 C-Z gates may be implemented efficiently by means of the circuit diagrams in Figs. 1 (a) and (c), without using $|h^{(1)}\rangle$'s. This is because $|h^{(1)}\rangle$'s, as chains of physical qubits without verification, are somewhat noisy.

(iii) *Concatenated cluster construction:* The level-$(l+1)$ hexa-cluster $|h^{(l+1)}\rangle$ is constructed in Fig. 1 (d) by combining the level-l clusters $|h^{(l)}\rangle$, $|0^{(l)}\rangle$, $|+^{(l)}\rangle$ with bare C-Z gates, where the $\oplus$ qubit includes the connection and transfer of $|0^{(l)}\rangle$ as $|+^{(l)}\rangle$ through one-bit teleportation similarly to the $\bullet$ qubit. These C-Z gates are combined in such ways that each qubit has at most one bare C-Z connection (wavy line), and that the output qubits ($\bigcirc$) as $|h^{(l+1)}\rangle$ are doubly verified. The level-$(l \quad 1)$ qubits, except $\bigcirc$'s, are measured to implement the computation. Then, if all the level-l syndromes are correct, the 6×7 level-$(l \quad 1)$ $\bigcirc$ qubits survive as the verified $|h^{(l+1)}\rangle$ passing the post-selection. That is, the level-l gate operation to be implemented with $|h^{(l+1)}\rangle$ has been verified beforehand. As noted previously, the level-2 construction may be implemented efficiently with the circuit diagrams in Figs. 1 (a) and (c) for the verified level-1 C-Z gates.

The level-1 $|0^{(1)}\rangle$ and $|+^{(1)}\rangle$ are prepared through verification by the stabilizer measurement and syndrome extraction with the usual method [1]. Then, the level-$(l+1)$ $|+^{(l+1)}\rangle$ and $|0^{(l+1)}\rangle$ are encoded with the level-l fundamental clusters via one-way computation in cluster diagrams, such as Fig. 1 (d) for the $|h^{(l+1)}\rangle$, which include suitably the verified C-Z gates (Figs. 1 (a) and (c)).

(iv) *Universal computation:* The fundamental clusters are constructed through verification up to the highest logical level $\bar{l}$ to achieve the fidelity required for a given computation size. Then, the desired Clifford computation is implemented by combining the highest-level hexa-clusters with bare C-Z gates. The preparation of $|\pi \quad 8\rangle = \cos(\pi \quad 8)|0\rangle + \sin(\pi \quad 8)|1\rangle$ is also needed for universal computation [4]. The $|\pi \quad 8\rangle$ is first encoded by the usual method [2] at the level 1, and concatenated up to the highest level $\bar{l}$ similarly to the $|+^{(l)}\rangle$. Then a high fidelity $|\pi \quad 8^{(\bar{l})}\rangle$ is obtained by using the magic state distillation with Clifford operations [6].

The errors on the qubits and Pauli frames should be considered properly to estimate the measurement errors and noise threshold.

(i) *Homogeneous errors on qubits:* The level-2 fundamental clusters are first constructed, and their constituent level-1 qubits are doubly verified. Then, it is reasonably expected that the level-0 qubits encoded in these verified level-1 qubits contain independently and identically distributed (homogeneous) depolarization errors in the leading order [7]. The homogeneous error probabilities ε_A $(A = X \ Y \ Z)$ of these level-0 qubits are determined essentially by the error probabilities p_{AB} of the physical gates which are used transversally for the level-1 double verification; $\varepsilon_X = p_{XI}$, $\varepsilon_Y = p_{YI}$, $\varepsilon_Z = 2p_{ZI}$ from the circuit diagram in Fig. 1 (c). In the level-$(l+1)$ construction, as described in Fig. 1 (d), *any operations are not implemented directly on the output* $\bigcirc$ *qubits, but the entanglement by the verified C-Z gates is transferred via one-way computation of the level-*$(l \quad 1)$ *qubits.* Hence, these output level-l qubits inherit transversally the homogeneous errors ε_A of the constituent level-0 qubits after the level-1 verification. The prepared level-$(l+1)$ clusters are further used for the level-$(l+2)$ construction, and some pairs of level-l qubits in these clusters are connected by bare C-Z gates. Then, extra errors are added to the constituent level-0 qubits through the bare C-Z gate as $\varepsilon'_X = \varepsilon_X + p_{XB}$, $\varepsilon'_Y = \varepsilon_Y + p_{YB}$, $\varepsilon'_Z = \varepsilon_Z + \varepsilon_X + \varepsilon_Y + p_{ZB}$ (summed over $B = I \ X \ Y \ Z$).

(ii) *Errors in measurements and threshold:* By the measurements of level-$(l \quad 1)$ qubits to construct the level-$(l+1)$ clusters, as in Fig. 1 (d), the level-$(l \quad 1)$ Pauli frames of the neighboring qubits are updated. The output level-l qubits to form the level-$(l+1)$ clusters are, however, doubly verified, and hence the propagation of the preceding measurement errors as Pauli frame errors is prohibited by post-selection. The fundamental clusters are therefore prepared to be free from the Pauli frame errors (up to the higher orders) through the concatenation. As a result, the error probability $p_q^{(l)}$ to measure solely a level-l qubit is reduced transversally to the level-0 $p_q^{(0)}$ on the 7-qubit code with distance 3 as

$$p_q^{(l)} \simeq {}_7\mathrm{C}_2 (p_q^{(l \ 1)})^2 \simeq ({}_7\mathrm{C}_2 p_q^{(0)})^{2^l} \ {}_7\mathrm{C}_2 \tag{1}$$

The level-0 qubits with the homogeneous errors ε'_A through the bare C-Z connection ($\varepsilon_A < \varepsilon'_A$) are measured in the X basis with the error probability $p_q^{(0)} = \varepsilon'_Z + \varepsilon'_Y + p_M$ (p_M is the error probability of the physical measurement). Then, the noise threshold is given from Eq. (1) as

$$p_q^{(0)} = Dp_e < 1 \ {}_7\mathrm{C}_2 \rightarrow p_{\mathrm{th}} = ({}_7\mathrm{C}_2 D) \ ^1 \tag{2}$$

where p_e represents the mean error probability of physical operations ($D \sim 1$). It is estimated as $p_{\mathrm{th}} = 0\ 042$ ($D = 17 \ 15$) typically with $p_{AB} = (1 \ 15)p_e$ for ε'_A and $p_M = (4 \ 15)p_e$ [2]. We have made a numerical simulation and confirmed the above estimates concerning the errors on the qubits and Pauli frames in the concatenation.

The resources required for the present scheme is calculated and compared with the conventional schemes [5]. The present scheme really consumes much less resources than the Steane's QEC scheme for $p_e \quad 10^{\ 3}$. We also find that compared with the C_4 C_6 scheme with post-selection (or with error-correction) [2], the present scheme provides a comparable threshold, requiring much less (or comparable) resources. The Fibonacci scheme such as C_4 C_6 based on the 4-qubit error-detecting code may be applied to improve especially the resources.

ACKNOWLEDGMENTS

This work was supported by JSPS Research Fellowships for Young Scientists No. 20 2157.

REFERENCES

1. A. M. Steane, Phys. Rev. Lett. **78**, 2252 (1997); A. M. Steane, Fortschr. Phys. **46**, 443 (1998); A. M. Steane, Nature **399**, 124 (1999); A. M. Steane, Phys. Rev. A **68**, 042322 (2003).
2. E. Knill, Nature **434**, 39 (2005); E. Knill, Phys. Rev. A **71**, 042322 (2005).
3. R. Raussendorf and H. J. Briegel, Phys. Rev. Lett. **86**, 5188 (2001); R. Raussendorf, D. E. Browne, and H. J. Briegel, Phys. Rev. A **68**, 022312 (2003);
4. K. Fujii and K. Yamamoto, *Proc. 8th Int. Conf. QCMC*, p. 105 (NICT Press 2007); M. Silva, V. Danos, E. Kashefi, and H. Ollivier, New J. Phys. **9**, 192 (2007).
5. K. Fujii and K. Yamamoto, arXiv:0802.4137 (2008).
6. S. Bravyi and A. Kitaev, Phys. Rev. A **71**, 022316 (2005).
7. B. Eastin, Phys. Rev. A **75**, 022301 (2007).

Two-photon multiqubit hyperentangled/cluster states

G. Vallone*, R. Ceccarelli*, A. Rossi*, F. De Martini*,† and P. Mataloni*

*Dipartimento di Fisica dell'Università "La Sapienza" and
Consorzio Nazionale Interuniversitario per le Scienze Fisiche della Materia, Roma, 00185 Italy
†Accademia Nazionale dei Lincei, via della Lungara 10, Roma 00165, Italy

Abstract. We describe the realization of a 6-qubit hyperentangled state by entangling two photons in three degrees of freedom. These correspond to the polarization, the longitudinal momentum and the indistinguishable emission produced by a 2-crystal system operating with Type I phase matching in the spontaneous parametric down conversion regime. A similar scheme involving a single crystal is also discussed.

Keywords: hyperentanglement, spontaneous parametric down conversion, multiqubit photon states
PACS: 03.67.Bg, 42.50.Dv, 42.65.Lm

INTRODUCTION

In the usual applications of quantum information, each photon carries one qubit of information encoded in a single degree of freedom (DOF), e.g. polarization, linear or transverse momentum and time-energy. However, by using more degrees of freedom for each photon, it's possible to encode more qubits in the same particle. If the particles are entangled in different degrees of freedom we commonly refere to them as *hyperentangled (HE) states* [1, 2, 3]. Increasing the number of qubits in a photon state is a fundamental issue for many quantum information (QI) processes: for example, more qubits we have more powerful will be the computation of a quantum computer. Moreover the violation of the local realism increases with the dimension of the Hilbert space.

We have developed a parametric source which enables the generation of a 4-qubit hyperentangled states of based on two photons that are entangled both in polarization and linear momentum (**k** modes) [2]. These states are easily transformed into high fidelity ($F = 0.880 \pm 0.013$) 2-photon 4-qubit *linear cluster states* [4], by entangling (with an half-wave plate) the polarization and momentum qubits of one of the two photons. By using this state we realized several algorithms [5, 6] in the framework of one-way quantum computation [7, 8]. Cluster states with more then 4 qubits can be obtained by adding more degrees of freedom for each photon. The first possibility is to use the time-energy entanglement by exploiting the uncertainty in the emission time of the photon pair. The second possibility is to consider more **k** emission modes of the Type-I crystal: by selecting, beside polarization, four possible **k** modes per photon, we can generate 6-qubit hyperentangled (or cluster) states. In this way, each photon encodes two qubits in the momentum and one qubit in the polarization DOF. We here discuss the new perspectives and the first results obtained in the generation of 2-photon 6-qubit hyperentangled states.

CP1110, *Quantum Communication, Measurement and Computing (QCMC), The Ninth International Conference,* edited by A. Lvovsky

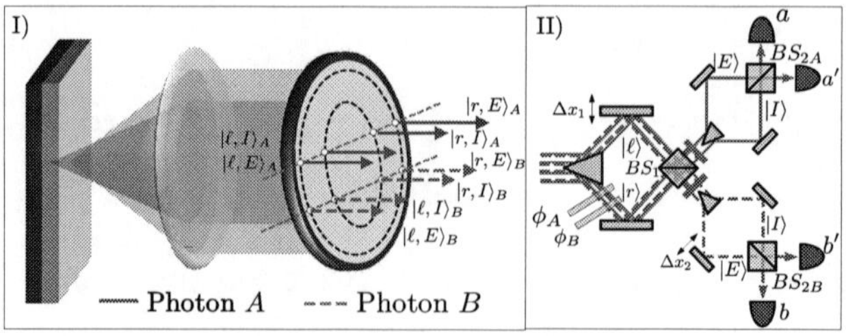

FIGURE 1. Experimental setup. I) The 2-crystal system generates photon pairs with equal probability over the internal (I) and external (E) emission cone, respectively corresponding to the two crystals. Polarization entanglement arises from double passage of the pump beam through the crystals and the combined action of a spherical mirror and a $\lambda/4$ wave-plate (not shown in the figure). The emission is selected by a single 8-hole screen. II) Measurement setup: the first interferometer (ending with BS_1) measures momentum entanglement in the basis $\frac{1}{\sqrt{2}}(|\ell\rangle + e^{i\phi_{A,B}}|r\rangle)$. The subsequent interferometers, corresponding to BS_{2A} and BS_{2B}, perform the measurement separately for the Alice and Bob photon in the basis $\frac{1}{\sqrt{2}}(|I\rangle \pm |E\rangle)$.

POLARIZATION-MULTIMODE ENTANGLEMENT

In this section we present two schemes for the experimental realization of a 6-qubit HE state based on the simultaneous triple entanglement of two photons.

Two-crystal scheme

In the first scheme we adopted a 2-crystal geometry to generate the 6-qubit HE state [9]. Besides polarization (π) entanglement, two photons experience a further double entanglement in the longitudinal momentum DOF. Precisely, two Type I β-barium-borate (BBO) crystal slabs, cut at different phase matching angles, (see Figure 1-I), aligned one behind the other, were operating under double excitation (back and forth) of a vertically (V) polarized continuous wave (CW) laser at wavelength λ_p. The polarization entangled Bell state $|\Phi^-\rangle = \frac{1}{\sqrt{2}}(|H\rangle_A|H\rangle_B - |V\rangle_A|V\rangle_B)$ was created by spatial superposing the two perpendicularly polarized emission cones of each Type I crystal, in agreement with a method already described by us in previous papers [2].

Photon pairs were created with equal probability by either one of BBO crystals. We refer to them as the "internal" (I) and the "external" (E) modes, corresponding to the first and the second crystal emission, respectively. Double longitudinal momentum entanglement was created by selecting four pairs of correlated modes with a 8-hole screen. It is worth to remember that only two photons, namely A (Alice) and B (Bob), are generated with equal probability by either one of the crystals. We label the corresponding mode emission as ℓ (r) by referring to the left (right) side of each cone and as I (E) by considering the internal (external) cone (see Figure 1-I)).

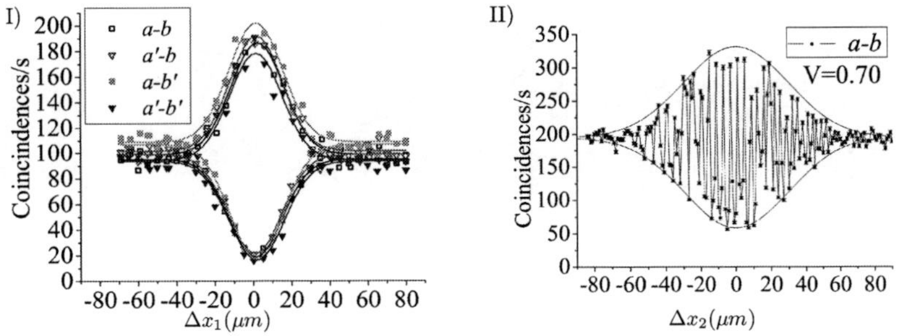

FIGURE 2. I) Coincidence rates a-b, a'-b, a-b', a'-b' vs Δx_1 for the state (1) showing the typical dip-peak feature with average visibility $V = 0.815 \pm 0.022$ (dip) and $V = 0.877 \pm 0.031$ (peak). The dip (peak) corresponds to a $-$ (+) relative sign between the state $|\ell\rangle_A|r\rangle_B$ and $|r\rangle_A|\ell\rangle_B$. The corresponding experimental fits are also shown. FWHM $\simeq 30nm$ of the interference patterns is in agreement with the expected value for a filter bandwidth of 6 nm. II) Coincidence rate a-b vs Δx_2 for the state (1) with visibility $V \simeq 0.70$ and FWHM $\simeq 60$ nm. The oscillations are due to the phase changing between the states $|I\rangle_A|I\rangle_B$ and $|E\rangle_A|E\rangle_B$. In both cases the time average of each experimental point is $3s$.

The generated state is given by the product of one polarization entangled state and two longitudinal momentum entangled states and is expressed as a 6-qubit HE state:

$$\frac{1}{\sqrt{2}}(|H\rangle_A|H\rangle_B + |V\rangle_A|V\rangle_B) \otimes \frac{1}{\sqrt{2}}(|\ell\rangle_A|r\rangle_B + |r\rangle_A|\ell\rangle_B) \otimes \frac{1}{\sqrt{2}}(|I\rangle_A|I\rangle_B + |E\rangle_A|E\rangle_B) \quad (1)$$

The measurement setup used to test the triple entanglement was given by the chained interferometric apparatus sketched in Figure 1-II). It is divided in two parts: the first one performs the measurement of the momentum entanglement existing within each emission cone. In the second part of the apparatus, measuring momentum entanglement between the two crystal emissions, the left (right) BS_1 output modes belonging to the Alice (Bob) photon are then injected into another interferometer where the I and E modes are matched onto the BS_{2A} (BS_{2B}) beam splitter. In Figure 2 we show the interference patterns obtained by varying the path delay on different arm.

We verified the presence of entanglement between the two photons for each separate DOF. This was done by tracing the other DOF's and evaluating separately an entanglement witness for π, **k** and **c** entanglement. Here we label as **k** and **c** the $\ell - r$ and $E - I$ DOF. The negative experimental values, $W_\pi = -0.6298 \pm 0.0080$, $W_{\mathbf{k}} = -0.7987 \pm 0.0055$ and $W_{\mathbf{c}} = -0.4101 \pm 0.0082$ demonstrate the entanglement in the three DOF's.

One-crystal scheme

We present here a different setup used to generate a double entanglement in the linear momentum DOF. It is based on the continuum **k**-emission of a single Type I crystal. By adding to this setup the a standard procedure to generate polarization entanglement [2], a 6-qubit hyperentangled state 1 can be generated. In the one-crystal scheme we select four

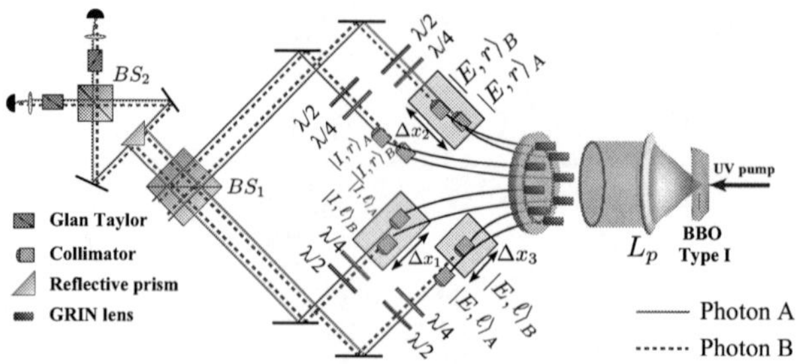

FIGURE 3. Scheme of the experimental setup used for the generation and measurement of the 8-mode entanglement. Photon A (B) is represented by a continuous red (dashed blue) line. The SPDC conical emission is transformed into a cylindrical one by the lens L_p, collected by eight GRIN lenses and then injected into the interferometer by eight corresponding single mode fibers. The beam splitter BS_1 acts as the BS_1 of Figure 1, while the BS_2 acts as the BS_{2A} and BS_{2B} of Figure 1 for both photons. Interference is observed by varying the path delays Δx_i ($i = 1,2,3$). Waveplates $\lambda/4$ and $\lambda/2$ are used for polarization restoration.

pairs of correlated modes within the conical emission of the crystal [10]. Precisely, we are able to couple the SPDC radiation into eight different single mode fibers (SMF's) by using an equal number of graded-index lenses (GL) that are glued each to the input of a SMF. Then, the eight GL-SMF systems are glued, after careful alignment, in an especially prepared eight-hole brass mask. The whole system represent a single integrated device allowing to intercept the multipath entangled parametric radiation. In the best alignment conditions we measured without any interferometric setup nearly $7 \cdot 10^3 coinc/s$ over a bandwidth $\Delta\lambda = 4.5nm$ for each pair of correlated modes. In the standard experimental conditions we put the 8-fiber system in an insulating box for temperature stabilization. By the interferometric apparatus shown in Figure 3, we demonstrated the multipath entanglement over the entire set of **k**-modes identified by the GL-SMF systems.

REFERENCES

1. P. G. Kwiat, *J. Mod. Opt.* **44**, 2173–2184 (1997).
2. M. Barbieri, C. Cinelli, P. Mataloni, and F. De Martini, *Phys. Rev. A* **72**, 052110 (2005).
3. J. T. Barreiro, N. K. Langford, N. A. Peters, and P. G. Kwiat, *Phys. Rev. Lett.* **95**, 260501 (2005).
4. G. Vallone, E. Pomarico, P. Mataloni, F. De Martini, and V. Berardi, *Phys. Rev. Lett.* **98**, 180502 (2007).
5. G. Vallone, E. Pomarico, F. De Martini, and P. Mataloni, *Phys. Rev. Lett.* **100**, 160502 (2008).
6. G. Vallone, E. Pomarico, F. De Martini, and P. Mataloni, *Phys. Rev. A* **78**, 042335 (2008).
7. R. Raussendorf, and H. J. Briegel, *Phys. Rev. Lett.* **86**, 5188 (2001).
8. P. Walther, K. J. Resch, T. Rudolph, E. Schenck, H. Weinfurter, V. Vedral, M. Aspelmeyer, and A. Zeilinger, *Nature (London)* **434**, 169 (2005).
9. F. D. M. P. M. G. Vallone, R. Ceccarelli, Hyperentanglement of two photons in three degrees of freedom (2008), preprint [arXiv:0810.4461].
10. P. M. F. D. M. A. Rossi, G. Vallone, Complete path entanglement of two photons (2008), preprint [arXiv:0809.4575].

CHAPTER 6

CONTINUOUS-VARIABLE QUANTUM OPTICS

Continuous-variable quantum information processing with squeezed states of light

Akira Furusawa

Department of Applied Physics, School of Engineering, The University of Tokyo, 7-3-1 Hongo, Bunkyo-ku, Tokyo 113-8656, Japan

Abstract. Our research activities on continuous-variable quantum information processing with squeezed states of light are briefly reviewed.

Keywords: quantum information, squeezing
PACS: 03.67.Lx, 03.67.Mn, 42.50.Dv

INTRODUCTION

We have been working on continuous-variable (CV) quantum teleportation since 1998 [1] and the fidelity has reached to $F = 0\,83$ [2]. Recently the concept of quantum teleportation is extended to *generalized* quantum teleportation both in qubit and CV regime [3, 4]. The scheme is shown in Fig. 1. The essence of the scheme is *off-line*, i.e., entanglement between the input $|\psi\rangle$ and an ancilla, and 'measurement and feedforward'.

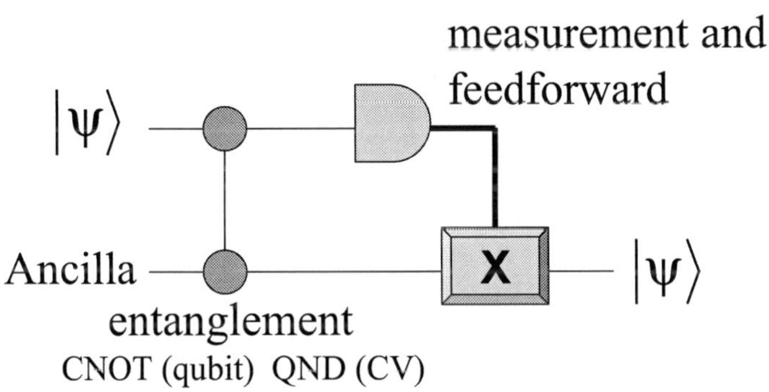

FIGURE 1. Generalized quantum teleportation.

OFF-LINE QUANTUM INFORMATION PROCESSING

The generalized quantum teleportation can be applied to off-line quantum information processing. One example is shown in Fig. 2.

In this scheme, the ancilla is a specific state $\hat{U}|c\rangle$, where unitary operation $\hat{U}$ is the one we want to make on the input. The advantage of the scheme is that the difficulty of operation is confined in state preparation of the ancilla, since the fidelity of teleportation

CP1110, *Quantum Communication, Measurement and Computing (QCMC), The Ninth International Conference,* edited by A. Lvovsky

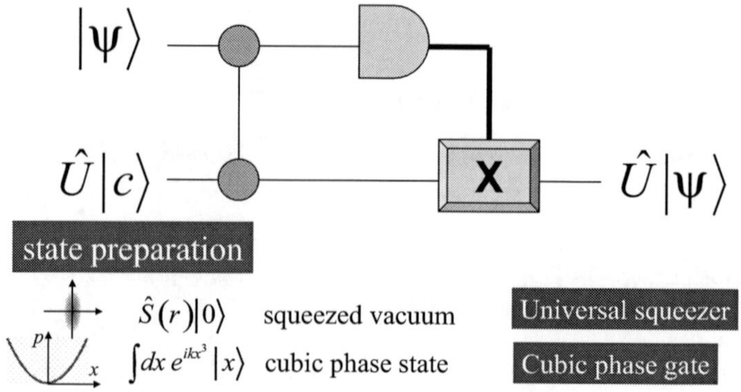

FIGURE 2. Off-line quantum information processing with the generalized quantum teleportation.

itself is rather high. Moreover, we do not have to make the unitary operation for an arbitrary input, instead it is enough to make it on a particular state $|c\rangle$, which is much easier than the case for an arbitrary input.

There are two important quantum gates with this scheme, a universal squeezer [5, 6] and a cubic phase gate [7], where the ancillae are a squeezed state and a cubic phase state, respectively. We demonstrated the universal squeezer [6] and a quantum non-demolition (QND) gate with the squeezers [8]. Here the QND gate is CV version of a controlled NOT gate and very important for CV quantum information processing.

ONE-WAY QUANTUM COMPUTATION WITH CLUSTER STATES

Another example of application of the generalized quantum teleportation is one-way quantum computation with cluster states both in qubit and CV regime [4, 9]. The scheme is shown in Figs. 3 and 4. Here $\hat{x}$ and $\hat{p}$ are quadrature operators and $\hat{a} = \hat{x} + i\hat{p}$ $(\hbar = 1\ \ 2)$.

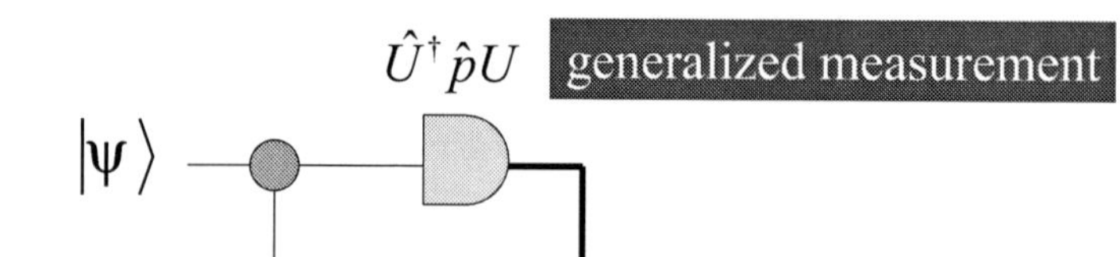

FIGURE 3. Generalized quantum teleportation is applied to one-way quantum computation with cluster states. A measurement is generalized as $\hat{U}^\dagger \hat{p} \hat{U}$.

The essence of the scheme is shown in Fig. 3. The difference from the scheme shown in Fig. 2 is that the ancilla is always $|p = 0\rangle$ here. Instead, a measurement is generalized as $\hat{U}^\dagger \hat{p} \hat{U}$ which corresponds to projection onto eigenstates of the operator. For example, operation of $\hat{U} = e^{ik\hat{x}^2}$ can be realized by homodyne measurement with local oscillator

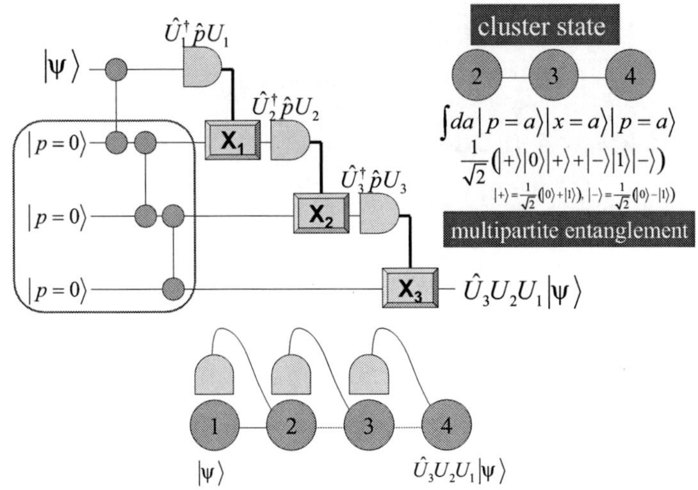

FIGURE 4. One-way quantum computation with cluster states. This example is on a three-mode *linear* cluster state.

phase tuning [4];

$$\begin{aligned} \hat{U}^\dagger \hat{p} \hat{U} &= e^{\ ik\hat{x}^2} \hat{p} e^{ik\hat{x}^2} \\ &= \hat{p} + k\hat{x} \\ &= \frac{\hat{p}\cos\theta \quad \hat{x}\sin\theta}{\cos\theta} \end{aligned} \tag{1}$$

where $\theta = \tan^{\ 1}(\ k)$ and it can be tuned by the local oscillator phase.

The scheme of Fig. 3 can be cascaded like the one shown in Fig. 4. In this case, one can prepare some special entangled state of many parties (optical beams) called a cluster state in advance as shown in Fig. 4. Then one makes projection onto the eigenstates of $\hat{U}_i^\dagger \hat{p} \hat{U}_i$ at each party (optical beam) according to desired operation. Finally, the measurement results are fed forward and the last party (optical beam) becomes the desired output state $\hat{U}_3 \hat{U}_2 \hat{U}_1 |\psi\rangle$.

Toward this goal, we created various four-mode cluster states [10]. Fig. 5 shows the cluster states.

CONCLUSION

Our research activities on continuous-variable quantum information processing with squeezed states of light are briefly reviewed especially from the viewpoint of application of generalized quantum teleportation.

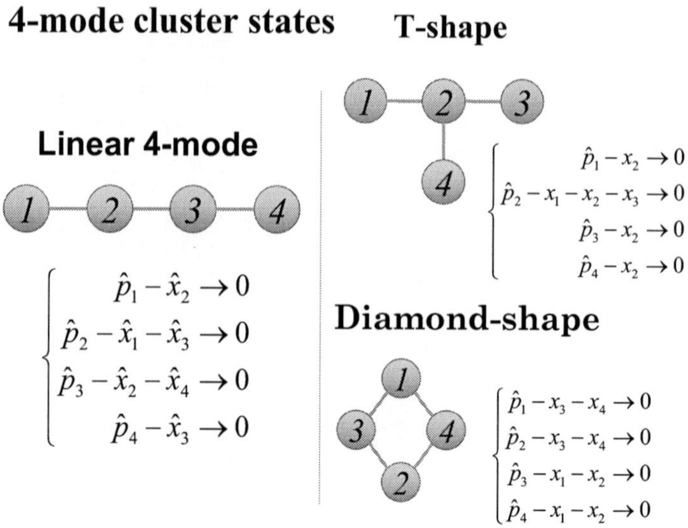

FIGURE 5. Four-mode linear, T-shape, and diamond-shape cluster states. These are simultaneous eigenstates of the operators listed near the figures.

ACKNOWLEDGMENTS

The author acknowledges all members of Furusawa group in Department of Applied Physics, The University of Tokyo for experimental achievements. Gratitude is extended to all collaborators including H. J. Kimble, S. L. Braunstein, P. van Loock, U. L. Andersen, E. H. Huntington, and M. Sasaki.

This work was partly supported by SCF, GIA, and G-COE commissioned by the MEXT of Japan, CREST, Japan Science and Technology (JST) Agency, and the Research Foundation for Opt-Science and Technology.

REFERENCES

1. A. Furusawa, J. L. Sørensen, S. L. Braunstein, C. A. Fuchs, H. J. Kimble, and E. S. Polzik, Science **282**, 706 (1998).
2. M. Yukawa, H. Benichi and A. Furusawa, Phys. Rev. A **77**, 022314 (2008).
3. X. Zhou, D. W. Leung, and I. L. Chuang, Phys. Rev. A **62**, 052316 (2000).
4. N. C. Menicucci, P. van Loock, M. Gu, C. Weedbrook, T. C. Ralph, and M. A. Nielsen, Phys. Rev. Lett. **97**, 110501 (2006).
5. R. Filip, P. Marek, and U. L. Andersen, Phys. Rev. A **71**, 042308 (2005).
6. J. Yoshikawa, T. Hayashi, T. Akiyama, N. Takei, A. Huck, U. L. Andersen, and A. Furusawa, Phys. Rev. A **76**, 060301(R) (2007).
7. D. Gottesman, A. Kitaev, and J. Preskill, Phys. Rev. A **64**, 012310 (2001).
8. J. Yoshikawa, Y. Miwa, A. Huck, U. L. Andersen, P. van Loock, and A. Furusawa, arXiv:0808.0551[quant-ph], Phys. Rev. Lett. accepted.
9. R. Raussendorf and H. J. Briegel, Phys. Rev. Lett. **86**, 5188 (2001).
10. M. Yukawa, R. Ukai, P. van Loock, and A. Furusawa, Phys. Rev. A 78, 012301 (2008).

Nondeterministic Noiseless Linear Amplification of Quantum Systems

T.C.Ralph* and A.P.Lund†

*Department of Physics, University of Queensland, Brisbane 4072, QLD, Australia
†Centre for Quantum Computer Technology, School of Science, Griffith University, QLD, Australia,

Abstract. We introduce the concept of non-deterministic noiseless linear amplification. We propose a linear optical realization of this transformation that could be built with current technology. We discuss the application of the device to distillation of continuous variable entanglement. We demonstrate that highly pure entanglement can be distilled from transmission over a lossy channel.

Keywords: quantum optics, distillation, amplification
PACS: 42.50.Dv, 03.67.Hk, 42.65.Yj

It is well known that a linear or phase insensitive amplifier acting on a quantum optical field, or more generally on any harmonic oscillator state, must introduce noise [1]. This noise enforces the no-cloning principle [2], the uncertainty principle for simultaneous measurements [3] and limits signal to noise in quantum limited communication and metrology protocols [4] as well as guaranteeing security in continuous variable quantum key distribution [5]. Never-the-less, we show here that a non-deterministic, but heralded, noiseless linear amplifier is possible.

We consider the following transformation

$$\hat{T}|\alpha\rangle = c|g\alpha\rangle \tag{1}$$

where g is a real number obeying $|g| > 1$, $|\alpha\rangle$ is a coherent state of the field or oscillator with complex amplitude α, and c is a complex number obeying $0 < |c| < 1$. The transformation of Eq.1 will only be valid if $\hat{T}$ is a non-deterministic (non-unitary) transformation. This will be physically allowed provided, on average, the distinguishability of the amplified states is not increased. This in turn implies that actually $|c| \leq 1/g$ [6]. Such a transformation, if heralded, might still be very useful. In the following we propose an explicit construction of such a non-deterministic, noiseless linear amplifier (NLA) using current quantum optics technology and discuss an application and the limits to the fidelity of the device under practical conditions.

The NLA is shown schematically in Fig.1. The optical mode to be amplified is divided evenly between N different paths using beamsplitters. Each path is then interacted with a single photon ancilla as shown in the inset to Fig.1. This interaction is a generalization of the quantum scissors introduced by Pegg et al [7]. The interaction is successful if one and only one photon is counted at the indicated ports. The paths are then recombined interferometrically with the inverse to the arrangement of beamsplitters used to split up the original mode. In the absence of the single photon interactions all the light would emerge in the original mode. Photon counters are placed at all the other outputs.

CP1110, *Quantum Communication, Measurement and Computing (QCMC), The Ninth International Conference,* edited by A. Lvovsky

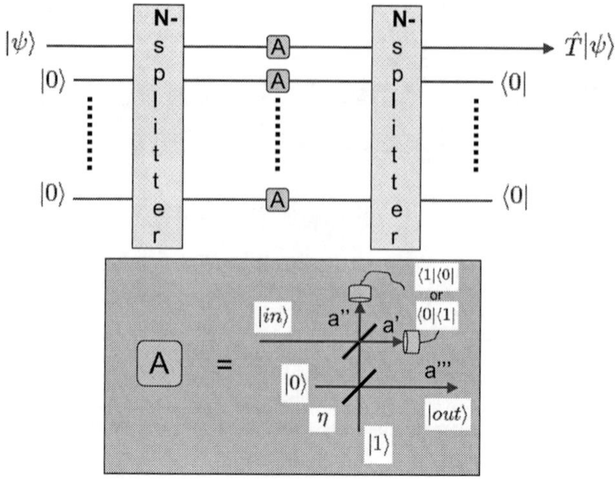

FIGURE 1. Schematic of the noiseless linear amplifier. The N-splitter is an array of beamsplitters that evenly divides the input beam. The second N-splitter coherently recombines the beams and is considered to have succeeded if no light exits through the other ports, as determined by photon counters. The interaction labelled "A" interacts a single photon ancilla with an input beam as shown in the gray inset. The upper beamsplitter is 50:50 whilst the lower beamsplitter has transmission η as shown. The interaction succeeds if a single count is recorded at a' and no count at a'', or vice versa.

Successful operation of the device is heralded when all these photon counters register no counts, given that all the ancilla interactions registered single counts.

We first calculate the effect of this device on an input coherent state, i.e. $|\psi\rangle = |\alpha\rangle$. The N-splitter divides the coherent state into the product state $|\alpha'\rangle|\alpha'\rangle|\alpha'\rangle...$, with $\alpha' = \alpha/\sqrt{N}$. Hence we can consider the effect of the interactions with the single photons individually for each mode. The action of the generalized quantum scissor is to truncate the coherent state to first order and simultaneously amplify it. Specifically, detection of a single photon at output port a' and zero photons at output port a'', or detection of a single photon at output port a'' and zero photons at output port a', produces the transformation

$$|\alpha'\rangle_{a'} \rightarrow e^{-\frac{|\alpha'|^2}{2}}\sqrt{\frac{\eta}{2}}(1 \pm \sqrt{\frac{1-\eta}{\eta}}\hat{a}^\dagger \alpha')|0\rangle_{a'''} \tag{2}$$

where the plus sign corresponds to the former case and the minus sign to the latter case. If the latter case occurs the phase flip can be corrected by feeding forward to a phase shifter. In the original quantum scissors $\eta = 0.5$ was used so the truncated state was not amplified [7].

Coherently recombining the modes at the second N-splitter and postselecting mode a on the basis that all other modes register zero photon counts leads to

$$e^{-\frac{|\alpha|^2}{2}}\eta^{\frac{N}{2}}(1+\sqrt{\frac{1-\eta}{\eta}}\hat{a}^{\dagger}\frac{\alpha}{N})^N|0\rangle_a \tag{3}$$

In the limit of N large, i.e. $N >> g|\alpha|$, we have the relationship

$$\lim_{N\to\infty}(1+g\hat{a}^{\dagger}\frac{\alpha}{N})^N|0\rangle = e^{g\hat{a}^{\dagger}\alpha}|0\rangle \tag{4}$$

where $g = \sqrt{(1-\eta)/\eta}$. We recognize the RHS of Eq.4 as being proportional to a coherent state with amplitude $g\,\alpha$. Putting everything together we conclude that in the large N limit the device of Fig.1 affects the transformation

$$|\alpha\rangle \to \eta^{\frac{N}{2}} e^{-\frac{(1-g^2)|\alpha|^2}{2}}|g\,\alpha\rangle \tag{5}$$

For $\eta < 1/2$ we have $g > 1$, and hence we achieve noiseless linear amplification as per Eq.1. The probability of success is given by the norm of the state

$$P = \eta^N e^{-(1-g^2)|\alpha|^2} \tag{6}$$

which is state dependent and also decreases with increasing N. This indicates that the cost of a better approximation to the amplified state is reduced probability of success. We will examine this trade-off shortly, but first let us consider a useful transformation we might achieve with the NLA.

We consider the action of the NLA when applied to one arm of an entangled state. We assume the arm that the NLA is applied to has experienced loss. We will show that on the occasions when the NLA is successful a more entangled state that has apparently suffered less loss is produced. We will consider Einstein, Podolsky, Rosen (EPR) type entanglement. First suppose no loss, i.e. the line transmission, $\varepsilon = 1$. The EPR (or two-mode squeezed) state is pure and given in the number basis by

$$|EPR\rangle = \sqrt{1-\chi^2}\sum_{n=0}^{\infty}\chi^n|n\rangle|n\rangle \tag{7}$$

where the strength of the entanglement is given by the the parameter χ, with $\chi = 0$ corresponding to no entanglement and $\chi = 1$ corresponding to maximal entangement. Using the number basis expansion of coherent states and Eq.5 it is straightforward to show that the effect of the NLA (in the large N limit) on an incident number state is the transformation

$$|n\rangle \to \eta^{\frac{N}{2}} g^n|n\rangle \tag{8}$$

Thus applying the NLA to one arm of the EPR state leads to the transformation

$$|EPR\rangle \to \eta^{\frac{N}{2}}\sqrt{1-\chi^2}\sum_{n=0}^{\infty}\chi^n g^n|n\rangle|n\rangle \tag{9}$$

We see that the post-selected state still has the form of an EPR state, however the effective value of the entanglement has changed to $\chi' = g\chi$. For $|g| > 1$ the entanglement has been increased, i.e. a more entangled state has been distilled. Notice again that an unphysical (unnormalizable) state is obtained if $|\chi g| \geq 1$. As before this indicates that the state does not converge for large N.

Now consider the case of non-zero loss, i.e. $\varepsilon < 1$. The EPR state is now mixed and given by the density operator

$$\rho = Tr_l\{|EPR'\rangle\langle EPR'|\} \tag{10}$$

where the global state is given by

$$|EPR'\rangle = \sqrt{1-\chi^2}\sum_{n=0}^{\infty}\chi^n\sum_{k=0}^{n}\sqrt{\tbinom{n}{k}}(1-\varepsilon)^{k/2}\varepsilon^{(n-k)/2}|n-k\rangle|n\rangle|k\rangle_l \tag{11}$$

and the trace is taken over the loss mode, labelled by the subscript l. Applying the NLA (as shown in Fig.2(b)) transforms the global state such that

$$\begin{aligned}|EPR'\rangle &\rightarrow \eta^{\frac{N}{2}}\sqrt{1-\chi^2}\sum_{n=0}^{\infty}\chi^n g^{n-k}\sum_{k=0}^{n}\sqrt{\tbinom{n}{k}}(1-\varepsilon)^{k/2}\varepsilon^{(n-k)/2}|n-k\rangle|n\rangle|k\rangle_l\\ &= \eta^{\frac{N}{2}}\sqrt{1-\chi^2}\sum_{n=0}^{\infty}\chi'^n\sum_{k=0}^{n}\sqrt{\tbinom{n}{k}}(1-\varepsilon')^{k/2}\varepsilon'^{(n-k)/2}|n-k\rangle|n\rangle|k\rangle_l \end{aligned}\tag{12}$$

Again we find that the post selected state can be written in a form equivalent to the initial EPR state but now with the substitutions

$$\chi' = \chi\sqrt{1+(g^2-1)\varepsilon}\,,\ \varepsilon' = \frac{g^2\varepsilon}{1+(g^2-1)\varepsilon}. \tag{13}$$

For $\chi < \chi\sqrt{1+(g^2-1)\varepsilon} < 1$ a physical state is obtained for which both the effective entanglement and line transmission are increased. Suppose that Alice holds the entanglement source and can control its strength, whilst Bob receives one of the entangled modes from Alice through a lossy line but then acts on this mode with the NLA. In such a scenario Alice and Bob may post select entanglement that can be stronger or more pure than is possible deterministically.

Up till this point we have considered the idealized limit in which $N \rightarrow \infty$. As previously pointed out this is impractical as the probability of success will tend rapidly to zero in this limit. We now examine the behavior of the NLA for finite N. Using Eq.3 we find the the exact state transformation for an input number state is

$$|n\rangle \rightarrow \eta^{\frac{N}{2}}\frac{N!}{(N-n)!N^n}g^n|n\rangle \tag{14}$$

From Eq.14 probabilities of success and fidelities can straightforwardly be calculated for various input states by expanding them in the number basis. Fig.3 shows, for an input coherent state, fidelities of the post-selected state with a target coherent state

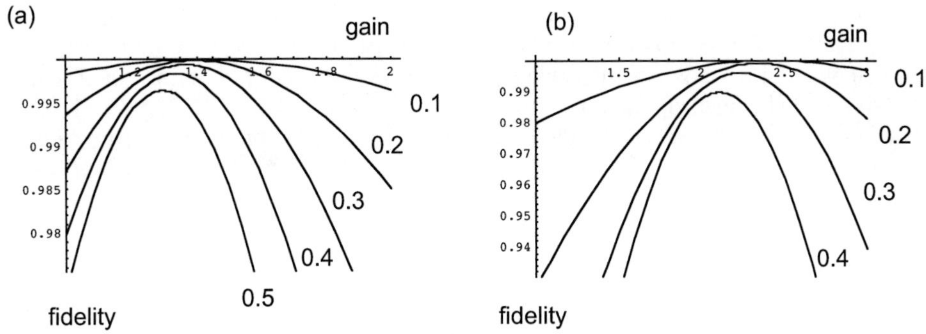

FIGURE 2. Fidelities of NLA output states versus target states with coherent input states and $N = 5$. In (a) $\eta = 1/3$ and in (b) $\eta = 1/7$. The amplitudes, α, of the input states are as indicated on the figures. The amplitudes of the target state is the gain times α.

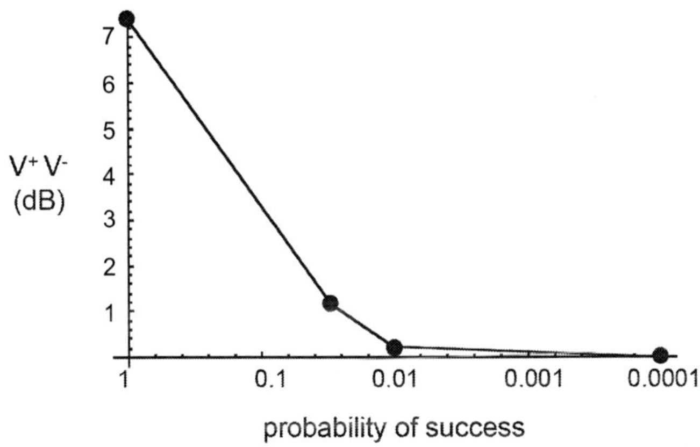

FIGURE 3. Purity of NLA output states versus probability of success with EPR state inputs. The purity is quantified by the uncertainty product of the squeezed and anti-squeezed quadrature correlations of the post-selected EPR state.

as a function of the gain between the input and the target for $N = 5$. In Fig.3(a) the ideal gain would be $g^2 \approx 2$ and in Fig.3(b) it is $g^2 \approx 6$. In both cases high fidelity is obtained however, due to the effect of finite N, we notice that the gain saturates and the peak fidelity reduces somewhat as larger inputs are chosen and that this effect is more pronounced when higher gains are targeted. The probabilities of success are 0.5% for gain 2 and 0.01% for gain 6. Notice that the inputs analyzed are all quite small. In order to achieve high fidelity for larger input amplitudes larger N's are required with subsequent reduction in success probability. In spite of these restrictions, useful outcomes can be achieved when amplifying entanglement as we now discuss.

Consider first pure entanglement. We find that useful boosts of entanglement can be achieved with realistic scenarios. For example with $N = 2$ and $\eta = 0.05$, an initial EPR state with squeezing parameter of $r = 0.1$ is transformed with high fidelity (F=0.993) into another EPR state with squeezing parameter $r = 0.4$. This represents an effective halving of the squeezing variance from $exp[-2r] \approx 0.82 \rightarrow 0.45$. The probability of success is 0.2%. Distillation of EPR states has previously been proposed by Browne and Eisert et al [8]. In their scheme multiple EPR sources are required. As well as being technologically easier, distillation via the NLA performs better in terms of probability of success and fidelity as a function of the number of single photon ancilla employed.

Of greater interest is the ability to purify entanglement that has been distributed through a decohering channel. Fig.4 shows the trade-off between probability of success and purification for transmission through a lossy channel. The channel is assumed to have 50% transmission efficiency. By Alice sending different strengths of entanglement and Bob varying the gain of the distiller they can distribute entanglement where the strength of the correlations is kept constant ($r = 0.4$) but the purity is varied. For all points the fidelity is above 0.99. High purity can be obtained, as quantified by the uncertainty product between the squeezed and anti-squeezed quadrature correlations of the EPR state but, as is typical of distillation schemes, this is at the expense of lower probability of success. The ability to perform such purification would be of direct benefit to continuous variable dense coding [9] and quantum key distribution [5] protocols.

We have introduced non-deterministic noiseless linear amplification and proposed a linear optics plus photon counting construction. Here we have discussed the application of the NLA to distillation however we might also expect applications for various other protocols in quantum communication and metrology.

ACKNOWLEDGMENTS

We thank E.H.Huntington for useful discussions. This work was supported by the Australian Research Council.

REFERENCES

1. C.M.Caves, Phys.Rev.D **23**, 1693 (1981).
2. W.K.Wootters and W.H.Zurek, Nature **299**, 802 (1982).
3. E. Arthurs and M. S. Goodman, Phys. Rev. Lett. **60**, 2447 (1988)
4. H-A.Bachor and T.C.Ralph, *A guide to experiments in quantum optics* (2nd Edition, Wiley-VCH, Weinheim 2004).
5. F.Grosshans, P.Grangier, Phys.Rev.Lett. **88**, 057902 (2002).
6. A.P.Lund and T.C.Ralph, in preparation (2008).
7. David T. Pegg, Lee S. Phillips, and Stephen M. Barnett, Phys. Rev. Lett. **81**, 1604 (1998)
8. D.E.Browne, J.Eisert, S.Scheel, and M.B.Plenio Phys. Rev. A **67**, 062320 (2003); J.Eisert, D.E.Browne, S.Scheel and M.B.Plenio, Annals of Physics, **311**, 431 (2004).
9. T.C.Ralph and E.H.Huntington, Phys.Rev.A 66, 042321 (2002).

A nonlocal single-mode amplification process using linear optics

Yu-Li Dong, Xu-Bo Zou and Guang-Can Guo

Key Laboratory of Quantum Information, University of Science and Technology of China, Chinese Academy of Sciences, Hefei, Anhui 230026, China

Abstract. We first present a scheme to implement a nonlocal single-mode amplification process using linear optics. Then by employing this process, cloning coherent state with known phases can be realized. With the help of ancillary squeezed states, the cloning fidelity can reach a high level.

Keywords: single-mode amplification, linear optics, cloning coherent state with known phase
PACS: 03.67.Hk, 03.65.Ta

INTRODUCTION

In optical quantum communication and computation, the single- and two-mode amplifications are very important operations. Based on these operations, various quantum clonings have been researched in the continuous-variable (CV) domain, such as optimal Gaussian cloning of coherent states [1, 2], non-Gaussian cloning of coherent states [3], cloning of phase-conjugate input (PCI) states [4] and cloning of coherent states with known phases [5, 6]. However, the coupling between signal modes via a nonlinear medium is in a low efficiency. Recently, a feasible implementation of local single-mode amplification [7] based entirely on linear optics, homodyne detection and off-line squeezed vacuum state was proposed, and it could be easily completed with the present technology. We also note that two-mode amplification based on linear optics has been proposed and experimentally realized in Ref. [8]. Subsequently, Zhang *et al.* [9] further presented a scheme for implementing nonlocal two-mode amplification by means of genuine four-mode entanglement.

In this paper, we first suggest a method that realizes a nonlocal single-mode amplification process using linear optics. Then we demonstrate a high fidelity of cloning of coherent states with known phases [10] can be achieved by employing this amplification process.

METHOD

Here, we consider a nonlocal single-mode amplification process using an ancillary EPR state. This process allows two distant parties to perform a single-mode amplification

CP1110, *Quantum Communication, Measurement and Computing (QCMC), The Ninth International Conference,* edited by A. Lvovsky

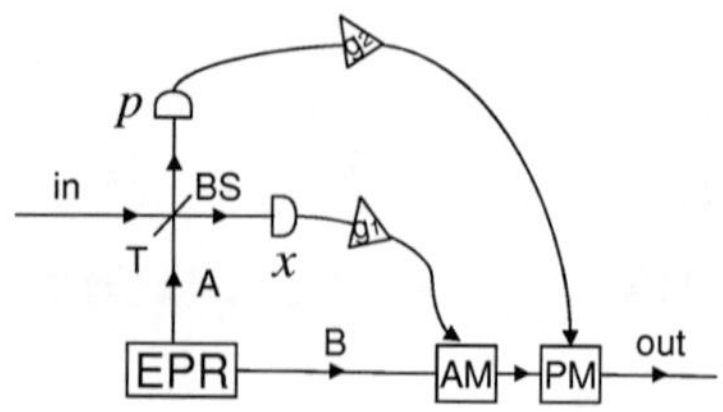

FIGURE 1. Scheme for nonlocal single-mode amplification using linear optics and homodyne detection. BS, beam splitter with transmission coefficient T; AM, amplitude modulator; PM, phase modulator; g_i, scaling factor of classical channels.

employing only linear optical elements and classical communication. The scheme for the nonlocal single-mode amplification process is depicted in Fig. 1. One half of the EPR state is mixed with the signal mode by a beam splitter (BS) with transmission coefficient T. The conjugate quadrature amplitudes of the two output beams are simultaneously measured by homodyne detections. The measurement results are labeled as

$$x = \sqrt{T}\hat{x}_{\text{in}} - \sqrt{1-T}\hat{x}_A, \tag{1}$$

$$p = \sqrt{1-T}\hat{p}_{\text{in}} + \sqrt{T}\hat{p}_A. \tag{2}$$

According these results, we can modify the amplitude and phase of another half of the EPR state via two independent modulators with the scaling factors g_1 and g_2, respectively, and the output field can be expressed as

$$\hat{x}_{\text{out}} = g_1\sqrt{T}\hat{x}_{\text{in}} + \hat{x}_B - g_1\sqrt{1-T}\hat{x}_A, \tag{3}$$

$$\hat{p}_{\text{out}} = g_2\sqrt{1-T}\hat{p}_{\text{in}} + \hat{p}_B + g_2\sqrt{T}\hat{p}_A. \tag{4}$$

When choosing the scaling factors $g_1 = 1/\sqrt{1-T}$ and $g_2 = 1/\sqrt{T}$, the output field reads

$$\hat{x}_{\text{out}} = \sqrt{\tfrac{T}{1-T}}\hat{x}_{\text{in}} + \hat{x}_B - \hat{x}_A, \tag{5}$$

$$\hat{p}_{\text{out}} = \sqrt{\tfrac{1-T}{T}}\hat{p}_{\text{in}} + \hat{p}_B + \hat{p}_A, \tag{6}$$

which achieves the single-mode amplification process for input signal.

APPLICATION

Then we consider the nonlocal cloning of coherent states with known phases. The schematic setup is shown in Fig. 2. The state to be copied is the coherent state $|\alpha\rangle$ with known phases. It can be characterized by two canonical conjugate quadratures, *i.e*, amplitude quadrature $\hat{x}$ and phase quadrature $\hat{p}$, which satisfy the canonical commutation relation $[\hat{x}, \hat{p}] = 2i$. Because we have known the phases of the coherent states, we set the amplitude α real without loss of generality. During the clone process, additional noise will be induced, which can be represented by the variances of $\hat{x}$ and $\hat{p}$. It has been

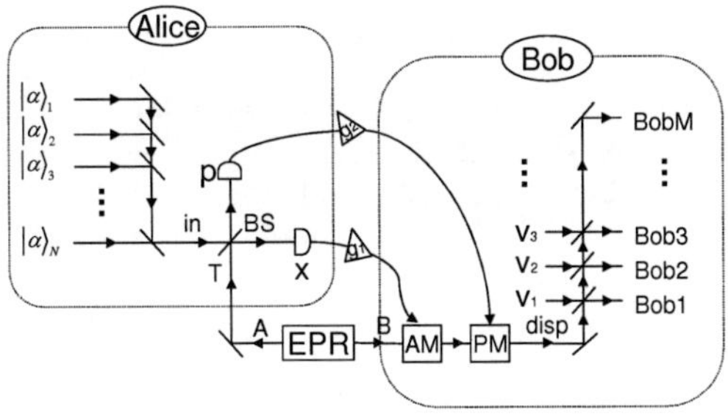

FIGURE 2. Scheme for nonlocal $N \Rightarrow M$ cloning of coherent states with known phases using linear optics and homodyne detection. BS, beam splitter with transmission coefficient T; AM, amplitude modulator; PM, phase modulator; v_i, auxiliary mode; g_i, scaling factor of classical channels.

shown that when the cloning process is restricted to the Gaussian operation, the fidelity quantifying the overlap between the input state and the output state can be written as

$$F_{\text{clone}} = \frac{2}{\sqrt{(1+\Delta^2 X_{\text{clone}})(1+\Delta^2 P_{\text{clone}})}}. \tag{7}$$

As shown in Fig. 2, we assume that Alice holds the N input states rather than N senders each of whom owns one of the N replicas [9], and Bob will distribute the states to be copied to M receivers. The protocol consists of three steps: (i) Alice performs a joint measurement of the unknown states and one mode of bipartite entangled state. (ii) Alice transfers the measurement results to Bob via a classical channel. (iii) According to the results, Bob modulates his part of entangled mode using amplitude and phase modulators and eventually distributes it to M receivers.

Let us analyze the above scenario in details. Firstly, the EPR entangled state can be prepared by mixing two individual squeezed vacuum modes on a 50-50 BS. Then one of the output beams is sent to Alice, and the other one is sent to Bob. Subsequently, Alice performs a Bell measurement on her mode A and a collected mode of N input states by an unbalance BS and two homodyne detectors. As we has mentioned above, by choosing the right transmission coefficient T and the scaling factors, a single-mode amplification process for the input signal can be achieved and the output field is given as Eq. (5, 6). Taking $T = M/(M+N)$, we have

$$\hat{x}_{\text{dis}} = \sqrt{\frac{M}{N}}\hat{x}_{\text{in}} + \hat{x}_B - \hat{x}_A, \tag{8}$$

$$\hat{p}_{\text{dis}} = \sqrt{\frac{N}{M}}\hat{p}_{\text{in}} + \hat{p}_B + \hat{p}_A. \tag{9}$$

In the third step, Bob distributes the states to M receivers using a series of BSs and $M-1$ squeezing vacuum states with the same squeezing parameter $e^{2r} = \sqrt{(1+1/N)/(1+N/M^2)}$. Then the variances of the quadrature noise of any clone are given as

$$\Delta^2 X_{\text{clone}} = \frac{1}{N} + \frac{M-1}{M}e^{2r} + \frac{1}{M}e^{-2r_{EPR}}, \tag{10}$$

$$\Delta^2 P_{\text{clone}} = \frac{N}{M^2} + \frac{M-1}{M} e^{-2r} + \frac{1}{M} e^{-2r_{EPR}}, \tag{11}$$

where r_{EPR} is the squeezing parameter of the two-mode entanglement state. It is obviously that if modes A and B are ideal EPR states ($r_{EPR} \rightarrow \infty$), the above equations combining with Eq. (7) will give the fidelity of cloning as

$$F_{N,M}^{(kp)} = \frac{2M\sqrt{N}}{\sqrt{N}(M-1) + \sqrt{(1+N)(M^2+N)}}, \tag{12}$$

which reaches a high level. For example, in the case of 1-to-2 cloning, the cloning fidelity is as high as 0.96. Finally, it has to be emphasized here that in the second step of this telecloning, Alice transfers to Bob not only the measurement results, but also the numbers of input and output modes, *i.e.*, N and M, so Bob can adjust the parameters of g_1, g_2, and the squeezing parameter r with respect to N and M to achieve the telecloning.

CONCLUSION

We have proposed a scheme to achieve nonlocal single-mode amplification using linear optics and homodyne detection. The whole process only relies on off-line two-mode squeezed state, which could be easily completed with present technology. Due to its nonlocality the proposed method might be extensively applied in a CV quantum information network. One of the applications of this process is the cloning of coherent states with known phases as shown in Fig. 2. With the help of squeezed vacuum states, this cloning can gain a high fidelity.

ACKNOWLEDGMENTS

This work was supported by National Fundamental Research Program, by the National Natural Science Foundation of China (Grants No. 10674128 and No. 60121503), by the Innovation Funds and "Hundreds of Talents" program of the Chinese Academy of Sciences, and by the Doctoral Foundation of the Education Ministry of China (Grant No. 20060358043).

REFERENCES

1. J. Fiurášek, Phys. Rev. Lett **86**, 4942 (2001).
2. Samuel L. Braunstein, Nicolas J. Cerf, Sofyan Iblisdir, Peter van Loock, and Serge Massar, Phys. Rev. Lett. **86**, 4938 (2001).
3. N. J. Cerf, O. Krüger, P. Navez, R. F. Werner, and M. M. Wolf, Phys. Revs. Lett. **95**, 070501 (2005).
4. N. J. Cerf and S. Iblisdir, Phys. Rev. Lett. **87**, 247903 (2001).
5. P. T. Cochrane, T. C. Ralph, and A. Dolinska, Phys. Rev. A **69**, 042313 (2004).
6. Yuli Dong, Xubo Zou, Shangbin Li, and Guangcan Guo, Phys. Rev. A **76** 014303 (2007).
7. Radim Filip, Petr Marek, and Ulrik L. Andersen, Phys. Rev. A **71**, 042308 (2005).
8. Vincent Josse, Metin Sabuncu, Nicolas J. Cerf, Gerd Leuchs, and Ulrik L. Andersen, Phys. Rev. Lett. **96**, 163602 (2006).
9. Jing Zhang, Changde Xie, and Kunchi Peng, Phys. Rev. A **76**, 064301 (2007).
10. Yu-Li Dong, Xu-Bo Zou, and Guang-Can Guo, Phys. Rev. A **77**, 034304 (2008).

Generation of Highly Squeezed Light at 860nm

Genta Masada[a,b,c], Koyo Nagashima[a,c] , Hajime Arao[a,c] , Yuishi Takeno[a,c] , Hidehiro Yonezawa[a,c] and Akira Furusawa[a,c]

[a]*Departmen of Applied Physics, School of Engineering, The University of Tokyo, 7-3-1 Hongo, Bunkyo-ku, Tokyo 113-8656, Japan*
[b]*Tamagawa University Research Institute, 6-1-1 Tamagawagakuen, Machida-city, Tokyo 194-8610, Japan*
[c]*CREST, Japan Science and Technology Agency, 5 Sanbancho, Chiyoda-ku, Tokyo 102-0075, Japan*
e-mail: *g.masada@lab.tamagawa.ac.jp*

Abstract. We report generation of squeezed light at 860nm by utilizing a subthreshold optical parametric oscillator with a periodically poled $KTiOPO_4$ crystal as a nonlinear medium. At the pump power of 130mW we observed the squeezing level of -9.55 dB and antisqueezing level of +16.36 dB.

Keywords: Quantum optics, Squeezed state, Optical parametric oscillator
PACS: 03.65.Ta, 03.67.-a.

INTRODUCTION

Squeezed state of optical fields is important resources for quantum information processing with continuous variables [1]. For example, quadrature squeezed vacuum states are applied to realize quantum teleportation [2] which is one of the most important protocols in quantum information processing. The fidelity of such protocols is limited directly by squeezing level. So it is important to generate highly squeezed state to achieve better performance.

EXPERIMENTAL SETUP

Degenerate parametric down conversion process in nonlinear optical crystals is often used for generation of squeezed states of light. We utilized a subthreshold optical parametric oscillator (OPO) which has a bow tie type ring configuration. High nonlinearity and low absorption losses are preferable as a nonlinear medium in order to reduce an oscillation threshold of the OPO and to generate highly squeezed light. We used periodically poled $KTiOPO_4$ (PPKTP) crystals which has rather low pump induced losses [3].

Figure 1 shows a schematic diagram of experimental setup. We use a continuous-wave Ti:Sapphire laser at 860 nm to accomplish this experiment. The setup mainly consists of three bow tie cavities, an OPO, an optical frequency doubler, and a mode cleaning cavity, and a balanced homodyne detector. A part of the beam from the

CP1110, *Quantum Communication, Measurement and Computing (QCMC), The Ninth International Conference,* edited by A. Lvovsky

Ti:Sapphire laser is introduced into the OPO. Another part of the beam is introduced into the frequency doubler to generate second harmonic as a pump beam for the OPO. The OPO cavity has two spherical mirrors whose radius of curvature is 50mm and two flat mirrors. One of the flat mirrors has a transmittance of 12.3% at 860nm and is used as an output coupler, while the others are highly reflectivity coated. All the mirrors have high transmittance at 430nm. A 10mm-long PPKTP crystal whose temperature is controlled at 50degrees is placed between the two spherical mirrors. The cavity length is stabilized actively controlled by the FM sideband locking method [4] using a piezo actuator.

The other part of the beam is introduced into the mode cleaning cavity in order that a local oscillator (LO) beam has a same spatial mode as the OPO output. Squeezed light from the OPO is combined with a LO beam at a half beam splitter and measured by a balanced homodyne detector. The spatial mode matching between the squeezed light and the LO beam is an important factor for the homodyne efficiency.

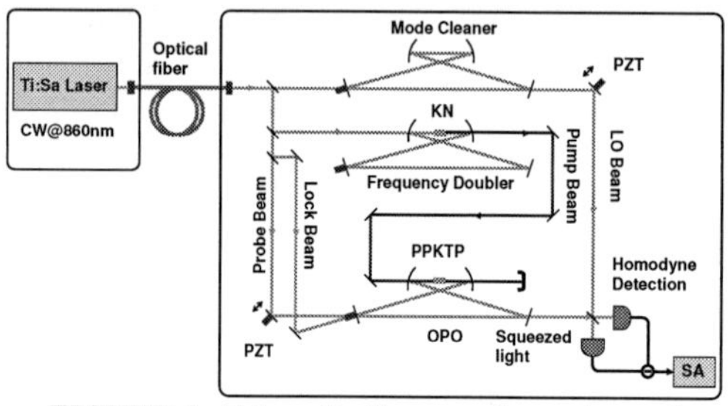

FIGURE 1. Schematic diagram of experimental setup.

In previous works -7.2 dB of squeezing level has been reported by Suzuki *et al.* [5] and -9.0 dB by Takeno *et al.* [6]. As is indicated in these references, one of the major factor which degrades the observed squeezed level is the intracavity losses. Another main factor is the fluctuation of relative phase between the squeezed light and the LO beam in homodyne measurement. In this work we have more focused on the searching a crystal position where the intracavity losses are as low as possible, and suppression of the phase fluctuation by improving a mechanical stability of experimental setup. An optical system is built as compactly as possible and a beam height is set as low as 2 inches in order to make a stable setup. The Ti:Sapphire laser is placed on a different optical table in order to avoid mechanical vibration and heat transfer to the optical system. The beam is introduced to the optical system by using a single mode fiber. A thick plastic box is placed over the optical system to eliminate a fluctuation caused by air flow. The improvement of mechanical design is expected to lead the reduction of the phase fluctuation.

RESULTS AND DISCCUSSION

Figure 2 shows one of the typical results of the observed quantum noise levels with the pump power of 130 mW. We locked the LO phase at the squeezed and antisqueezed quadratures and succeeded in obtaining -9.55 dB squeezing and +16.36

dB antisqueezing level, respectively. The setting of the spectrum analyzer is zero span mode at 1 MHz with 30kHz of resolution band width and 300Hz of video bandwidth.

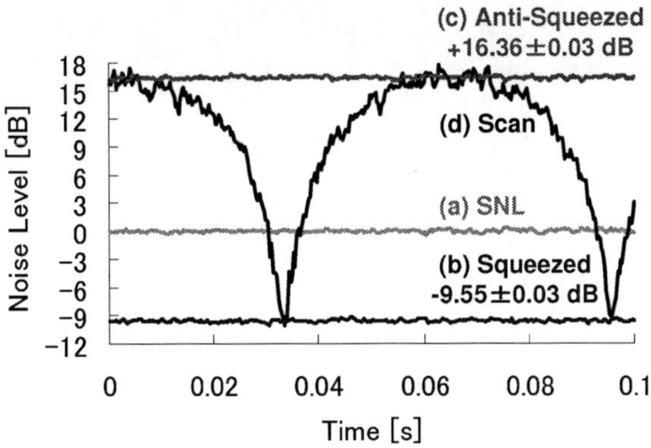

FIGURE 2. Power levels of quantum noise. (a) Shot noise level. (b) LO phase is locked at the squeezing quadrature. (c) LO phase is locked at antisqueezing quadrature. (d) LO phase is scanned. These are normalized to make the shot noise level 0 dB. Measurement frequency: 1 MHz, resolution band width: 30 kHz and video bandwidth: 300 Hz. Traces (a), (b) and (c) are averaged 20 times.

We repeated the above measurement at various pump powers up to 165mW. The results are summarized in figure 3. The squeezing level saturates at the higher pump power level. This fact could be due to the phase fluctuation which has a large effect on the observed squeezing level through mixing of the highly antisqueezed quadrature as follows.

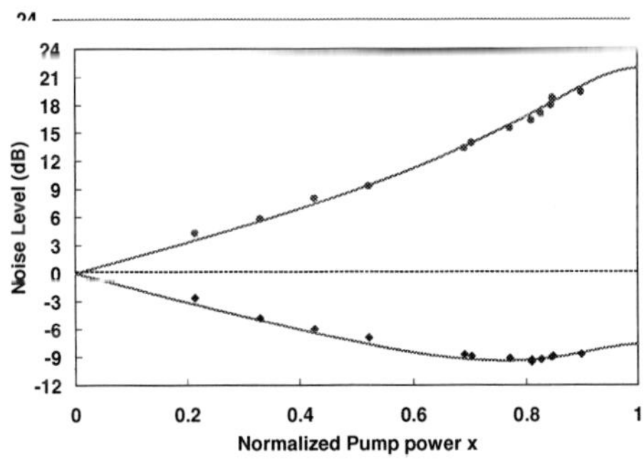

FIGURE 3. Pump power dependence of the squeezing and antisqueezing levels. Squares (circles) indicate the observed squeezing (antisqueezing) levels. Solid curves indicate theoretically calculated results by using eq.(2) in the condition of the phase fluctuation as 1.4 degrees.

The noise levels $R_{\pm}$ for the antisqueezed (+) and squeezed (-) quadratures can be modeled as [7]

$$R_{\pm} = 1 \pm \alpha\rho \frac{4x}{(1 \mp x)^2 + 4f^2} \tag{1}$$

where α is the detection efficiency, ρ is the escape efficiency of the OPO cavity, x is the normalized pump papameter, and f is the detuning factor. The escape efficiency is written as $\rho = T/(T+L)$ where T is transmittance of the output coupler, L is the intracavity losses. The normalized pump parameter is given as $x = \sqrt{P/P_{th}}$ where P is pump power, and P_{th} is an oscillation threshold of the OPO. In the current setup, α = 0.964, T = 0.123, P_{th} = 205mW and f = 0.154. Taking account of the phase fluctuation $\tilde{\theta}$, the theoretical squeezing level R'_- and antisqueezing level R'_+ are calculated as follws [8]

$$R'_\pm \approx R_\pm \cos^2\tilde{\theta} + R_\mp \sin^2\tilde{\theta}\,. \qquad (2)$$

Table 1 summarizes the comparison of experimental parameters, experimental results, and theoretically calculation for the previous and present work. The detector noise which is -20dB compared to the shot noise level is considered in calculation. We could find a good spot on the crystal where the intracavity losses were rather low. This fact yielded the improvement of the escape efficiency. And we could suppress the phase fluctuation to $\tilde{\theta}$ =1.4 degrees which was consistent with the increase of x_{opt} (optimum normalized pump parameter).

TABLE 1. Comparison of the experimental parameter, the experimental results, and the theoretical calculation results with the previous work (ref.6).

	L	$\tilde{\theta}$	x_{opt}	R_{-ex}(dB)	$R'_{-theory}$ (dB)
Previous work	0.0038	1.5	0.745	-9.01	-9.21
Present work	0.0033	1.4	0.810	-9.55	-9.30

CONCLUSION

We have demonstrated generation of squeezed light at 860nm by utilizing a subthreshold optical parametric oscillator with a periodically poled $KTiOPO_4$ crystal as a nonlinear medium. At the pump power of 130mW we observed the squeezing level of -9.55 dB and antisqueezing level of +16.36 dB. The main reason of the improvement on the observed squeezing levels may be considered as a good spot on the crystal and improvement on the phase stability. By improving the phase stability, more squeezing level will be expected.

REFERENCES

1. S.L.Braunstein and P.van Look, Rev. Mod. Phys. **77**, 513 (2005).
2. A.Furusawa, J.L.Sorensen, S.L.Braunstein, C.A.Fuchs, H.J.Kimble, and E.S.Polzik, Science **282**, 706 (1998).
3. S.Wang, V.Pasiskevicius, and F.Laurell, J. Appl. Phys. 96,2023 (2004)
4. R.W.P.Drever *et.al.*, Appl. Phys. B **31** 97 (1983)
5. S.Suzuki, H.Yonezawa, F.Kannari, M.Sasaki, and A.Furusawa, Appl. Phys. Lett. **89**, 061116 (2006).
6. Y.Takeno, M.Yukawa, H.Yonezawa, and A.Furusawa, Opt. Express **15**, 4321 (2007).
7. M.J.Collet and C.W.Gardiner, Phys.Rev. A**30**,1386 (1984)
8. T.C.Zhang, K.W.Goh, C.W.Chou, P.LOdahl, and H.J.Kimble, Phys. Rev.A **67**, 033802 (2003)

Time-resolved pulsed homodyne detector and its application to measurement of pulsed squeezed states

Ryuhi Okubo*, Mayumi Hirano*, Yun Zhang† and Takuya Hirano*,**

*Department of Physics, Gakushuin University, 1-5-1 Mejiro, Toshima-ku, Tokyo, 171-8588, Japan
†Department of Electronic-Engineering, The University of Electro-Communications, 1-5-1 Chofugaoka, Chofu, Tokyo 182-8585, Japan
**CREST, Japan Science and Technology Agency, 1-9-9 Yaesu, Chuou-ku, Tokyo, 103-0028, Japan

Abstract. We report a pulsed homodyne detector that operates in the time domain and can measure individual pulses at a repetition rate of 76 MHz. The individuality of the measurement is interrogated by correlation coefficient between adjacent pulses. The homodyne detector is composed of a low-noise high-speed operational amplifier. We observed 2.3 dB of squeezing in both the time and frequency domains.

Keywords: homodyne detector, squeezed state
PACS: 42.50.Dv, 03.67.Hk

INTRODUCTION

A Homodyne detector (HD) is one of the most widely-used apparatus in both quantum optics and quantum information with continuous variables [1], and it is used for measuring the quadrature amplitude of a light field [2]. With the development of quantum optics and quantum information fields, time-domain detection has become more and more important. Recently, time-domain HD has been successfully applied to quantum tomography experiments [3] and generating Schrodinger cat states [4]. In time-domain pulsed-homodyne detection, the HD should yield a single value of a field quadrature for each pulse [5]. Another important requirement is the ability to measure pulse trains at high repetition rates. In order to realize high-speed communications, a high-repetition-rate pulsed laser is usually employed as the light source [6]. However, the implementation of a fast-response HD is more technically challenging.

In this paper, we present an efficient pulsed homodyne system that operates in the time domain and can measure individual pulses at a repetition rate of 76 MHz, which is almost two orders of magnitude higher than previous experiments using a charge sensitive amplifier [7]. This system has a usable bandwidth of 250 MHz in the frequency domain [8]. Pulse-resolved measurement was evaluated by the correlation coefficient (CC) between adjacent pulses. In order to test quantum-limited operation, a squeezed state was measured and 2.3 dB of squeezing was observed in both the time and frequency domains.

CP1110, *Quantum Communication, Measurement and Computing (QCMC), The Ninth International Conference,* edited by A. Lvovsky

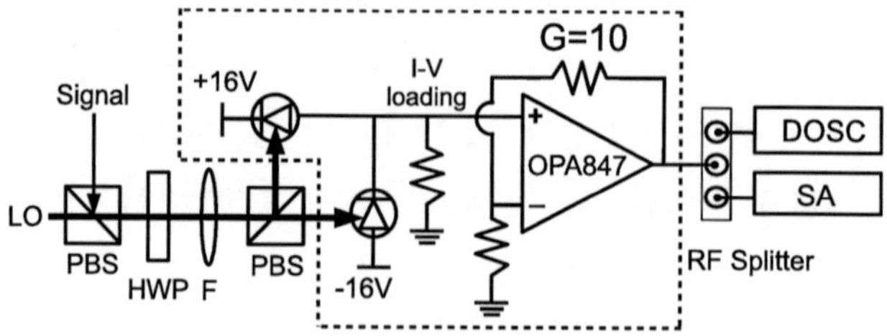

FIGURE 1. An electro-optic schematic of homodyne detection(see text for details).

HOMODYNE DETECTOR

Figure 1 presents an electro-optic schematic of our homodyne detection system. A local oscillator (LO) beam and a signal beam, which are orthogonally polarized to each other, are overlapped using a polarizing beam splitter (PBS), and then propagate collinearly to the balanced HD. A half-wave plate (HWP) is inserted to rotate the polarization by 45° and a second PBS is used to split the beam into two beams of equal intensity. The two beams are then focused on a pair of InGaAs photodiodes (PDs; Fermionics, FD150W) by a 100-mm-focal-length lens, which is placed immediately after the second HWP.

The two PDs are connected to each other and biased oppositely to generate a difference photocurrent. The difference photocurrent of the PDs is converted to a voltage signal by a load resistor and then amplified with a gain of G=10 by an operational amplifier (Texas Instruments, OPA847) in a non-inverting configuration. The quantum efficiencies of the two PDs were measured to be 0.86 and 0.84. Lead accumulators were used for the amplifier power supply.

The HD output is split into two signals by a RF splitter. One signal is monitored by a spectrum analyzer (SA), the other signal is recorded by a digital oscilloscope. The scope has an analog bandwidth of 4 GHz, a sampling rate of 5 G Samples/s, and a storage depth of 500 k points.

TIME-DOMAIN ANALYSIS

In the time-domain measurement, the quadrature value is obtained by integrating HD output over the time interval corresponding to the separation between two successive pulses. In our case, the time interval (13 ns) is the reciprocal of the laser pulse repetition rate of 76 MHz. A time sequence of 500 k points, covering time windows of 100μs is numerically integrated to produce quadrature values of 7600 pulses.

Figure 2(a) shows the shot-noise variance in the time domain as a function of the LO power. Subtracting the electronic noise reveals a linear dependence between the LO power and the shot-noise variance up to 2.0 mW. Our experiment was performed with an LO power of less than 2.0 mW to avoid saturating the detector at high power. The SNL at an LO power of 1.6 mW is 7.5 dB above the electronic noise of the HD.

To ensure individual measurement of each laser pulse, we measured the CC (which indicates the degree of linearity between two samples) between measured quadrature

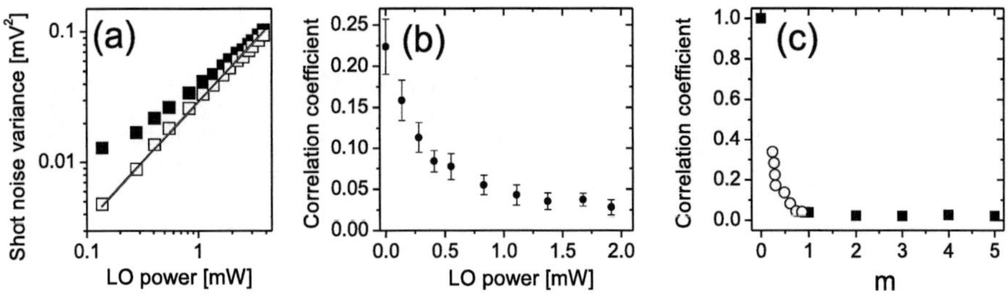

FIGURE 2. (a) Shot-noise variance measured in time domain as a function of LO power. The filled squares represent the measured noise variance. The open squares were obtained from these values by subtracting the amplifier noise. (b) CC between adjacent pulses as a function of the LO power. (c) CC between n-th and n+m-th pulses at an LO power of 1.6 mW. Open circles represent the CC between n-th and n+1-th by integrating for several point of interval.

sequences starting at two nearby laser pulses. To ensure that a single pulse yields one quadrature value, the interval integration is started at the point that maintains the minimum CC between two quadrature sequences corresponding the n-th pulse and the n+1-th pulse, respectively.

Figure 2(b) shows the CC between adjacent pulses as a function of the LO power. Each point is the average of ten measurements, and error bar is standard deviation of ten times measurements. At the low LO power, the correlation exhibits due to low frequency noise. Figure 2(c) depicts the CC between two measured quadrature sequences corresponding to the n-th pulse and the n+m-th pulse at an LO power of 1.6 mW. Each point is the average of ten measurements. This figure clearly shows complete correlation when m= 0 and negligible dependence for m> 1. The value of CC between the quadrature of n-th pulse and the quadrature of n+1-th pulse is 0.04. The open circles represent the CCs between two quadrature sequences produced by the n-th interval and the n+1-th interval when an inappropriate starting point for interval integration is used. A correlation clearly exists because one pulse is split across two adjacent integration intervals.

To demonstrate an application of our HD system in the time domain, we also measured squeezed states. Figure 3(a) gives the measured quadrature amplitude distribution and the best-fitting Gaussian distribution for squeezed quadrature, antisqueezed quadrature, and SNL. The quadrature amplitude distributions are constructed from 7600 pulses at the corresponding LO phase, which is monitored by a SA in the frequency domain. The frequency-domain results are presented in Fig. 3(b). The results obtained in both the frequency and time domains demonstrate a squeezing of −2.3 dB below the SNL and antisqueezing of +7.5 dB above the SNL. In the time domain measurement, a CC of 0.04 between adjacent squeezed pulses was obtained.

CONCLUSION

In conclusion, we have described an efficient pulsed homodyne system that operates in the time domain and can measure individual pulses at high repetition rates. The value of

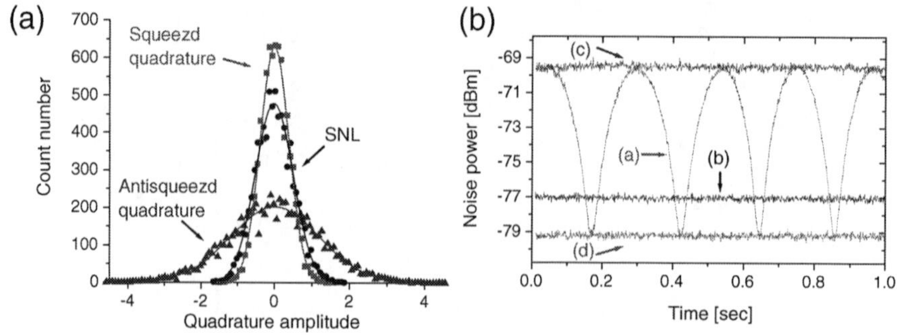

FIGURE 3. Measured quadrature amplitudes for squeezed quadrature, antisqueezed quadrature, and SNL both in the time domain and the frequecny domain. Fig. 3(b): Noise power for the SNL and the pulsed squeezed light at a fixed RF of 10 MHz. Trace (a): LO phase scanned, trace (b): SNL, trace (c): LO phase is fixed in the antisqueezed quadrature, trace (d): LO phase is fixed in the squeezed quadrature. The resolution bandwidth of the SA is 1 MHz, and its video bandwidth is 300 Hz.

CC between the quadrature of n-th pulse and the quadrature of n+1-th pulse is 0.04. As a demonstration of an application of our HD, a squeezed state with squeezing of −2.3 dB was measured in both the time and frequency domains.

REFERENCES

1. S. L. Braunstein and A. K. Pati eds., *Quantum Information with Continuous Variables* (Kluwer Academic, Dordrecht, 2003).
2. H. P. Yuen and V. W. S. Chan, Opt. Lett. **8**, 177 (1983).
3. A. Zavatta, S. Viciani, and M. Bellini, Science **306**, 660 (2003).
4. A. Ourjoumtsev, R. Tualle-Brouri, J. Laurat, and P. Grangier, Science **312**, 83 (2006).
5. D. T. Smithey, M. Beck, A. Faridani, and M. G. Raymer, Phys. Rev. Lett. **70**, 1244 (1993).
6. A. Zavatta, M. Bellini, P. L. Ramazza, F. Marin, and F. T. Arecchi, J. Opt. Soc. Am. B **19**, 1189 (2002).
7. H. Hansen, T. Aichele, C. Hettich, P. Lodahl, A. I. Lvovsky, J. Mlynek, and S. Schiller, Opt. Lett. **26**, 1714 (2001).
8. R. Okubo, M. Hirano, Y. Zhang, T. Hirano, Opt. Lett. **33**, 1458 (2008).
9. Y. Zhang, T. Furuta, R. Okubo, K. Takahashi, and T. Hirano, Phy. Rev. A **76**, 012314 (2007).
10. R. E. Slusher, P. Grangier, A. LaPorta, B. Yurke, and M. J. Potasek, Phys. Rev. Lett. **59**, 2566 (1987).
11. T. Hirano and M. Matsuoka, Opt. lett. **15**, 1153 (1990).
12. M. E. Anderson, M. Beck, M. G. Raymer, and J. D. Bierlein, Opt. Lett. **20**, 620 (1995).
13. J. Wenger, R. Tualle-Brouri, and P. Grangier, Opt. Lett. **29**, 1267 (2004).
14. Y. Eto, T. Tajima, Y. Zhang, and T. Hirano, Opt. Lett. **32**, 1698 (2007).

Generation of Squeezed Vacuum States via Polarization Self-Rotation

G. Campbell, C. Healey, J. Appel, K.-P. Marzlin and A. I. Lvovsky

Institute for Quantum Information Science, University of Calgary, Alberta T2N 1N4, Canada

Abstract. Recently, the possibility of generating squeezed vacuum states via polarization self-rotation in hot atomic vapour has been debated. We present experimental results confirming that vacuum squeezed states can be generated using this method.

Keywords: squeezing, atomic vapour, non-linear optics
PACS: 42.50.Dv

INTRODUCTION

Squeezed vacuum states [1], in which the noise for some quadrature values drops below the quantum noise limit, are frequently used as a quantum resource in quantum communication, processing and cryptographic protocols. Despite the widespread use of squeezed vacuum states, producing them is typically not a trivial task. Using self-rotation in atomic vapour provides a simpler method to generate squeezed vacuum states [2] than many currently used.

The effect of squeezing via polarization self-rotation arises from cross-phase modulation (XPM) induced by non-linear interaction with the pump field. An elliptically polarized field in a Kerr medium will rotate by $\phi = gl\varepsilon$ where $\varepsilon \ll 1$ is the ellipticity, l the cell length and g the rotation parameter (Fig.1).

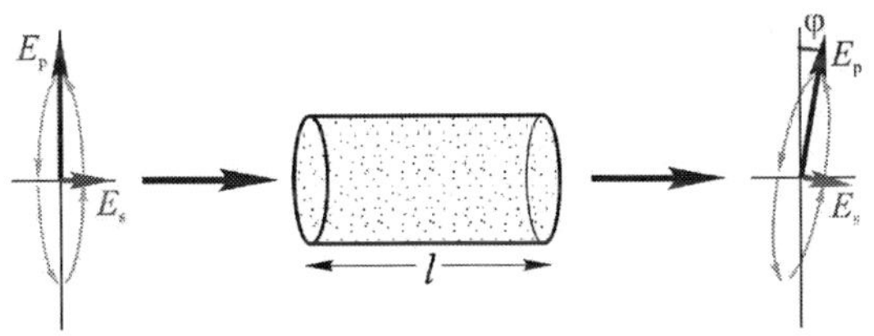

FIGURE 1. Propagation through the medium causes a polarization rotation of elliptically polarized light.

Suppose now the light entering the Kerr medium is linearly, vertically (y) polarized. Vacuum field noise in the orthogonal (x) polarization component leads to microscopic ellipticity, which is given by $\varepsilon = P_x/X_y$, where (X,P) are the field quadratures in and out of phase with the macroscopic field, respectively. Polarization self-rotation results in the macroscopic y-polarized field making a contribution to the X quadrature of the orthogonal polarization component, leading to the following transformation:

$$X_x' \rightarrow X_x + glP_x \tag{1}$$

CP1110, *Quantum Communication, Measurement and Computing (QCMC), The Ninth International Conference,* edited by A. Lvovsky

$$P_x' \to P_x \tag{2}$$

The mapping results in a linear shear of the Wigner function of the field state in the horizontal polarization, leading to quadrature squeezing (Fig.2).

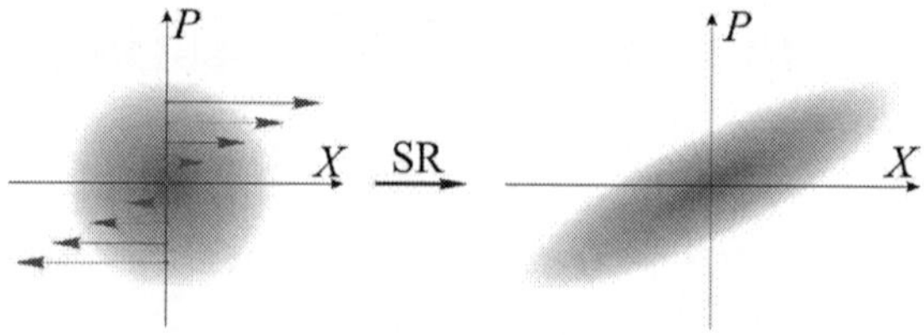

FIGURE 2. Transformation of the vacuum Wigner function due to the XPM.

Following the proposal, a 2003 experiment [3] demonstrated 0.7 dB of squeezing below the quantum noise limit using this technique. The result, however, was contested on the grounds that noise due to spontaneous emission into the signal mode would overwhelm the squeezed state and prevent a reduction of noise below the quantum limit [4]. The claim was backed by two experiments that were unable to reproduce the 2003 result.

The discrepancy in the experiments demands further investigation such that the matter can be resolved. A more detailed theoretical analysis than what has been presented thus far is underway; however, this paper is focused on the presentation of clear experimental results demonstrating a squeezed vacuum state.

We have repeated the experiment and confirmed the earlier results that squeezing can be obtained. Our investigation has yielded noise reduction of up to 0.8dB below the quantum limit with a bandwidth of approximately 20 MHz.

EXPERIMENT

^{87}Rb, heated to 80°C in a glass vapour cell, is used. The setup requires only a single, linearly polarized laser field to be sent through an atomic vapour near resonance. The input field is generated by an external cavity diode laser with a pump power of up to 30mW and a beam diameter of 0.8 μm. Detection of squeezing is done by frequency-domain balanced homodyne detection. The vertically polarized light is split off from the squeezed vacuum signal using a polarizing beam splitter, and its polarization is rotated such that it can then be used as the local oscillator for homodyne detection (Fig. 3).

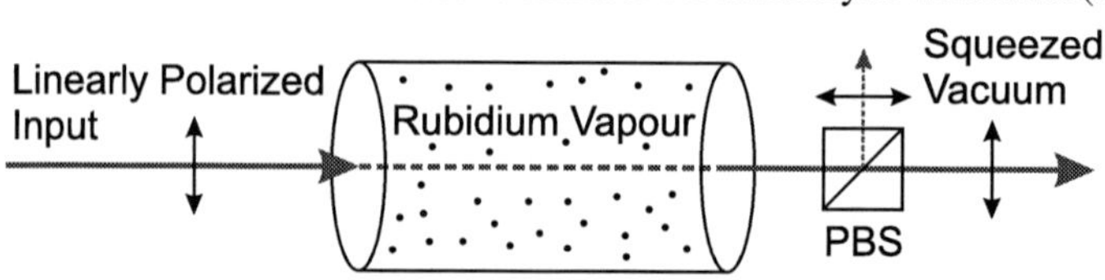

FIGURE 3. A simplified experimental diagram of the setup.

We were able to reproduce the results of the original 2003 experiment and obtained squeezing below the quantum noise limit in a range of 1 MHz to 20 MHz (Fig. 4) with a maximum of 0.8 dB at roughly 3 MHz. Once squeezing was observed we attempted to eliminate any possibility of a noise reduction due to sources other than atomic squeezing. The noise characteristics of the detector were investigated in the absence of the signal field. We found the the shot noise scaled linearly with the pump power, a signature of shot noise measurement, to a correlation of 0.99. We also confirmed that the laser was within 0.2 dB of being shot noise limited by checking noise addition on the two detectors. The common-mode rejection ratio of the detector was ≈ 40 dB. Artefacts due to electronic noise or saturation of the detectors were also ruled out.

Squeezing was observed using two homodyne detectors of considerably different designs (one of which was that used in Ref. [3]), each confirmed to operate properly. We also obtained squeezing using the vapour cell that was used in Ref. [4] as well as two of our own vapour cells, ruling out impurities in one of the cells as a source of the discrepancy. Squeezing was also revealed to be relatively insensitive to initial ellipticity and to imperfect subtraction. Both effects lead to a displaced squeezed state but little reduction in squeezing.

Discussion with the authors of Ref. [4] have not yielded a conclusion regarding what caused the extra noise; however, some possibilities have been put forward. A possible source of extra noise that has been identified is the active frequency stabilization systems in the Ti:Sapphire lasers used in the experiments of Ref. [4], since all successful experiments have used diode laser systems. Furthermore, the loss-sensitive nature of the squeezed state requires excellent transmission through all optics; condensation of rubidium vapour of the cell's windows presents a problem for squeezing using this method. To mitigate this effect we employed an air cooling system to keep the cell side walls cooler than the windows.

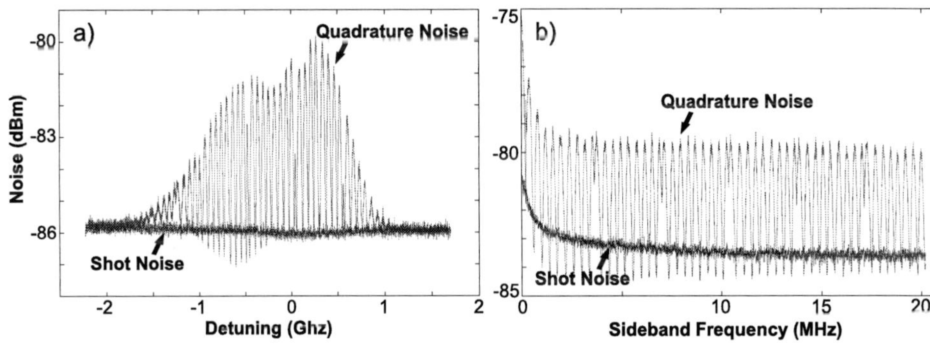

FIGURE 4. (a) Squeezing as a function of detuning at a sideband of 3MHz. Here the detuning is scanned slowly while the quadrature phase is scanned rapidly. (b) Spectrum of squeezed light. The sideband frequency is scanned slowly, the quadrature phase quickly and the detuning is held constant where maximum squeezing is observed.

CONCLUSION

We have observed 0.8 dB of squeezing below the shot noise limit using cross-phase modulation via polarization self-rotation. A recent experimental paper [5] has also reported 0.9 dB of squeezing via polarization self-rotation at low sideband frequencies (<1 MHz). Also a recent theoretical work [6] with detailed theoretical analysis does not rule out the possibility of squeezing in a system similar to ours. Theoretical work is under way to examine the noise in a more rigorous manner than has previously been done and preliminary results support the latest experimental findings.

ACKNOWLEDGMENTS

We would like to thank Magnus Hsu and Ping Koy Lam for fruitful discussions and collaboration. We are grateful to NSERC, AIF, CFI, Quantum*Works* and CIFAR for funding opportunities.

REFERENCES

1. D. F. Walls (Nature 306, 141-146, 1983).
2. A. B. Matsko, I. Novikova and G.R. Welch, (Phys. Rev. A **66**, 043815, 2002).
3. J. Ries, B. Brezger and A. I. Lvovsky, (Phys. Rev. A **68**, 025801, 2003).
4. M. T. L. Hsu *et al.*, (Phys. Rev. A **73**, 023806, 2006).
5. E. Mikhailov and I. Novikova (Opt. Lett. **33**, 1213 2008).
6. A. Lezama *et al.*, (Phys. Rev. A **77**, 013806 (2008).

Continuous Variable Entanglement Distillation of Non-Gaussian States

Mikael Lassen*, Ruifang Dong*, Joel Heersink*, Christoph Marquardt*, Radim Filip†, Gerd Leuchs* and Ulrik L. Andersen**

*Institut für Optik, Information und Photonik, Max-Planck Forschungsgruppe, Universität Erlangen-Nürnberg, Günther-Scharowsky-Str. 1, 91058, Erlangen, Germany
†Department of Optics, Palacký University, 17. listopadu 50, 77200 Olomouc, Czech Republic
**Department of Physics, Technical University of Denmark, Building 309, 2800 Lyngby, Denmark

Abstract. We experimentally demonstrate distillation of continuous variable entangled light that has undergone non-Gaussian attenuation loss. The continuous variable entanglement is generated with optical fibers and sent through a lossy channel, where the transmission is varying in time. By employing simple linear optical components, a measurement induced operation and classical communication, we demonstrate that the entanglement is probabilistically increased.

Keywords: Entanglement,distillation, quantum information
PACS: 42.50.Dv,42.50.Ex

1. INTRODUCTION

Quantum information technology is a promising platform for developing new optical communication systems with absolute security and higher information rates. However, the technological difficulty is to distribute highly entangled states over long distances, which due to the unavoidable coupling with the environment, leads to losses and decoherence of the entangled states. Fortunately, this can be overcome by using entanglement distillation, which is a protocol that extracts from an ensemble of less entangled states a subset of states with a higher degree of entanglement [1]. This protocol has sofar only been experimentally demonstrated for discrete variable systems [2, 3]. The implementation of a scheme that is capable of distilling entanglement of continuous variable (CV) systems has remained an experimental challenge. However, it has been shown theoretically that entanglement distillation can be done by performing highly nonlinear, thus difficult, operations no matter whether it is a pure or a mixed state [4, 5]. Fortunately, in many practical scenarios the transmitted quantum state will be non-Gaussian: for example the transmission of light through a turbulent atmospheric channel, where the attenuation coefficient will fluctuate in time, results in a non-Gaussian quantum state. As we will show it is therefore possible to distill entanglement that has undergone such non-Gaussian noise by means of a simple measurement induced operation and classical communication.

CP1110, *Quantum Communication, Measurement and Computing (QCMC), The Ninth International Conference*, edited by A. Lvovsky

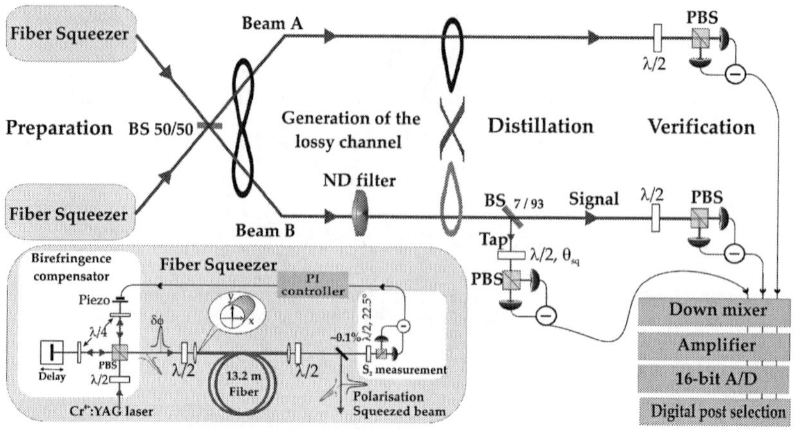

FIGURE 1. Schematics of the experimental setup for the preparation, distillation and verification of entanglement.

2. THE EXPERIMENTAL REALIZATION

The distillation of corrupted entangled states consists of three parts as schematically illustrated in Fig. 1: the preparation, distillation and verification. We generate polarization entanglement by interfering two independent polarization squeezed beams A and B with the same noise properties on a 50/50 beam splitter (BS). The squeezing is generated by single passing two linearly polarized light pulses with identical intensity through the fiber along the orthogonal polarization axes, which generates two quadrature squeezed states, independently. The emerging pulses are overlapped with a $\pi/2$ relative phase shift after the fiber, thus generating a circular polarized light pulse. With this method polarization squeezing is generated [6]. The non-Gaussian mixture of polarization entangled states is implemented by sending one of the entangled beams, e.g. beam B, through a controllable neutral density filter (ND).

The distillation operation consists of a tap measurement of $\hat{X}$ on a small portion (7%) of the mixed entangled beam, and a probabilistic heralding process where the remaining state is kept or discarded conditioned on the measurement outcomes: if the outcome of the weak measurement is larger than some threshold value, X_{th}, the state is kept. The verification measurement are accomplished simultaneously in two independent Stokes measurements, see Fig. 1 .

The channel we are testing here is a discrete information channel. The channel alternates between full transmission and 25% transmission with each realization occurring at a probability of 0.5. After transmission through such a channel the resulting state is a mixture of a highly entangled state and a weakly entangled state. The statistics of beam B is contaminated with the attenuated entangled state thus producing non-Gaussian statistics. For this state we measure the correlations in $\hat{X}$ and $\hat{P}$ to be above the shot noise level by 5.5 ± 0.3 dB and 5.6 ± 0.3 dB, respectively, and the Gaussian logarithmic negativity (LN) to be -1.63 ± 0.02. Note that the LN, is only valid as an entanglement monotone

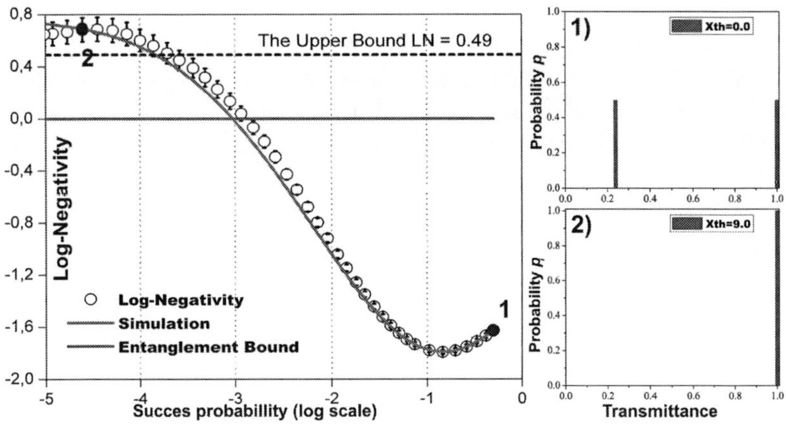

FIGURE 2. Experimental results for distillation of an entangle state transmitted through a discrete lossy channel. The experimental results are marked by circles and the theoretical prediction is plotted by the red solid line. The bound for Gaussian entanglement is given by the blue line, and the upper bound for total entanglement before distillation is given by the black dashed line. The weight of the two constituents in the mixed state after distillation for various threshold values is shown in the plots labelled by 1 and 2.

if the state is Gaussian, which is not the case for all stages of our experiment. However, the Gaussian LN is a good measure of entanglement useful for Gaussian operations. We find that the Gaussian entanglement is completely lost as a result of the introduction of such time-dependent loss. This is in contrast to the scenario where only stationary loss ,corresponding to Gaussian loss, is inflicting the entangled states. In the stationary case, a certain degree of Gaussian entanglement will always survive, although it will be small for high loss level.

The LN as a function of the success probability (postselection threshold) is shown in Fig. 2 by black open circles. The probability distributions of the two superimposed states in the mixture after distillation are shown for different postselection thresholds, which corresponds to $X_{th} = 0.0, 9.0$, labelled by 1 and 2, respectively. As the threshold increases, the Gaussian LN increases, ultimately approaching the LN of the input entanglement and the probability distribution tends to a single valued distribution, therefore the mixture of two Gaussian entangled states reduces to a single highly entangled Gaussian state, thus demonstrating the act of Gaussification; however, the amount of distilled data or success probability decreases, causing an increase in the statistical error on the distillable entanglement. Based on the experimental parameters, the theoretical simulation is shown by red curve and it shows a very good agreement with the experimental results. To show that the total entanglement of the state has increased as a result of distillation, we compute an upper bound for the LN before distillation given by: $LN = Log_2||\rho||_1$, where ρ is the partial transposed density matrix of the state. The upper bound of the LN without the Gaussian approximation is computable from the LN of each Gaussian state in the mixture and we find $LN_{upper} = 0.49$, which is shown in Fig. 2 by the dashed black line. We see that for a success probability around 10^{-4} the Gaussian LN crosses the upper bound for entanglement, and since the state at this point is perfectly Gaussified we may conclude that the total entanglement of the state has increased as a result of the distillation protocol. Fig. 2-2 gives the explicit explanation by showing that the probability contribution from the 75% attenuated data reaches 0 when the post selection threshold is set to $X_{th} = 9.0$, this corresponds to the distilled entanglement of $LN_S^{dist} = 0.67 \pm 0.08$ with a success probability $P_S = 1.69 \times 10^{-5}$. However, we see that even a small contribution from the 75% attenuated data will submerge the useful entanglement for Gaussian operations.

3. SUMMARY

We have experimental realized distillation of CV entanglement. Simply by implementing a weak measurement based on a beam splitter, a homodyne detector and classical communication. The protocol was successfully demonstrated for a discrete transmission channel. We would like to note that in Ref.[7] different kinds of transmission channels has been investigated.

ACKNOWLEDGMENTS

This work was supported by the EU project COMPAS (no.212008), the Deutsche Forschungsgesellschaft and the Danish Agency for Science Technology and Innovation (no. 274-07-0509). ML acknowledges support from the Alexander von Humboldt Foundation and RF acknowledges MSM 6198959213, LC 06007 of Czech Ministry of Education and 202/07/J040 of GACR.

REFERENCES

1. Bennett, C. H. *et al.* Purification of noisy entanglement and faithful teleportation via noisy channels. *Phys. Rev. Lett.* **76**, 722-725 (1996).
2. Pan, J.-W., Gasparoni, S., Ursin, R., Weihs, G. & Zeilinger, A. Experimental entanglement purification of arbitrary unknown states. *Nature*, **423**, 417-422 (2003).
3. Zhao, Z., Yang, T., Chen, Y.-A., Zhang, A.-N. & Pan, J.-W., Experimental realization of entanglement concentration and a quantum repeater. *Phys. Rev. Lett.* **90**, 207901(2003).
4. Eisert, J., Scheel, S. & Plenio, M. B. Distilling Gaussian states with Gaussian operations is impossible. *Phys. Rev. Lett.* **89**, 137903(2002).
5. Fiurasek, J., Gaussian transformations and distillation of entangled Gaussian States. *Phys. Rev. Lett.* **89**, 137904 (2002).
6. Dong, R. *et al.* Experimental evidence for Raman-induced limits to efficient squeezing in optical fibers. *Opt. Lett.* **33**, 116-118(2008).
7. R. Dong, M. Lassen, J. Heersink, C. Marquardt, R. Filip, G. Leuchs and U. L. Andersen, Experimental entanglement distillation of mesoscopic quantum states, *Nature physics*, **4**, 919 - 923 (2008).

Pulsed homodyne detection of quadrature entanglement at telecommunication wavelength

Yujiro Eto*,†, Akihiro Nonaka*,†, Yun Zhang** and Takuya Hirano*,†

*Department of Physics, Gakushuin University, 1-5-1 Mejiro, Toshima-ku, Tokyo, 171-8588, Japan
†CREST, Japan Science and Technology Agency, 1-9-9 Yaesu, Chuou-ku, Tokyo, 103-0028, Japan
**Department of Electronic Engineering, The University of Electro-Communications, Chofu 182-8585, Japan

Abstract. We report on the generation of quadrature entangled light pulses at the telecommunication wavelength of 1.535 μm by combining two squeezed pulses. The quadrature phase amplitudes of the entangled light pulses are measured by using two time-domain pulsed homodyne detectors, which yield one pair of measurement values by single-shot measurement for each pulse. The entanglement is confirmed with a sufficient criterion $\langle \Delta[\hat{x}_a + \hat{x}_b]^2 \rangle + \langle \Delta[\hat{p}_a - \hat{p}_b]^2 \rangle = 0.64 < 1$.

Keywords: quadrature entanglement, pulsed homodyne detection, telecommunication wavelength, ring interferometer
PACS: 03.67.Hk, 42.50.Dv, 42.65.Wi

INTRODUCTION

Entanglement is a most important resource in the field of continuous-variable quantum communications (CVQCs) [1]. Many groups reported on the generation of quadrature entanglement at various wavelengths [2-12]. Entanglement at telecommunication wavelengths is very useful for fiber-based CVQCs because the existing optical fiber has a minimum-loss window for the telecommunication wavelengths. Silberhorn *et al.* reported on the generation of entanglement at telecommunication wavelength employing the $\chi^{(3)}$ nonlinearlity [5], and they observed entangled light pulses in the frequency domain using a spectrum analyzer. For the practical applications of CVQCs, the time-domain measurement for individual pulses [13, 14] would be preferable to the measurement in the frequency domain [9].

Quadrature entanglement can be generated by combining two squeezed beams on a beam splitter [4]. The degree of entanglement significantly depends on the relative phase between two squeezed beams. Therefore the stability and the controllability of the relative phase between two squeezed beams are required for the prolonged generation of high-quality entanglement.

In this paper, we report on the pulsed homodyne detection of quadrature entanglement at telecommunication wavelength, which is generated by utilizing a novel scheme using a ring interferometer. In this scheme, the long-term stability and the controllability of the relative phase between two squeezed beams can be achieved simultaneously.

CP1110, *Quantum Communication, Measurement and Computing (QCMC), The Ninth International Conference,* edited by A. Lvovsky

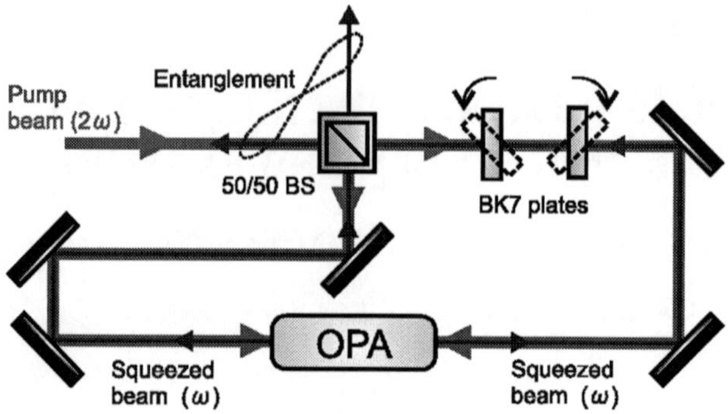

FIGURE 1. Generation of entanglement by using a ring interferometer. The heavy line (the thin line) indicate the light pulse at frequency of 2ω (ω).

GENERATION OF ENTANGLEMENT USING A RING INTERFEROMETER

Figure 1 describes the novel scheme to generate the quadrature entanglement using a ring interferometer, which consists of a dual-wavelength 50/50 beam splitter (50/50 BS) at frequencies of both ω and 2ω, an optical parametric amplifier (OPA), and dispersive media (two BK7 plates). The pump beam at frequency 2ω, which is employed to pump the OPA, is divided into half by 50/50 BS. The two squeezed beams at ω are generated by bidirectional pumping, and they are combined on the 50/50 BS. By virtue of the common-path arrangement of the ring-type interferometer, The relative phase between two squeezed beams θ is kept stable. The beams at different frequencies ω and 2ω pass bidirectionally through BK7. Thanks to the refractive-index dispersion of BK7, the relative phase can be changed by symmetrical tilting two BK7 plates.

In the actual experiment, the light source is a passively Q-switched erbium doped glass laser emitting light pulses at wavelength of 1.535 μm with a duration of 3.9 ns and a repetition rate of 2.7 kHz (Cobolt model Tango). The second harmonic pulse at 767 nm is used as the pump pulse for the OPA [15, 16]. The two squeezed beams are generated in a periodically poled MgO:$LiNbO_3$ waveguide (NGK Insulators Ltd.). The each of generated entangled beam is interrogated in the pulsed homodyne detector.

OBSERVATION OF QUANTUM CORRELATIONS

The amplitude quadratures (x_a and x_b) and the phase quadratures (p_a and p_b) of the entangled beams are measured, after θ is set to $\pi/2$. Figure 2(a) shows the measured values of x_a and x_b for 30 entangled pulse pairs. Although the quadrature values of x_a or x_b seem to be randomly distributed, the simultaneously measured x_a and x_b are anticorrelated, namely $x_a \simeq -x_b$. Figure 2(b) shows the measured values of p_a and p_b. p_a and p_b are correlated ($p_a \simeq p_b$). In this measurment output pump average powers from PPLN2 are 71 μW and 58 μW (the peak powers are 6.9 W and 5.7 W) in

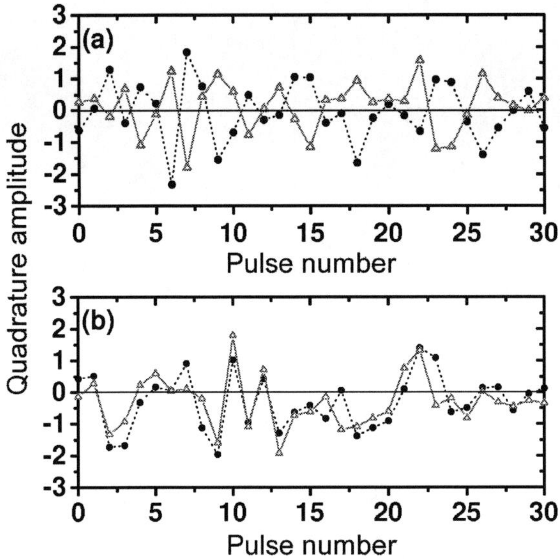

FIGURE 2. (a) The measured values of x_a and x_b for 30 entangled pulse pairs. (b) The measured values of p_a and p_b for 30 entangled pulse pairs.

the clockwise and the counterclockwise direction. The quantum efficiencies of two homodyne detectors are 0.80 and 0.84. The electronic noise levels of two homodyne detectors are 11.8 dB and 12.6 dB below the shot noise level respectively.

Figure 3(a) shows the scatter plot of the measured x_a against the data of the measured x_b. The number of total pulse pairs is 2000, and the total measuring time is about 45 seconds. The quantum correlation of $\langle\Delta[\hat{x}_a+\hat{x}_b]^2\rangle = 0.32$ (-2.0 dB compared to the corresponding vacuum variances) is obtained after correcting for the electronic noise. The scatter plot for $\hat{p}_a$ and $\hat{p}_b$ is shown in Fig. 3(b). The quantum correlation of $\langle\Delta[\hat{p}_a-\hat{p}_b]^2\rangle = 0.32$ (-1.9 dB) is obtained. These results show that a sufficient criterion for entanglement [17, 18] is satisfied $\langle\Delta[\hat{x}_a+\hat{x}_b]^2\rangle + \langle\Delta[\hat{p}_a-\hat{p}_b]^2\rangle = 0.32+0.32 = 0.64 < 1$.

CONCLUSION

We have reported the pulsed homodyne detection of the quadrature entanglement at telecommunication wavelength. The entangled light pulses are generated by utilizing our proposed method based on the use of the ring interferometer. By using our proposed method, the relative phase between squeezed beams is controllable easily and is kept stable. We believe that our experimental results will be useful for practical applications of CVQCs.

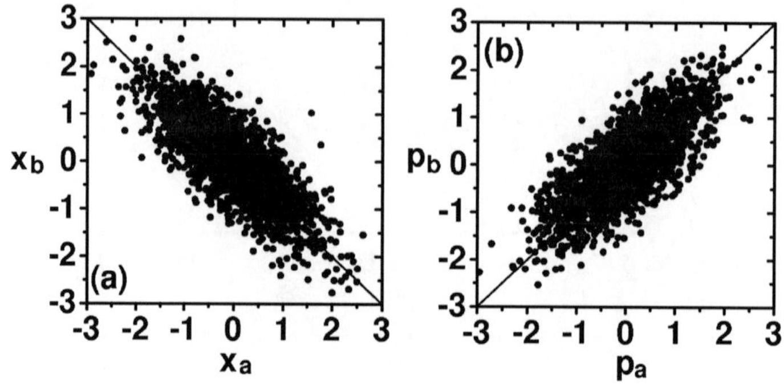

FIGURE 3. (a) The scatter plot of the measured x_a against the data of the measured x_b. (b) The scatter plot of the measured p_a against the data of the measured p_b.

ACKNOWLEDGMENTS

The authors gratefully acknowledge the contribution of Takashi Tajima. This work is partly supported by Grants-in-Aid for Scientific Research from the Ministry of Education, Culture, Sports, Science, and Technology of Japan. One of the authors (Y. E.) is supported by the Research Fellowships of the Japan Society for the Promotion of Science for Young Scientists.

REFERENCES

1. S. L. Braunstein and P. van Loock, Rev. Mod. Phys. **77,** 513 (2005).
2. Z. Y. Ou, S. F. Pereira, H. J. Kimble, and K. C. Peng, Phys. Rev. Lett. **68,** 3663 (1992).
3. Z. Y. Ou, S. F. Pereira, and H. J. Kimble, Appl. Phys. B **55,** 265 (1992).
4. A. Frusawa *et al.*, Science **282,** 706 (1998).
5. Ch. Silberhorn *et al.*, Phys. Rev. Lett. **86,** 4267 (2001).
6. W. P. Bowen, R. Schnabel, P. K. Lam, and T. C. Ralph, Phys. Rev. Lett. **90,** 043601 (2003).
7. W. P. Bowen, R. Schnabel, P. K. Lam, and T. C. Ralph, Phys. Rev. A **69,** 012304 (2004).
8. J. Laurat *et al.*, Phys. Rev. A **71,** 022313 (2005).
9. J. Wenger, A. Ourjoumtsev, R. Tualle-Brouri, and P. Grangier, Eur. Phys. J. D **32,** 391 (2005).
10. N. Takei *et al.* Phys. Rev. A **74,** 060101(R) (2006).
11. K. Yoshino, T. Aoki, and A. Furusawa, Appl. Phys. Lett. **90,** 041111 (2007).
12. Y. Zhang *et al.*, Phys. Rev. A **76,** 012314 (2007).
13. D. T. Smithey, M. Beck, M. G. Raymer, and A. Faridani, Phys. Rev. Lett. **70,** 1244 (1993).
14. R. Okubo, M. Hirano, Y. Zhang, and T. Hirano, Opt. Lett. **33,** 1458 (2008).
15. Y. Eto, T. Tajima, Y. Zhang, and T. Hirano, Jpn. J. Appl. Phys., Part 2 **45,** L821 (2006).
16. Y. Eto, T. Tajima, Y. Zhang, and T. Hirano, Opt. Lett. **32,** 1698 (2007).
17. L. M. Duan, G. Giedke, J. I. Cirac, and P. Zoller, Phys. Rev. Lett. **84,** 2722 (2000).
18. R. Simon, Phys. Rev. Lett. **84,** 2726 (2000).

Levels of Quantum Correlations in the Continuous Variable Regime: Review and Experimental Illustrations with OPOs

V. D'Auria [a], G. Keller [a], J. Laurat [a], N. Treps [a], T. Amri [a], T. Coudreau [b], C. Fabre [a]

[a] *Laboratoire Kastler Brossel, Université P. et M. Curie, Ecole Normale Supérieure, CNRS, 4 place Jussieu 75252 Paris, France*

[b] *Laboratoire Matériaux et Phénomènes Quantiques, Université Denis Diderot, CNRS, 10 rue Alice Domon et Léonie Duquet 75205 Paris, France*

Abstract. We discuss different levels of quantum correlations in the continuous variable regime and illustrate them by means of experimental results with an optical parametric oscillators. More specifically, we report the first experimental generation of bright frequency-degenerate EPR beams with a type II OPO. Degeneracy is obtained by introducing a birefringent plate inside the cavity, resulting in phase locking. Correlation is characterized by homodyne detection.

Keywords: Quantum Optics, quantum fluctuations, entanglement, Optical parametric oscillator.
PACS: 42.50.Dv; 42.50.Lc

THE SELF-PHASE LOCKED OPO

Beyond a fundamental significance, entanglement has proven to be a necessary resource for quantum information protocols. In quantum optics, in parallel to the photon-counting regime, an active direction focuses on continuous variables of light. Parametric interaction in an optical cavity provides the most efficient technique to generate entanglement. An OPO consists in a non-linear crystal inserted inside an optical cavity. By pumping it with a laser beam with power above a certain threshold, a type II OPO emits bright twin beams orthogonally polarized (the *signal* (A1) and the *idler* (A2)). These beams exhibit correlations in amplitude and anti-correlations in phase, namely entanglement.

In our experiment, the OPO is based on a type II non-linear crystal and it is forced to work at frequency degeneracy. In order to achieve the degenerate operation, a birefringent plate is inserted inside the OPO cavity; when tilted with respect to the crystal axes, the plate projects a small fraction of the idler on the signal and vice versa, thus introducing a linear coupling between the two beams. As an effect of the linear coupling, the beams tend to synchronize their oscillations [1]. For a well-defined range of parameters, the frequency-locking phenomenon occurs and can be maintained during more than one hour. In experiments, the plate orientation is fixed with respect

CP1110, *Quantum Communication, Measurement and Computing (QCMC), The Ninth International Conference,* edited by A. Lvovsky

to the crystal axes and induce the locking phenomenon by acting on the crystal temperature and on the pump frequency. For very small coupling (0.02°), this configuration preserves the strong entanglement between the signal and idler modes predicted in an OPO above threshold. Such an OPO is called a self phase locked OPO [2].

In experiments we set the OPO to frequency degeneracy operation and detect together with the signal and idler beams two auxiliary modes (sum/difference of the entangled fields A±). A single homodyne detector measures the quadrature noise of each beam, with local oscillator at the degenerate frequency. Considering the noise spectrum of the sum or difference of signal and idler fluctuations is strictly equivalent to considering the noise spectrum of the rotated modes. For entangled modes A1 and A2, the noise fluctuations of the modes A± are squeezed below the standard quantum limit (normalized to 1 in the following), respectively on the phase quadrature (Y)) and the orthogonal quadrature (X)). By measuring the squeezing, we obtain information on the entanglement.

Different Level of quantum correlation for the OPOs.

Different criteria have been developed to prove entanglement [3]. A first level is given by the Mancini inequality and its corollary, the Duan inequality, both expressed in terms of the variances of modes A± respectively as: $\Delta^2Y)\,\Delta^2X))\,)$ and) $0.5(\Delta^2Y) + \Delta^2X)))\,)$. These inequalities state a sufficient condition for A1 and A2 to be non-separable. A stricter condition is stated by the EPR criterion, which characterizes the correlations described in 1935 by Einstein, Podolsky and Rosen [4]. This criterion states that the beams are EPR-entangled if, by performing a measurement on one of them, it is possible to retrieve the value of an observable on the other beam within the quantum noise. More specifically the product of the conditional variances $V(X_1|X_2)$ and $V(Y_1|Y_2)$, on the quadratures of A_1, obtained by measuring the quadratures of A_2, must verify: $V(X_1|X_2)V(Y_1|Y_2)<1$. The quantities $V(X_1|X_2)$ and $V(Y_1|Y_2)$ are experimentally retrieved by measuring the individual signal and idler modes as well as their combinations $A_\pm$ [3].

In our experiments we consider the generation of bright entangled beams from a self-phase locked OPO above threshold; the quantum correlations are characterized according to Mancini, Duan and EPR criteria. In the ideal case, the signal and idler beams generated by the OPO exhibit perfect intensity correlations and phase anti-correlations, or, in other words, a perfect noise suppression on the phase quadrature of the fields sum Y) and simultaneously on the orthogonal quadrature of the fields difference X). Nevertheless, above threshold, the presence of classical phase noise on the pump beam can seriously affect the phase anti-correlations [5]. Accordingly EPR criterion is demonstrated in the spectral region where pump noise becomes negligible.

EXPERIMENTAL RESULTS ABOVE THRESHOLD

In our setup, the OPO pump is provided by a continuous wave laser @532nm (Diabolo-Innolight, frequency doubled). In order to reduce the phase noise, the pump is filtered in a triangular cavity. The OPO is a semi-monolithic one employing a KTP non-linear type II crystal. The homodyne local oscillator (LO) is provided by the coherent laser output at 1064 nm (see figure 1). The noise spectrum of the homodyne signal is measured and the noise properties quantified for different frequencies. Thanks to their orthogonal polarizations, the signal and idler at the OPO output beams can be optically combined, by using few polarization components; a half-wave plate followed by a polarizing beam splitter enables to choose the mode to be detected: the signal A1, the idler A2, or the ±45° rotated modes A±. The selected beam is sent to an homodyne detector, measuring its amplitude and phase quadratures. The local oscillator (LO) is provided by the coherent laser output at 1064 nm. The spectrum of the homodyne signal is evaluated and the noise properties quantified for different analyzing frequencies.

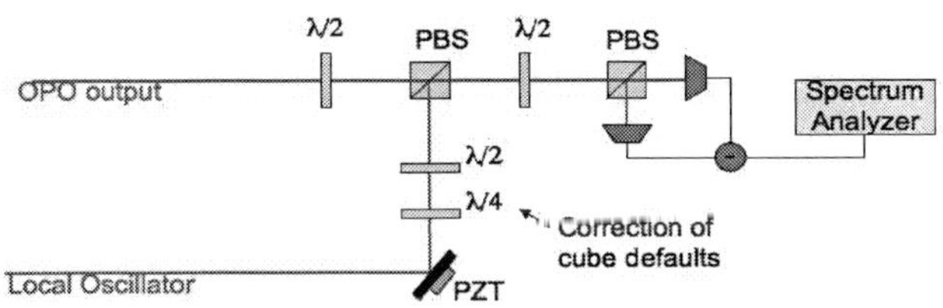

FIGURE 1. : Scheme of the detection. The output of the type II OPO is sent to a system of an half wave plate and a polarizing beam splitter. By turning the plate, it is possible to choose the beam to be detected, the signal A1, the idler A2, or the ±45° rotated modes A±. The selected beam is sent to an homodyne detector, measuring its amplitude and phase quadratures. The local oscillator (LO) is provided by the coherent laser output at 1064 nm.

When scanning the local oscillator phase, one can obtain the whole quadrature noise properties of the modes A±. In order to evaluate the entanglement criteria, one can hold the local oscillator phase so that the measured quadrature would be the one with the minimum noise. Figure 2 a) presents the minimum quadrature noises of the rotated modes, normalized to the shot noise, for an analyzing frequency of 20 MHz, where it has been measured that the pump excess noise can be neglected. Both quadratures are squeezed: the noise suppression is -0.8±0.7 dB for A), and -1.7±0.8 dB for A). The Mancini and the Duan criterion can be directly evaluated from these measurements (cf. Figure 2 b): one finds a separability of 0.7±0.1<1 and 0.55± 0.10 < 1 for the

Mancini criterion: the beams are indeed entangled. In order to check for the more restrictive EPR criterion, the fields A1 and A2 are also detected, as individual noises have to be taken into account to determine conditional variances. As theoretically predicted, the noise on these fields is phase-insensitive and they are found to be approximately shot noise limited. These results correspond to a value for the EPR criterion of 0.85±0.10<1, thus proving the EPR type correlation between the signal and idler beams [6].
In conclusion, the self-locked OPO realizes an extremely compact source of bright EPR beams.

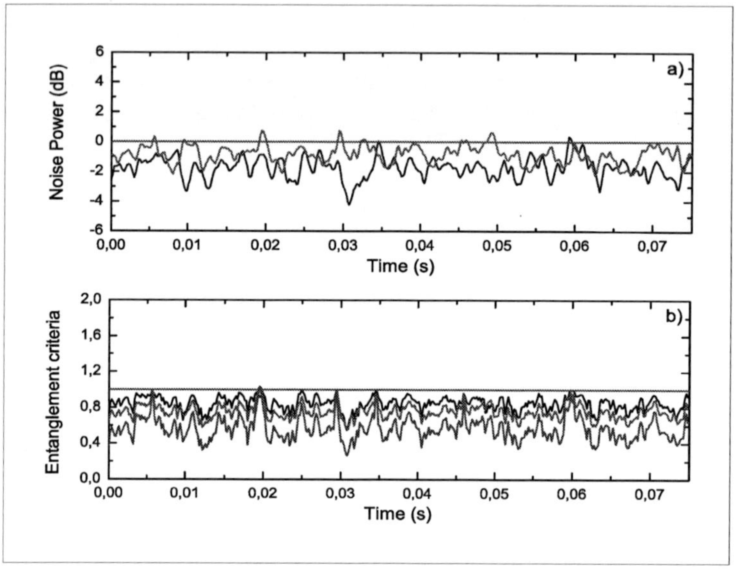

FIGURE 2. : Top: Normalized noise variances @20 MHz of for the Y+ (green curve) and X- (blue). Bottom: Inseparability (violet-Mancini, green-Duan) and EPR (black) criteria. Resolution bandwidth, 100 kHz, video bandwidth 300 Hz.

REFERENCES

1. C. Fabre et al., *Opt. Commun.* **170**, 299-307 (1999)
2. L. Longchambon, *Eur. Phys. J. D* **30**, 279-285 (2004), L. Longchambon, *Eur. Phys. J. D* **30**, 287-293 (2004)
3. N. Treps, C. Fabre, *Laser Physics* **15**, 187 (2005)
4. A. Einstein, B. Podolsky, N. Rosen, *Phys. Rev.* **47**, 777-780 (1935)
5. J. Laurat et al., *Optics Letters* ***30,*** 1177 (2005)
6. G.Keller et al. *Opt. Expr.* **16,** 9351-9356 (2008).

Experimental Demonstration of a Quantum Nondemolition Gate

Yoshichika Miwa[*,†], Jun-ichi Yoshikawa[*,†], Alexander Huck[**,‡], Ulrik L. Andersen[**,‡], Peter van Loock[‡,§] and Akira Furusawa[*,†]

*Department of Applied Physics, School of Engineering, The Unibersity of Tokyo, 7-3-1 Hongo, Bunkyo-ku, Tokyo 113-8656, Japan
†CREST, Japan Science and Technology Agency, 5 Sanbancho, Chiyoda-ku, Tokyo 102-0075, Japan
**Department of Physics, Technical University of Denmark, Building 309, 2800 Lyngby, Denmark
‡Institut für Optik, Information und Photonik, Max-Planck Forshungsgruppe, Universität Erlangen-Nürnberg, Günther-Scharowsky Straße 1, 91058 Erlangen, Germany
§OQI Group, Institute of Theoretical Physics I, Universität Erlangen-Nürnberg, Staudtstr. 7/B2, 91058 Erlangen, Germany

Abstract. We experimentally demonstrate a quantum nondemolition (QND) gate which is a continuous variable analogue of a qubit controlled-NOT gate. The QND gate is shown to entangle two initially separable coherent states, which will find uses in cluster state computation.

Keywords: quantum optics, quantum computation, entangle
PACS: 03.67.Lx, 03.67.Mn

INTRODUCTION

A quantum nondemolition (QND) gate is a two-mode entangling gate for continuous variable (CV) quantum information processing. Unlike a beam splitter, the QND gate is capable of entangling two input coherent states. As applications, the QND gate can be used for CV cluster computation [1, 2, 3] and CV error correction [4, 5].

The QND gate is characterized by the interaction Hamiltonian $\hat{H}_{\text{int}} \propto \hat{x}_1 \hat{p}_2$, where the subscripts "1" and "2" denote two independent modes. The input-output relations in Heisenberg picture are as below:

$$\begin{aligned} \hat{x}_1^{\text{out}} &= \hat{x}_1^{\text{in}} & \hat{x}_2^{\text{out}} &= \hat{x}_2^{\text{in}} + G\hat{x}_1^{\text{in}} \\ \hat{p}_1^{\text{out}} &= \hat{p}_1^{\text{in}} \quad G\hat{p}_2^{\text{in}} & \hat{p}_2^{\text{out}} &= \hat{p}_2^{\text{in}} \end{aligned} \tag{1}$$

where G is the interaction gain which is variable in our experiment. We examined these input-output relations. Furthermore we verified entanglement in the output states. The performance for the QND measurements and the results for $G = 1\ 0$ are discussed in Ref. [6]; hence, here we will concentrate on the experimental details for $G = 1\ 5$.

CP1110, *Quantum Communication, Measurement and Computing (QCMC), The Ninth International Conference,* edited by A. Lvovsky

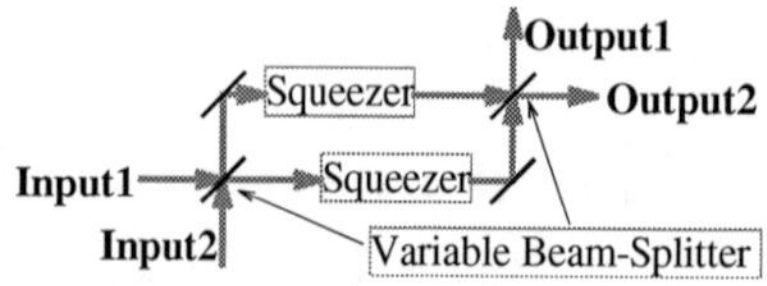

(a)Schematic of the QND gate.

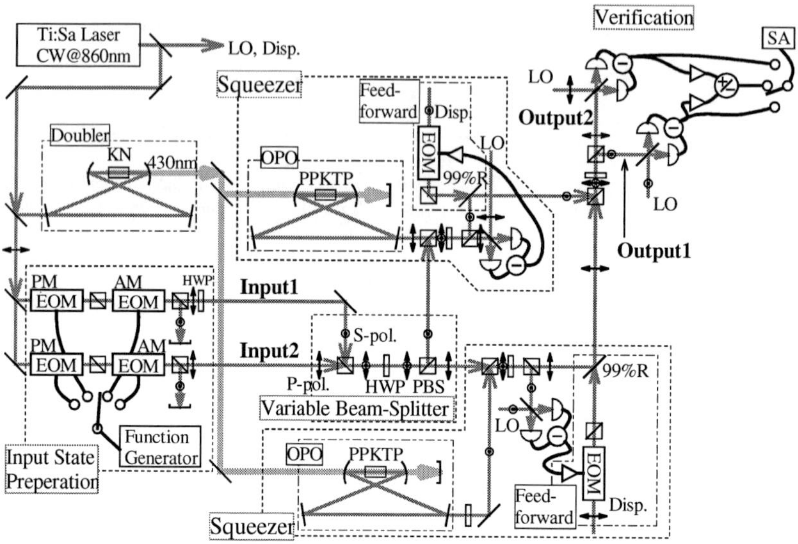

(b)Detailed experimental setup.

FIGURE 1. Schematic of our experimental setup. OPO: optical parametric oscillator, EOM: electro-optic modulator, AM: amplitude modulation, PM: phase modulation, PBS: polarized beam splitter, HWP: half wave plate, ⊙: s-polarized, ↕: p-polarized, 99% R: beam splitter with 99% reflectivity, Disp.: Auxiliary beam for a phase space displacement, LO: optical local oscillator, and SA:spectrum analyzer.

EXPERIMENT

Fig. 1 is a schematic of our experimental setup. The light source is a continuous-wave Ti:sapphire laser with the wave length of 860nm. Quantum information is encoded in the sidebands at 1.25MHz relative to the carrier frequency. As shown in Fig. 1(a), our setup is basically a Mach-Zehnder interferometer with a squeezer in each arm. The interaction gain G in Eq. (1) is determined by the four variable beam splitters, two of them constituting the Mach-Zehnder interferometer and the other two determining squeezing levels of the squeezers. These beam splitters are composed of polarizing beam splitters (PBS) and half wave plates (HWP); thus, the beam-splitting ratios are controlled via the rotating angle of the HWPs.

As squeezers, we use teleportation-based squeezing gates [7, 8] which utilize offline resources of squeezed vacuum states, balanced homodyne measurements, and feedfor-

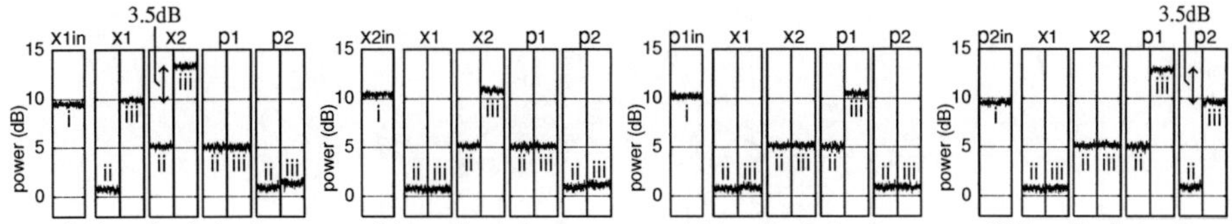

(a)Amplitude in x_1^{in} (b)Amplitude in x_2^{in} (c)Amplitude in p_1^{in} (d)Amplitude in p_2^{in}

FIGURE 2. Second moments of the quadrature components of the input and output states with respect to four different input states, for $G = 1\ 5$. The vertical axes are normalized to the shot noise limit. (i) The excited input quadratures. (ii) The output quadratures without coherent excitation. (iii) The output quadratures with the coherent excitation in the input.

ward. Each squeezed vacuum is generated by a subthreshold optical parametric oscillator (OPO) which is composed of a bow-tie shaped cavity and a periodically-poled KTiOPO$_4$(PPKTP) crystal as a nonlinear optical medium [9], see Fig. 1(b). Using such a teleportation-based approach, we avoid lossy direct-coupling of an information carrier beam to a nonlinear optical medium. Besides, teleportation-based squeezers have good control of the squeezing level which simply relies on the reflectivity of the beam splitter, and an appropriate feedforward gain. Therefore, teleportation-based approach is suitable for an efficient QND gate.

Feedforward in each squeezer is a phase space displacement proportional to the homodyne results. An auxiliary beam is phase-modulated by an electro-optic modulator (EOM) and subsequently combined with the information carrier beam at an asymmetric beam splitter (99:1).

In the experiment, we use coherent states as inputs. The coherent amplitude is generated by modulating the amplitude or phase of the bright carrier using an EOM.

Fig. 2 is the experimental results that examine input-output relations in Eq. (1) for $G = 1\ 5$. The output quadratures with (ii) and without (iii) coherent amplitude in the input are compared with each other, for four individual input amplitudes. From Fig. 2, each input amplitude is well preserved in the output state; besides, the amplitude in x_1^{in} (p_2^{in}) is added to x_2^{out} (p_1^{out}) with the gain close to 1 5 (about 3 5 dB in a logarithmic scale of power). Therefore, these results agree with the theoretical prediction in Eq. (1).

Next, we measure noise correlations between output quadratures, the results of which are shown in Fig. 3. We subtract x_2 from x_1 (add p_1 to p_2) with several gains. The subtraction (addition) gain g is changed electrically by a step attenuator. For the verification of entanglement, we use the Duan-Simon criterion [10, 11], yielding a sufficient condition for entanglement, as follows:

$$\langle(\hat{x}_1^{out} \quad g\hat{x}_2^{out})^2\rangle + \langle(\hat{p}_2^{out} + g\hat{p}_1^{out})^2\rangle < 4|g| \qquad (2)$$

where g is the subtraction (addition) gain. As Fig. 3, the experimental data (ii) go below the lines $2g$ (iv) simultaneously in $\langle(\hat{x}_1^{out} \quad g\hat{x}_2^{out})^2\rangle$ and $\langle(\hat{p}_2^{out} + g\hat{p}_1^{out})^2\rangle$; thus, Eq. (2) is satisfied in the domain $0\ 3 < g < 1\ 1$. Therefore, the output states are entangled even though the input states are separable coherent states.

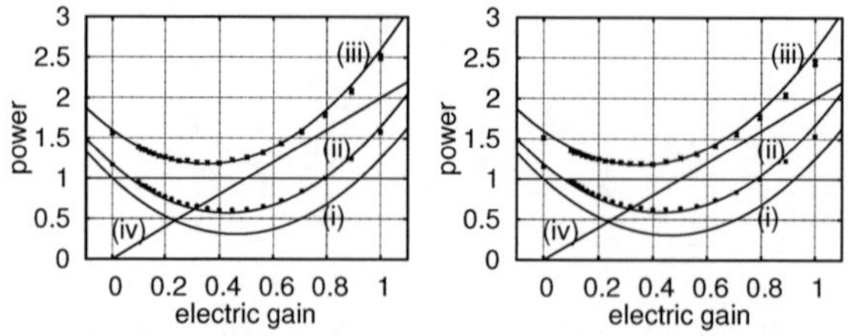

(a)Variance of $(\hat{x}_1^{out}\text{-}g\hat{x}_2^{out})$ (b)Variance of $(\hat{p}_2^{out}+g\hat{p}_1^{out})$

FIGURE 3. Noise correlation between output quadratures, for $G = 1\ 5$, normalized to the shot noise limit. (i) Ideal interaction. (ii) Experimental data and theoretical curve with squeezed ancilla (about 5 dB relative to the shot-noise level). (iii) Experimental data and theory with vacuum ancilla, (iv) Sufficient condition for entanglement.

CONCLUSION

We have experimentally demonstrated a QND gate whose input-output relations agree well with the theory. Furthermore, the entanglement in the output states is witnessed, for the input of two independent coherent states.

ACKNOWLEDGMENTS

This work was partly supported by SCF, GIA, and G-COE commissioned by the MEXT of Japan, and the Research Foundation for Opt-Science and Technology. ULA and AH acknowledge financial support from the EU under project No. 212008 (COMPAS) and the Lundbeck foundation. PvL acknowledges support from the Emmy Noether programme of the DFG in Germany.

REFERENCES

1. S. L. Braunstein and P. van Loock, *Rev. Mod. Phys.* **77**, 513 (2005).
2. J. Zhang and S. L. Braunstein, *Phys. Rev. A* **73**, 032318 (2006).
3. N. C. Menicucci, P. van Loock, M. Gu, C. Weedbrook, T. C. Ralph, and M. A. Nielsen, *Phys. Rev. Lett.* **97**, 110501 (2006).
4. S. L. Braunstein, *Phys. Rev. Lett.* **80**, 4084 (1998)
5. S. Lloyd and J. -J. E. Slotine, *Phys. Rev. Lett.* **80**, 4088 (1998).
6. J. Yoshikawa, Y. Miwa, A. Huck, U. L. Andersen, P. van Loock, and A. Furusawa, arXiv:0808.0551[quant-ph], *Phys. Rev. Lett.* accepted.
7. R. Filip, P. Marek, and U. L. Andersen, *Phys. Rev. A* **71**, 042308 (2005).
8. J. Yoshikawa, T. Hayashi, T. Akiyama, N. Takei, A. Huck, U. L. Andersen, and A. Furusawa, *Phys. Rev. A* **76**, 060301(R) (2007).
9. S. Suzuki, H. Yonezawa, F. Kannari, M. Sasaki, A. Furusawa, *Appl. Phys. Lett.* **89**, 061116 (2006).
10. L. -M. Duan, G. Giedke, J. I. Cirac, and P. Zoller, *Phys. Rev. Lett.* **84**, 2722 (2000).
11. R. Simon, *Phys. Rev. Lett.* **84**, 2726 (2000).

Probing Quantum Rules By The Experimental Implementation Of Single-Photon Creation And Annihilation Operators

A. Zavatta[a], V. Parigi[a], M.S. Kim[a] and M.S. Bellini[a]

[a]*Istituto Nazionale di Ottica Applicata (INOA-CNR) Largo E. Fermi 6, 50125 Firenze (Italy).*
[b]*Physics Dep., University of Florence, Via G. Sansone 1, 50019 Sesto Fiorentino – Firenze (Italy).*
[c]*School of Mathematics and Physics, The Queen's University, Belfast, BT7 1NN (UK).*
[d]*European Laboratory for Nonlinear Spectroscopy (LENS), University of Florence, Via N. Carrara 1, 50019 Sesto Fiorentino – Firenze (Italy).*

Abstract. We have realized experimental schemes to implement the action of single photon creation and annihilation operators on single-mode light fields. In particular, we apply alternated sequences of creation and annihilation operators to completely classical and fully incoherent thermal light states. By verifying that the resulting states depend on the order in which the two quantum operators are applied, we provide the most direct experimental verification of the non-commutation of bosonic creation and annihilation operators.

Keywords: Nonclassical states of light, quantum homodyne tomography, parametric downconversion, single-photon addition, single-photon subtraction.
PACS: 03.65.Wj, 42.50.Dv

INTRODUCTION

In the last few years a growing interest in the quantum information field has been reserved to the implementation of quantum operations like addition and subtraction of a single photon to and from a state of light. Several experiments have demonstrated that through the implementation of such basic quantum operators it is possible to generate nonclassical and non-Gaussian states of light and increase the entanglement amount of a bipartite state [1-5]. The manipulation of such kind of quantum operations also leads to the implementation of novel schemes for quantum communication as well as to violations of the Bell's inequality in the continuous variable domain [6,7]. Here we show the experimental application of single-photon operator sequences onto a completely incoherent state of light as the thermal state. The addition and subtraction of single photons is obtained by the conditional implementation of the creation and annihilation operators $\hat{a}^\dagger$ and $\hat{a}$ using the parametric down-conversion process [2,8] and a beam splitter [9]. Three different operator modules have been used to easily implement alternated sequences of the two quantum operations on the light state. The single-photon-creation module is placed between two annihilation modules as shown in FIGURE 1. The desired operator sequence is obtained by choosing the right coincidence pattern of conditional signals, and then homodyne detection is performed on the resulting final state.

CP1110, *Quantum Communication, Measurement and Computing (QCMC), The Ninth International Conference,* edited by A. Lvovsky

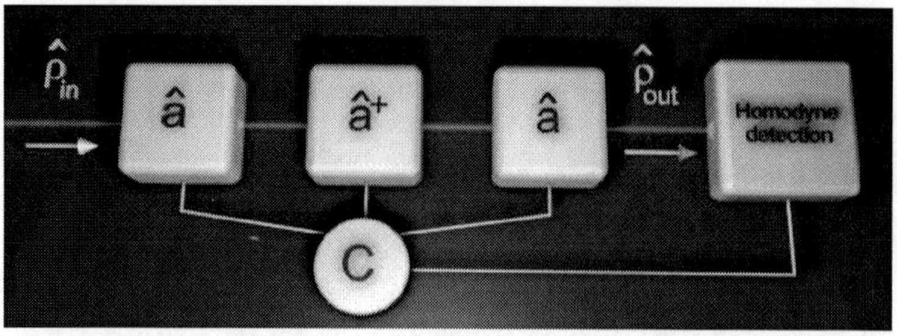

FIGURE 1. Implementation of the single photon creation and annihilation operators in arbitrary combinations.

EXPERIMENTAL SETUP AND RESULTS

The experimental setup is shown in FIGURE 2. Here conditional parametric amplification in a nonlinear crystal is used for single-photon excitation of the thermal states using the single-photon addition scheme as recently demonstrated in Ref. [10].

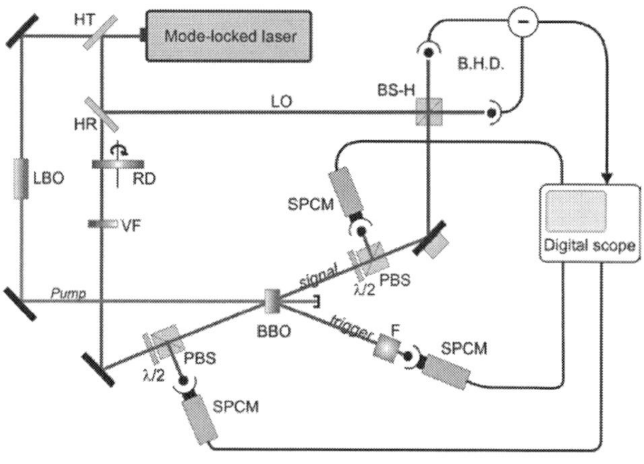

FIGURE 2. Experimental setup.

The second harmonic beam of a picosecond mode-locked Ti:Sa laser is used to pump the parametric amplifier which consists in a 3mm long BBO crystal. If no input fields are injected into the parametric amplifier, spontaneous down-conversion takes place and photon pairs are emitted along the signal and trigger channels. If a thermal field is injected along the idler channel, the generation of the single-photon added thermal state occurs in the signal channel when a single photon is emitted in the trigger channel. In order to select a pure state, a couple of spatial and spectral filters (F) are placed in front of the single-photon detector (SPCM) placed along the trigger channel.

The thermal state is generated by a small portion of the main laser output impinging on a rotating round glass disk (RD); its amplitude is adjusted by a variable attenuator (VF).
The conditional attenuation by a high-transmissivity beam-splitter is employed for photon annihilation where a single-photon detector is placed in the reflection channel [5]. A variable reflectivity beam-splitter is obtained by a half-wave plate (λ/2) followed by a polarizing beam-splitter (PBS). A click from the single-photon detector assures that a successful single-photon subtraction has taken place on the input field.
If the parametric gain of the crystal and the reflectivity of the subtracting beam-splitter are low enough, the above operations are a good approximation of the ideal processes of photon creation and annihilation on the signal mode. Both conditions are always well satisfied in our experiments.
A single click from the single-photon detectors conditions the generation of a photon-added or photon-subtracted state, while a double click can either produce a first-subtracted-then-added thermal state or vice versa, depending on the sequence of the two clicks.
We finally use ultrafast pulsed homodyne detection (BHD) and quantum tomography to investigate the resulting states in the signal channel [11,12]. Homodyne measurements on the signal channel are only performed when the right combination of two clicks arrives from the detectors. After the acquisition of about 10^5 quadrature measurements for each investigated state with random phases, we have reconstructed the diagonal density matrix elements by means of an iterative maximum likelihood algorithm [13,14], correcting for the finite detection efficiency [1]. The reconstructed density matrix elements have then been used to obtain the corresponding Wigner functions.

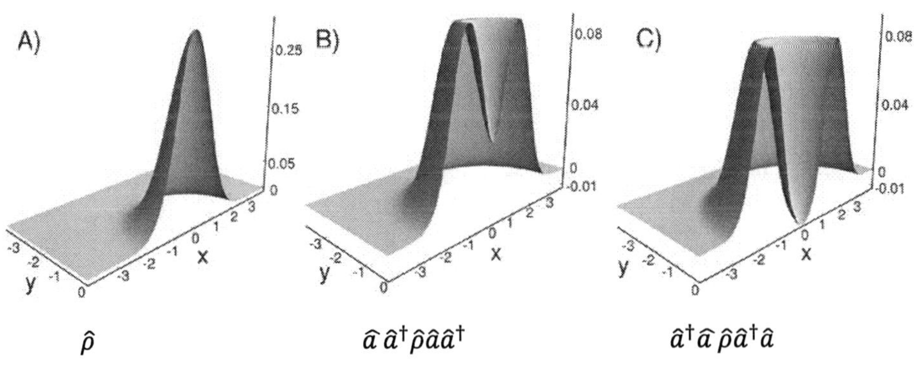

FIGURE 3. Reconstructed Wigner functions: A) Initial thermal state; B) added and then subtracted state; C) subtracted and then added thermal state.

In FIGURE 3 we report the obtained Wigner functions. We have first measured the initial thermal state and then the state resulting from both the addition-subtraction and subtraction-addition sequences. The Wigner function for added and subtracted sequences exhibits a clear central dip, which is absent in the Wigner function of the thermal field; such a dip reaches negative values for the subtract-then-add sequence, whereas it stays well in the positive region for the add-then-subtract sequence. It is evident that the two final states are different from each other and from the original thermal state. This provides a direct experimental verification of the noncommutativity of the quantum bosonic creation and annihilation operators.

CONCLUSIONS

We clearly demonstrate that different states result from the action of the two inverse sequences of creation and annihilation operators, thus showing that $[\hat{a}, \hat{a}^{\dagger}] \neq 0$.
Besides the fundamental importance of these results, demonstrating the non-commutation of creation and annihilation operators, the experimental implementation of such a sequence of basic quantum operations has strong implications for future quantum applications, as it can also be seen as a first step towards a full-scale engineering of a quantum light state.

ACKNOWLEDGMENTS

This work was partially supported by Ente Cassa di Risparmio di Firenze and by CNR-RSTL Projects.

REFERENCES

1. V. Parigi, A. Zavatta, M. S. Kim, and M. Bellini, *Science* **317**, 1890 (2007).
2. A. Zavatta, S. Viciani, and M. Bellini, *Science* **306**, 660 (2004).
3. A. Ourjoumtsev, R. Tualle-Brouri, J. Laurat, and P. Grangier, *Science* **312**, 83 (2006).
4. A. Zavatta, M. D'Angelo, V. Parigi, and M. Bellini, *Phys. Rev. Lett.* **96**, 020502 (2006).
5. A. Ourjoumtsev, A. Dantan, R. Tualle-Brouri, and P. Grangier, *Phys. Rev. Lett.* **98,** 030502 (2007).
6. M. D'Angelo, A. Zavatta, V. Parigi, and M. Bellini, *Phys. Rev.* A **74**, 052114 (2006).
7. R. García-Patrón, J. Fiurášek, N. J. Cerf, J. Wenger, R. Tualle-Brouri, and P. Grangier, *Phys. Rev. Lett.* **93**, 130409 (2004).
8. A. Zavatta, S. Viciani, and M. Bellini, *Phys. Rev. A* **72**, 023820 (2005).
9. J. Wenger, R. Tualle-Brouri, and P. Grangier, *Phys. Rev. Lett.* **92**, 153601 (2004).
10. A. Zavatta, V. Parigi, and M. Bellini, *Phys. Rev. A* **75**, 052106 (2007).
11. A. Zavatta, M. Bellini, P. L. Ramazza, F. Marin, F. T. Arecchi, *J. Opt. Soc. Am. B* **19**, 1189 (2002).
12. A. Zavatta, S. Viciani, and M. Bellini, *Phys. Rev. A* **70**, 053821 (2004).
13. A. I. Lvovsky, *J. Opt. B: Quantum Semiclassical Opt.* **6**, S556 (2004).
14. Z. Hradil, D. Mogilevtsev, and J. Rehacek, *Phys. Rev. Lett.* **96**, 230401 (2006).

Measurement of Entanglement in Time Domain —— an Experiment towards Demonstration of EPR paradox with Continuous Variable using Laser Pulses

Yun Zhang, R. Okubo, M. Hirano and T. Hirano

Department of physics, Gakushuin University, 1-5-1 Mejiro, Toshima-ku, Tokyo, 171-8588, Japan, and CREST, Japan Science and technology agency, 1-9-9, Yaesu, Chuou-ku, Toky, 103-0028, Japan

Abstract. We report on a measurement of entanglement in the time domain. The entanglement is examined by the Duan-Simon criteria. We obtained Duan-Simon criteria of $\Delta^2(x_1 - x_2) + \Delta^2(p_1 + p_2) = 1.4 < 2$ without optimal gain and $\Delta^2(x_1 - x_2) + \Delta^2(p_1 + p_2) = 1.2 < 2$ with optimal gain, respectively. This opens the way to realize both efficient and truly causal EPR paradox.

Keywords: Entanglement, optical parametric amplifier, EPR paradox
PACS: 42.50.Dv, 03.67.Hk

INTRODUCTION

In the development of quantum information, such as quantum teleportation, entanglement is the central [1]. It is at the heart of the Einstein-Podolsky-Rosen (EPR) paradox, which demonstrated the incompatibility between local realism and the completeness of quantum mechanics. The originally devised entanglement had been experimentally realized both in the discrete variable and in continuous variable, and the demonstration of EPR paradox is also carried out both in the discrete variable and continuous variable [2, 3, 4, 5]. Causal separation was first achieved in discrete variable leaving the low detector efficiencies loophole. In contrast, high detector efficiencies can be easily achieved by measuring continuous variable with a homodyne detector. However, these experiments have been limited to situations of poor or non-existent causal separation. A conclusive experiment would require both high efficiency and locality. Such loophole free experiments for the Bell inequality are yet to be carried out. In recent years, several proposals have been proposed to demonstration of loophole-free EPR paradox. Among them, Ralph et al. proposed a scheme to measure the photon correlation using homodyne detection [6]. Nha et al and Garcia-Patron et al proposed a scheme using non-Gaussian entangled state, which is generated by subtracting one photon from two-mode squeezed state, as the source of nonlocal correlation [7, 8]. Following the proposal of Reid et al. [9], we pursue to demonstrate a loophole EPR paradox with continuous variable using laser pulses. The causal separation can be obtained by quantum correlation between a pair of laser pulses, which are one pair of a series of individual laser pulse train, and high efficiency maintains by employing a homodyne detector. Here we report the result of the first step of our pursuit, measurement of entanglement in the time domain.

CP1110, *Quantum Communication, Measurement and Computing (QCMC), The Ninth International Conference,* edited by A. Lvovsky

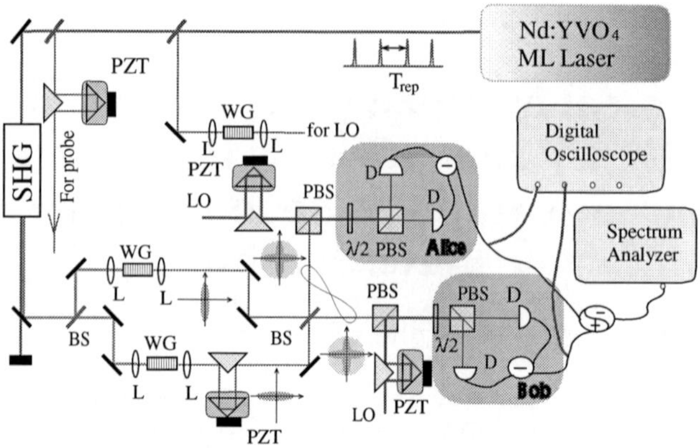

FIGURE 1. Experimental setup. SHG: second harmonic generator, WG: waveguide, BS: 50:50 beam splitter, L: lens, PBS: polarizing beam splitter, Ds: photodiodes, PZTs: piezoelectric transducers, $\lambda/2$: half-wave plate, and LO: local oscillator.

EXPERIMENTAL SETUP

The schematic of our experimental setup is shown in Fig. 1. In order to generate squeezed states, single passed optical parametric amplifiers (OPAs) are used. Single-mode PPLN waveguides (WGs) are employed as nonlinear optical media. The choice for a single-mode PPLN waveguide has the benefit of the transverse confinement and of the possibility of spatial mode control. We use a cw mode-locked Nd:YVO4 laser, which is pumped by a semiconductor laser diode, operating at 1064 nm with a duration of 7 ps and a pulse repetition rate of 76 MHz, as a light source. The fundamental was divided into tree parts. A major fraction of this light was sent to a single pass through second harmonic generator (SHG) to generate 532 nm pump source for pumping the OPAs. Another small fraction of the fundamental beam is taken out to serve as a probe to study classical parametric amplification occurring in the OPAs. A local oscillator (LO) beam, which passed through a similar PPLN waveguide to ensure proper spatial mode matching to the signal, is also separated from the fundamental wave. Each OPA was pumped by a second harmonic beam obtained from SHG. The down-conversion signal produced in the OPAs propagates to a 50:50 beam splitter (BS), where they will be spatial and temporal overlap and interference to form the entanglement state. The phase of one of two squeezed vacua is locked so that the squeezing ellipses are perpendicular to each other. Under this condition, the entanglement state is generated. Each of entangled beams is measured on a homodyne detector by mixing with a LO light. The output of homodyne detector is split into two signals by a RF splitter. The sum or difference of two outputs of homodyne detectors is monitored by a spectrum analyzer, which is also able to analyze the information in the frequency domain. The other signal is recorded by a digital oscilloscope, which records and analyzes the information in the time domain.

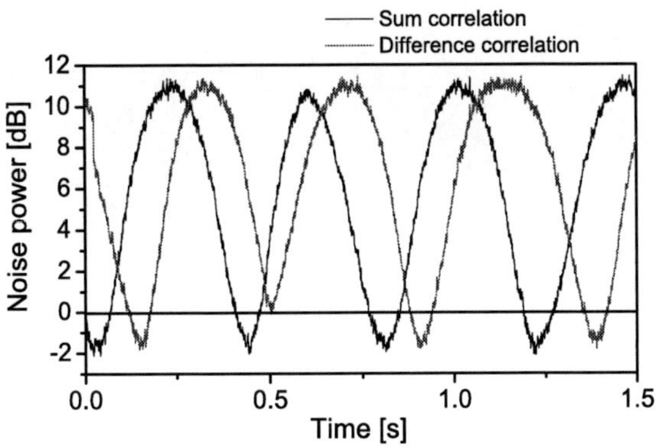

FIGURE 2. Entanglement correlation between two entangled beams in the frequency domain.

EXPERIMENTAL RESULTS

Figure 2 was a typical recorded noise power spectrum by scanning the LO phase of one homodyne detector and electronically adding or subtracting the photocurrents of two sets of homodyne detectors. The noise power was normalized to shot noise level, which was obtained by blocking the output of OPAs. The noise reduction of about 2.0 dB below the shot noise level for two beams was observed. We obtained Duan-Simon criteria of $\Delta^2(x_1 - x_2) + \Delta^2(p_1 + p_2) = 0.63 + 0.66 = 1.29 < 2$ [10, 11]. This result attests for the quantum entanglement of the pulsed state generated by combing two squeezed vacua on beam splitter.

Figure 3 gives a direct representation of the quantum correlation between two entangled states and separable state. The total number of pulses of each color was 7600. Each point corresponded to one simultaneous measurement of entangled state or separable state. These results show clear correlations in the relevant quadratures. To show the quantum entanglement, after the data were stored we sampled the variance of recorded data. We determined the entanglement criteria by normalizing the sum or different variance of entangled beams to the sum or different variance of separable state. We obtained Duan-Simon criteria of $\Delta^2(x_1 - x_2) + \Delta^2(p_1 + p_2) = 1.4 < 2$ without optimal gain and $\Delta^2(x_1 - x_2) + \Delta^2(p_1 + p_2) = 1.2 < 2$ with optimal gain, respectively. The measured entanglement level without optimal gain is a little small than that of in the frequency domain. We believe the reason is that the electronic noise of the homodyne detector cannot be calibrated in the time domain.

Demonstration of the EPR paradox relies on quantum correlation between a pair of non-commuting observable, so that measurement of either observable in one subsystem 1 allows the interference of that variable in subsystem 2 to better than the quantum noise limit. Between the amplitude and phase quadratures of a pair of optical beams this is quantified by the product of conditional variance, we therefore define the demonstration of EPR paradox for $\Delta^2 x_{1|2} \Delta^2 p_{1|2}$. This is a striking contrast to the

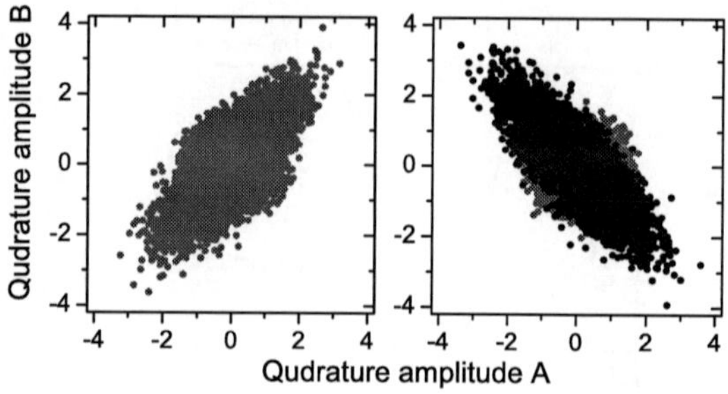

FIGURE 3. Quantum correlation between two entangled beams in the time domain. (a) amplitude quadrature (b) phase quadrature

Duan-Simon criterion. The reason for this difference is that the Duan-Simon criterion is independent of the purity of the entanglement (ie. independent of the variance of the sub-system), the EPR paradox criterion is very sensitive to. In our present result, a condition variance of $\Delta^2 x_{1|2} = 1.13$ for amplitude quadrature and $\Delta^2 x_{1|2} = 1.16$ for phase quadratrue without optimal gain were obtained, respectively. An EPR paradox of $\Delta^2 x_{1|2}\Delta^2 p_{1|2} = 1.31$ can be directly obtained. Unfortunately, this result did not show a demonstration of EPR paradox. We believe that the EPR paradox criterion can be further improved by improving the squeezing level of the squeezed vacua.

CONCLUSION

In conclusion, we have experimentally demonstrated generation and characterization pulsed entanglement both in the time domain and frequency domain. The measured conditional variance shows we pave the way to demonstrate both efficient and truly causal EPR paradox

REFERENCES

1. S. L. Braunstein and P. van Loock, *Rev. Mod. Phys.* **77**, 513 (2005).
2. M. D. Reid and P. Drummond, *Phys. Rev. Lett.* **60**, 2731(1988).
3. M. D. Reid, *Phys. Rev. A* **40**, 913 (1989).
4. Z. Y. Ou, S. F. Pereira, H. J. Kimble, and K. C. Peng, *Phys. Rev. Lett.* **68**, 3663(1992).
5. B. P. Bowen, R. Schnabel, P. K. Lam, and T. C. Talph, *Phys. Rev. Lett.* **90**, 043601(2003).
6. T. C. Ralph, W. J. Munro, and R. E. S. Polkinghorne, *Phys. Rev. Lett.* **85**, 2035(2000).
7. R. Garcia-Patron, J. Fiura?ek, N. J. Cerf, J. Wenger, R. Tualle-Brouri, and Ph. Grangier, *Phys. Rev. Lett.* **93**, 130409 (2004).
8. H. Nha and H.J. Carmichael, *Phys. Rev. Lett.* **93**,020401(2004).
9. M. D. Reid, P. D. Drummond, E. G. Cavalcanti, W. P. Bowen, P. K. Lam, H. A. Bachor, U. L. Andersen, G. Leuchs, arXiv:0806.0270.
10. L. M. Duan G. Giedke, J. I. Cirac, and P. Zoller, *Phys. Rev. Lett.* **84**, 2722 (2000).
11. R. Simon, *Phys. Rev. Lett.* **84**, 2726 (2000).

Quantum Tomography of the Single-Photon State Generated by Down Conversion in a Periodically Poled KTP Crystal

Nitin Jain*, Sun-Hyun Youn† and A. I. Lvovsky*

*Institute of Quantum Information Science, University of Calgary, Canada, T2N1N4
†Department of Physics, Chonnam National University, Gwangju, Korea, 500-757

Abstract. We report the preparation of the single-photon Fock state $|1\rangle$ by conditional measurements on the bi-photon generated in a periodically poled KTP crystal. We employ time-resolved homodyne detection to measure the quadrature fluctuations and characterize the single photon state with an overall efficiency of 53%. The photon production rate is over 9000 s^{-1}, which is much higher than in previous experiment of the kind. Additionally, the experiment employs a new balanced detector whose bandwidth permits time-resolved measurement at the repetition rate of the mode-locked master laser, without the need for a pulse picker.

Keywords: Single Photon, Homodyne Tomography, Parametric Down Conversion
PACS: 03.65.Wj, 42.50.Dv

Quantum nature of light was discovered over a hundred years ago as a tool to explain the spectrum of blackbody radiation. Since then, the field of quantum optics has witnessed rapid growth, playing a vital role in applications ranging from quantum teleportation, cryptography to QND measurements, violation of Bell's inequalities. A workhorse of quantum-optical information technology is parametric down conversion (PDC), employed in *creation* of photon pairs. Combined with *manipulation* techniques such as conditional measurements, PDC has allowed us to generate a plethora of interesting quantum states of light [1, 2, 3]. Quantum state *characterization*, i.e. obtaining complete information about its density operator $\hat{\rho}$, is achieved by multiple measurements performed on identically prepared copies of the state. One of the state-reconstruction methods, homodyne tomography is based on measuring the noise statistics of the electromagnetic field quadratures at different phases [4].

Conditional preparation of pure optical states requires tight spatio-temporal filtering of trigger photons [5] which in turn results in low state preparation rates, especially for multiphoton states. An additional challenge is associated with time-domain homodyne detection, which must resolve individual local oscillator pulses. Most modern pulsed homodyne detectors feature a bandwith of only a few MHz [6]. On the other hand, pulse repetition rates of commercial mode-locked lasers are on a scale of at least ~80 MHz. Attempts to increase the bandwidth of detection electronics lead to a reduction of the signal-to-noise ratio [7], which is equivalent to a lower detection efficiency [8]. One way to address this issue is employing a pulse picker, which however renders the whole setup bulky and complicated [9]. Alternatively, a cavity dumped pulsed laser can be used, in which case, the laser pulse rate and thus the state production rate is significantly reduced [10]. Our experiment overcomes these issues by performing PDC in a periodically poled

CP1110, *Quantum Communication, Measurement and Computing (QCMC), The Ninth International Conference,* edited by A. Lvovsky

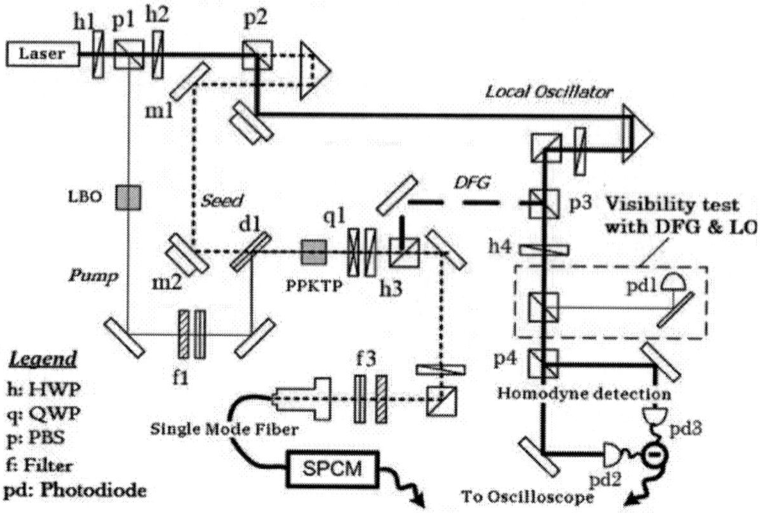

FIGURE 1. Experimental setup

KTP crystal (PPKTP) and employing fast, time-domain based homodyne detection. We achieve photon production rates of the order of 9000 s^{-1} which can be attributed to the higher nonlinearity offered by this crystal. Further, the homodyne detector features a bandwidth of around 90 MHz and a signal-to-noise ratio of $\sim$12 dB, allowing us to detect every pulse emitted by the mode-locked laser, thereby eliminating the need for a pulse picker. As a demonstration of our apparatus' capability, we perform an experiment on conditional preparation and homodyne tomography of the single-photon state, along the lines of Ref. [9]. Figure 1 shows a basic schematic of the experimental apparatus. A mode-locked Ti:Sapphire laser (Coherent MIRA 900) emits a pulse train with the wavelength 790 nm, repetition rate about 76 MHz and pulse width around 2 ps. A major portion of the laser output is frequency-doubled in a single pass through a 3-mm LBO crystal (type I) yielding $\sim$80 mW of average power at 395 nm and then passed on to a PPKTP crystal [11, 12] for a type II collinear down conversion. The generated photon pair is separated into two emission channels according to the photons' polarizations. Detecting a photon in one of these channels (labelled 'trigger') projects the quantum state in the remaining channel (labelled 'signal') into a single photon state. However, given the multi-mode photon production and intrinsic spectral and spatial correlation of the modes of the two photons, the signal photon is actually in an incoherent, mixed state [5, 12]. A higher degree of coherence is established by projecting the trigger photon into a single mode by spectral selection (using a 0.3-nm bandwidth interference filter) and spatial filtering (with an optical fiber). This fiber-coupled photon triggers a single photon counting module (SPCM-AQR-14-FC, photon-detection efficiency $\eta \simeq$45 % at 790 nm), while the 'signal' photon forms the quantum state to overlap with the local oscillator at the input of the homodyne detector.

The degree of spatio-temporal mode matching between the single photon and local

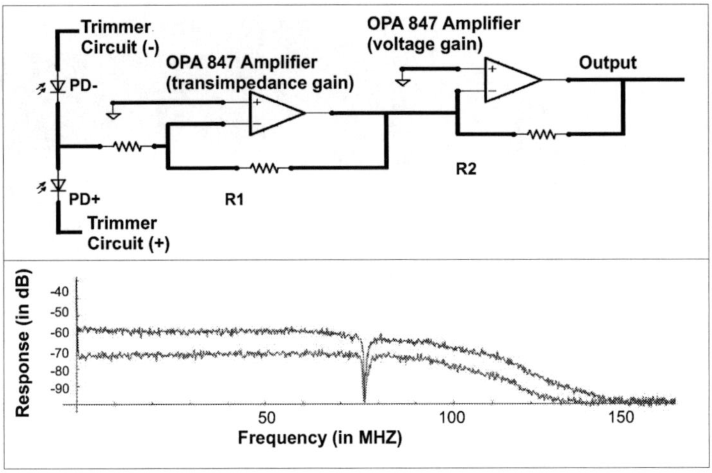

FIGURE 2. (Above) Basic schematic of the circuit and (below) spectral response of our homodyne detector. The (lower/red) curve is the electronic noise, the (upper/blue) curve is the optical response (Shot noise) at an average local oscillator power of ~ 25 mW, the suppression is due to a notch filter tuned at the laser repetition rate (76.3 MHz).

oscillator has a direct impact on the reconstruction efficiency and thus is a critical component of the experiment. The mode matching is achieved by employing a *seed* beam (fig. 1) which is coupled to the fiber in the trigger channel. The difference frequency generated from the pump and seed beams is then matched to the local oscillator [5, 9].

The experiment employed a new homodyne detector with a 3-dB bandwidth around 90 MHz (fig. 2), which exceeds the pulse repetition rate of the laser, and hence permits a time-resolved measurement. The detector employs two Hamamatsu S5972 photodiodes with a quantum efficiency of 85%, followed by two-stage amplification in a circuit consisting of two OPA 847 operational amplifiers (fig. 2). The amplified signal features a ratio between the shot noise and electronic noise of 10-12 dB, which corresponds to a quantum efficiency of 90–96% [8]. Trimmer-based circuits allow varying the reverse bias across each of the photodiodes on the fly. This is useful in achieving balanced detection, compensating for non-identical characteristics of the photodiodes. Careful design and board-layout practices as well as suitably chosen components help minimizing instability and oscillations associated with parasitic capacitances.

The function of the detector is illustrated in fig. 3(b), which displays multiple oscilloscope traces of its output signal. One can see time-resolved measurements of the field quadratures in individual signal pulses. The oscilloscope is triggered by the detection of trigger photons, so the central peak corresponds to the single-photon Fock state. Other peaks correspond to the vacuum state and are used to calibrate the detector.

As discussed in Ref. [9], all inefficiencies associated with the experiment (linear losses, dark counts, photodiode inefficiencies etc.) manifest themselves as the admixture of the vacuum state to the single-photon state. The resulting measured state is:

$$\hat{\rho}_{meas} = \eta_f |1\rangle\langle 1| + (1-\eta_f)|0\rangle\langle 0| \tag{1}$$

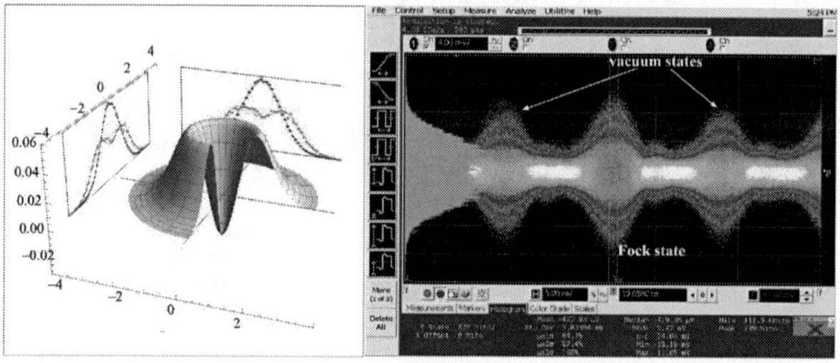

a b

FIGURE 3. (a) Marginal distributions for the vacuum (red) and $\hat{\rho}_{meas}$ (green) and the reconstructed Wigner Function with the 85% detector efficiency taken into account. (b) Oscilloscope screenshot showing multiple traces of the homodyne detector output. The central peak corresponds to the state $\hat{\rho}_{meas}$, and side peaks to the vacuum state.

with η_f being the overall efficiency of our experiment. The Wigner function of this state is given by

$$W_{\eta_f}(x,p) = \frac{2}{\pi}\left(4\eta_f(x^2+p^2)+1-2\eta_f\right)e^{-2(x^2+p^2)})). \tag{2}$$

A plot of the acquired phase-independent quadrature histograms and the corresponding Wigner function is shown in fig. 3(a). The observed distribution corresponds to a quantum efficiency of $\eta_f = 0.53$, which means that the reconstructed state's Wigner function is negative even without accounting for imperfect detection efficiency.

In summary, we have described the homodyne tomography of a single photon state, created conditionally at the output of a PPKTP crystal, in a real-time fashion by using a high-bandwidth detection scheme. We hope to accomplish a higher value of η_f with a better mode-matching, elimination of linear losses and implementing more efficient photodiodes in the homodyne detector circuit. Equipped with these high-rate preparation and fast real-time acquisition systems, we will be able to perform multi-photon quantum state characterization experiments.

REFERENCES

1. A.I. Lvovsky and J. Mlynek, Phys. Rev. Lett. **88**, 25 (2002)
2. A.I. Lvovsky and S. A. Babichev, Phys. Rev. A. **66**, 011801 (R) (2002)
3. A. Ourjoumtsev *et al.*, Science **312**, 83-86 (2006)
4. A.I. Lvovsky and M.G. Raymer, quant-ph/0511044 (2008)
5. T. Aichele, A.I. Lvovsky, S. Schiller, Eur. Phys. J. D **18**, 237-245 (2002)
6. H. Hansen *et al.*, Opt. Lett **26**, 21 (2001)
7. A. Zavatta *et al.*, J. Opt. Soc. Am. B **19**, 5 (2002)
8. J. Appel *et al.*, Phys. Rev, A **75**, 035802 (2007)
9. A.I. Lvovsky *et al.*, Phys. Rev. Lett. **87**, 5 (2001)
10. J. Wenger *et al.*, Opt. Lett **29**, 11 (2004)
11. C. Kuklevicz *et al.*, Phys. Rev. A 69, 041801 (2004); B. Shi et al, J. Opt. Soc. Am. B, **21**, 12 (2002)
12. S. Castelletto and R. Scolten, Eur. Phys. J. Applied Physics **41**, 181-194 (2008)

Near-optimal quantum receivers for the coherent state discrimination

Masahiro Takeoka, Kenji Tsujino, and Masahide Sasaki

National Institute of Information and Communications Technology, 4-2-1, Nukuikitamachi, Koganei-city, Tokyo 184-8795, Japan
CREST, Japan Science and Technology Agency, 5 Sanbancho, Chiyoda-ku, Tokyo 102-0075, Japan

Abstract.
We propose practical near-optimal quantum receivers for the binary phase shift keyed coherent signals to overcome the homodyne limit in all areas of the signal power. Our schemes consist of Gaussian operations and photon detection and do not require realtime electrical feedback. We numerically show that the scheme is feasible with the current technology.

Keywords: optimal state discrimination, coherent state, quantum receiver
PACS: 03.67.Hk, 42.50.Dv

INTRODUCTION

Coherent state is one of fundamental information carriers in optical and quantum communication networks. Since coherent states are nonorthogonal to each other, one cannot discriminate them without errors. In particular, the conventional coherent communication technology is based on coherent signals and homodyne or heterodyne measurements. For example, binary phase shift keyed (BPSK) signals $\{|\alpha\rangle, |-\alpha\rangle\}$ are discriminated by a homodyne receiver with the average error probability of $P_{err} = \mathrm{erfc}[\sqrt{2}|\alpha|]/2$, in ideal, which is often called the shot noise limit or the homodyne limit, falling short of the error free criterion (10^{-9}) when $|\alpha|^2 < 10$ (in the following, we assume equal prior probabilities for a sake of simplicity).

It is well known that one can greatly surpass the above limit in principle [1]. The quantum mechanically optimized measurement for the BPSK signal could attain $P_{err} = (1-\sqrt{1-e^{-4|\alpha|^2}})/2$ which is often called the Helstrom bound. To implement such a measurement in real world, Kennedy proposed a simple near-optimal quantum receiver whose error rate is only twice of the Helstrom bound [2]. For surpassing the BPSK homodyne limit in practice with this Kennedy receiver, the visibility of the displacement must be higher than $1-P_{err}$ and a photon detector with high quantum efficiency and low dark counts is required [3]. In particular, at present technology, the error probability would be saturated at $|\alpha|^2 \sim 1$ due to the visibility imperfection, while the performance of the Kennedy receiver is comparable or worse than the homodyne limit for $|\alpha|^2 < 1$ even in ideal. This is in principle overcome by introducing realtime adaptive feedback as proposed by Donlinar [4]. Although its concept was demonstrated recently [5], the Dolinar's scheme is too complex to really exceed the homodyne limit with the current technology.

CP1110, *Quantum Communication, Measurement and Computing (QCMC), The Ninth International Conference,* edited by A. Lvovsky

SIMPLE NEAR-OPTIMAL QUANTUM RECEIVERS

Here we propose new simple and practical quantum receivers that circumvent the problem of the Kennedy receiver. In the Kennedy receiver, the BPSK signal $\{|\alpha\rangle, |-\alpha\rangle\}$ is shifted to $\{|2\alpha\rangle, |0\rangle\}$ by the displacement operation $\hat{D}(\alpha) = \exp(\alpha\hat{a}^\dagger - \alpha^*\hat{a})$, where $\hat{a}$ and $\hat{a}^\dagger$ are annihilation and creation operators, respectively, and then detected by an on/off type photon detector which discriminates zero or non-zero photons. It is well known that $\hat{D}(\alpha)$ can be realized by using a beamsplitter with the transmittance $\tau \to 1$ and the coherent LO $|\alpha/\sqrt{\tau}\rangle$. Our proposed receivers are depicted in Fig. 1(a), where the displacement $\hat{D}(\alpha)$ is replaced by a single-mode general Gaussian unitary operation $\hat{U}_G$. We will seek its optimal structure.

An on/off detector is described by the POVM $\{\hat{\Pi}_{\text{off}}, \hat{\Pi}_{\text{on}}\}$ with

$$\hat{\Pi}_{\text{off}} = e^{-\nu} \sum_{m=0}^{\infty} (1-\eta)^m |m\rangle\langle m|, \quad \hat{\Pi}_{\text{on}} = \hat{I} - \hat{\Pi}_{\text{off}}, \tag{1}$$

where $|m\rangle$ is an m-photon state, η is the quantum efficiency, and ν is the dark counts. The Gaussian unitary operation $\hat{U}_G$ consists of phase shift, displacement, and squeezing, while one can omit the phase shift since the on/off detection is insensitive to the global phase. Then the average error probability is calculated from

$$P_e = \frac{1}{2}\left(\langle\alpha|\hat{U}_G^\dagger \hat{\Pi}_{\text{off}} \hat{U}_G|\alpha\rangle + \langle-\alpha|\hat{U}_G^\dagger \hat{\Pi}_{\text{on}} \hat{U}_G|-\alpha\rangle\right), \tag{2}$$

where $\hat{U}_G = \hat{D}(\beta)\hat{S}(\zeta)$, $\hat{S}(\zeta) = \exp[\frac{1}{2}(\zeta^*\hat{a}^2 - \zeta\hat{a}^{\dagger 2})]$ is the squeezing operator, and $\zeta = re^{i\varphi}$ is the complex squeezing parameter.

After some calculation, one can find that r, φ, and β have the extreme at the same point, where $\varphi = 0$, β is real, and the optimal displacement β_{opt} and squeezing r_{opt}, are given by β and r satisfying

$$\frac{8\eta\alpha\beta}{1-e^{4r}} = \left\{\frac{4\eta(\alpha^2+\beta^2)}{1-e^{4r}} - \frac{\eta+(2-\eta)e^{-2r}}{\eta+(2-\eta)e^{2r}}\right\}\tanh\left(\frac{4\eta\alpha\beta}{\eta+(2-\eta)e^{-2r}}\right), \tag{3}$$

$$\alpha = \beta\tanh\left(\frac{4\eta\alpha\beta}{\eta+(2-\eta)e^{-2r}}\right), \tag{4}$$

simultaneously. The optimized average error probability is then given by

$$\begin{aligned} P_e^{DS} &= \frac{1}{2} - \frac{2e^{-\nu}}{\sqrt{(\eta+(2-\eta)e^{2r_{\text{opt}}})(\eta+(2-\eta)e^{-2r_{\text{opt}}})}} \\ &\quad \times \exp\left[-\frac{2\eta(\alpha^2+\beta_{\text{opt}}^2)}{\eta+(2-\eta)e^{-2r_{\text{opt}}}}\right]\sinh\left[\frac{4\eta\alpha\beta_{\text{opt}}}{\eta+(2-\eta)e^{-2r_{\text{opt}}}}\right]. \end{aligned} \tag{5}$$

In the following, we call it as the Type-I receiver. When one is restricted to use only linear unitary operation, that is the displacement $\hat{D}(\gamma)$, the conditions in Eqs. (3) and (4) are simplified as

$$\alpha = \gamma\tanh(2\eta\alpha\gamma), \tag{6}$$

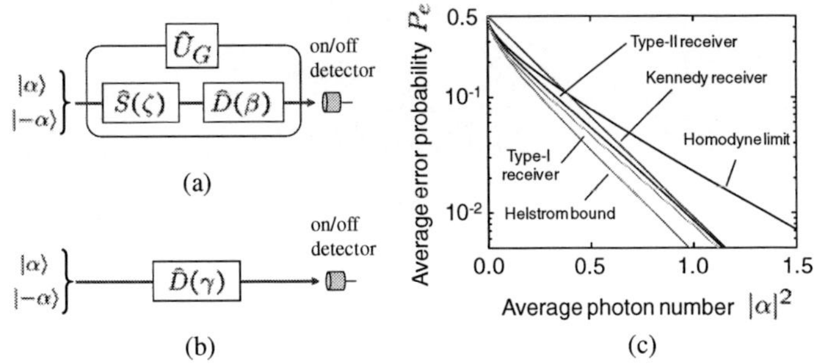

FIGURE 1. (a), (b) Schematic of the proposed receivers. $\hat{S}(\zeta)$ represents a squeezing operation. (c) Average error probabilities for several receivers.

The schematic is shown in Fig. 1(b) which we call the Type-II receiver. Its average error probability is given by

$$P_e^D = \frac{1}{2} - e^{-\nu-\eta(\alpha^2+\gamma_{\rm opt}^2)} \sinh\left(2\eta\alpha\gamma_{\rm opt}\right), \tag{7}$$

where $\gamma_{\rm opt}$ is the γ satisfying Eq. (6). Note that its physical setup is the same as that of the Kennedy receiver. However, we stress that $\gamma_{\rm opt} \neq \alpha$ in general and thus the conventional Kennedy receiver is easily improved by using $\hat{D}(\gamma_{\rm opt})$ instead of $\hat{D}(\alpha)$. Figure 1(c) plots the average error probabilities for the Type-I, Type-II, and Kennedy receivers, the homodyne limit, and the Helstrom bound. It is shown that the error probabilities for both the Type-I and II receivers are better than the homodyne limit for any $|\alpha|^2$. Note that a proof-of-principle experiment of the Type-II receiver has been demonstrated recently [6].

TOWARD REALIZATION OF QUANTUM RECEIVERS BEYOND THE HOMODYNE LIMIT

Let us finally discuss the practical perspective of these quantum receivers, particularly, the Type-II receiver. The superiority of the Type-II receiver rather than the Kennedy receiver in $|\alpha|^2 \leq 1$ is significant to beat the homodyne limit in realistic experiments. It is known that the Kennedy-type receiver is not robust against thermal noise or dark counts. Moreover, even without environmental noises, the mode mismatch between the signal and LO causes additional dark counts. As mentioned above, the displacement $\hat{D}(\beta)$ is realized by interfering the signal with the coherent state LO $|\beta/\sqrt{1-\tau}\rangle$ via the beamsplitter of the transmittance τ. The effect of mode mismatch can be characterized by introducing the mode match factor ξ $(0 \leq \xi \leq 1)$ representing the overlap between the signal and LO pulse areas. Since these two pulses are in a coherent state, the average

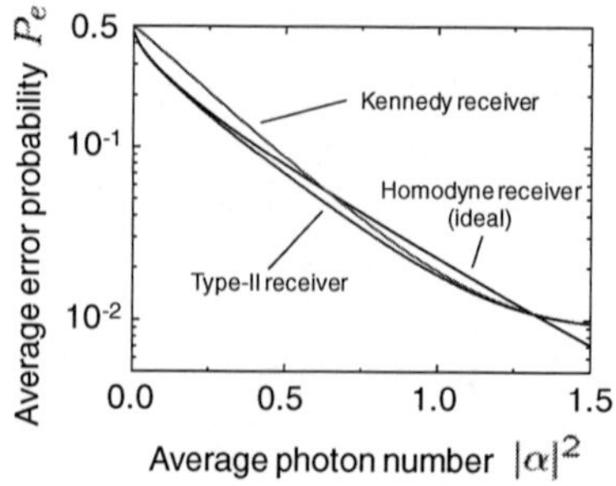

FIGURE 2. Average error probabilities for the ideal homodyne receiver and the Type-II and the Kennedy receivers with practical imperfections, $\tau = 0.99$, $\eta = 0.9$, $\nu = 10^{-3}$, and $\xi = 0.995$.

intensity of the signal field after the interference is simply given by

$$I = (1-\xi)\left(\tau|\alpha|^2 + |\beta|^2\right) + \xi\left|\pm\sqrt{\tau}\alpha + \beta\right|^2. \tag{8}$$

Due to its Poissonian photon number distribution, the average discrimination error including τ and ξ at the on/off detector is described as

$$\tilde{P}_e^D = \frac{1}{2} - e^{-\nu-\eta(\tau\alpha^2+\tilde{\gamma}_{\rm opt}^2)} \sinh\left(2\eta\xi\sqrt{\tau}\alpha\tilde{\gamma}_{\rm opt}\right), \tag{9}$$

where $\tilde{\gamma}_{\rm opt}$ fulfills the optimality condition

$$\xi\sqrt{\tau}\alpha = \tilde{\gamma}_{\rm opt}\tanh(2\eta\xi\alpha\tilde{\gamma}_{\rm opt}). \tag{10}$$

An example of the average error probabilities including the imperfections is shown in Fig. 2 which clearly shows the advantage of our proposed receiver would be crucial to experimentally observe the gain of the quantum receiver beyond the homodyne limit. Although the requirement for η in the weaker signal is still high, recent experimental progress in this field (e.g. [7]) is rather promising for near-future realization of the quantum receivers that outperform the homodyne limit.

REFERENCES

1. C. W. Helstrom, *Quantum Detection and Estimation Theory* (Academic Press, New York, 1976).
2. R.S. Kennedy, RLE, MIT, QPR, **108**, 219 (1973)
3. M. Takeoka and M. Sasaki, Phys. Rev. A **78**, 022320 (2008).
4. S. Dolinar, RLE, MIT, QPR, **111**, 115 (1973)
5. R. L. Cook, P. J. Martin, and J. M. Geremia, *Nature* **446**, 774 (2007).
6. C. Wittmann, M. Takeoka, K. N. Cassemiro, M. Sasaki, G. Leuchs, and U. L. Andersen, Phys. Rev. Lett. **101**, 210501 (2008).
7. D. Rosenberg, A E. Lita, A. J. Miller, and S.-W. Nam, Phys. Rev. A **71**, 061803(R) (2005).

CHAPTER 7

PHOTONS

Entanglement and Decoherence in a Microscopic-Macroscopic system

Nicolò Spagnolo*, Chiara Vitelli*, Fabio Sciarrino*,† and Francesco De Martini*,**

*Dipartimento di Fisica della Sapienza Università di Roma, Roma, 00185 Italy and Consorzio Nazionale Interuniversitario per le Scienze Fisiche della Materia, Roma, 00185 Italy
†Centro di Studi e Ricerche "Enrico Fermi", Via Panisperna 89/A,Compendio del Viminale, Roma 00184, Italy
**Accademia Nazionale dei Lincei, via della Lungara 10, Roma 00165, Italy

Abstract. We investigate the quantum features, entanglement and resilience to losses, of multiphoton fields generated by optical amplification of photon entangled states. We first report the realization of a class of Microscopic-Macroscopic entangled states of more than 5×10^4 photons, based on parametric amplification of single-photon and two-photons states. A non-locality CHSH-type test has been performed on this system. Then, we apply a decoherence model to this Micro-Macroscopic entangled state, showing their high resilience to photon-loss. The theoretical results are in full agreement with many experimental observations.

Keywords: Macroscopic States, Decoherence, Entanglement, Non-Locality
PACS: 03.65.Ta,03.67.Pp

INTRODUCTION

In recent years two fundamental aspects of quantum mechanics have attracted a great deal of interest, namely the investigation on the irreducible nonlocal properties of Nature implied by quantum entanglement and the physical realization of the "Schroedinger Cat" [1, 2]. The last concept, by applying the nonlocality property to a combination of a microscopic and of a macroscopic systems, enlightens the concept of the quantum state, the dynamics of large systems and ventures into the most intriguing philosophical problem, i.e. the emergence of quantum mechanics in the real life. In this context, the coupling with the enviroment [3, 4, 5] is responsible of the very short persistence of coherence in these quantum system. This phenomenon, well known as *decoherence*, represents a fundamental aspect of quantum physics and also the main limitation to quantum computation.

In this paper we report the realization of a new class of quantum states, involving a number of particles N in excess of 5×10^4. The basic idea is to investigate quantum phenomena with large size systems generated in the microscopic world and then transferred in the macroscopic one through an amplification process. This approach allows the direct observation of entanglement and non-locality between a microscopic and a macroscopic photonic state and shows a high resilience to decoherence by coupling with environment [6, 7]. This property is directly demonstrated in this paper by a numerical simulation of the effects of a lossy channel in these Macroscopic states [8]. The present work shows

CP1110, *Quantum Communication, Measurement and Computing (QCMC), The Ninth International Conference,* edited by A. Lvovsky

that the amplification process applied to a microscopic system is a natural approach to enlighten the quantum-to-classical transition and to investigate the persistence of quantum phenomena into the "classical" domain in quantum systems of increasing size.

ENTANGLEMENT AND NON-LOCALITY IN A MICRO-MACROSCOPIC SYSTEM

The realization of a Macro-state [6] consisting of $N = 3.5 \times 10^4$ photons in a quantum superposition and entangled with a far apart single-photon state (Micro-state) has been obtained with the following scheme. Precisely, an entangled photon pair was created by a nonlinear optical process, then one photon of the pair was injected into an optical parametric amplifier (OPA) operating for any input polarization state, i.e. into a phase covariant cloning machine [9, 10, 11, 12, 13]: Fig.1. The generated Micro-Macro state

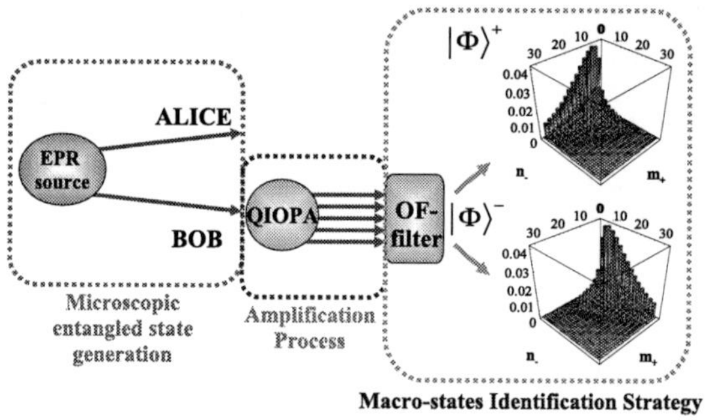

FIGURE 1. Scheme of the setup used for the generation of a Micro-Macroscopic entangled state. One of the two photons of an EPR pair is amplified by a collinear OPA in a Macroscopic field of $N = 3.5x10^4$ particles. The measurement scheme is then based on an Orthogonality Filter which allows to treat the amplified field as a Macro-qubit.

can be written in the form:

$$|\Sigma^-\rangle_{AB} = \frac{1}{\sqrt{2}}\left(|\pi\rangle_A|\Phi^{\pi_\perp}\rangle_B - |\pi_\perp\rangle_A|\Phi^{\pi}\rangle_B\right) \quad (1)$$

In this expression, $|\pi\rangle$ are single-photon states with $\vec{\pi}$ polarization, and $|\Phi^{\pi,\pi_\perp}\rangle_A$ are the single-photon amplified state of the OPA. The interaction Hamiltonian of this device takes the form $\widehat{\mathscr{H}_{coll}} = \imath\hbar\chi\widehat{a}_H^\dagger\widehat{a}_V^\dagger + \text{h.c.}$ in the $\{\vec{\pi}_H, \vec{\pi}_V\}$ polarization basis, and $\widehat{\mathscr{H}_{coll}} = \frac{\imath\hbar\chi}{2}e^{-\imath\phi}\left(\widehat{a}_\phi^{\dagger 2} - e^{\imath 2\phi}\widehat{a}_{\phi_\perp}^{\dagger 2}\right)$ for any "equatiorial" basis $\{\vec{\pi}_\phi, \vec{\pi}_{\phi\perp}\}$ on the Poincaré sphere where: $\vec{\pi}_\phi = \frac{1}{\sqrt{2}}\left(\vec{\pi}_H + e^{\imath\phi}\vec{\pi}_V\right)$. In order to demonstrate the entanglement feature of the biparitte system, the fundamental problem to overcome has been to find a viable generalized measurement scheme. This scope has been achieved by identifying the amplified macro-state as a macro-qubit system. Hence we have introduced a probabilistic measurement able to discriminate between orthogonal macro-state wavefunctions. The output radiation of the Macroscopic field was analyzed in polarization and revealed by

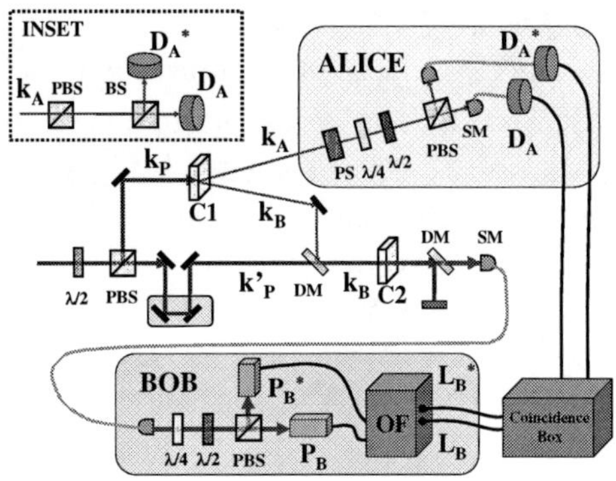

FIGURE 2. Optical configuration of the QI-OPA apparatus. The first NL crystal :C1, generates an EPR entangled state. The NL crystal 2: C2, realizing the optical parametric amplification (OPA), is cut for collinear type II phase matching. INSET: Two photon π−states over the mode $\mathbf{k}_A$ are detected through a fiber coupled beam-splitter *BS* by a pair of single-photon detectors D_A and D_A^*.

linear detectors, namely, photomultipliers. Through an electronic device dubbed "Orthogonality Filter" (OF), we locally filtered the detected signal exploiting the macroscopic difference existing between the probability distributions of the photon numbers associated with the orthogonal Macro-states. With this POVM scheme we demonstrated the non-separability of the bipartite system by adopting an entanglement criterion introduced for qubit-qubit correlations [14].

A further investigation on the persistence of quantum effects in large multi-particle systems was carried out by performing a non-locality test implying the violation of a Bell's inequality [7]. To carry out such a test higher correlations among the fields realized on the $\mathbf{k}_A$ and $\mathbf{k}_B$ modes are required. Hence we adopted the QI-OPA amplification of a 4-photon entangled state. In this second experiment, the NL crystal 1 generated by SPDC two simultaneous entangled photon couples, i.e. a 4-photon entangled state. The 4 photons were emitted over the two output modes $\mathbf{k}_i$ $(i = A, B)$ in the *singlet* state of two spin-1 subsystems

$$|\Psi_2^-\rangle_{AB} = \frac{1}{\sqrt{3}}(|2H\rangle_A|2V\rangle_B - |H;V\rangle_A|H;V\rangle_B + |2V\rangle_A|2H\rangle_B) \tag{2}$$

The 2 photons generated over the mode $\mathbf{k}_B$ were injected into the collinear QI-OPA amplifier. After the amplification process the overall output state can be expressed in any "equatorial" polarization basis $\left\{\vec{\pi}_\phi, \vec{\pi}_\phi^\perp\right\}$ on the Poincaré sphere as

$$|\Omega\rangle_{AB} = \frac{1}{\sqrt{3}}\left(|2\phi\rangle_A\left|\Phi^{2\phi^\perp}\right\rangle_B - |\phi;\phi^\perp\rangle_A\left|\Phi^{\phi,\phi^\perp}\right\rangle_B + |2\phi^\perp\rangle_A\left|\Phi^{2\phi}\right\rangle_B\right) \tag{3}$$

where $|\Phi^\pi\rangle_B$ stands for the amplified field generated by injecting the 2-photon state $|\pi\rangle_B$: $|\Phi^\pi\rangle_B = \widehat{U}\,|\pi\rangle_B$. In order to demonstrate the non-locality of the micro-macro state we have defined two sets of dichotomic observables for A and B which can be measured by adopting different polarization basis sets. Both partners performed measurements in the equatorial basis $\left\{\vec{\pi}_{\phi}, \vec{\pi}_{\phi}^{\perp}\right\}$ as follows. A associated to the detection of the 2-photons with polarization $\vec{\pi}_{\phi}$, i.e. of $|2\phi\rangle_A$, the observable $+1$ and to the detection of the 2-photons with polarization $\vec{\pi}_{\phi\perp}$, i.e. of $|2\phi^{\perp}\rangle_A$, the observable -1. The two possible output values $\{\mathbf{a}, \mathbf{a}'\}$ corresponded to $\phi_a = \frac{\pi}{4}$ and $\phi_a' = -\frac{\pi}{4}$. B associated to the detection of the $|\Phi^{2\phi}\rangle_B$ the observable $+1$ and to the detection of the $|\Phi^{2\phi\perp}\rangle_B$ the observable -1. The two possible values $\{\mathbf{b}, \mathbf{b}'\}$ correspond to $\phi_b = 0$ and $\phi_b' = \frac{\pi}{2}$. In order to achieve a high discrimination among $\left\{|\Phi^{2\phi}\rangle_B, |\Phi^{2\phi\perp}\rangle_B\right\}$ the OF-filter was adopted. We obtained:

$$S = 2.256 \pm 0.049 \tag{4}$$

The experimental obtained value for the parameter S results: $S > 2 = S_{CHSH}$ [15], and hence implies the non-locality of the micro-macro photon system.

COHERENCE IN MACROSCOPIC QUANTUM SUPERPOSITIONS

Let us define now an useful parameter that can be used to quantify the degree of coherence in a MQS. Our starting point is the concept of "fidelity" [16] $\mathscr{F}$ between two generic density matrices $\widehat{\rho}$ and $\widehat{\sigma}$, defined as: $\mathscr{F}(\widehat{\rho}, \widehat{\sigma}) = \mathrm{Tr}\left(\sqrt{\widehat{\rho}^{\frac{1}{2}}\widehat{\sigma}\widehat{\rho}^{\frac{1}{2}}}\right)$. From this quantity, we can define the "Bures distance" [17, 18]:

$$D(\widehat{\rho}, \widehat{\sigma}) = \sqrt{1 - \mathscr{F}(\widehat{\rho}, \widehat{\sigma})} \tag{5}$$

which is a metric in the quantum state space. The notion of distance adopted is strictly related to the probability $\mathscr{F}^2(\widehat{\rho}, \widehat{\sigma})$ to obtain a wrong result in measurements which are performed to discriminate between two non-orthogonal quantum states [19]. If we calculate this quantity between two macroscopic states $|\phi_1\rangle$ and $|\phi_2\rangle$, the distance $D(|\phi_1\rangle, |\phi_2\rangle)$ represents their *distinguishability*. Let us now define the two corresponding MQS's: $|\phi^{\pm}\rangle = \frac{\mathscr{N}_{\pm}}{\sqrt{2}}(|\phi_1\rangle \pm |\phi_2\rangle)$. We consider the distance $D(|\phi^+\rangle, |\phi^-\rangle)$ as the "*Visibility*", i.e. "degree of orthogonality" of the MQS's. Indeed, the value of the MQS visibility depends exclusively on the relative phase of the component states: $|\phi_1\rangle$ and $|\phi_2\rangle$. In presence of decoherence, the relative phase in the MQS will progressively randomize, and the visibility will decrease from the initially unitary value to $D \to 0$ as $|\phi^{\pm}\rangle$ will both evolve to the same mixed state $\widehat{\rho} = \widehat{\rho}_1 + \widehat{\rho}_2$. In this expressions, $\widehat{\rho}_{(1,2)}$ are the density matrices of the component states $|\phi_1\rangle$ and $|\phi_2\rangle$ after losses. The choice of the "classical" states $|\phi_{1,2}\rangle$ with respect to $|\phi^{\pm}\rangle$ is determined considering the quantum states which exhibit higher resilience to losses as the components of the superposition. This choice directly depends on the concept of *envariance* and *einselection* in fundamental quantum theory [20]. Finally, a simple beam-splitter model allows to simulate the effects of a lossy channel and to calculate how the distance between two initially orthogonal MQS decrease under decoherence processes.

The decoherence model presented in this paper has been applied to QI-OPA equatorial single-photon amplified states:

$$|\Phi^{\phi}\rangle = \hat{U}|\phi\rangle = \sum_{i,j=0}^{\infty} \gamma_{ij}|(2i+1)\phi,(2j)\phi_{\perp}\rangle \tag{6}$$

where $\gamma_{ij} = \frac{1}{C^2}\left(e^{-\imath\varphi}\frac{\Gamma}{2}\right)^i\left(-e^{\imath\varphi}\frac{\Gamma}{2}\right)^j\frac{\sqrt{(2i+1)!}\sqrt{(2j)!}}{i!j!}$. In these expressions $C = \cosh g$ and $\Gamma = \tanh g$, where g is the non linear gain of the amplifier. Due to the unitary evolution in the crystal, the superposition form of the input single-photon qubit is maintained during the amplification process. The experimental results obtained in [6], where the resilience to losses allowed to verify the entanglement, can be understood evaluating numerically how the distance between two initially orthogonal states decreases with an increasing value of the losses parameter R. We choose $\{|\Phi^+\rangle, |\Phi^-\rangle\}$ as initial states, where the $\vec{\pi}_{\pm}$ basis corresponds to a value of $\phi = 0$ in the definition. We note that the following results still hold for all orthogonal equatorial Macro-States due to the phase-covariance of the process. In Fig.3 we report the visibility as a function of the average number of lost photons $R\langle n\rangle$ for different values of the gain. We compare this class of MQS's with the quantum superposition of coherent states [21, 22, 23, 24]. The same model of decoherence used in this paper has been applied in [8] on the latter class of states, obtaining an "universal" curve, valid for all values of $\langle n\rangle$, reported in Fig.3 as the dotted line. The single-photon amplified states, differently from the coherent MQS states, become more robust to losses with an increasing average number of generated photons. This resilience depends on the unbalancement in the stimulated emission regime be-

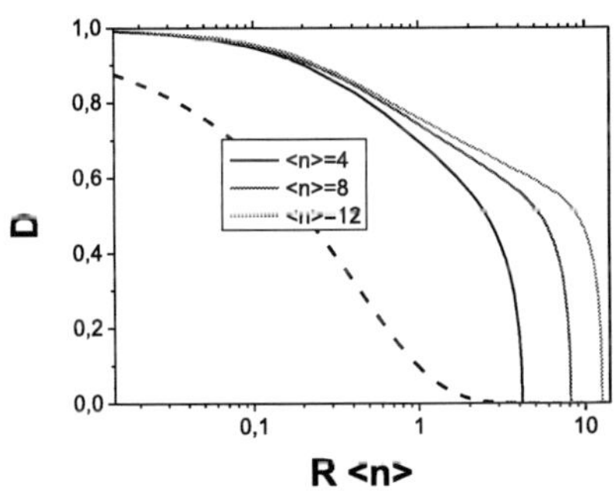

FIGURE 3. Plot of the MQS "Visibility" as a function of the average number of lost photons $R\langle n\rangle$. Dashed blue line corresponds to the superposition of coherent states for all values of α. Black curve corresponds to QIOPA amplified states with g=0.8 ($\langle n\rangle \sim 4$), red line to g=1.1 ($\langle n\rangle \sim 8$) and green line to g=1.3 ($\langle n\rangle \sim 12$).

tween the number of photons generated in the two orthogonal polarization, depending on the injected qubit. This feature is enlightened by the analysis of the photon-number distribution for the $|\Phi^+\rangle$ and $|\Phi^-\rangle$ states for different values of the parameter R. The two distributions remain distinguishable even with a high value of lost photons and then can be discriminated through the POVM scheme described previously.

CONCLUSIONS

In this paper we addressed the problem of entanglement and decoherence in a Microscopic-Macroscopic system. First, we reported the realization of a Micro-Macroscopic entangled state obtained by phase-covariant amplification of single photon-states. With the same device, the amplification of a two-photon state lead to the violation of Bell's inequalities in the CHSH form. Then, we defined a quantitative criterium that identifies the persistance of coherence in MQS under the propagation in a lossy channel. The chosen model has been applied to the amplified Macro-qubits, which show a high resilience to decoherence. This property allows experimental realization and implementation of quantum information protocols.

ACKNOWLEDGMENTS

Work supported by PRIN 2005 of MIUR and INNESCO 2006 of CNISM.

REFERENCES

1. A. Einstein, B. Podolski, and N. Rosen, *Phys. Rev.* **47**, 777 (1935).
2. E. Schroedinger, *Naturwissenschaften* **23**, 807 – 812, 823 – 828, 844 – 849 (1935).
3. M.A. Nielsen and I.L.Chuang, Quantum Information and Quantum Computation (Cambridge University Press, 2000)
4. W. Zurek, Physics Today, October 1991, pag.36; *Rev. Mod. Phys.* **75**, 715 (2003); *Progr. Math. Phys.* **48**, 1 (2007)
5. W. Dur, J. Simon, and J.I. Cirac, *Phys. Rev. Lett.* **89**, 210402 (2002) ; W. Dur, and H.J. Briegel, *Phys. Rev. Lett.* **92**, 180403 (2004)
6. F. De Martini, F. Sciarrino, and C. Vitelli, *Phys. Rev. Lett.* **100**, 253601 (2008).
7. F. De Martini, F. Sciarrino, and C. Vitelli, quant-ph/0804.0341/v1.
8. F. De Martini, F. Sciarrino, and N, Spagnolo, quant-ph/arxiv:0809.3889
9. F. De Martini, *Phys. Rev. Lett.* **81**, 2842 (1998); *Phys. Lett. A* **250**, 15 (1998).
10. F. De Martini, F. Sciarrino, and V. Secondi, *Phys. Rev. Lett.* **95**, 240401 (2005).
11. F. De Martini, and F. Sciarrino, *Prog. Quant. Elect.* **29**, 165 (2005).
12. F. Sciarrino, and F. De Martini, *Phys. Rev. A* **72**, 062313 (2005)
13. E. Nagali, T. De Angelis, F. Sciarrino, and F. De Martini, *Phys. Rev. A* **76**, 042126 (2007).
14. H.S. Eisenberg, G. Khoury, G. A. Durkin, C. Simon, and D. Bouwmeester, *Phys. Rev. Lett.* **93**, 193901 (2004).
15. J.F. Clauser, M.A. Horne, A. Shimony, and R.A. Holt, *Phys. Rev. Lett.* **23**, 880 (1969)
16. R. Jozsa, *J. Mod. Opt.* **41**, 2315 (1994)
17. D. Bures, *Trans. Am. Math. Soc.* **135**, 199 (1969)
18. M. Hubner, *Phys. Lett. A* **163**, 239 (1992); *Phys. Lett. A* **179**, 226 (1993)
19. B. Huttner, A. Muller, J. D. Gautier, H. Zbinden, and N. Gisin, *Phys. Rev. A* **54**, 3783 (1996)
20. W.H. Zurek, *Phys. Rev. Lett.* **90**, 120404 (2003) and references therein.
21. W. Schleich, M. Pernigo, and Fam Le Kien, *Phys. Rev. A* **44**, 2172 (1991)
22. M. Brune, S. Haroche, J.M. Raimond, L. Davidovich, and N. Zagury, *Phys. Rev. A* **45**, 5193 (1992)
23. J.M. Raimond, M. Brune, and S. Haroche, *Rev. Mod. Phys.* **73**, 565 (2001)
24. A. Ourjoumtsev, R. Tualle-Brouri, J. Laurat, and P. Grangier, *Science* **382**, 83 (2006); A. Ourjoumtsev, H. Jeong, R. Tualle-Brouri, and P. Grangier, *Nature* **448**, 784 (2007)

Simple class of nonclassical states generated by seeded down-conversion and imperfect photon detection

Jonas Söderholm and Shuichiro Inoue

Institute of Quantum Science, Nihon University, 1-8-14 Kanda-Surugadai, Chiyoda-ku, Tokyo 101-8308, Japan

Abstract. We have analyzed a scheme employing seeded nondegenerate down-conversion and a single-photon threshold detector with non-unit quantum efficiency and finite dark-count probability. The single-mode state generated conditioned on the click of the detector is found to be easily expressed using the negative binomial distribution.

Keywords: Nondegenerate down-conversion, Fock seed, Negative binomial distribution
PACS: 42.50.-p, 42.50.Dv, 42.50.Ar, 42.65.Lm

INTRODUCTION

Nondegenerate down-conversion allows one to generate approximate single-photon states probabilistically [1]. As the down-converted photons are generated in pairs, with one photon each in the signal and idler modes, the detection of a photon in one mode signals the presence of a photon in the other. By decreasing the pump power, the effects of multiple photon pairs can be controlled at the expense of a lower success probability. The quality of the heralded single-photon setup also depends on the detector parameters. Experimentally, approximate single-photon states with negative measured Wigner functions have been achieved, and thus also demonstration of a non-Gaussian operation [2]. Such operations are necessary in addition to linear optics and squeezers in order to achieve universal quantum computation with continuous variables [3].

In more recent experiments [4, 5], non-Gaussian operations have also been realized with seeded nondegenerate down-conversion through photon addition. Here, we analyze the seeded heralded single-photon scheme, considering losses and detector imperfections. We find that the generated state can be easily expressed using the negative binomial distribution.

THE NEGATIVE BINOMIAL DISTRIBUTION

Using the gamma function, the negative binomial distribution can be written as

$$p_{\text{nb}}(s,\mu,n) = \frac{\Gamma(s+n)\, s^s \mu^n}{\Gamma(s)\, n!\, (s+\mu)^{s+n}}, \tag{1}$$

CP1110, *Quantum Communication, Measurement and Computing (QCMC), The Ninth International Conference,* edited by A. Lvovsky

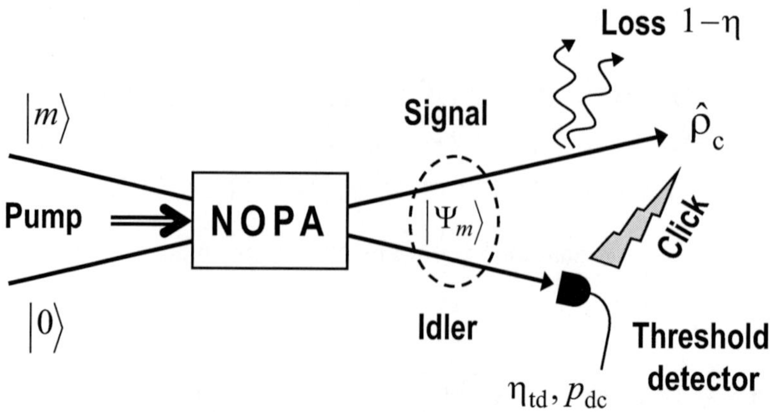

FIGURE 1. The considered setup consists of a nondegenerate optical parametric amplifier (NOPA), whose idler mode is seeded with an m-photon state, and a single-photon threshold detector. The single-mode state $\hat{\rho}_c$ is generated when the threshold detector clicks. Its quantum efficiency and dark-count probability are denoted as η_{td} and p_{dc}, respectively.

where s and the expectation value $\mu = \sum_{n=0}^{\infty} n p_{nb}(s,\mu,n)$ are positive. Using a different parametrization, the distribution can be expressed using the binomial coefficient with a negative argument, which explains its name. The negative binomial distribution keeps its form under linear loss, and the thermal and Poisson distributions are special cases. The thermal distribution

$$p_{th}(\mu,n) = p_{nb}(1,\mu,n) = \frac{\mu^n}{(\mu+1)^{n+1}} \tag{2}$$

is obtained for $s = 1$, whereas the Poisson distribution

$$p_{poi}(\mu,n) = \lim_{s\to\infty} p_{nb}(s,\mu,n) = \frac{e^{-\mu}\mu^n}{n!} \tag{3}$$

corresponds to the limit $s \to \infty$.

SEEDED NONDEGENERATE DOWN-CONVERSION

Nondegenerate down-conversion seeded with an m-photon state in one mode will generate photon pairs with a negative binomial distribution [6]. The two-mode state generated by the nonlinear crystal in the setup depicted in Fig. 1 is therefore of the form

$$|\Psi_m\rangle = \sum_{n=0}^{\infty} e^{i\varphi_n} \sqrt{p_{nb}(m+1,[m+1]\mu_0,n)}\,|n,m+n\rangle. \tag{4}$$

Ignoring dark counts, the probability for the threshold detector not to click is $p_0 = (1-\eta_{td})^m/(1+\eta_{td}\mu_0)^{m+1}$, where η_{td} is the quantum efficiency. Hence, the probability of a photon-induced click is $1-p_0$ and the probability for a click induced by dark count is $p_{dc}p_0$, where p_{dc} is the dark-count probability. In both cases, the resulting state in the signal mode is a statistical mixture of Fock states, where the probability for n photons takes the form

$$P(\lambda,s,\mu_1,\mu_2,n) = (1-\lambda)p_{nb}(s,\mu_1,n) + \lambda p_{nb}(s,\mu_2,n) \tag{5}$$

with

$$s = m+1, \qquad \mu_1 = \eta\mu_0(m+1), \qquad \text{and} \qquad \mu_2 = \frac{\eta\mu_0(m+1)(1-\eta_{td})}{1+\eta_{td}\mu_0}. \tag{6}$$

Since the negative binomial distribution keeps its form under loss, $1-\eta$ can be the total loss in the signal mode, not necessarily limited to losses before the detector click. In the case of a photon-induced click, we have

$$\lambda = \left[1 - \frac{(1+\eta_{td}\mu_0)^{m+1}}{(1-\eta_{td})^m}\right]^{-1}, \tag{7}$$

whereas $\lambda = 1$ in the case corresponding to a dark count. As we cannot distinguish the two cases, the conditional state $\hat{\rho}_c$ in the signal mode corresponds to

$$\lambda = \left[1 - \frac{(1+\eta_{td}\mu_0)^{m+1}}{(1-p_{dc})(1-\eta_{td})^m}\right]^{-1}. \tag{8}$$

The condition for the probability distribution (5) to be sub-Poissonian can be written as

$$\lambda < -\frac{\sqrt{(s+1)^2(\mu_1^2-\mu_2^2)-(s^2-1)\mu_1\mu_2}}{2s(\mu_1-\mu_2)}. \tag{9}$$

Hence, whenever Eq. (8) satisfies condition (9), the generated state $\hat{\rho}_c$ must be nonclassical.

Finally, we note that the heralded single-photon scheme corresponds to $m = 0$, in which case the negative binomial distribution simplifies to the thermal distribution.

ACKNOWLEDGMENTS

This work was supported by the National Institute of Information and Communications Technology.

REFERENCES

1. C. K. Hong and L. Mandel, *Phys. Rev. Lett.* **56**, 58–60 (1986).
2. A. I. Lvovsky, H. Hansen, T. Aichele, O. Benson, J. Mlynek, and S. Schiller, *Phys. Rev. Lett.* **87**, 050402 (2001).
3. S. Lloyd and S. L. Braunstein, *Phys. Rev. Lett.* **82**, 1784–1787 (1999).
4. A. Zavatta, S. Viciani, and M. Bellini, *Science* **306**, 660–662 (2004).
5. A. Zavatta, V. Parigi, and M. Bellini, *Phys. Rev. A* **75**, 052106 (2007).
6. G. S. Agarwal, *Phys. Rev. A* **45**, 1787–1792 (1992).

Numerical Analysis of Parametric Downconversion

Wolfgang Mauerer and Christine Silberhorn

Max Planck Research Group for Optics, Information, and Photonics
University Erlangen-Nürnberg
Günther-Scharowsky-Straße 1/Bau 24
91058 Erlangen, Germany

Abstract. Parametric downconversion (PDC) is a popular technique to produce twin beams of photons that are entangled in multiple degrees of freedom. The generated states form the basis for numerous applications that require entanglement. An exact quantification of this resource is therefore essential, for instance for quantum cryptography that relies on a complete knowledge of the correlation contained in the state. While the determination of an entanglement monotone for the PDC process is only possible analytically in special cases, an exact calculation must usually be performed numerically. Recent work by Mikhailova et al. [2] analyses a certain class of PDC states for which the concurrence entanglement measure can be obtained by an analytical approximation. In this contribution, we analyse the validity of the approximation by comparison with exact numerical methods.

Keywords: Parametric downconversion, entanglement, concurrence, large spectral correlation
PACS: 42.50.-p, 02.60.Cb

REVIEW OF PARAMETRIC DOWNCONVERSION

Parametric downconversion (PDC) exploits the second-order nonlinearity $\chi^{(2)}$ of a dielectric to produce strictly photon number correlated twin beams. The interaction Hamiltonian of the process is given by $\hat{H}_{\text{int}}(t) = \int_V \mathrm{d}^3\vec{x}\chi^{(2)}\hat{E}_p^{(+)}(\vec{x},t)\hat{E}_s^{(-)}(\vec{x},t)\hat{E}_i^{(-)}(\vec{x},t) + \text{H.c.}$, where the subscripts "p", "s", and "i" denote pump, signal, and idler, respectively. Physically, the operator describes the annihilation of a pump photon impinging on the crystal and the creation of two outgoing photons – signal and idler. Since the coupling strength $\chi^{(2)}$ is small, it usually suffices to consider the resulting state in first-order perturbation theory for quantum-mechanical considerations. Additionally, we can assume a classical pump because $\hat{E}_{\text{p}}$ is much stronger than both signal and idler field, which leads to a state of the form [1]

$$|\psi\rangle \approx |0\rangle_{\text{MM}} + C\iint \mathrm{d}\omega_1\mathrm{d}\omega_2 f(\omega_1,\omega_2)\hat{a}^\dagger(\omega_1)\hat{b}^\dagger(\omega_2)\,|0\rangle_{\text{MM}}. \tag{1}$$

where $|0\rangle_{\text{MM}}$ denotes a multi-mode vacuum, and $f(\omega_s,\omega_i)$ is the spectral distribution function (SDF) that controls the correlations between signal and idler in the frequency domain (for the sake of simplicity, we omit any polarisation degrees of freedom).

The exact form of $f(\omega_1,\omega_2)$ depends on the properties of the dielectric. Phase-matching conditions ensure conservation of energy and momentum. They have been investigated by numerous authors, cf. Ref. [1] and references therein. For so-called type-

CP1110, *Quantum Communication, Measurement and Computing (QCMC), The Ninth International Conference,* edited by A. Lvovsky

I processes, signal and idler are emitted into the same spatial mode and carry the same polarisation. In the degenerate case, it is possible that the SDF is identical for signal and idler.

Consider a $LiIO_3$ crystal, for which the SDF is given by [2]

$$\psi_I(\nu_1,\nu_2) \propto \exp\left(-\frac{(\nu_1+\nu_2)^2\tau^2}{8\ln 2}\right) \times \operatorname{sinc}\left(\frac{L}{2c}\left(A(\nu_1+\nu_2) - B\frac{(\nu_1-\nu_2)^2}{\omega_0}\right)\right), \quad (2)$$

where $\nu = 2\pi c/\lambda - \pi c/\lambda_0$, τ is the pulse duration of the Gaussian pump, and $\lambda_0 =$ 400 nm. The constants $A = 0.17, B = 0.069$ are determined by the dispersive properties of the crystal (where an expansion up to second polynomial order has been used), and $L = 0.05$ cm denotes the crystal length. We concentrate on this process in the following.

For numerous applications, it is of particular interest that signal and idler are spectrally entangled. To quantify the amount of entanglement, we re-write the SDF using a *Schmidt decomposition* defined by

$$f(\omega_1,\omega_2) = \sum_{n=0}^{N} \sqrt{\lambda_n}\xi_n^{(1)}(\omega_1)\xi_n^{(2)}(\omega_2) \quad (3)$$

$\xi_i(\omega_{1/2})$ is a bi-orthogonal set of complex functions called Schmidt functions, and $\sqrt{\lambda_n}$ are real expansion coefficients. Notice that the decomposition is unique for every well-behaved function $f(\omega_s,\omega_i)$. Computing an analytical representation of the Schmidt decomposition (3) for a particular SDF is usually impossible, so numerical methods are required. Once the decomposition is known, however, the coefficients $\sqrt{\lambda_n}$ readily allow for computing two quantities: The *logarithmic negativity* S and *concurrence* K. They are defined by $S = -\sum_{n=0}^{N}\lambda_n \log\lambda_n$, $K = \frac{1}{\sum_{n=0}^{N}\lambda_n^2}$. Both measure the amount of entanglement contained in the spectral degree of freedom [3].

The concurrence K can be computed approximately in the limit of long and short pump pulses, as was shown in Ref. [2]. A formula for all pump durations can be obtained by quadratic interpolation between the boundary cases, and reads as $K_{\text{approx}} = \sqrt{\frac{75.5^2}{\eta} + (44\eta)^2}$, where $\eta = 2\tau/(L/v_g^{(p)} - L/v_g^{(s)})$. $v_g^{(p,s)}$ gives the group velocities for pump and signal. In the following section, we will analyse how well the proper value of K is approximated by this technique.

NUMERICAL METHODS

Various numerical alternatives to perform a Schmidt decomposition are available; they differ greatly in their numerical convenience and power (see Ref. [4] for a complete discussion). One particularly scalable and well-behaved approach is to first discretise the SDF into a matrix $C \in \mathbb{C}^{N\times M}$ and then perform a singular value decomposition (SVD) that transforms C into $C = U \cdot S \cdot V^{\mathrm{T}}$.

The matrices $U \in U(N)$ and $V \in U(M)$ are unitaries, and $S \in \mathbb{R}^{N\times M}$ is a real diagonal matrix that contains the singular values s_i. It can be shown that $s_i = \sqrt{\lambda_n}$ holds, that is, the singular values are immediately connected to the Schmidt coefficients. Likewise, the (discretised) Schmidt basis is contained in the row and column vectors of U and V^{T}, respectively.

The dimension of C required to obtain convergent results depends on the amount of entanglement contained in the state. For many downconversion configurations, only small matrices with dimensions of around 1000×1000 are required, which does not present any computational difficulties. Since the SDF defined in Eq. 2 is very narrow, the situation is different in this case. Depending on the pump length, matrices of size 20000×20000 and larger are necessary, so more refined computational methods involving parallel techniques must be employed. Since we focus on the results of the computations here, the reader is again referred to Ref. [4] for more details about the underlying process.

RESULTS

The matrices resulting from a discretisation of Eq. (2) are thinly populated; typically less than 1 percent of their entries are non-zero. Physically, this equates to a high-dimensional Hilbert space that contains a large amount of correlation. We have therefore resorted to sparse matrices and a corresponding algorithm (Lanczos bi-diagonalisation with partial re-orthogonalisation) for the SVD calculation. Any quantity obtained from the decomposition will only exhibit the proper value if a grid of sufficiently large size is chosen for the discretisation. This is illustrated for the logarithmic negativity in Figure 1. Notice that for different pump widths (and thus for different SDFs), different grid sizes are required to achieve convergence.

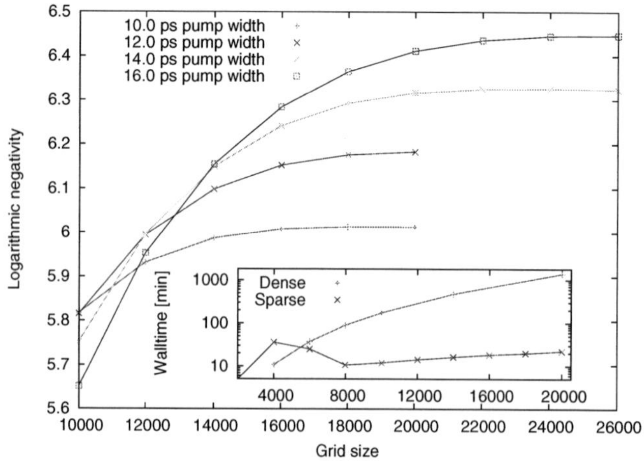

FIGURE 1. Convergence of logarithmic negativity depends on the chosen grid size. The inset illustrates the computational complexity of the problem. Crosses etc. represent points on the grid for which the calculation was performed. The connecting lines have no significance and are just used to guide the eye.

To ensure that the sparse representation provides all relevant information, we have also performed a cross-check with dense matrices (using bi-diagonal reduction and QR iteration for the SVD) that did not unearth any significant differences. However, this method requires significantly increased computational effort, as is shown in the inset of Figure 1. This emphasises the importance of choosing the proper methods and algorithms – the class of treatable physical situations is much enlarged by proper techniques.

The results of our numerical calculations are shown in Figure 2. The agreement between approximation and numerical solution is excellent for temporal pump widths larger than 6 ps, and again for pump widths smaller than 0.5 ps. A noticeable difference of up to 15% arises for pumps at around 1.5 ps. This is the region handled by quadratic interpolation between long and short pump pulses in the approximation. Notice, however, that the amount of entanglement is never over-, but only underestimated in this region.

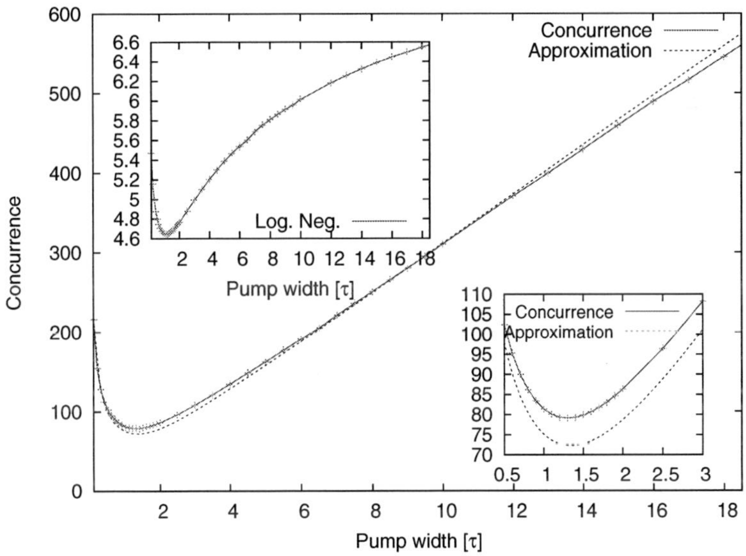

FIGURE 2. Comparison between concurrence approximated by analytical methods and the numerical solution. The bottom inset shows the region where most disagreement between both approaches can be observed. The upper inset displays the logarithmic negativity, a second entanglement monotone that can only be inferred from the numerical calculation, not from the analytical approximation.

ACKNOWLEDGMENTS

We acknowledge valuable conversations with A. Christ and A. Eckstein about PDC in general, and helpful discussions with Mikhail Fedorov on the type-I process.

REFERENCES

1. W. P. Grice, and I. A. Walmsley, *Phys. Rev. A* **56**, 1627–1634 (1997).
2. Y. M. Mikhailova, P. A. Volkov, and M. V. Fedorov, *arXiv:0801.0689* (2008).
3. S. Parker, S. Bose, and M. B. Plenio, *Physical Review A* **61**, 32305 (2000).
4. W. Mauerer, M. Avenhaus, and C. Silberhorn, *to appear in J. Las. Phys.* (2009).

Local transformation of two EPR photon pairs into a three-photon W state

T. Tashima*, T. Wakatsuki*, Ş. K. Özdemir†, T. Yamamoto*, M. Koashi* and N. Imoto*

*Division of Materials Physics, Department of Materials Engineering Science, Graduate School of Engineering Science, Osaka University, Toyonaka, Osaka 560-8531, Japan, CREST Photonic Quantum Information Project
†Division of Materials Physics, Department of Materials Engineering Science, Graduate School of Engineering Science, Osaka University, Toyonaka, Osaka 560-8531, Japan, CREST Photonic Quantum Information Project, ERATO Nuclear Spin Electronics Project

Abstract. We propose and demonstrate a local transformation of two EPR photon pairs to three-photon W state. The key issue of our scheme is that it requires only one polarization dependent beamsplitter to perform the necessary local operation. Using a quantum state tomography, the fidelity of the prepared three-photon state was estimated as 0.778 ± 0.043. The expectation value of the witness operator of W_3 was -0.111 ± 0.043.

Keywords: multipartite entanglement, W state, LOCC
PACS: 03.67.-a, 42.50.-p, 03.67.Mn

Entanglement is a powerful resource for quantum information processing. In bipartite systems, the properties are well-understood: any two-qubit state can be prepared from an EPR pair by local operation and classical communication (LOCC). On the other hand, entanglement of multipartite systems contains a rich and complex structure which is still under intense research. In particular, the relation between different multipartite entangled states and their inter-convertibility under LOCC has attracted much attention. It has been shown that there are inequivalent classes of genuine multipartite entanglement, simplest of which are the tripartite W state and Greenberger-Horne-Zeilinger(GHZ) states represented in photonic systems as $|\mathrm{W}_3\rangle = \frac{1}{\sqrt{3}}(|\mathrm{HHV}\rangle + |\mathrm{HVH}\rangle + |\mathrm{VHH}\rangle)$ and $|\mathrm{GHZ}_3\rangle = \frac{1}{\sqrt{2}}(|\mathrm{HHH}\rangle + |\mathrm{VVV}\rangle)$, with $|\mathrm{H}\rangle$ and $|\mathrm{V}\rangle$ denoting the horizontal and vertical polarization respectively. These two states cannot be converted to each other by LOCC with nonzero success probability [1].

Recently, there have been a number of experimental demonstrations for the preparation of tripartite GHZ and W states [2, 3, 4]. One particularly interesting demonstration is the preparation of a tripartite GHZ state from two EPR pairs [2] and then the local conversion of this GHZ state to an approximate W state using a generalized measurement and classical communication[3]. The latter shows a trade off between the probability of success and the fidelity of the final state to the W state: The higher is the fidelity, the lower is the success probability. One naturally expects that if we prepare a tripartite W states directly from two EPR pairs via LOCC, we can avoid this trade-off and obtain unit fidelity.

CP1110, *Quantum Communication, Measurement and Computing (QCMC), The Ninth International Conference,* edited by A. Lvovsky

In this study, we demonstrate the local transformation of two EPR photon pairs shared among three parties to a three-photon W state with LOCC. The key issue of the proposal and its demonstration is the use of a polarization dependent beamsplitter (PDBS) which has different transmission/reflection characteristics for H and V polarized photons: This PDBS acts as a 50:50 beamsplitter (BS) for the V-polarized photons while as an unbalanced BS with transmission efficiency of $(7+\sqrt{17})/16$ and reflection efficiency of $(9-\sqrt{17})/16$ for H-polarized photons. We assume that Alice, Bob and Charlie who are spatially separated share two EPR photon pairs $|\Psi\rangle_{12}\otimes|\Psi\rangle_{34} = (|H\rangle_1|H\rangle_2 - |V\rangle_1|V\rangle_2)/\sqrt{2}\otimes(|H\rangle_3|H\rangle_4 - |V\rangle_3|V\rangle_4)/\sqrt{2}$ where modes 2 and 3 belong to Charlie, and Alice and Bob holds the modes 1 and 4. After the action of the PDBS on two photons on Charlie's side and the selection of H-polarzied photon in one of the outputs (mode 5) of the PDBS, the postselected state in the remaining modes 1, 4 and 6 becomes

$$\frac{\sqrt{9-\sqrt{17}}}{8\sqrt{2}}\left[|HHV\rangle + |HVH\rangle + \frac{\sqrt{7+\sqrt{17}}}{\sqrt{9-\sqrt{17}}}|VHH\rangle\right]_{146},$$

where we have rotated the polarization of the photon in mode 6 of the PDBS by a half wave plate (HWP) on Charlie's side. Finally, by introducing local polarization dependent loss in mode 1 (Alice's side), we obtain

$$\sqrt{3(9-\sqrt{17})/128}|W_3\rangle_{146}. \tag{1}$$

This means that local transformation of two EPR photon pairs into a three-photon W state is achieved with unit fidelity at a success probability of $3(9-\sqrt{17})/128 \sim 11.4\%$.

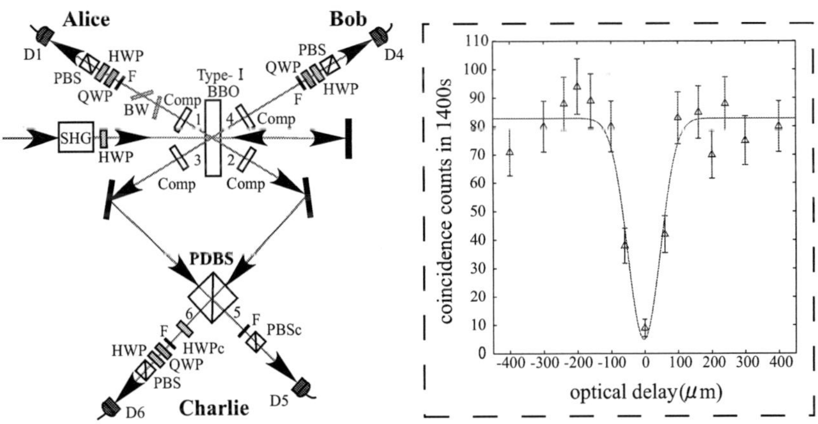

FIGURE 1. Experimental setup and observed two photon interference by recording coincidence of four V-polarized photons.

In our scheme shown in Figure 1, the ultraviolet pulse (wavelength 395nm; power 380mW) with diagonal polarization prepared from mode-locked Ti:Sapphire laser (wavelength 790nm; pulse width 90fs; repetition rate 82MHz) makes two passes through a pair of Type-I BBO crystals each with thickness 1.5mm stacked together with their optical axes orthogonal to each other to produce two EPR photon pairs. To compensate walk off effects of H- and V-polarized photons, we used compensation BBOs (comp) with thickness 1.65mm. Successful events are postselected by four-photon coincidences using single photon counting detectors in modes 1, 4, 5, and 6. Charlie performs the local operation by the PDBS and polarizing beamsplitter (PBSc) by detecting a H-polarized photon in mode 5 and changing the polarization on mode 6 by HWPc. Alice performs the local operation by a set of glass plates in mode 1 which forms a Brewster window (BW) to equalize the amplitude. Quarter wave plates (QWP), HWPs and PBSs in front of the detectors in mode 1, 4 and 6 are used for quantum state tomography. Fixing the polarization of photons in modes 2 and 3 to V-polarization, we adjust their temporal overlap by a two photon interference at the PDBS by changing the optical delay and recording four-fold coincidences. Since the PDBS acts as a 50:50 BS for V-polarized photons, when time delay is properly set, the coincidence detection approaches to zero. The result is shown in Fig. 1. The visibility of the dip was 0.885 ± 0.056. The following experiments were performed at the center of the interference dip of the solid Gaussian fitting curve.

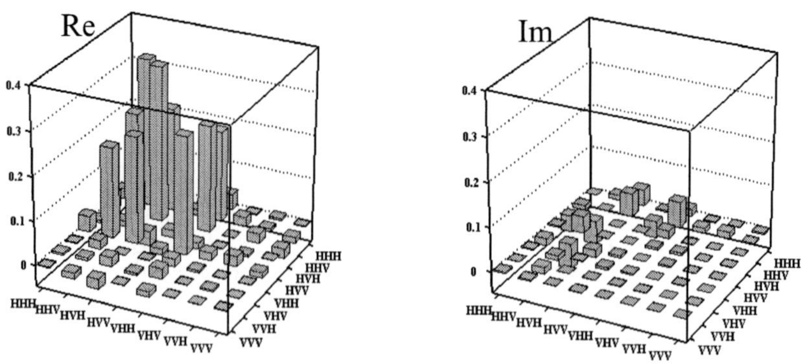

FIGURE 2. 3-dimensional matrix of real part and imaginary part of the reconstructed density matrix of a prepared three-photon W state.

First, by quantum state tomography using 16 tomographic bases, we confirmed that initially prepared photon pairs in modes 1 and 2, and modes 3 and 4 are highly entangled. Their estimated fidelity values were $F_{12} = 0.976 \pm 0.002$ and $F_{34} = 0.967 \pm 0.002$. Next, we reconstructed the three-photon density matrix (ρ_{exp}) from the acquisition data set of 64 tomographic measurements by using the iterative maximum-likelihood reconstruction [5]. The result is shown in Fig. 2. The density matrix for the ideal W_3 state consists of only nine real elements. It has three elements of $1/3$ height corresponding to $|\mathrm{HHV}\rangle$, $|\mathrm{HVH}\rangle$ and $|\mathrm{VHH}\rangle$ and six maximal positive coherences between them. It can be clearly seen from the data that the experimentally obtained density matrix is dominated by the same nine elements although some extra terms and imaginary elements with very small weights are seen. From the result, we estimated the fidelity and the witness of the prepared states. The fidelity of the converted state to the ideal W_3 state is $\mathscr{F}_{\mathscr{W}} = \langle \mathrm{W}_3 | \rho_{\mathrm{exp}} | \mathrm{W}_3 \rangle = 0.778 \pm 0.043$. The witness operators for distinguishing between the W_3 state and biseparable states (including separable states) are given as $\mathscr{W}_W = \frac{2}{3}\mathbf{1} - |W_3\rangle\langle W_3|$ [4]. The obtained expectation values have $\mathrm{Tr}(\mathscr{W}_W \rho_{exp}) = -0.111 \pm 0.043$. This values show that the prepared state is not a biseparable state. Finally, we observe whether the marginal bipartite states are entangled or not, because presence of entanglement in the marginal bipartite states is an important property of a W state. We evaluated the entanglement of the marginal bipartite state using the Pres-Horodecki criterion [6, 7] and concurrence [8] which, respectively, have the values of -0.206 and 0.667 for a marginal entangled state of an ideal three-photon W state. The results of these analyses are given in Table. 1, which clearly shows the entanglement in the marginal bipartite states. These results together with the density matrix in Fig 2 prove that we have achieved the transformation of two EPR photon pairs to a three-photon W state using only LOCC.

TABLE 1. Peres-Horodecki criterion and Concurrence of the prepared marginal bipartite states.

	Peres-Horodecki criterion	Concurrence
modes 1-4	-0.091 ± 0.026	0.322 ± 0.073
modes 1-6	-0.143 ± 0.030	0.421 ± 0.066
modes 4-6	-0.123 ± 0.027	0.415 ± 0.068

ACKNOWLEDGMENTS

This work was partially supported by JSPS Grant-in-Aid for Scientific Research(C) 20540389 and by MEXT Grant-in-Aid for the Global COE Program and Young scientists(B) 20740232.

REFERENCES

1. W. Dür, G. Vidal, and J. I. Cirac, Phys. Rev. A. **62,** 062314 (2000).
2. K. J. Resch, P. Walther, and A. Zeilinger, Phys. Rev. Lett. **94,** 070402 (2005).
3. P. Walther, K. J. Resch, and A. Zeilinger, Phys. Rev. Lett. **94,** 240501 (2005).
4. M. Bourennane, et al., Phys. Rev. Lett. **92,** 087902(2004).
5. D. F. V. James, P. G. Kwiat, W. J. Munro, and A. G. White, Phys. Rev. A. **64,** 052312 (2001).
6. A. Peres, Phys. Rev. Lett. **77,** 1413 (1996).
7. M. Horodecki, P. Horodecki, and R. Horodecki, Phys. Lett. A **223,** 1 (1996).
8. V. Coffman, J. Kundu, and W. K. Wootters, Phys. Rev. A. **61,** 052306 (2000).

Measurement of the cross-Kerr nonlinearity induced by a single-photon-level coherent pulse in a photonic crystal fiber

Nobuyuki Matsuda[a,b], Ryosuke Shimizu[c], Yasuyoshi Mitsumori[a,b], Hideo Kosaka[a,b] and Keiichi Edamatsu[a,b]

[a]*Research Institute of Electrical Communication, Tohoku University, Sendai, Miyagi 980-8577, Japan.*
[b]*CREST, Japan Science and Technology Agency, Kawaguchi, Saitama 330-0012, Japan*
[c]*PRESTO, Japan Science and Technology Agency, Kawaguchi, Saitama 330-0012, Japan*

Abstract. We demonstrate the first observation of the cross-Kerr nonlinearity induced by a single-photon-level coherent optical pulse in a photonic crystal fiber. Taking advantage of large nonlinearity and managed dispersion of the optical fiber, we have successfully measured very small (10^{-7}) conditional phase shifts.

Keywords: quantum optics, cross-phase modulation, optical fibers, optical Kerr effect.
PACS: 42.50.-p

INTRODUCTION

Control of light by light plays an important role of optical information processing and communication technologies. All-optical switching, which is the typical application of such technologies in classical communications, requires large optical nonlinearity. Waveguide media and cavity structures are often utilized as the nonlinear media [1-3] because they have strong confinement of light and long interaction time. Meanwhile, control of one photon by another, which is the quantum analog of such optical switching technology, is required in photonic quantum information processing (QIP) [4]. Although typical optical Kerr nonlinearity induced by a single photon is extremely weak for that purpose, cavity quantum electrodynamics-based devices have performed nonlinear Kerr phase shift up to $\pi/4$ at the single-photon level [5,6]. In addition, a recent proposal opened the way to utilizing moderately weak nonlinearity to realize the quantum-gate operation via strong coherent light [7]. Thus, exploring the availability of the single-photon nonlinearity in various media is attracting great interest. In this paper, we report the first successful measurement of the single-photon-level Kerr nonlinearity in a photonic crystal fiber (PCF) [8].

CP1110, *Quantum Communication, Measurement and Computing (QCMC), The Ninth International Conference,* edited by A. Lvovsky

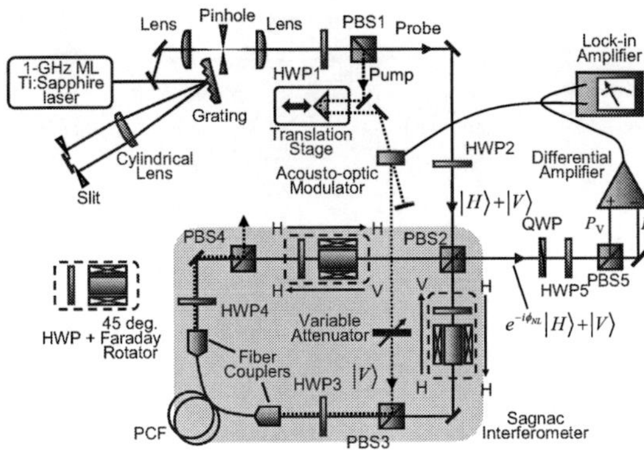

FIGURE 1. Experimental setup. PBS, polarizing beam splitters, HWP, half-wave plates, QWP, quarter-wave plates, PCF, photonic crystal fiber. H and V are horizontal and vertical polarizations, respectively.

EXPERIMETNAL METHOD

The experimental setup based on the pump-probe technique is shown in Fig. 1. It is essential for our measurement to reduce the probe shot noise, which is proportional to the inverse of the square root of the probe photon number used for the measurement. It is also necessary to reduce the probe peak power because the self-Kerr effect of the probe pulses increases the probe phase noise. We used a mode-locked femtosecond Ti:Sapphire laser with a high repetition rate (1 GHz) as a light source. The output beam is spectrally filtered to have a spectral width of approximately 4.5 nm. We used a 5 m long PCF (NL-2.4-800, Blaze Photonics), whose zero-dispersion wavelength is 802 nm, to which the center wavelength of the laser spectrum was tuned.

The probe pulses are introduced to a polarization-division Sagnac interferometer, which enables the high-sensitivity measurement of the phase difference between both counter-propagating beams. The clockwise probe pulses undergo nonlinear phase shifts ϕ_{NL} induced by the pump beam in the PCF via the cross-Kerr effect. Output probe beam from the interferometer has the state of polarization $\exp(-\phi_{NL})|H\rangle+|V\rangle$. Subsequent quarter-wave plate (QWP) converts the state into $\exp(-\phi_{NL})|R\rangle+|L\rangle$; the linear polarization of which the plane of polarization is rotated by $\phi_{NL}/2$. ϕ_{NL} is measurable by projecting the probe state of polarization. See Ref. 9 for more detail.

RESULTS

To analyze the probe polarization state with large ϕ_{NL}, we plotted the interference fringe of the Sagnac interferometer by measuring optical power P_V reflected by polarizing beam splitter (PBS) 5 as a function of the rotation angle of half-wave plate (HWP) 5. Figure 2 shows the measured interference fringe for various pump powers indicated. The shifts of the interference fringes in the presence of the pump indicate

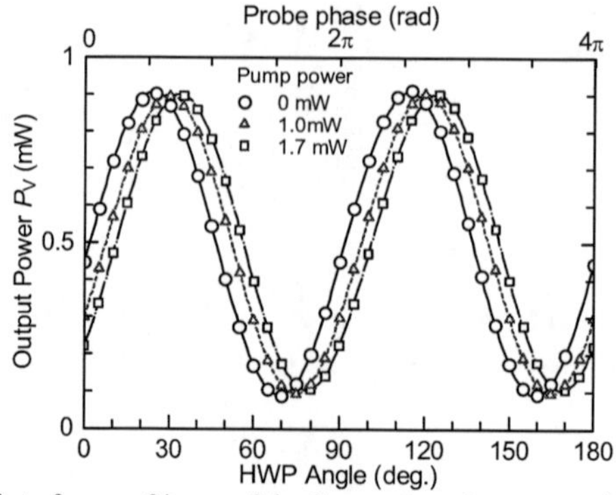

FIGURE 2. Observed interference fringes of the Sagnac interferometer with the PCF. As the pump power increases, the fringe exhibits a shift to the right.

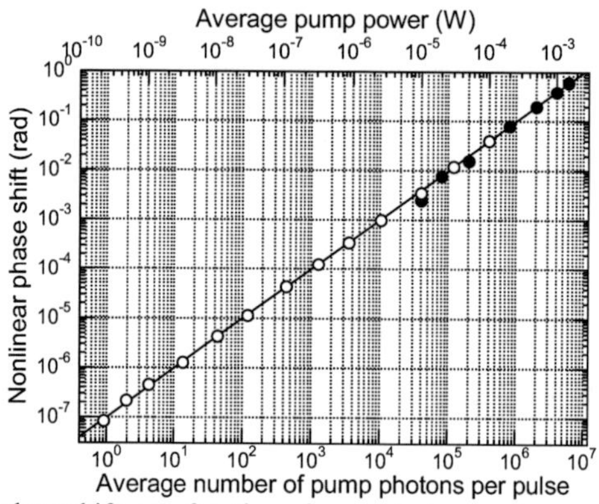

FIGURE 3. Nonlinear phase shifts as a function of mean photon number per pump pulse. The closed and open circles are the phase shift obtained from the interference fringe of the Sagnac interferometer and the optical bridge technique, respectively.

the nonlinear phase shift $\phi_{NL}/2$ [9]. It can be seen that the phase shift increases with increasing pump power.

To measure the small phase shift induced by the pump pulse that possesses a few photons, we utilized a polarization-sensitive optical bridge technique [10]. The differential optical power ΔP at the balanced receiver for the bridge is $\Delta P = P_V - P_H \propto \phi_{NL} P_0$ when $\phi_{NL} \ll 1$, where P_0 is the total power of the interfering component of the probe pulses [10]. We adopted lock-in detection synchronized with the pump pulse modulation by an acousto-optic modulator (AOM) to eliminate lower-frequency phase noise originating from the laser noise and the index fluctuation of the PCF. All the measurement was carried out at room temperature, without cooling the PCF. Figure 3 shows the measured phase shift as a function of the mean photon number in a pump pulse. The closed and open circles are the phase shift obtained from the interference fringe of the Sagnac interferometer and the optical bridge technique, respectively. We have successfully measured approximately 10^{-7} phase shift induced by the pump pulses which contains one photon in a pulse in average. It is noteworthy that the measured value is induced by every single-photon-level coherent pulse, although this result was carried out by repeated measurement with 3 s lock-in time constant. To our knowledge, this is the first observation of the single-photon-level Kerr nonlinearity in an optical fiber.

For the implementation of QIP, however, the measurement in a single shot is required. Assuming the probe photon number of ~ 10^7 per pulse used in our measurement, the measurable phase shift in the single-shot measurement is ϕ_{NL} of ~ 10^{-4}. The value would be obtainable if optical fibers a few kilometers long with the same nonlinearity as ours combined with reduced loss of 1 dB/km and flattened group-velocity dispersion (GVD) were available. Those demands were recently realized independently [11,12]. However, the increase of fiber length (or nonlinearity) at the same time increases the phase noise originating from self-Kerr induced antisqueezing, so it may remain difficult to realize the QOIP in the extrapolation of current fiber technology. However, the technology developed in the present study will be a useful tool for not only proof-of-principle experiments of QOIP but also further advanced studies of quantum nonlinear phenomena not yet revealed.

ACKNOWLEDGMENTS

The authors are grateful to Professors K. Koshino and H. Ishihara for valuable discussions. This research was supported in part by a Grant-in-Aid for Creative Scientific Research (No. 17GS1204) and a Grant-in-Aid for JSPS Fellows (No. 20009351) from the Japan Society for the Promotion of Science.

REFERENCES

1. K. Kitayama, Y. Kimura, K. Okamoto and S. Seikai, *Appl. Phys. Lett.* **46**, 623 (1985).
2. J. E. Sharping, M. Fiorentino, P. Kumar and R. S. Windeler, *IEEE Photonics Technol. Lett.* **14**, 77 (2002).
3. Z.-H. Zhu, W.-M. Ye, J.-R. Ji, X.-D.Yuan and C. Zen, *Opt. Express* **14**, 1783 (2006).
4. M. A. Nielsen and I. L. Chuang, *Quantum Computation and Quantum Information* (Cambridge University Press, Cambridge, 2000), p. 287.
5. I. Fushman, D. Englund, A. Faraon, N. Stoltz, P. Petroff and J. Vučković, *Science* **320**, 769 (2008).
6. Q. A. Turchette, C. J. Hood, W. Lange, H. Mabuchi and H. J. Kimble, *Phys. Rev. Lett.* **75**, 4710 (1995).
7. W. J. Munro, K. Nemoto and T. P. Spiller, *New J. Phys.* **7**, 137 (2005).
8. P. Russell, *Science* **299**, 358 (2003).
9. N. Matsuda, R. Shimizu, Y. Mitsumori, H. Kosaka and K. Edamatsu, *Appl. Phys. Lett.* **91**, 171119 (2007).
10. J. J. Baumberg, S. A. Crooker, D. D. Awschalom, N. Samarth, H. Luo and J. K. Furdyna, *Phys. Rev. B.* **50**, 7689 (1994).
11. W. Reeves, J. Knight, P. Russell and P. Roberts, *Opt. Express* **10**, 609 (2002).
12. K. Tajima, J. Zhou, K. Nakajima and K. Sato, *J. Lightwave Technol.* **22**, 7 (2004).

Observation of time correlation of photon pairs created in photonic crystal fiber

T. Nakanishi[*,†], K. Sakami[*], H. Kobayashi[*], K. Sugiyama[*,†] and M. Kitano[*,†]

[*]*Department of Electronic Science and Engineering, Kyoto University, Kyoto 606-8501, Japan*
[†]*CREST, Japan Science and Technology Agency, Tokyo 102-0075, Japan*

Abstract. We observe two-photon interference of photon pairs from a photonic crystal fiber with the help of an unbalanced Mach-Zehnder interferometer. Two-photon interference fringes with 83% visibility, which exceed the classical limit of 50%, are obtained. The correlation time of the photon pairs is roughly estimated to be 0.2 ps

Keywords: photon pair, four-wave mixing, two-photon interference, photonic crystal fiber
PACS: 42.50.Dv, 42.50.St, 42.65.Lm, 03.65.Ud

INTRODUCTION

Two-photon interference is one of the purely quantum phenomena attributed to time correlations between a pair of photon. The two-photon interference fringe has half the period of a one-photon interference fringe [1]. Two-photon interference can be applied to high-resolution lithographic technology that overcomes the classical diffraction limit [2]. Recently, four-wave mixing (FWM) in $\chi^{(3)}$ nonlinear material has been studied as an effective method to produce entangled photon pairs [3, 4, 5, 6, 7, 8]. In three-wave mixing, a pump photon is split into a pair of photons, while in FWM, two photons in the pumping beam are converted into a pair of time-correlated photons. Hence, in FWM, the resolution of two-photon interference is twice that of one-photon interference produced by the pumping light. High nonlinearity can be achieved by confining the optical field to small area. Therefore, photonic crystal fibers (PCFs) with small core diameters are expected to be useful.

In this paper, we introduce an unbalanced Mach-Zehnder interferometer with a long arm and a short arm to verify the time correlation of photon pairs created in a PCF. In this interferometer, two quantum states corresponding to the case where both photons traverse the same arm contribute to the two-photon interference. With precise coincidence counting for eliminating the possibility that each photon follows different paths, we obtained definite two-photon interference fringes with 83% visibility exceeding the classical limit, which provides clear evidence for the time correlation of the photon pairs. We examined the correlation time of the photon pairs by inserting an acrylic plate into one arm, and roughly estimated it to be 0.2 ps.

CP1110, *Quantum Communication, Measurement and Computing (QCMC), The Ninth International Conference,* edited by A. Lvovsky

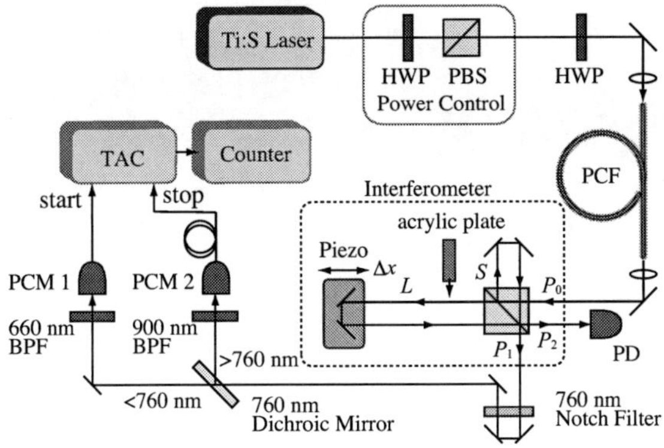

FIGURE 1. Experimental setup for two-photon interference. HWP, half-wave plate; PBS, polarizing beam splitter; BPF, band-pass filter; PCM, photon counting module; TAC, time-to-amplitude converter.

TWO-PHOTON INTERFERENCE IN UNBALANCED MACH-ZEHNDER INTERFEROMETER

For observation of two-photon interference, we use an unbalanced Mach-Zehnder interferometer comprising a long arm "L" with length L_L and a short arm "S" with L_S as shown in Fig. 1. The photon pairs are fed to the input port P_0, and they exit from one of the two output port P_1 and P_2, after taking the long path or the short path. We represent the photon state $|k_a, P_j\rangle$ with the wavenumber k_a $(a = \mathrm{s}, \mathrm{i})$ located at port $P_j (j = 0, 1, 2)$. We calculate the probability amplitude $|\Psi_1\rangle$ with which we observe the signal and idler photons exiting from the same output port P_1: $|\Psi_1\rangle = \iint dk_\mathrm{s} dk_\mathrm{i} \Psi(k_\mathrm{s}, k_\mathrm{i}) |k_\mathrm{s}, P_1\rangle |k_\mathrm{i}, P_1\rangle$. The state $|k_a, P_1\rangle$ is a superposition of states in which the photon from P_0 follows either path S or path L: $|k_a, P_1\rangle = \frac{1}{\sqrt{2}} e^{ik_a L_\mathrm{L}} |k_a, P_0\rangle + \frac{1}{\sqrt{2}} e^{ik_a L_\mathrm{S}} |k_a, P_0\rangle$. Thus, we obtain

$$|\Psi_1\rangle = \frac{1}{2} \iint dk_\mathrm{s} dk_\mathrm{i} \Psi(k_\mathrm{s}, k_\mathrm{i})\, e^{i2k_\mathrm{p} L_S} (e^{i2k_\mathrm{p}\Delta L} + e^{ik_\mathrm{s}\Delta L} + e^{ik_\mathrm{i}\Delta L} + 1) |k_\mathrm{s}, P_0\rangle |k_\mathrm{i}, P_0\rangle, \quad (1)$$

where we use the relations $k_\mathrm{s} + k_\mathrm{i} = 2k_\mathrm{p}$ $(k_\mathrm{p} = \omega_\mathrm{p}/c)$ and $\Delta L = L_\mathrm{L} - L_\mathrm{S}$.

If ΔL is larger than the coherent length of each photon, the probability of finding both the photons at port P_1 is given as $\langle \Psi_1 | \Psi_1 \rangle \propto 1 + \frac{1}{2}\cos(2k_\mathrm{p}\Delta L)$. The visibility of two-photon interference is only 50%, which is equal to the classical limit [9], because the second and third terms in Eq. (1) don't contribute. These terms correspond to the states where each photon follows a different path, so it is possible to eliminate them by counting only the photon pairs within a time interval shorter than the propagation time difference. In other words, we can postselect the first and fourth terms as follows:

$$|\Psi'_1\rangle = \frac{1}{2} \iint dk_\mathrm{s} dk_\mathrm{i} \Psi(k_\mathrm{s}, k_\mathrm{i})\, e^{i2k_\mathrm{p} L_S} (e^{i2k_\mathrm{p}\Delta L} + 1) |k_\mathrm{s}, P_0\rangle |k_\mathrm{i}, P_0\rangle. \quad (2)$$

The probability is $\langle \Psi'_1 | \Psi'_1 \rangle \propto 1 + \cos(2k_\mathrm{p}\Delta L)$. All the terms in Eq. (2) contribute to the interference and produce the interference fringes with 100% visibility, which cannot be explained by the classical theory.

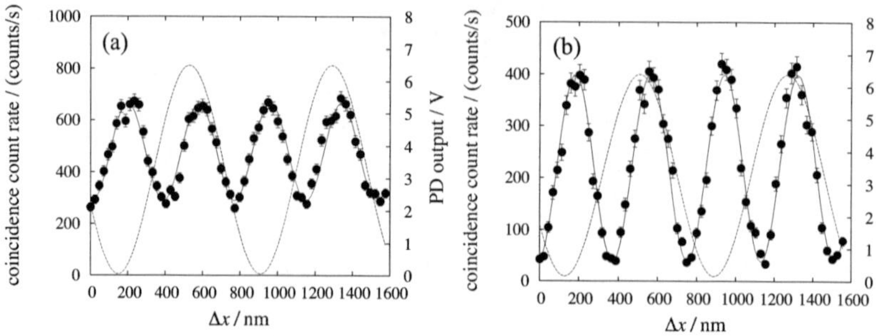

FIGURE 2. The circles represent the coincidence count rates for (a) $T = 6\,\text{ns}$ and (b) $T = 1.5\,\text{ns}$. The one-photon interference pattern of the pump light, which is detected by the PD, is superimposed for reference (dotted line).

EXPERIMENTS

A schematic of the experimental setup is shown in Fig. 1. We use a polarization-maintaining PCF with the zero-dispersion wavelength of $\lambda_0 = 760\,\text{nm}$ (Crystal Fiber, NL-PM-760) and the length of 1.93 m. In order to obtain broad photon-pair spectrum, we set the pump wavelength slightly higher than λ_0. The photon pairs from the PCF are sent to the unbalanced Mach-Zehnder interferometer. For elimination of the pumping light, the output light from P_1 is separated into two paths by a dichroic mirror after passing through a notch filter twice. The signal photon at 660 nm and the idler photon at 900 nm are detected by the respective photon counting modules. The band-pass filters with an FWHM of 10 nm passband are placed in front of the detectors. We use a time-to-amplitude converter (TAC) and an electronic counter for coincidence counting. The counter counts the number of photon pairs within a specific time interval T, which can be controlled by adjusting the discrimination levels of the counter. We set the path difference $\Delta L \sim 60\,\text{cm}$, which corresponds to the optical delay of $\tau = 2\,\text{ns}$. The piezo actuator attached to a mirror in the interferometer provides a variation of Δx in ΔL. The pump laser is operated at a power of 4 mW and the wavelength of 760.4 nm, which produces about 2,000 pairs per second at the detectors without the interferometer. The rate at which photon pairs are produced is sufficient for the two-photon interference experiment, and the accidental coincidence rate is negligibly small.

We first measured one-photon interference for the signal and idler photons by individually counting the photons. As expected, there is no clear interference fringe, because ΔL is much larger than the coherent length of each photon. We performed two-photon interference experiments by changing the gate time T for coincidence counting. First, we set $T = 6\,\text{ns}\,(> \tau)$ to measure the coincidence count rate that involves all the photon pairs passing through the interferometer. Figure 2(a) shows the one-photon interference pattern of the pump light (dotted line) as a reference and the two-photon interference pattern of the photon pairs (circles). The period of the two-photon interference fringe is half that of the one-photon interference fringe for the pump light. The visibility is only 40%. Next, we set $T = 1.5\,\text{ns}\,(< \tau)$ to eliminate the possibility of the signal and idler

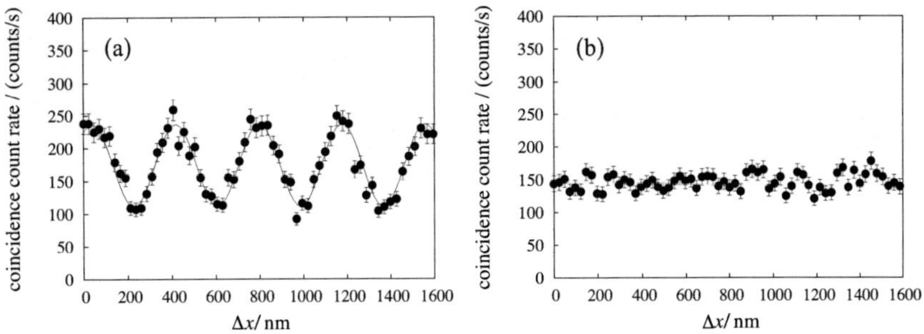

FIGURE 3. Two-photon interference when optical delay of (a) 0.12ps or (b) 0.24ps is introduced to path *L*.

photons following different paths. The results are shown in Fig. 2(b). We observed a two-photon interference fringe with 83% visibility, which exceeds the classical limit of 50%. The result shows that the photon pairs generated in the PCF have time correlations that can be explained only by the quantum theory.

For the gate time $T = 1.5\,\text{ns}$, we examined correlation time of the photon pairs by inserting an acrylic plate into path *L*. The plate thickness is 3 mm, which induces group delay of 0.12 ps between the signal and idler photon. When the photons in path *L* passed through the acrylic plate once, the visibility of two-photon interference decreased to 40% as shown in Fig. 3(a). When the plate was placed in a way so that the photons pass through it twice, the two-photon interference almost vanished as shown in Fig. 3(b). The interference disappears when the delay between the photon pair exceeds the correlation time of the pair, so we roughly estimated the correlation time to be about 0.2 ps.

ACKNOWLEDGMENTS

This research is supported by the global COE program "Photonics and Electronics Science and Engineering," at Kyoto University and a Grant-in-Aid for Young Scientists (B), No. 18740247, from the Japan Society for the Promotion of Science.

REFERENCES

1. K. Edamatsu, R. Shimizu, and T. Itoh, *Phys. Rev. Lett.* **89**, 213601 (2002).
2. A. N. Boto, P. Kok, D. S. Abrams, S. L. Braunstein, C. P. Williams, and J. P. Dowling, *Phys. Rev. Lett.* **85**, 2733 (2000).
3. K. Edamatsu, G. Oohata, R. Shimizu, and T. Itoh, *Nature* **431**, 167 (2004).
4. M. Fiorentino, P. Voss, J. Sharping, and P. Kumar, *IEEE Photonics Technol. Lett.* **14**, 983 (2002).
5. X. Li, P. L. Voss, J. E. Sharping, and P. Kumar, *Phys. Rev. Lett.* **94**, 53601 (2005).
6. J. Chen, X. Li, and P. Kumar, *Phys. Rev. A* **72**, 33801 (2005).
7. J. G. Rarity, J. Fulconis, J. Duligall, W. J. Wadsworth, and P. S. J. Russell, *Optics Exp.* **13**, 534 (2005).
8. J. Fulconis, O. Alibart, J. L. O'Brien, W. J. Wadsworth, and J. G. Rarity, *Phys. Rev. Lett.* **99**, 120501 (2007).
9. K. Sanaka, K. Kawahara, and T. Kuga, *Phys. Rev. Lett.* **86**, 5620 (2001).

CHAPTER 8

INTERFACING QUANTUM INFORMATION BETWEEN LIGHT AND MATTER

Matter-matter entanglement for quantum networks

Julien Laurat *, Kyung Soo Choi†, Hui Deng† and H. J. Kimble†

*Laboratoire Kastler Brossel, Universite P. et M. Curie, Ecole Normale Superieure and CNRS, UPMC, Case 74, 4 place Jussieu, 75252 Paris Cedex 05, France.
†Norman Bridge Laboratory of Physics 12-33, California Institute of Technology, Pasadena, California 91125, USA.

Abstract. Developments in quantum information science rely critically on entanglement, as its distribution between different parties enables quantum communication protocols, such as quantum key distribution or teleportation. This talk focused on two different ways to generate heralded entanglement between matter systems, a critical requirement for scalable quantum networking.

Keywords: Entanglement, Quantum memory, Quantum repeater
PACS: 03.67.Mn, 03.65.Yz, 03.67.Hk, 42.50.Dv

INTRODUCTION

Quantum control of entanglement between material systems is an essential capability for quantum networks and scalable quantum communication architectures [1, 2]. In recent years, significant advances have been achieved in the control of the quantum states of atomic systems, including entanglement of trapped ions [3, 4] and between macroscopic quantum spins [5]. By following the seminal paper of Duan, Lukin, Cirac and Zoller (*DLCZ*) [6], entanglement between single collective excitations stored in two remote atomic ensembles has also been demonstrated [7]. In the *DLCZ* protocol [6], entanglement is created in a probabilistic but heralded way from quantum interference in the measurement process [8, 9, 10]. The detection of a photon from one or the other atomic ensemble in an indistinguishable fashion results in an entangled state with one collective spin excitation shared coherently between the ensembles. The first section will describe such generation [11].

However, as it will be shown later, an inherent drawback of this approach is the compromise between the amount of entanglement and its preparation probability, leading intrinsically to low count rate for high entanglement. In the second section, we will present a protocol where entanglement between two atomic ensembles is created by coherent mapping of an entangled state of light [12]. By splitting a single-photon and subsequent state transfer, we separate the generation of entanglement and its storage. After a programmable delay, the stored entanglement is mapped back into photonic modes with overall efficiency of 17%. Improvements of single-photon sources [13] together with our protocol will enable "on-demand" entanglement of atomic ensembles.

CP1110, *Quantum Communication, Measurement and Computing (QCMC), The Ninth International Conference,* edited by A. Lvovsky

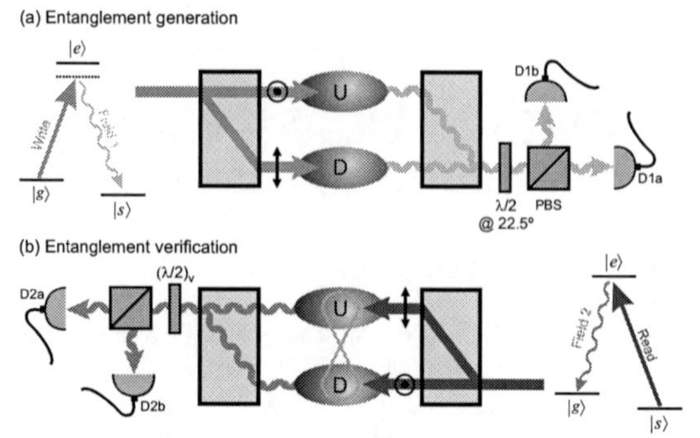

FIGURE 1. (a) A weak write pulse is split into two paths separated by 1 mm by a birefringent crystal and excites simultaneously two atomic samples, U,D. The resulting fields $1_{U,D}$ are combined at the polarizing beamsplitter (PBS) and sent to the single-photon detectors $D_{1a,1b}$. A detection event at D_{1a} or D_{1b} heralds the creation of entanglement [7]. (b) After a user-controlled storage time τ, entanglement is verified by mapping the atomic state to propagating light fields $2_{U,D}$ by way of strong read pulses. Tomography of the fields $2_{U,D}$ is then achieved. The atomic cloud is initially prepared in the ground state $|g\rangle$. $\{|g\rangle, |s\rangle, |e\rangle\}$ denote the Cesium hyperfine levels $\{|6S_{1/2}, F=4\rangle, |6S_{1/2}, F=3\rangle, |6P_{3/2}, F=4\rangle\}$, respectively. Note that the fields $1_{U,D}$ and $2_{U,D}$ are detected with a small angle relative to the classical beams, which is not represented here for the sake of simplicity.

MEASUREMENT-INDUCED ENTANGLEMENT

The generation of entanglement following the DLCZ approach is illustrated in Fig. 1. A single cloud of cesium atoms in a magneto-optical trap is used; two ensembles are defined by different optical paths 1 mm apart . This separation is obtained by the use of birefringent crystals close to the cloud, which separate orthogonal polarizations. At 40 Hz, the trap magnetic field is switched off for 7 ms. After waiting 3 ms for the magnetic field to decay, the two samples are simultaneously illuminated with 30 ns-long and 10 MHz red-detuned write pulses, at a rate of 1.7 MHz. Given the duty cycle of the experiment, the effective rate is 180 kHz. Spontaneous Raman scattered fields induced by the write beams are collected into single-mode fibers, defining for each ensemble optical modes that we designate as fields $1_{U,D}$ with 50 μm waist and a 3° angle relative to the direction of the write beams [14]. Fields $1_{U,D}$ are frequency filtered to block spontaneous emission from atomic transitions $|e\rangle \rightarrow |g\rangle$, which do not herald the creation of a collective excitation. After this stage, and before detection, fields $1_{U,D}$ are brought to interfere on a polarizing beam-splitter. A detection event at $D_{1a,1b}$ that arises indistinguishably from either of the fields $1_{U,D}$ projects the atomic ensembles into an entangled state where, in the ideal case, one collective excitation is coherently shared between the U,D ensembles [6, 7].

A model-independent determination of entanglement based upon quantum tomography of the fields $2_{U,D}$ obtained by sending a read pulse has been developed in the Supplemental Information accompanying Ref. [7]. The model consists of reconstructing a reduced density matrix, $\tilde{\rho}_{2_U,2_D}$, obtained from the full density matrix by restriction to the subspace with no more than one photon per mode. We also assume that all off-diagonal

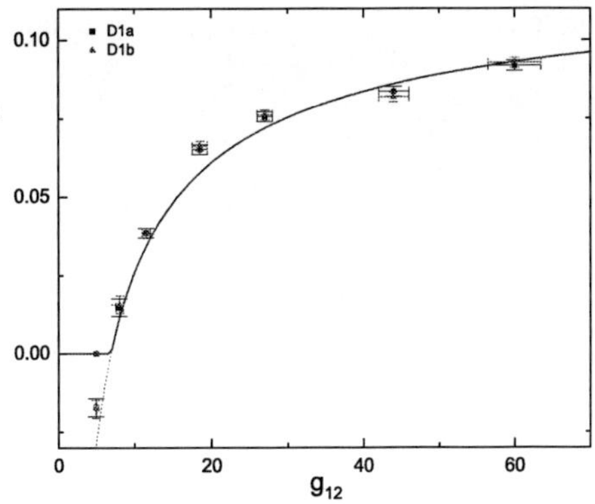

FIGURE 2. Concurrence C as a function of the normalized cross correlation function g_{12}, for the two possible heralding events (detection at D_{1a} or D_{1b}).

elements between states with different numbers of photons vanish. The model thus leads to a lower bound for entanglement. As detailed in Ref. [7], $\tilde{\rho}_{2_U,2_D}$ can be written in the photon-number basis $|n\rangle|m\rangle$ with $\{n,m\} = \{0,1\}$ as follows:

$$\tilde{\rho}_{2_U,2_D} = \frac{1}{P}\begin{pmatrix} p_{00} & 0 & 0 & 0 \\ 0 & p_{01} & d & 0 \\ 0 & d^* & p_{10} & 0 \\ 0 & 0 & 0 & p_{11} \end{pmatrix}, \quad (1)$$

Here, p_{ij} is the probability to find i photons in mode 2_U and j in mode 2_D; d is the coherence term between the $|1\rangle|0\rangle$ and $|0\rangle|1\rangle$ states; and $P = p_{00} + p_{01} + p_{10} + p_{11}$. From $\tilde{\rho}_{2_U,2_D}$, one can calculate the concurrence C, which is a convenient monotone measurement of entanglement ranging from 0 for a separable state to 1 for a maximally entangled state:

$$C = max(0, C_0)\ with\ C_0 = \frac{1}{P}(2|d| - 2\sqrt{p_{00}p_{11}}). \quad (2)$$

In the regime of low excitation and detection probabilities in which the current experiment is performed, the vacuum component p_{00} can be approximated by $p_{00} \sim 1 - p_c$, while the terms p_{01} and p_{10} are given by $p_{10} = p_{01} = p_c/2$. p_c is the conditional probability of detecting a photon in field 2 from one ensemble following a detection event for field 1.

To investigate the scaling of entanglement with excitation probability χ, we determined C for various values of χ for fixed memory time $\tau = 200$ ns (Figure 2). As χ increases, higher order terms cannot be neglected, precisely as in parametric down conversion. A convenient parameter to assess the excitation regime of each ensemble is the normalized intensity cross correlation function g_{12} between field 1 and field 2 [14], defined as $g_{12} = p_{12}/(p_1 p_2)$ with p_{12} the joint probability for detection events from field 1 and 2 in a given trial and p_i the probability for unconditional detections in field i. In the ideal case, this function is related to the excitation probability χ by $g_{12} = 1 + 1/\chi$, where $g_{12} > 2$ defines the nonclassical border in the ideal case and $g_{12} \gg 2$ being the single-excitation regime for the ensembles.

MAPPING PHOTONIC ENTANGLEMENT INTO AND OUT

The drawback of the previous measurement induced protocol is that entanglement and count rate are inherently correlated. Here, we demonstrate the reversible mapping of an entangled state into a quantum memory. The mapping is obtained by using adiabatic passage based upon dynamic Electromagnetically Induced Transparency (EIT) [15, 16]. Storage and retrieval of optical pulses have been demonstrated previously, for both classical pulses [17], and for single photon pulses [18, 19]. Adiabatic transfer of a collective excitation has been demonstrated between two ensembles coupled by a cavity mode [20], which can provide a suitable approach for generating on-demand entanglement over short distances.

In our experiment, entanglement between two atomic ensembles L_a, R_a is created by first splitting a single photon into two modes L_{in}, R_{in} to generate an entangled state of light. This entangled field state is then coherently mapped to an entangled matter state for L_a, R_a. On demand, the stored atomic entanglement for L_a, R_a is reversibly converted back into entangled photonic modes L_{out}, R_{out}. As opposed to the original *DLCZ* scheme, our approach is inherently deterministic, suffering principally from the finite efficiency with which single excitations can be mapped to and from an atomic memory, with efficiency $\simeq 50\%$ having been achieved. Moreover, the contamination of entanglement for the L_a, R_a ensembles from processes involving 2 excitations can be arbitrarily suppressed (independent of the mapping probabilities) with continuing advances in on-demand single photon sources [13]. Our experiment thereby provides a promising avenue to distribute and store entanglement deterministically over remote atomic ensembles for quantum networks.

The experimental setup is depicted in Fig. 3. Our single photon source is based on Raman transitions in an optically thick cesium ensemble [6, 14]. This system generates 28 ns-long single photons (resonant with the $6S_{1/2}, F = 4 \leftrightarrow 6P_{3/2}, F' = 4$ transition) in a heralded fashion [14]. The single photons are polarized at 45° from the eigen-polarizations of the beam displacer BD_1 (Fig. 1b), which splits them into entangled optical modes L_{in}, R_{in} (called the signal modes) to produce, in the ideal case, the state in the following form :

$$\frac{1}{\sqrt{2}}(|0_{L_{in}}\rangle|1_{R_{in}}\rangle + e^{i\phi_{rel}}|1_{L_{in}}\rangle|0_{R_{in}}\rangle). \tag{3}$$

The next stage consists in coherently mapping photonic entanglement for L_{in}, R_{in} into atomic ensembles L_a, R_a (called the memory ensembles) within a single cloud of cold cesium atoms in a magneto-optical trap (MOT) (Fig. 1c). Ensembles L_a, R_a are defined by the well-separated optical paths of the entangled photonic modes L_{in}, R_{in}. To avoid dissipative absorption for the fields in modes L_{in}, R_{in} for our choice of polarization [18], we initially spin-polarize the atomic ensemble into $|F = 4, m_F = 0\rangle$. A synchronous clock governs the trials of both the single photon source and memory

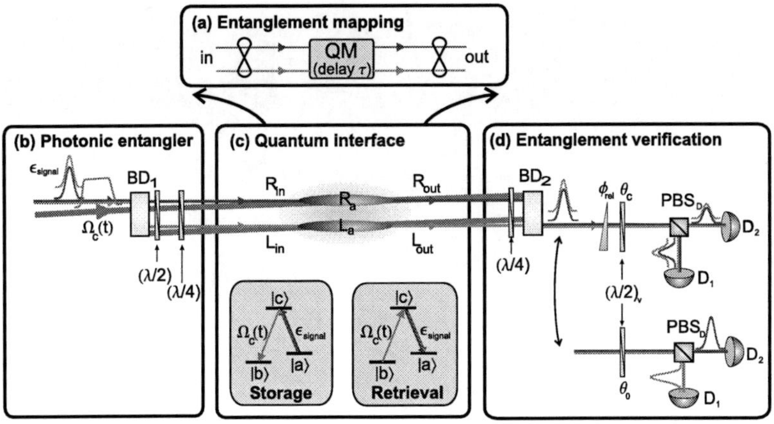

FIGURE 3. (a) Mapping of an entangled state of light into and out of a quantum memory (QM) with storage time τ. (b) Photonic "entangler": A beam displacer BD_1 splits an input single photon into two orthogonally polarized, entangled modes L_{in}, R_{in}, which are spatially separated by 1 mm. With waveplates $\lambda/2$ and $\lambda/4$, the signal fields ε_{signal} for L_{in}, R_{in} and control fields $\Omega_c^{(L,R)}(t)$ are transformed to circular polarizations with the same helicity along each path L, R, and copropagate with angle of 3°. (c) Photonic entanglement between L_{in}, R_{in} is coherently mapped into the memory ensembles L_a, R_a by switching $\Omega_c^{(L,R)}(t)$ off adiabatically. After a programmable storage time, the atomic entanglement is reversibly mapped back into optical modes L_{out}, R_{out} by switching $\Omega_c^{(L,R)}(t)$ on. Relevant energy diagrams for the storage and retrieval processes are shown in the insets. States $|a\rangle$, $|b\rangle$ are the hyperfine ground states $F = 4$, $F = 3$ of $6S_{1/2}$ in atomic cesium; state $|c\rangle$ is the hyperfine level $F' = 4$ of the state $6P_{3/2}$. (d) Verification : After a $\lambda/4$ plate, the beam displacer combines modes L_{out}, R_{out} into one beam with orthogonal polarizations. With $(\lambda/2)_v$ at $\theta_c = 22.5°$ before the polarization beamsplitter (PBS_D), single photon interference is recorded at detectors D_1, D_2 by varying the relative phase ϕ_{rel} by a Berek compensator. With $(\lambda/2)_v$ at $\theta_0 = 0°$, photon statistics for each mode L_{out}, R_{out} are measured independently.

ensembles with a period of 575 ns. Initially, the strong control fields $\Omega_c^{(L,R)}$ (resonant with the $6S_{1/2}, F = 3 \leftrightarrow 6P_{3/2}, F' = 4$ transition) open transparency windows $\Omega_c^{(L,R)}(0)$ in L_a, R_a for the signal modes. As the wavepacket of the signal field propagates through each ensemble, the control fields $\Omega_c^{(L,R)}(t)$ are turned off in 20 ns by an electro-optical intensity modulator, thus coherently transforming the fields of the respective signal modes to collective atomic excitations within L_a, R_a. This mapping leads to heralded entanglement between ensembles L_a, R_a. After a user-defined delay, chosen here to be 1.1 μs, the atomic entanglement is converted back into entangled photonic modes by switching on the control fields $\Omega_c^{(L,R)}(t)$. Figure 4 gives the entaglement measurement before and after storage. A transfer efficiency, defined as the ratio of input and output concurrence, of 20% is demonstrated.

In conclusion, we have reported two different ways of generating heralded entanglement between atomic ensembles. Our second protocol alleviates the significant drawback of probabilistic protocols [6], where low preparation probabilities prevent its potential scalability. Our strategy can be incorporated into recent extensions of [6] for efficient scaling for high fidelity quantum communication.

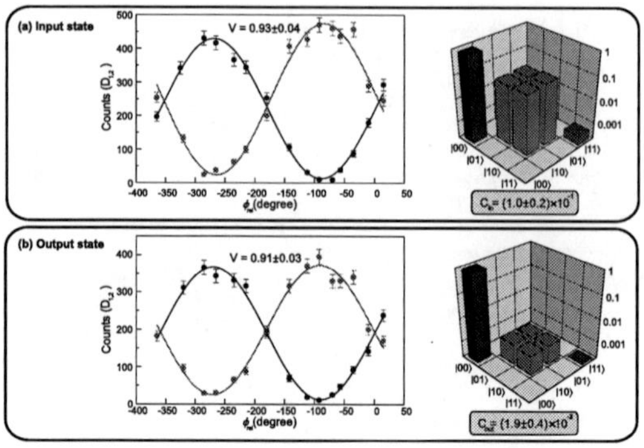

FIGURE 4. Entanglement for the input (a) and output (b) optical modes. The figure shows the interference fringes and the reconstructed density matrices (in log scale) for the photonic modes **a,** at the input of the memory and **b,** at the output after storage and retrieval. The estimated concurrence is given in each case. Each point of the fringe is taken for 20,000 (100,000) heralding events for the input (output) state.

ACKNOWLEDGMENTS

We acknowledge our ongoing collaboration with S. J. van Enk and discussions with D. Felinto, H. de Riedmatten and C.W. Chou. This research is supported by the DTO and by the NSF. J. L. acknowledges financial support from the EU (Marie Curie fellowship) and H. D. from the CPI at Caltech.

REFERENCES

1. P. Zoller *et al.*, Eur. Phys. J. D **36**, 203 (2005).
2. H.J. Briegel *et al.*, Phys. Rev. Lett. **81**, 5932 (1998).
3. H. Häffner *et al.*, Appl. Phys. B **81**, 151 (2005).
4. C. Langer *et al.*, Phys. Rev. Lett. **95**, 060502 (2005).
5. B. Julsgaard, A. Kozhekin, E.S. Polzik, Nature **413**, 400 (2001).
6. L.M. Duan *et al.*, Nature **414**, 413 (2001).
7. C.W. Chou *et al.*, Nature **438**, 828 (2005).
8. R.H. Dicke, Am. J. Phys. **49**, 925 (1981).
9. C. Cabrillo *et al.*, Phys. Rev. A **59**, 1025 (1999).
10. S. Bose *et al.*, Phys. Rev. A **83**, 5158 (1999).
11. J. Laurat, K.S. Choi, H. Deng, C.W. Chou, H.J. Kimble, Phys. Rev. Lett. **99**, 180504 (2007).
12. K.S. Choi, J. Laurat, H. Deng, H.J. Kimble, Nature **452**, 67 (2008).
13. B. Lounis, M. Orrit, Rep. Prog. Phys. **68**, 1129-1179 (2005).
14. J. Laurat, J. *et al.*, Opt. Express **14**, 6912-6918 (2006).
15. S.E. Harris, Phys. Today **50**, 36-40 (1997).
16. M. Fleischhauer, M.D. Lukin, Phys. Rev. Lett. **84**, 5094-5097 (2000).
17. L.V. Hau, S.E. Harris, Z. Dutton, C.H. Behroozi, Nature **397**, 594-598 (1999).
18. T.Chaneliére, *et al.*, Nature **438**, 833-836 (2005).
19. M.D. Eisaman,*et al.*, Nature **438**, 837-841 (2005).
20. J. Simon, H. Tanji, S. Ghosh, V. Vuletić, Nature Physics **3**, 765-769 (2007); advance online publication, 23 September 2007 (DOI 10.1038/nphys726).

Coherent spin state transfer from light to electrons in a semiconductor

H. Kosaka[a,b], H. Shigyou[a], Y. Mitsumori[a,b], Y. Rikitake[b,c]
H. Imamura[b,c], T. Kutsuwa[b] and K. Edamatsu[a]

[a]*Research Institute of Electrical Communication, Tohoku University, Sendai 980-8577, Japan*
[b]*CREST, Japan Science and Technology Agency (JST), Saitama 332-0012, Japan*
[c]*Department of Information Engineering, Sendai National College of Technology, Sendai 989-3128, Japan*
[d]*National Institute of Advanced Industrial Science and Technology (AIST), Tsukuba 305-8568, Japan*

Abstract. We demonstrate that the superposition of light polarization states is coherently transferred to electron spins in a semiconductor quantum well. By using time-resolved Kerr rotation we observe the initial phase of Larmor precession of electron spins whose coherence is transferred from light. To break the electron-hole spin entanglement, we utilized the big discrepancy between the transverse g-factors of electrons and light holes. The result encourages us to make a quantum media converter between flying photon qubits and stationary electron spin qubits in semiconductors.

Keywords: photon, polarization, electron, spin, coherence, qubit

PACS: 72.25.Fe, 78.67.-n, 85.35.Ds

INTRODUCTION

Electron spin coherence originates in the quantum coherence between up ($\uparrow_e$) /down ($\downarrow_e$) spin states. Likewise, light polarization coherence is based on photon pseudo-spin coherence between right (σ^{+}_{ph})/left (σ^{-}_{ph}) circular polarization states. It is known that the electron spin states can be programmed or prepared by injecting the circularly polarized photons [1,2]. However, we cannot obtain the coherent superposition of up and down spin states in the conventional preparation methods due to the entanglement between the electron spin state and the simultaneously created hole spin state. In this letter we demonstrate spin coherence transfer from light to electrons. We not only show the projective up/down spin states or spin population as reported before [3-6] but also show that an arbitrary coherent superposition state can be transferred from light polarization to electron spins [7,8]. Although we used classical light and a spin ensemble, the spin coherence transfer is the first demonstration of the conditions needed for the quantum state transfer from a photon to an electron.

CP1110, *Quantum Communication, Measurement and Computing (QCMC), The Ninth International Conference,* edited by A. Lvovsky

THOERY

We begin by summarizing the operating principle of our Quantum media conversion (QMC) scheme (Fig. 1). Any light polarization state is represented by a state vector, called a Stokes vector, in the Poincaré sphere (PS) spanned by the basis vectors $|\sigma^+\rangle$ and $|\sigma^-\rangle$, corresponding to right and left circular polarizations. In the same manner, any electron spin state is represented by a Bloch vector in the Bloch sphere (BS) spanned by the basis vectors $|\uparrow\rangle_e$ and $|\downarrow\rangle_e$, corresponding to up and down spin polarizations. Both the PS and BS thus are the equivalent special unitary group, SU(2) Hilbert spaces connected via selection rules. The selection rules for a light hole (LH) exciton are given by $|\sigma^+\rangle \rightarrow |\uparrow\rangle_e \otimes |\downarrow\rangle_{lh}$ and $|\sigma^-\rangle \rightarrow -|\downarrow\rangle_e \otimes |\uparrow\rangle_{lh}$ [9]. An arbitrary light polarization state $\alpha|\sigma^+\rangle + \beta|\sigma^-\rangle$ is thus transferred to the exciton spin state $\alpha|\uparrow\rangle_e \otimes |\downarrow\rangle_{lh} - \beta|\downarrow\rangle_e \otimes |\uparrow\rangle_{lh}$, where the electron spin state is entangled with the hole spin state. The coherence time of a hole spin, however, is so short that the entangled electron spin might be instantly converted into incoherent mixture of spin states projected along the z axis, and an incoherent mixture cannot be used as a qubit. The problem can be solved by applying an in-plane (transverse) magnetic field B_x to configure a V-shaped three-level system shown in Fig. 1(b). The magnetic field lifts the Kramers degeneracy of the LH and reconfigures the eigenstates as $|\pm x\rangle_{lh} = |\downarrow\rangle_{lh} \pm |\uparrow\rangle_{lh}$ while keeping the degeneracy of the electron with the smaller g-factor. Via the resonant transition between the $|-x\rangle_{lh}$ state and the degenerated electron states, the same selection rules lead to the transition from $\alpha|\sigma^+\rangle + \beta|\sigma^-\rangle$ to $(\alpha|\uparrow\rangle_e + \beta|\downarrow\rangle_e) \otimes |-x\rangle_{lh}$. Since the transferred electron spin state is not entangled with the LH state, its coherence time is not limited by the short LH spin coherence time or the fast relaxation to the ground heavy hole (HH) state. This is the key point of our QMC scheme. It is essential to use the LH states since we cannot sufficiently lift the Kramer's degeneracy of the HH state by applying an in-plane magnetic field because of the very small transverse g factor [10]. The electron coherence time is limited by the electron-hole exchange interaction, which can be greatly extended by extracting a hole while leaving an electron in a quantum well [11] or a quantum dot [2,12].

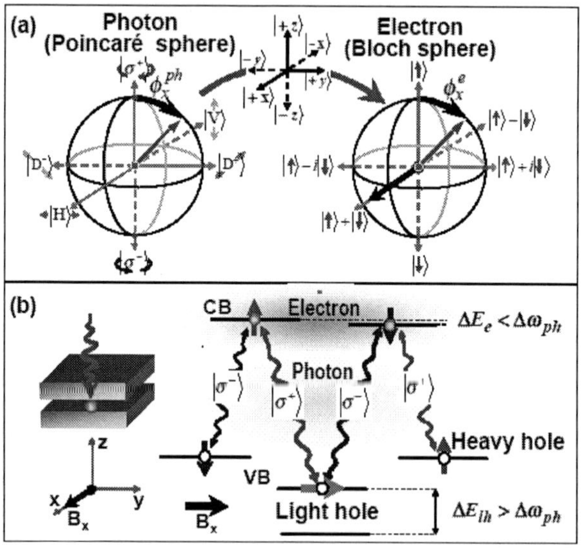

FIGURE 1. (a) The scheme of quantum media conversion from photon to electron spin. (b) The relevant energy levels under an in-plane magnetic field B_x forming V-shaped system.

SAMPLE AND SETUP

The sample we used contained a high-quality 11-nm-thick undoped GaAs quantum well embedded in undoped $Al_{0.3}Ga_{0.7}As$ grown by molecular beam epitaxy on a semi-insulating GaAs substrate. The LH exciton is energetically well separated from the HH exciton. The quantum well was designed to have a relatively small electron transverse g-factor ($g_{e,\perp} = -0.21$) and a relatively large LH transverse g-factor ($g_{lh,\perp} = -3.5$) under the in-plane magnetic field B_x used to configure the V-shaped three-level system. We chose the non-zero g-factor for an electron intentionally in order to be able to observe the initial phase of the spin precession under a moderate B_x. A mode-locked Ti:Sapphire laser was used to pump and probe the sample through a wavelength-tunable filter consisting of a diffraction grating (FWHM = 0.38 nm) in each path and a variable delay line in only the probe path. All the measurements were made at 5 K and at normal incidence in the Voigt geometry (Fig. 1(b)). The pump power dependence of the typical Kerr rotation (KR) signals is linear up to 1 mW, which corresponds to the exciton density of 5×10^9 cm^{-2} or an average exciton spacing of 160 nm, which is much greater than the exciton Bohr radius (< 12 nm). Below this power level, all the temporal responses were unchanged. All the measurements in the work reported here were made in this linear range where created excitons were well separated from each other.

EXPERIMENATAL RESULTS

Spin coherence of an electron is often indicated by the quantum beats, which is usually visualized by time-resolved KR measurements [13]. Although the KR provides information only about the spin projection or population along the growth axis, we can infer the spin orientation from the initial phase of the spin precession. Here we show how the inferred initial phase of the generated electron spin is related to the pump light polarization. We selected {$|\sigma^{\pm}\rangle$ circular, $|D^{\pm}\rangle$ diagonal and $|H/V\rangle$ horizontal/vertical} light polarization and measured the temporal evolution of the KR angle under B_x = 7 T (Fig. 2a). The lower-energy side of the LH

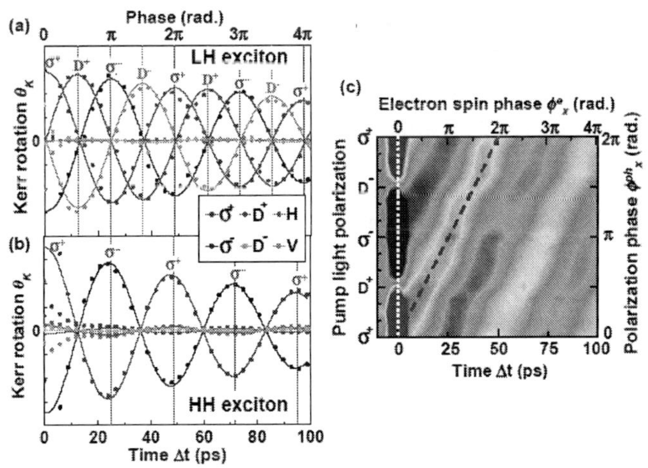

Fig. 2. Time-dependent KR signal in the case of (a) light-hole excitation and (b) heavy-hole excitation. (c) Coincidence of the phase angle between the pump light polarization and the created electron spin.

exciton peak was excited with one of the above-mentioned polarizations and probed with the H polarization. Four of the states perpendicular to the x axis ($|\sigma^{\pm}\rangle$, $|D^{\pm}\rangle$) provided approximately the same amplitude, while the phases were sequentially shifted by the angle of $\pi/2$. These results suggest that the $|D^{\pm}\rangle$ states, which are coherent superpositions of $|\sigma^{+}\rangle$ and $|\sigma^{-}\rangle$ states $|\sigma^{+}\rangle \pm i\ |\sigma^{-}\rangle$, create equivalent superpositions of the $|\uparrow\rangle_e$ and $|\downarrow\rangle_e$ states $|\uparrow\rangle_e \pm i\ |\downarrow\rangle_e$. It is natural that the states along the x axis ($|H/V\rangle$) provided negligible amplitude because those state vectors are parallel to the field. In contrast to the LH excitation, the HH excitation gave the conventional result (Fig. 2b), where the $\sigma^{\pm}$ pump corresponding to the projected states $|\uparrow/\downarrow\rangle_e$ provided significant amplitude but the others did not. The lack of the amplitude seen with $D^{\pm}$ pumping indicates that superposition of the $|\uparrow\rangle_e$ and $|\downarrow\rangle_e$ states could not be created by the HH excitation. We verified that the electron phases ϕ_x^{e} inferred from the peaks well coincide with the pump-light phases ϕ_x^{ph} (Fig. 2c).

CONCLUSIONS

In conclusion, we have demonstrated coherent polarization transfer from light to electron spins in a semiconductor. From the phase of spin precession we inferred the initial phase that should correspond to the phase of light polarization. The electron-hole spin entanglement was broken by the big discrepancy between the g-factor of the electron and the g-factor of the light hole. Although we have demonstrated only a necessary condition of QMC, this is a step toward the realization of all-semiconductor quantum interfaces between a photon and an electron spin.

ACKNOWLEDGMENTS

This work was supported in part by the SCOPE program (No. 41402001) of MIC Japan, a grant from MEXT Japan (No. 16710061) and a grant from NEDO Japan.

REFERENCES

1. M. Kroutvar et al., Nature 432, 81 (2004).
2. R. Young et al., New J. of Phys. 9, 365 (2007).
3. T. Amand et al. Phys. Rev. Lett. 78, 1355 (1997)
4. A. P. Heberle, W. W. Ruhle and K. Ploog, Phys. Rev. Lett. 72, 3887 (1994).
5. M. Kroutvar et al., Nature 432, 81 (2004).
6. H. Kosaka et al., Y. Mitsumori, Y. Rikitake and H. Imamura, Appl. Phys. Lett. 90, 113511 (2007).
7. R. Vrijen and E. Yablonovitch, Physica E 10, 569 (2001).
8. H. Kosaka, et al., Phys. Rev. Lett. 100, 096602 (2008).
9. F. Meier and B. P. Zakharchenya, Eds., Optical Orientation (Elsevier, Amsterdam, 1984).
10. X. Marie et al., Phys. Rev. B 60, 5811 (1999).
11. J. M. Kikkawa and D. D. Awschalom, Nature 397, 139 (1999).
12. H. Kosaka et al., Phys. Rev. B 67, 045104 (2003).
13. J.J. Baumberg, D.D. Awschalom, Samarth, N., Luo, H. & Furdyna, J.K., Phys. Rev. Lett. 72, 717–720 (1994).

Electromagnetically-induced transparency and squeezed light

E. Figueroa,[1,2] J. Appel,[1,3] D. Korystov,[1] M. Lobino,[1] C. Kupchak,[1] and A. I. Lvovsky[1]

[1] *Institute for Quantum Information Science, University of Calgary, Calgary, Alberta T2N 1N4, Canada*
[2] *Present address:Max-Planck-Institut für Quantenoptik, Hans-Kopferman-Strasse 1, D-85748 Garching, Germany*
[3] *Present address: Niels Bohr Institute, Blegdamsvej 17 DK - 2100 Kobenhavn, Denmark*

Abstract. We report theoretical and experimental studies of the transmission and storage of squeezed vacuum under the conditions of electromagnetically induced transparency in rubidium vapor. A model based on a three-level atom allows us to evaluate the linear losses and extra noise that degrade nonclassical properties of the squeezed vacuum during the atomic interaction and eventually predict the quantum states of the transmitted light with a high precision. Experimentally, we show that squeezing is preserved after storage in rubidium vapor for 1 μs.

Keywords: Electromagnetically induced transparency, quantum memory, squeezing
PACS: 42.50.Gy, 03.67.-a, 42.50.Dv

Electromagnetically-induced transparency (EIT) was first introduced [1] as a technique to create a transparency window in an otherwise opaque medium. Years later, Fleischhauer and Lukin [2] suggested EIT as means for implementing quantum memory for light, i.e. transferring the state of the electromagnetic field to a collective atomic spin excitation. After the first proof-of-principle experimental realization of this protocol with classical light pulses [3], several experimental groups demonstrated that EIT-based memory can preserve nonclassical properties of light.

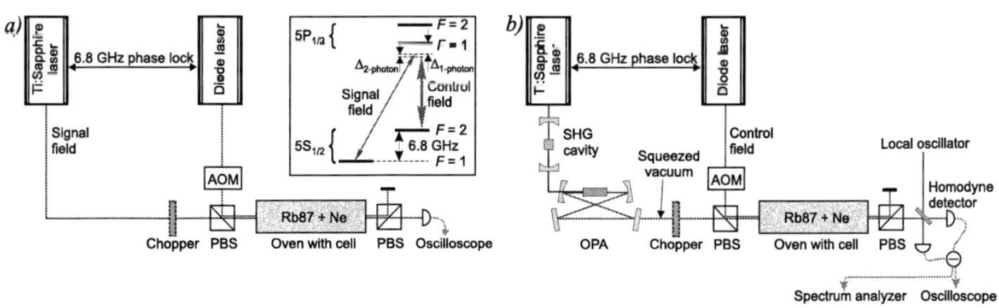

FIGURE 1. Experimental setup for classical (a) and quantum (b) measurements. SHG, second harmonic generation; PBS, polarizing beam splitter; AOM, acousto optical modulator. The elements used in experiments with pulsed light are shown with a dashed line. The inset displays the atomic energy level configuration.

CP1110, *Quantum Communication, Measurement and Computing (QCMC), The Ninth International Conference,* edited by A. Lvovsky

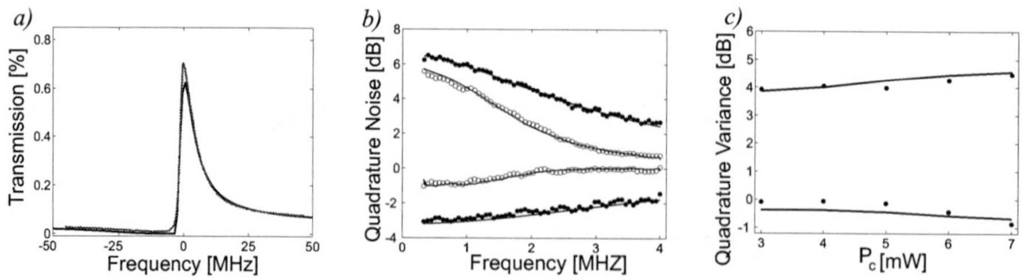

FIGURE 2. Propagation of squeezed vacuum through an EIT medium. (a) The EIT line with a theoretical fit; (b) spectrum of squeezing before (•) and after (○) the Rb cell along with theoretical curves; (c) squeezing in the pulsed regime as a function of the EIT control field power. In (b) and (c), both the squeezed and antisqueezed quadratures are shown.

A particularly important quantum state of light is squeezed vacuum. Here we report a theoretical and experimental study of slow propagation and storage of this state under EIT conditions. In our experiment, the EIT medium was an atomic ^{87}Rb vapor at 65°C, with a Λ energy level configuration formed by one of the hyperfine sublevels of the $5P_{1/2}$ state and two hyperfine sublevels of the $5S_{1/2}$ state (Fig. 1, inset). The classical signal field, resonant to the $|5S_{1/2}, F=1\rangle$–$|5P_{1/2}, F=1\rangle$ transition, was provided by a Ti:Sapphire laser at a wavelength of 795 nm [Fig. 1(a)]. The squeezed vacuum signal was obtained from a sub-threshold continuous-wave optical parametric amplifier (OPA) with a bow-tie cavity, employing a 20-mm-long periodically poled $KTiOPO_4$ crystal as the nonlinear element [4]. The OPA was pumped by the second harmonic of the master Ti:Sapphire laser and generated a squeezing of approximately 3 dB. The same laser was used as the local oscillator in homodyne detection [Fig. 1(b)]. When required, the signal was chopped into 600-ns (FWHM) pulses using a home-made mechanical chopper [5]. The control field, interacting with the $|5S_{1/2}, F=2\rangle$–$|5P_{1/2}, F=1\rangle$ transition, was obtained from an additional diode laser phase locked at 6.834 GHz to the master laser to ensure two-photon resonance with the signal. Orthogonal, linear polarizations for the control and signal were utilized. The two fields were spatially mode matched inside the Rb cell.

The classical transmission spectrum $|T(\omega)|^2$ of the EIT line was measured in order to characterize the absorption and dispersion of the medium. We fitted the spectrum with $|T(\omega)|^2 = |e^{i\frac{\omega}{2c}\chi L}|^2$ [Fig. 2(a)], where L is the cell length and

$$\chi(\Delta_p, \delta_2) \propto \frac{i\gamma_{bc} + \delta_2}{|\Omega|^2 - (i\gamma_{bc} + \delta_2)(\Delta_p + iW)} \tag{1}$$

is the linear susceptibility of a Doppler broadened gas of 3-level atoms under EIT conditions [6]. In the above equation, Ω is the control Rabi frequency, Δ_p is the one-photon detuning of the signal field, δ_2 is the two-photon detuning, γ_{bc} is the ground state dephasing rate, and W is the width of the Doppler-broadened line.

Based on the atomic parameters determined in this observation, we can predict how the propagation through the EIT medium influences continuous-wave squeezed light. The experimentally measurable quadrature noise spectrum is given by $V(\omega) =$

$\langle \hat{q}_\vartheta(\omega)\hat{q}_\vartheta(-\omega)\rangle$, where $\hat{q}_\vartheta(\omega)$ is the frequency-domain quadrature operator and ϑ is the local oscillator phase. We begin with the known expression for the initial squeezed vacuum, produced by a subthreshold OPA [7]: when the local oscillator phase is varied, the observed quadrature noise oscillates between the extremal values

$$V^{\pm}(\omega) = \frac{1}{2} \pm \eta \frac{2\sqrt{P/P_{th}}}{(\omega/\gamma)^2 + (1 \mp \sqrt{P/P_{th}})^2}, \tag{2}$$

where ω is the sideband frequency, η is the overall quantum efficiency, P/P_{th} is the ratio of the OPA pump and threshold powers, and 2γ is the cavity bandwidth.

Upon propagation through the EIT medium with amplitude transmissivity $T(\omega)$, the field operator undergoes the transformation $\hat{a}'(\omega) = T(\omega)\hat{a}(\omega) + \sqrt{1-|T(\omega)|^2}\hat{v}(\omega)$, where $\hat{v}(\omega)$ denotes vacuum. Substituting the field operator into the above expression for $V(\omega)$, we find for the highest and lowest quantum noise:

$$\begin{aligned} V'^{\pm}(\omega) &= \frac{|T(\omega)|^2 + |T(-\omega)|^2}{2}\left(V^+(\omega) + V^-(\omega)\right) \\ &+ |T(\omega)||T(-\omega)|\left(\pm V^+(\omega) \mp V^-(\omega)\right) + \frac{1-|T(\omega)|^2}{4} + \frac{1-|T(-\omega)|^2}{4} + V_{\text{noise}}(\omega). \end{aligned} \tag{3}$$

In the above equation, $V_{\text{noise}}(\omega)$ is the excess noise generated by atoms in the cell [8]. This contribution is a consequence of the population exchange of the rubidium ground states and has been investigated theoretically in Refs. [9, 10].

Experimental measurements of $V^{\pm}(\omega)$ and $V'^{\pm}(\omega)$ were performed by feeding the homodyne detector output to a spectrum analyzer, respectively without and with the EIT cell in the signal beam path. At each sideband frequency, the local oscillator phase was varied and the highest and lowest noise levels were recorded [Fig. 2(b)]. The measurement results were fitted with Eqs. (2,3). The agreement between theory and experiment shows that absorption and the extra noise are the main factors responsible for the degradation of squeezing.

We found that the squeezing is preserved better when the lasers are slightly (540 kHz) detuned from the two-photon resonance. This is because the noise reduction observed at a specific electronic frequency ω originates from quadrature entanglement between the spectral modes associated with the frequencies $\Omega \pm \omega$ of the optical signal field. Thus both modes need to be transmitted through the EIT transparency window.

Propagation of pulses of squeezed vacuum through the EIT medium was studied by means of time-domain homodyne tomography, using a procedure similar to that of Ref. [11]. The output of the homodyne detector was multiplied by the temporal mode function $W(t)$ and integrated over time, producing a single sample of the field quadrature. The weight function was equal to the square root of the transmitted classical pulse intensity. The local oscillator phase values associated with each quadrature were determined as described in Ref. [5], i. e. by measuring the degree of squeezing in the transmitted light with a spectrum analyzer running continuously. The 50,000 to 100,000 quadrature-phase pairs acquired in this manner were used to reconstruct the density matrix and the Wigner function of the transmitted, pulsed optical mode by means of the maximum-likelihood algorithm. The reconstructed optical state strongly resembles a

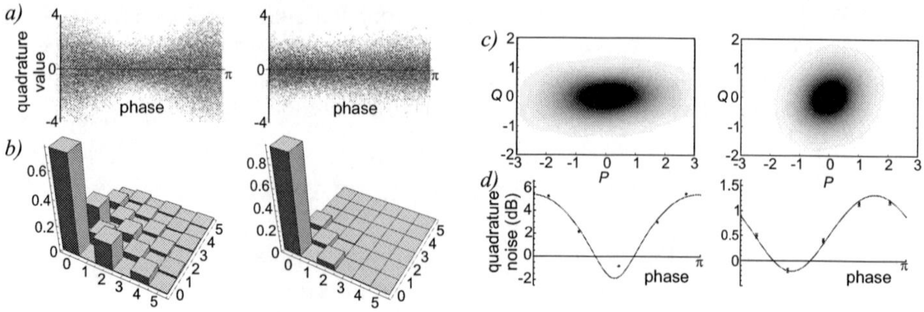

FIGURE 3. Homodyne tomography of the squeezed vacuum before (left column) and after (right column) storage. Raw samples of phase-dependent quadrature noise (a), maximum-likelihood reconstruction of the density matrices in the Fock basis b), Wigner functions (c) and quadrature noise variances (d) are displayed. In (d), the solid lines are calculated from the density matrices while the points with error bars show variances of binned quadrature values.

squeezed thermal state. The minimum and maximum noise levels observed at different control field powers are summarized in Fig. 2(c). The experimental data on classical transmission of the EIT line, the parameters of the OPA and the atomic excess noise spectra are sufficient to make theoretical predictions for the transmission of squeezed vacuum in the pulsed regime. For all control powers, over 99% fidelity between the theoretically predicted and experimentally observed states is reached.

Quantum memory was implemented by turning the control field off for one microsecond. The retrieved pulse was subjected to time-domain homodyne detection. The results of homodyne tomography measurements are shown in Fig. 3 for both the initial and retrieved ensembles. Density matrices and Wigner functions of these states were reconstructed using the maximum likelihood method and exhibit 1.86+/-05 and 0.21+/-.04 dB of squeezing, respectively. The quantum memory fidelity associated with this reconstruction equals 0.89, which is higher than the classical maximum of 0.73.

REFERENCES

1. K. J. Boller, A. Imamoglu, S.E. Harris, Phys. Rev. Lett. **66**, 2593 (1991)
2. M. Fleischhauer and M. D. Lukin, Phys. Rev. Lett. **84**, 5094, 2000; M. Fleischhauer and M. D. Lukin, Phys. Rev. A **65**, 022314.
3. D. F. Phillips *et al.*, Phys. Rev. Lett. **86**, 783 (2001).
4. J. Appel *et al.*, Phys. Rev. A **75**, 035802 (2007)
5. J. Appel *et al.*, Phys. Rev. Lett. **100**, 093602 (2008).
6. F. Figueroa *et al.*, Opt. Lett. **31**, 2625 (2006).
7. M. O. Scully and M. S. Zubairy, *Quantum Optics* (Cambridge University Press, 1997)
8. M. T. L. Hsu *et al.*, Phys. Rev. Lett. **97**, 183601 (2006).
9. A. Peng *et al.*, Phys. Rev. A **71**, 033809 (2005); G. Hétet *et al.*, Phys. Rev. A **74**, 059902(E) (2006).
10. G. Hétet *et al.*, Phys. Rev. A **77**, 012323 (2008);
11. J. S. Neergard-Nielsen *et al.*, Phys. Rev. Lett. **97**, 083604 (2006).

12. This work was supported by NSERC, CIFAR, AIF, CFI and Quantum*Works*.

Electromagnetically induced transparency experiments in cesium vapor - Towards single photon operation

D. Höckel, M. Scholz, O. Benson

Nano Optics, Institute of Physics, Humboldt-Universität zu Berlin, Hausvogteiplatz 5-7
10117 Berlin, Germany

Abstract. We describe first results with an experimental setup designed to store light pulses of ultra low intensity inside a cesium vapor cell by employing the effect of electromagnetically induced transparency. A phase-locked laser system is used as pump and probe laser and probe laser pulses of Gaußian shape are delayed by up to 280 ns. We outline the route to single photon operation of the setup, in particular by introducing a novel filtering concept.

Keywords: light coherence, electromagnetically induced transparency, EIT, optical phase locked loops, single photons, delayed light, Fabry-Pérot interferometer
PACS: 42.50.Gy, 42.65.-k, 07.60.Ly

INTRODUCTION

Transferring of quantum states between light and matter is an important topic in the field of quantum communication and computing. Promising tools for that task are systems based on electromagnetically induced transparency (EIT) which can be realized in an alkali gas at room temperature. So far many groups world-wide have studied EIT in rubidium vapor [1, 2, 3]. Cesium (Cs) however is particularly attractive for EIT, e.g. in order to establish all optical atomic clocks utilizing the 9.2 GHz hyperfine splitting of the ground state directly [4].

We use a phase-locked laser system that allows to perform EIT experiments in Cs vapor at room temperature [5]: An 894 nm extended-cavity diode laser (master laser) is stabilized to the ^{133}Cs D1 line by Doppler-free frequency modulation spectroscopy. A second extended-cavity diode laser is locked to the master laser using a digital-only optical phased-locked loop with a frequency offset of 9.2 GHz. From the fraction of power contained in the beat-note signal (98.4%), the residual phase noise of the system has been calculated to be less than 0.02 rad^2. In a magnetically shielded Cs vapor cell, EIT experiments are performed using these two stabilized lasers as probe and pump laser.

Here, we present experimental results of delayed light pulses that are a first step on the way to storage and manipulation of single photons inside the Cs vapor cell. Furthermore, the route to single photon operation of this setup will be outlined.

CP1110, *Quantum Communication, Measurement and Computing (QCMC), The Ninth International Conference,* edited by A. Lvovsky

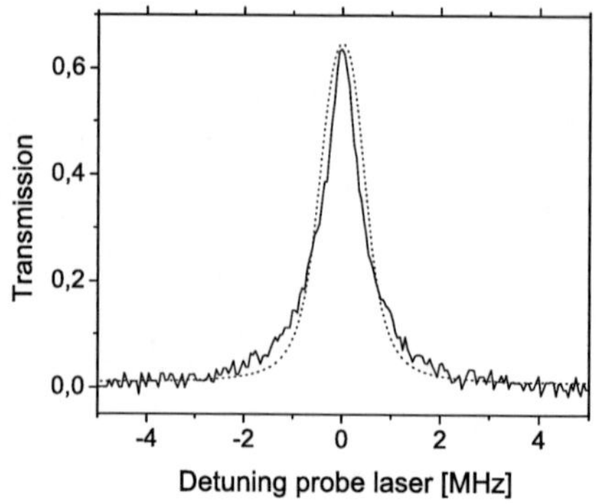

FIGURE 1. EIT: Scans across the probe laser resonance for a cell temperature of 40 °C. Measurement (solid line) compared to theoretical fit (dotted line) [5].

EXPERIMENTAL REALIZATION

In a first experiment we used the phase-locked laser system to demonstrate EIT in a Λ-shaped atomic system on the D1 line of ^{133}Cs. Two acousto-optic modulators were used to change the frequency of the probe laser allowing a scan over the $6^2S_{1/2}(F=3) \rightarrow 6^2P_{1/2}(F'=4)$ transition while the pump laser stayed locked to the $6^2S_{1/2}(F=4) \rightarrow 6^2P_{1/2}(F'=4)$ transition. A transmission of $>60\%$ for a transparency window width of 1.1 MHz was demonstrated. A theoretical description of the measured transmission relying on the dephasing rate as the only free parameter was presented (see Fig. 1) [5].

Delayed light pulses

The described experimental setup was then extended by an electro-optical modulator (EOM) allowing amplitude modulation of the probe laser (see Fig. 2). Pulses of Gaußian shape were generated and the delay associated with the steep dispersion within the EIT transparency window was measured (see Fig. 3). In a first step, delays of up to 280 ns were measured, corresponding to a speed of light of $c/2000$ inside the cell. These first promising results indicate, that the delay will increase with further optimization of the experimental parameters, mainly laser stability.

SINGLE PHOTON OPERATION

In order to reach true single photon operation in our experimental setup two major additions are required: A narrow-band single photon source and the filtering of probe photons with respect to the pump beam. Narrow-band single photons can be generated by cavity-enhanced parametric down-conversion by an optical parametric oscillator

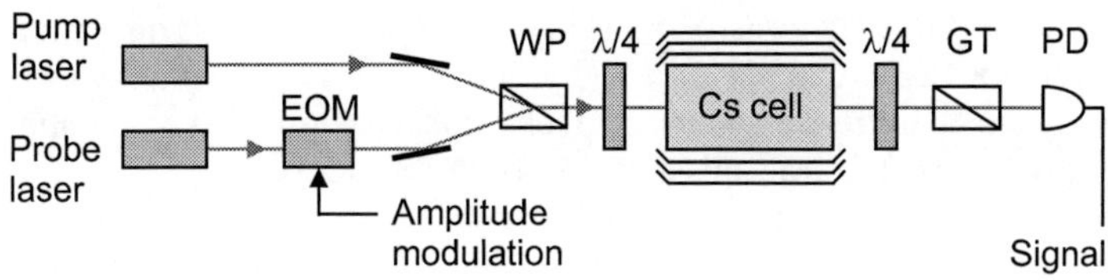

FIGURE 2. EIT setup: The power of the probe laser is modulated by an electro-optical modulator (EOM). Pump and probe laser are linearly polarized and superimposed by a Wollaston prism (WP). The polarization is changed to circular and both lasers copropagate through a magnetically shielded Cs cell. Behind the cell, the polarization is transformed back to linear. A Glan-Thompson prism (GT) then filters the pump beam before the probe beam is detected by a photo detector (PD).

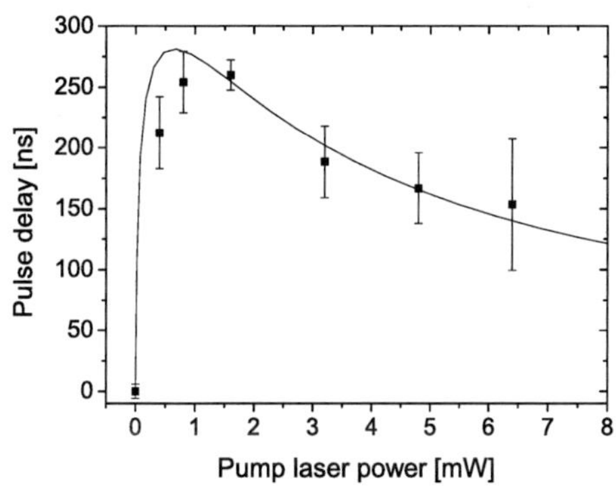

FIGURE 3. Delayed light pulses: A Gauß-shaped light pulse of 2 μs length is delayed by up to 280 ns, depending on the pump laser power. The continuous line is based on calculations described in [5].

(OPO) below threshold [6, 7, 8]. Our setup is already designed for such an approach: The probe laser is amplified, frequency-doubled, and used as the pump of the OPO. In a degenerate process the down-converted photons from the OPO are frequency-matched to the EIT window and can be slowed down and eventually brought to a hold. The suppression of the pump beam is necessary since several mW of pump power will copropagate with the single photon probe field. In our setup, a combination of high-quality polarization filters and a multi-pass etalon as spectral filter will be able to suppress the pump power by $\geq$12 orders of magnitude which is sufficient for single photon experiments. As the polarization filters alone allow a suppression of around $\sim 10^5$, more than 10^7 needs to be accomplished by the spectral filtering.

The spectral filter

A Fabry-Pérot filter or etalon is one of the most widely used optical filters. It basically consists of a resonant cavity formed by two parallel mirrors. Interference between the multiple reflections of light between the two reflecting surfaces leads to a transmission spectrum that shows peaks corresponding to resonances of the cavity. The wavelength separation between adjacent transmission peaks, i.e. the free spectral range (FSR), and the transmission linewidth $\Delta\nu$ define the finesse of the device $F = FSR/\Delta\nu$. In theory, the finesse is given by the mirror reflectivity (R) only, yielding $F_r = \pi\sqrt{R}/(1-R)$, but in real setups several material parameters like surface flatness of the optical substrates and mirror tilt together with F_r require to generalize to the so called *effective* finesse F_{eff}. The performance of the filter is described by the contrast $C = T_{max}/T_{min} = (1+4F_{eff}^2/\pi^2)$ and the peak transmission, which is no longer $T_{max} = 1$ as in the ideal case, but $T_{max} = F_{eff}/F_R < 1$ [9].

In our setup, a finesse of $F = 69$ together with a maximum transmission $T_{max} =$ 92% has been achieved. This was accomplished by carefully optimizing the relevant parameters, by using high-quality mirror substrates with a matched surface quality of $\lambda/200$, piezo-controlled spacers between the mirrors, stabilized to a length accuracy of $\Delta d = 1$ nm to remove the mirror tilt, and by using an optimized mirror reflectivity of $R =$ 95.6%. By tuning the mirror the FSR is adjusted to 18.4 GHz, i.e. twice the frequency offset between probe and pump laser. This allows a suppression of the pump laser power by a (calculated) contrast factor of $C \approx 1900$. To reach the required suppression of $> 10^7$, the etalon is used in *multi-pass* configuration. In an n-pass setup, transmission and contrast theoretically scale to the power of n, i.e. $C_n = C^n$ and $T_n = T^n$. The finesse scales $F_n = F_1/(2^{1/n}-1)^{1/2}$ [9]. A triple-pass setup will therefore fulfill the performance requirements. Such a setup is currently in preparation. First results with a double-pass configuration are very promising, $F_2 = 106$ and $T_2 = 76\%$ have been achieved already.

ACKNOWLEDGMENTS

This work was supported by the Deutsche Forschungsgemeinschaft DFG (Grant No. BE2224/5). D. Höckel acknowledges funding by the state of Berlin (NaFöG) and M. Scholz by Deutsche Telekom Stiftung.

REFERENCES

1. D. F. Phillips, A. Fleischhauer, A. Mair, R. L. Walsworth, and M. D. Lukin, *Phys. Rev. Lett.* **86**, 783–786 (2001).
2. E. Figueroa, F. Vewinger, J. Appel, and A. I. Lvovsky, *Opt. Lett.* **31**, 2625–2627 (2006).
3. I. Novikova, A. V. Gorshkov, D. F. Phillips, A. S. Sørensen, M. D. Lukin, and R. L. Walsworth, *Phys. Rev. Lett.* **98**, 243602 (2007).
4. S. Knappe, V. Shah, P. D. D. Schwindt, L. Hollberg, J. Kitching, L. Liew, and J. Moreland, *Appl. Phys. Lett.* **85**, 1460 (2004).
5. D. Höckel, M. Scholz, and O. Benson, *Appl. Phys. B* **in press** (2008).
6. Y. J. Lu, R. L. Campbell, and Z. Y. Ou, *Phys. Rev. Lett.* **91**, 163602 (2003).
7. C. E. Kuklewicz, F. N. C. Wong, and J. H. Shapiro, *Phys. Rev. Lett.* **97**, 223601 (2006).
8. J. S. Neergaard-Nielsen, B. M. Nielsen, H. Takahashi, A. I. Vistnes, and E. S. Polzik, *Opt. Exp.* **15**, 7940 (2007).
9. C. Roychoudhuri, and M. Hercher, *Appl. Opt.* **16**, 2514 (1977).

Testing Quantum Memories Via Entanglement Verification

Hauke Häseler and Norbert Lütkenhaus

Institute for Quantum Computing, University of Waterloo, Canada
Quantum Information Theory Group, University of Erlangen, Germany

Abstract. We present a general criterion for a quantum device such as a quantum memory to work in a non-classical regime. The criterion is based on the verification of effective entanglement and requires only a very small amount of resources. We apply the test to quantum memories for squeezed light and obtain results in terms of second moments of measured quadratures of the retrieved states. These results are checked against optimized classical storage strategies.

Keywords: quantum memory, entanglement verification, classical benchmarks
PACS: 03.67.Hk, 03.67.Mn, 42.50.Dv

INTRODUCTION

Recently, there has been increasing experimental interest in the storage and retrieval of light, and of nonclassical (e.g. squeezed) light in particular (see, e.g., [1, 2]). But how do we know that the same task could not have been achieved by a classical memory, i.e., a measure & re-prepare strategy? This question can be answered by theory, and many solutions for squeezed-state storage have been proposed [3, 4, 5]. They involve calculations of the maximum fidelity between input and output states that a classical strategy can achieve, averaged over an ensemble of input states.

We propose a different criterion which uses the fact that a measure & re-prepare strategy acts as an entanglement-breaking channel. Our criterion hinges on two known results: Firstly, any source of non-orthogonal states admits an entanglement-based description, so that verification of effective entanglement between source and memory output excludes classical storage strategies. Secondly, a discrete-continuous entanglement criterion is required, since the number of test states must be finite in any experiment, but the dimension of the output space is in general infinite. A suitable criterion for this situation is the Expectation Value Matrix (EVM) criterion [6, 7], which also removes the need for tomography on the output states. In fact, we will work in the minimal setting of two equiprobable test states and homodyne detection of the quadrature expectation values and variances. The resulting classical data allow for the direct evaluation of our criterion.

In order to check the strength of the criterion, we test it against optimized classical storage strategies.

CP1110, *Quantum Communication, Measurement and Computing (QCMC), The Ninth International Conference,* edited by A. Lvovsky

ENTANGLEMENT VERIFICATION

Pure test states: A source of two pure test states, say the squeezed vacua $|0;r\rangle$ and $|0;-r\rangle$ with the squeezing parameter $r \in \mathbb{R}$, can be described as the creation of an entangled state

$$|\psi^{src}\rangle = \frac{1}{\sqrt{2}}\left(|0\rangle_A|0;r\rangle_B + |1\rangle_A|0;-r\rangle_B\right), \tag{1}$$

followed by the projective measurement $\{|0\rangle_A\langle 0|, |1\rangle_A\langle 1|\}$. The EVM then contains the reduced density matrix $\rho_A = \mathrm{tr_B}(|\psi^{\mathrm{src}}\rangle\langle\psi^{\mathrm{src}}|)$ and the measurement outcomes on the retrieved states ρ_0^{out} and ρ_1^{out}, i.e., all available information on the system. For each set of possible measurement outcomes, the EVM criterion indicates whether the set originated from an effective entangled state, and it is for those outcomes that the memory works in the quantum domain.

In order to evaluate our criterion, let us consider typical measurement outcomes: Since the input states are not displaced, the same can be expected for the retrieved states, i.e., all first quadrature moments are zero. Also, we assume that the memory is oblivious of the choice of test state and of the measurement setting, i.e., $\mathrm{tr}(\rho_0^{out}\hat{x}^2) = \mathrm{tr}(\rho_1^{out}\hat{p}^2)$ and vice versa. Now, each set of measurement outcomes is specified by just two parameters, e.g., $\mathrm{Var}_0(\hat{\mathrm{x}})$ and $\mathrm{Var}_0(\hat{\mathrm{p}})$. Note that the EVM criterion does not rely on any assumptions about the output states, but the above parameterization helps to visualize the results.

The EVM can be constructed in accordance with Ref. [7] and be evaluated on the partially transposed state to form the desired criterion. Figure 1 shows which variance pairs originate from effective entangled states, for a particular choice of test state overlap s.

Mixed test states: The generation of pure squeezed vacuum states is in practice unfeasible, since any losses in the process will result in mixed states. As a first approach, we could simply attribute these losses to the memory and use the above method for pure test states. However, this would certainly lead to a suboptimal benchmark criterion. We therefore extend the EVM method to sources of mixed test states ρ_0^{in} and ρ_1^{in}. In general, there is no unique entanglement-based formulation for such sources and therefore no unique construction of an EVM exists. We will address the problem as follows: In analogy to the source state in Eq. (1), define

$$|\Psi^{src}\rangle = \frac{1}{\sqrt{2}}\left(|0\rangle_A|\phi_0\rangle_{BC} + |1\rangle_A|\phi_1\rangle_{BC}\right), \tag{2}$$

where the $|\phi_i\rangle_{BC}$ are *purifications* of the test states ρ_i^{in}, and the system C is a copy of system B. The simplest way to include this auxiliary system C in the EVM is to extend the measurements $\{|i\rangle\langle j|\}$, $i,j \in \{0,1\}$, on system A to the set $\{|i\rangle\langle j| \otimes \mathbb{1}_C\}$. The form of the EVM then remains unchanged, but s must be replaced by the overlap of the two purifications, $\langle\phi_0|\phi_1\rangle$. We can now exploit the unitary freedom inherent to purifications to optimize this overlap. It turns out that the given EVM leads to the strongest entanglement criterion when $\langle\phi_0|\phi_1\rangle$ takes its maximum value. According to Uhlmann's theorem, this maximum value is given by the fidelity $F(\rho_0^{in},\rho_1^{in})$.

An evaluation of the criterion is shown in Fig. 1 for squeezed and anti-squeezed thermal states as test states. The values of the squeezing parameter and the mixedness shown in Fig. 1 are experimentally viable (c.f. [1]) . The area of variance pairs in the quantum domain is bounded from above by the straight line $\mathrm{Var}(\hat{p}) = F(\rho_0^{in}, \rho_1^{in})/2$. For decreasing values of $\mathrm{Var}(\hat{x})$, the border of the quantum domain rises, up to the point $\mathrm{Var}(\hat{x}) = \mathrm{Var}(\hat{p}) = 1/2$, which corresponds to the trivial case of a completely lossy channel.

OPTIMAL CLASSICAL STORAGE

Mathematically, a measure & re-prepare strategy is expressed as an entanglement-breaking channel

$$\Lambda(\rho^{in}) = \sum_i \mathrm{tr}(\rho^{in}\pi_i)|\phi_i\rangle\langle\phi_i|, \tag{3}$$

which in our case acts on system B of either source state (1) or (2). The π_i form a POVM, and the two output states ρ_0^{out} and ρ_1^{out} will be mixtures of the $|\phi_i\rangle$. For each possible value of $\mathrm{Var}(\hat{x})$, we now seek the smallest possible $\mathrm{Var}(\hat{p})$, i.e.,

$$\min_{\pi_i, |\phi_i\rangle} \mathrm{Var}_0(\hat{p}) \tag{4}$$

$$\text{s.t. } \mathrm{Var}_0(\hat{x}) = c. \tag{5}$$

Since the objective function is a variance, all $|\phi_i\rangle$ should be minimum uncertainty states. Furthermore, since displacements do not affect variances, no generality is lost by restricting ourselves to the class of squeezed vacuum states. The minimization over $\{|\phi_i\rangle\}$ can now be performed over a set of squeezing parameters r_i.

For the pure state source, the optimal choice of $\{\pi_i\}$ turns out to be the minimum error discrimination POVM. For two mixed states, we can use a known result concerning the fidelity [8]:

$$F(\rho_0, \rho_1) = \min_{\pi_i} \sum_i \sqrt{\mathrm{tr}(\rho_0\pi_i)\mathrm{tr}(\rho_1\pi_i)}, \tag{6}$$

where the minimization runs over all POVMs. We can then transform our optimization problem to a form similar to the right-hand side of Eq. (6). The variances resulting from these optimized measure & re-prepare strategies are shown in Fig. 1.

DISCUSSION

There are two points to note about the above results: Firstly, for the pure test states the boundaries of the classical region and the quantum domain coincide. Hence, the EVM criterion detects all entangled states which can be witnessed in the given setup. All other measurement data are compatible with classical storage strategies. It is worthwhile to note that the observation of squeezing alone in the output light does not suffice to verify storage in the quantum domain.

The second point concerns the gap between the optimal classical storage strategy and the boundary of the quantum domain, which is present for the source of mixed

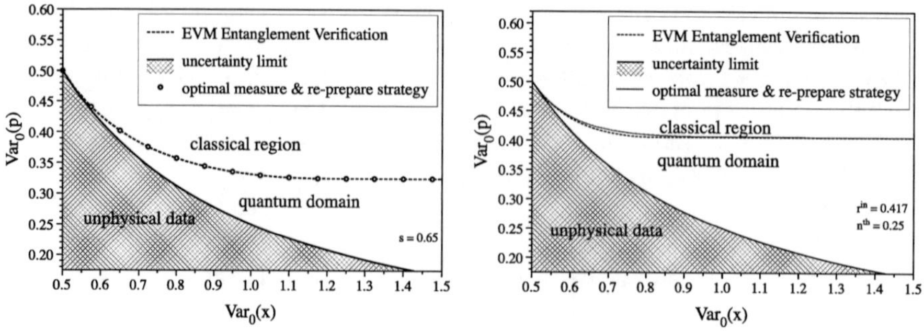

FIGURE 1. Left: Output data from pure test states. The criterion is tight for the whole output space. Right: Output data from mixed test states. Some data pairs in the quantum domain are not detected by the entanglement criterion, unless the EVM is enlarged.

states, but not for the pure-state case. Since the classical strategy is optimal, it forms the actual boundary of the quantum domain. The gap arises from states which are not detected by the entanglement criterion. The reason for this is that the EVM does not include all available information. From the source state (2), the reduced density matrix ρ_{AC} is known and fixed, but since it is in general infinite-dimensional, it cannot be fully included in the EVM. It is in fact possible to detect more entangled states by extending the EVM and thus adding more information about the purifying system C. Ultimately, there is a trade-off between the size of the EVM and the strength of the resulting benchmark criterion.

In conclusion, we presented a benchmark criterion for quantum memories and similar quantum devices or channels, which requires very few experimental resources. Its application to memories for squeezed light leads to an experimentally feasible way of verifying operation in the quantum domain.

ACKNOWLEDGMENTS

We would like to thank Alex Lvovsky and Mirko Lobino, Sarah Croke as well as Tobias Moroder for helpful discussions. This work was funded by the EU Integrated Project QAP, the Canadian NSERC Discovery Grant and by Quantum Works.

REFERENCES

1. J. Appel, E. Figueroa, D. Korystov, M. Lobino, and A. I. Lvovsky, *Phys. Rev. Lett.* **100**, 093602 (2008).
2. K. Honda, D. Akamatsu, M. Arikawa, Y. Yokoi, K. Akiba, S. Nagatsuka, T. Tanimura, A. Furusawa, and M. Kozuma, *Physical Review Letters* **100**, 093601 (2008).
3. K. Hammerer, M. M. Wolf, E. S. Polzik, and J. I. Cirac, *Phys. Rev. Lett.* **94**, 150503 (2005).
4. J. Calsamiglia, M. Aspachs, R. Munoz-Tapia, and E. Bagan, quant-ph/0807.5126v1 (2008).
5. R. Namiki, *Physical Review A (Atomic, Molecular, and Optical Physics)* **78**, 032333 (2008).
6. J. Rigas, O. Gühne, and N. Lütkenhaus, *Phys. Rev. A* **73**, 012341 (2006).
7. H. Häseler, T. Moroder, and N. Lütkenhaus, *Phys. Rev. A* **77**, 032303 (2008).
8. C. A. Fuchs, Ph.D. thesis, University of Mexico, Mexico (1995).

Rate degradation in quantum repeaters with imperfect memories

M. Razavi*, M. Piani† and N. Lütkenhaus†

*Institute for Quantum Computing and Department of Electrical and Computer Engineering, University of Waterloo, ON, Canada
†Institute for Quantum Computing and Department of Physics and Astronomy, University of Waterloo, ON, Canada

Abstract. Memory dephasing and its impact on the rate of entanglement generation in quantum repeaters is addressed. For systems that rely on probabilistic schemes for entanglement distribution and connection, we estimate the maximum achievable rate per employed memory for our optimized partial nesting protocol. We show that, the polynomial scaling of rate with distance, L, can only be achieved if quantum memories with infinitely long coherence times are available. For finite coherence times and for long distances, the above rate degrades *exponentially* in $\sqrt{L}$ regardless of the employed purification protocol.

Keywords: Quantum Communication, Quantum Memory, Quantum Key Distribution
PACS: 03.67.Hk, 03.67.Bg, 42.50.Dv, 03.67.Dd

Quantum repeaters have been proposed to remedy the ill effect of channel loss in quantum communication systems [1, 2, 3]. The essential elements for implementing such systems are quantum memory units that exhibit efficient coupling to light, scalability, and long coherence times comparable to the transmission delay between the two parties. In the ideal case of infinitely long coherence times, quantum repeaters can exchange entanglement with a rate that scales polynomially with distance, as opposed to direct entanglement-distribution links for which the rate commonly decays exponentially. This scenario may change drastically if we allow for imperfections in our quantum memories [2]. In this paper we analytically address this issue by employing an optimistic approach to estimating the throughput of quantum repeaters in the presence of memory dephasing. Our results show that, for the structure considered here, finite coherence times result in throughputs that, per memory used, decay exponentially in $\sqrt{L}$.

Consider a quantum repeater system of nesting level n, for which the total distance $L = 2^n L_0$ has been partitioned into segments of length L_0, see Figure 1. Suppose, at the two ends of such segments, there are $N \gg 1$ memories. In an ideal quantum repeater scenario, we first establish entanglement over segments of length L_0, and then extend this entanglement to the farthest ends by performing Bell-state measurements (BSMs) at all intermediate nodes. These measurements can be done simultaneously if BSMs are deterministic and error free, otherwise, at each nesting level, we should wait until we learn, via classical communication, which BSMs, from the previous level, have been successful. This waiting time depends on the nesting level of the corresponding BSM, and it doubles as we move on to the next level. To find the rate at which we create entanglement, we not only need to account for the probabilistic nature of entanglement distribution and connection procedures, but we must also account for the memory decay

CP1110, *Quantum Communication, Measurement and Computing (QCMC), The Ninth International Conference,* edited by A. Lvovsky

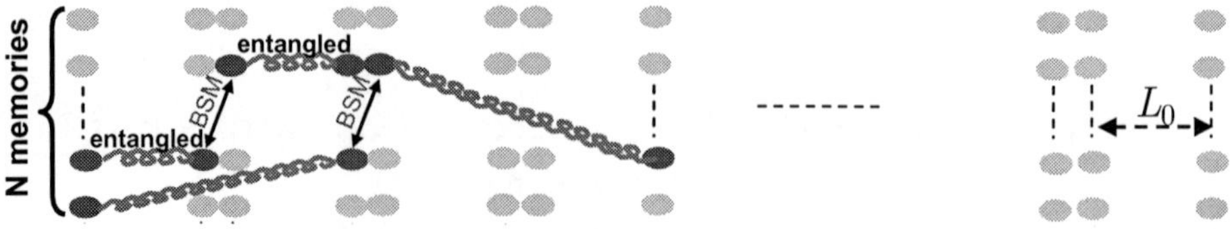

FIGURE 1. A quantum repeater with multiple quantum memories per node. At each round, we first employ an entanglement distribution protocol to entangle any unentangled memory pairs over shortest links. It takes at least $T_0 = L_0/c$ to learn about the success/failure of these attempts, which succeed with probability P_S, hence T_0 is the shortest period at which any protocol can run. At any such cycle, we also match up entangled pairs at different stations, according to our partial nesting protocol (see the text), to perform Bell-state measurements (BSMs), which succeed with probability P_M.

while we are waiting for the above classical signals.

We use a probabilistic framework to model entanglement distribution and connection procedures in quantum repeaters [2, 3]. We denote the probability of success for our employed entanglement distribution scheme over distance L_0 by P_S. The implicit assumption here is that our entanglement distribution scheme is heralding, i.e., there is a mechanism by which we can verify or become informed of whether or not our entangling attempt has succeeded. We also assume that the whole process of entanglement distribution from the time that it starts until we learn about its result takes $T_0 \equiv L_0/c$. This is the fundamental time period at which we can attempt successively to entangle two specific memories at distance L_0. We further assume that, in the absence of memory errors, a successful entangling attempt results in a Bell state $\hat{\pi}^+$, where $\hat{\pi}^\pm = |\psi^\pm\rangle\langle\psi^\pm|$ with $|\psi^\pm\rangle \equiv (|01\rangle \pm |10\rangle)/\sqrt{2}$. Similarly, we associate a success probability P_M to each BSM. This probability, in principle, depends on the state of the memories before performing the BSM. In our optimistic analysis, we assume $P_M < 1$ is the maximum probability of obtaining a conclusive result for a given BSM module. For instance, if we rely on a specific pattern of photodetection events for performing a BSM, the inefficiency of photodetectors may result in inconclusive outcomes, in which case we consider the BSM has failed.

The key to the polynomial scaling of the rate in quantum repeaters is in performing *informed*, rather than *blind*, BSMs. A BSM on memories B and C is considered to be informed if we know, at the time of measurement, that B and C are entangled with relevant memories in distant locations, otherwise the BSM is considered to be blind. Hence in order to do informed BSMs, we need to wait for classical information from distant nodes in the network, whereas blind BSMs can be done at any time. With informed BSMs, however, we can extend the entanglement in a nested way [1], step by step, doubling the range at each nesting level, such that the rate scales as $(P_M)^n$, whereas with blind BSMs all required BSMs must succeed simultaneously, which results in a rate that scales as $(P_M)^{2^n}$. Informed BSMs provide us with a better rate-versus-distance scaling, provided that memories with sufficiently long coherence times are available.

If our memories do not enjoy infinitely long coherence times, it may be beneficial to use blind BSMs in a few final nesting levels. That brings us to our mth-order partial nesting protocol (m-PNP), for $m = 1, \ldots, n$, which proceeds as follows. At each T_0-

long cycle, we first attempt to entangle unused memories of the shortest links. We learn about the success/failure of these attempts by the next cycle. At each cycle, intermediate stations also match up as many memories as possible in their two banks for performing BSMs. In our protocol, we perform *informed* BSMs at stations that correspond to nesting levels 1 to m, and *blind* BSMs at all other stations on memories known to be entangled up to nesting level m. These blind BSMs are performed as soon as such measurements can be done. Purification may or may not be used over the first m nesting levels.

We model the memory degradation by a dephasing process in which a qubit state $\hat{\rho}$ is mapped, after decaying for a time period t, to

$$\Gamma_t(\hat{\rho}) = p(t/2)\hat{\rho} + [1 - p(t/2)]\hat{\sigma}_z\hat{\rho}\hat{\sigma}_z, \tag{1}$$

where $\hat{\sigma}_z$ is the Pauli z-operator, $p(t) = [1 + \exp(-t/\tau_c)]/2$, and τ_c is the memory coherence time. The above dephasing process maps $\hat{\pi}^{\pm}$ to $p(t)\hat{\pi}^{\pm} + [1 - p(t)]\hat{\pi}^{\mp}$.

Using the above model, we estimate $R_m^{(n)}$, the generation rate of maximally entangled states, per employed memory, in the presence of dephasing errors and without restricting ourselves to any specific purification protocol. For any memories X and Y at distance L in the m-PNP, we overestimate the ratio between maximally entangled states that can be obtained out of many copies of the final degraded state of X and Y by its entanglement cost [4]. For a state $\hat{\rho}$, entanglement cost is the ratio M/K for starting with $(\hat{\pi}^+)^{\otimes M}$ and obtaining $\hat{\rho}^{\otimes K}$, in the limit of large K, using local operations and classical communication. We can then prove that, for the (m-PNP),

$$R_m^{(n)} \approx \frac{P_S P_M^{(2^{n-m+1}+m-2)}}{2L/c} E_C(T_m^{(n)}), \tag{2}$$

where $T_m^{(n)} = \max(T_0, 2^{m-2}T_0)$, $E_C(t) = H(\frac{1}{2} + \sqrt{p(t)[1-p(t)]})$, and $H(x) = -x\log_2 x - (1-x)\log_2(1-x)$. The fraction term in the above equation accounts for the probabilistic nature of our system, and $E_C(T_m^{(n)})$ accounts for memory dephasing.

For any given distance L, one can find the optimum values of m and n that maximize $R_m^{(n)}$. Figure 2 shows this optimum rate versus coherence time for $L = 1000\,\text{km}$. Here, we assume $P_S = 0.2 \times 10^{-0.01L_0}$, which is an appropriate choice for the schemes proposed in [3]. The underlying physical channel is assumed to be an optical fiber for which $c = 2 \times 10^8\,\text{m/s}$, hence $L/c = 5\,\text{ms}$. On several points on the graph, we have indicated pairs (n,m) that achieve the maximum achievable rate, with and without purification. For $\tau_c \ll L/c$, the partial nesting protocol favors using blind BSMs as they require memories with shorter coherence times. For lower values of P_M, one needs to use fewer nesting levels to attain the maximum achievable rate.

Figure 2 also demonstrates the rate behavior in both limits of long and short coherence times. First, it is noticeable that the rate, regardless of the employed purification scheme, approaches a constant value if $\tau_c \gg L/c$, in accord with what has been reported in [2]. In the other unexplored extreme of $L \gg c\tau_c$, however, the situation is much different. Because, to the first order, $E_C(t) \propto t\exp(-2t/\tau_c)$, for $t \gg \tau_c$, we can show that, at $L \gg c\tau_c$ and for $P_M < 1$ and $P_S \propto \exp(-\alpha L_0)$ with α being a constant, $R_m^{(n)}$ in Eq. (2) scales as a power of $\exp[-\sqrt{(L/c)/\tau_c}]$, at the optimum values of m and n.

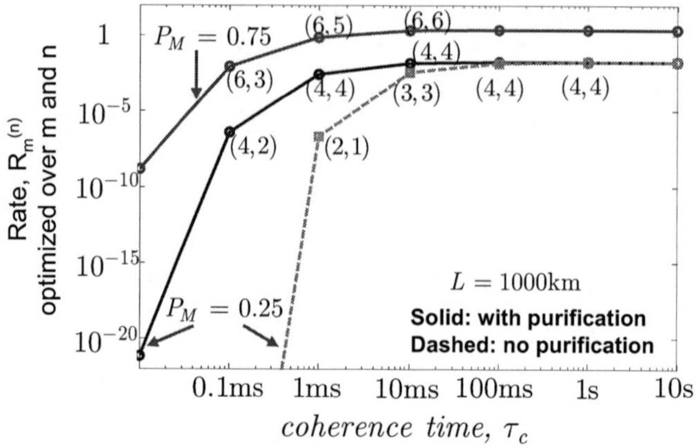

FIGURE 2. Optimistic estimates for the generation rate of maximally entangled states per employed memory, $R_m^{(n)}$, versus coherence time, optimized over m and n, for $P_S = 0.2 \times 10^{-0.01L_0}$, $L = 1000$km, and $L/c = 5$ms. For fixed values of τ_c, the curves scale exponentially in L if no purification is used and exponentially in $\sqrt{L}$ with purification. The two-tuples (n,m), on several points on the graph, show the optimum nesting levels, n, and the optimum numbers of informed BSMs, m, at those points.

The above analysis provides us with practical rate-versus-cost measures that help us estimate the minimal cost of quantum communication systems. It alarmingly underlines the importance of coherence times in quantum repeaters. For a constant target rate, our results imply that the cost of building such a system, under optimistic assumptions, explodes exponentially in $\sqrt{L}$, if the employed quantum memories do not supply coherence times far exceeding the transmission delay L/c. It is noticeable that this rate behavior is still superior to the fully exponential decay of rate with distance in quantum relays, which do not use quantum memories [5]. In the particular case of quantum key distribution [6], in order to create 1000 secure key bits/s over 1000km, using a quantum repeater, thousands of memories are needed with coherence times on the order of 10ms.

We would like to thank W. Munro for his suggesting the memory model. This work was sponsored by QuantumWorks, OCE, and NSERC Discovery Grant.

REFERENCES

1. H.-J. Briegel, W. Dür, J. I. Cirac, and P. Zoller, Phys. Rev. Lett. **81,** 5932 (1998).
2. L. Hartmann *et al.*Phys. Rev. A **75,** 032310 (2007); O. A. Collins, Phys. Rev. Lett. **98,** 060502 (2007).
3. L. M. Duan *et al.*, Nature **414,** 413 (2001); N. Sangouard *et al.*, Phys. Rev. A **76,** 050301(R) (2007).
4. G. Vidal, W. Dür, and J. I. Cirac, Phys. Rev. Lett. **89,** 027901 (2002); M. Horodecki, P. Horodecki, and R. Horodecki, Phys. Rev. Lett. **84,** 2014 (2000).
5. H. de Riedmatten *et al.*, Phys. Rev. Lett. **92,** 047904 (2004).
6. A. Ekert, Phys. Rev. Lett. **67,** 661 (1991).

Pair-wise Entanglement Characterizing a Quantum State of an Atomic Ensemble Storing a Squeezed Light

Kaoru Shimizu and Daisuke Hashimoto

NTT Basic Research Laboratories, NTT Corporation, 3-1Morinosato Wakamiya,Atsugi-shi, Kanagawa 243-0198, Japan

Abstract. This paper characterizes a pair-wise quantum spin correlation formed in an atomic ensemble that stores a phase or amplitude squeezed state of light on the framework of the electromagnetically induced transparency (EIT). The photonic quantum correlation in the squeezed light is transformed into an atomic quantum correlation between N atoms. If we regard the atomic state as a 1/2-spin and utilize the pair-wise spin entanglement to analyze the atomic quantum correlation, we can specify a certain entangled state of a spin pair as the signature of the squeezed light.

Keywords: Atomic ensemble quantum memory, Quantum correlation and entanglement, Squeezed light, EIT
PACS: 42.50.Dv, 03.67-a, 03.65.Ud

INTRODUCTION

Recent experiments based on the electromagnetically induced transparency (EIT) have made it possible for us to store a quantum state of signal light in an ensemble of atomic Λ-type three-level systems [1]. Just after the storing process, an atomic electron is coherently trapped in a certain superposition $a|1\rangle+b|0\rangle$ of the low-lying two states of the Λ-type system. We can retrieve the signal light from the ensemble by carrying out the inverse operation later.

If we consider the superposed state $a|1\rangle+b|0\rangle$ to be a 1/2-spin state, we can assign an appropriate quantum many-body spin system to the quantum state of the ensemble. If a phase or amplitude squeezed light is stored [2], therefore, the quantum correlation between photons is transformed into the quantum correlation between the spins of the atoms.

This paper clarifies the relationships between those two different kinds of the quantum correlations. First, we derive the two-point reduced density matrix ρ of an arbitrary spin pair belonging to the ensemble of N atoms. Then, on the basis of the framework for analyzing a pair-wise spin entanglement embedded in the density matrix ρ, we evaluate the concurrence $C(\rho)$ [3] and executed the Lewenstein-Sanpera (LS) decomposition for ρ [4]. The former quantifies the strength of the quantum correlation and the latter clarify the correlation form. As a result of the LS

CP1110, *Quantum Communication, Measurement and Computing (QCMC), The Ninth International Conference,* edited by A. Lvovsky

decomposition, we can identify a certain entangled state of a spin pair as the atomic signature of the squeezed light. The pair-wise entangled state specifies the intensity, phase and the squeeze parameter of the squeezed light stored in the ensemble.

TRANSFER OF PHOTONIC QUANTUM CORRELATION TO AN ATOMIC ENSEMBLE

The quantum state transfer from the signal light to the atomic ensemble is based on the adiabatic control of the dark-state polariton that is a coupled mode of the signal photon with an atomic electron trapped in the superposed state $a|1\rangle+b|0\rangle$. During the storing process, the relative weight of the photonic and atomic parts changes from 1:0 to 0:1 in the dark-state polariton. The initial quantum state of the whole system is a product state of an arbitrary photonic quantum state $\sum_{i=0}^{\infty} C_i\,|\,i\rangle_{\mathrm{PH}}$ of the signal light and the N-atomic state $|0;N\rangle$ that all atoms are in the spin down state $|0\rangle$. The final quantum state is a product state of the zero photon state $|0\rangle_{\mathrm{PH}}$ and the atomic quantum state $|\Psi\rangle_{\mathrm{AE}}=\sum_{i=0}^{\infty}(-1)^i C_i\,|\,i;N\rangle$. Here $|i;N\rangle$ is the symmetric entangled state of N spins that is given by the equally-superposition of the $N!/i!(N-i)!$-different states of arbitrary i up spins. The coefficient C_i is transformed from the number state $|i\rangle_{\mathrm{PH}}$ of photons to the symmetric state of N spins without disturbing the coherence [5].

The state $|\Psi\rangle_{\mathrm{AE}}$ must be a symmetric entangled state in general and is possible to be characterized with pair-wise quantum correlation characteristics. By taking the pair of spins A and B, we derive the density matrix $\rho_{(AB)} = \mathrm{Tr}_{\overline{AB}}\,(|\Psi\rangle\langle\Psi|)_{\mathrm{AE}}$, where the state $|\Psi\rangle_{\mathrm{AE}}$ suffers from the averaging for the quantum states of all the other spins. As the state $|\Psi\rangle_{\mathrm{AE}}$ is symmetric for an exchange of arbitrary spins, the density matrix is independent of the choice of the spin pair. We can calculate the density matrix ρ_{sq} of the atomic ensemble that stores a phase or amplitude squeezed state $|\xi\rangle$ of light if the coefficients C_i are knowledgeable for $|\xi\rangle$. To determine the coefficient, we can start from the following operational definition for the squeezed state $|\xi\rangle$ that is evolved from the initial coherent state $|\beta\rangle$; $|\xi\rangle= exp[R(\hat{a}^{+2}-\hat{a}^{2})]|\beta\rangle$. Here R means the squeeze parameter and the squeeze operation also results in amplification. The intensity of $|\xi\rangle$ is $|\alpha|^2$ if $\beta=\alpha e^{-2R}$. The detailed calculation of ρ_{sq} will be reported elsewhere [6].

For the coherent state $|\alpha\rangle$, we can easily show that the quantum state $|\Psi\rangle_{\mathrm{AE}}$ is represented by the product state of N parallel spins with the identical spin state; $-a|1\rangle+b|0\rangle$ provided that the atomic parameter $a^2(=\alpha^2/N)$ is far less than unity. The density matrix $\rho_{|\alpha\rangle}$ approaches to $|\Psi_0\rangle\langle\Psi_0|$, where $|\Psi_0\rangle=(-a|1\rangle+b|0\rangle)_A(-a|1\rangle+b|0\rangle)_B$. For the phase or amplitude squeezed state $|\xi\rangle$, the quantum state $|\Psi\rangle_{\mathrm{AE}}$ is an entangled state that is deviated from the product state $(-a|1\rangle+b|0\rangle)^{\otimes N}$. An intuitive consideration leads us to the above statement. The effect of the creation/annihilation operator on the photonic state is equivalent to the excitation/de-excitation in a 1/2-spin. As the squeeze operator $exp[R(\hat{a}^{+2} - \hat{a}^{2})]$ involves the creation/annihilation operator in a quadratic form, the spin state inversion occurs simultaneously for an arbitrary pair of the parallel spins. After all, the successive operation of $\hat{a}^{+2}$ and $\hat{a}^{2}$ changes the quantum state to an entangled state of N spins.

PAIR-WISE QUANTUM CORRELATION FORMED IN AN ATOMIC ENSEMBLE

Once the reduced density matrix ρ_{sq} is obtained for the atomic ensemble storing the squeezed state $|\xi\rangle$, we can evaluate the concurrence $C(\rho_{sq})$ to quantify the strength of the pair-wise quantum correlation formed in the ensemble. The concurrence is derived as follows as a function of the small squeeze parameter R and the number N of the atoms;

$$C(\rho_{sq}) \cong \frac{1}{N} 4|R| \left[1 - 2|R| + \frac{14}{3} R^2 - \frac{8}{3} |R|^3 + \cdots \right] . \quad (1)$$

The concurrence is independent of the atomic quantities $(a,b) = (\alpha/\sqrt{N}, \sqrt{1-\alpha^2/N})$ provided that the intensity α^2 of the squeezed light is far less than the number N of the atoms. This suggest that the concurrence relates only with the deformed quantum fluctuation characteristics around the classical point $(\alpha,0)$ in the q-p plane with regardless of the value of α. For the coherent state $|\alpha\rangle$, the concurrence asymptotically becomes zero for the limitation of $|R| \to 0$.

The pair-wise quantum correlation reflects all the properties of the squeezed light stored in the atomic ensemble. We can clarify this point by executing the Lewenstein-Sanpera (*LS*) decomposition [4] of the density matrix ρ_{sq}. The *LS* decomposition is the unique decomposition of the two-point reduced density matrix into the separable and inseparable parts. Suppose that we can find the possible maximal weight Λ for the separable part, we can determine the *LS* decomposition as follows by use of the normalized separable density matrix ρ_{sep} and the normalized entangled state $|\Psi e\rangle$ of a 1/2-spin pair;

$$\rho_{sq} = \Lambda \rho_{sep} + (1-\Lambda)|\Psi e\rangle\langle\Psi e|. \quad (2)$$

The pair-wise entangled state $|\Psi e\rangle$ determines the quantum correlation form of a spin pair formed in the atomic ensemble. All the properties regarding with the atomic quantum correlation must be involved in the inseparable part of the *LS* decomposition. Therefore, the weight $1-\Lambda$ and the pair-wise entangled state $|\Psi e\rangle$ reflect all the physical properties of the squeezed light.

In general, the normalized entangled state $|\Psi e\rangle$ can be expressed as $|\Psi e\rangle = s|11\rangle + t|10\rangle + u|01\rangle + v|00\rangle$. Moreover, the weight $1-\Lambda$ of the inseparable part can be related with the concurrence and $\{s,t,u,v\}$ by $1-\Lambda = C(\rho_{sq})/2|sv-tu| \leq 1$. Hence, our goal is to determine the set of $\{s,t,u,v\}$ with the normalization condition so that $|sv-tu|$ can be maximized. We can adopt the several physical constraints to simplify our problem. The derivation of the *LS*-decomposition will be reported elsewhere [6].

We here present our final result. We can obtain a pair of the simple expressions for the pair-wise entangled state $|\Psi e\rangle$;

$$
\begin{aligned}
|\Psi e\rangle &= (a^2+b^2x)|11\rangle - ab(1-x)(|10\rangle+|01\rangle) + (b^2+a^2x)|00\rangle \\
&\cong (1-x^2/2)\cdot(-a|1\rangle+b|0\rangle)_A(-a|1\rangle+b|0\rangle)_B + x\,(b|1\rangle+a|0\rangle)_A(b|1\rangle+a|0\rangle)_B
\end{aligned}
\tag{3}
$$

and the weight $1-\Lambda$;

$$1-\Lambda = C(\rho_{sq})/2|x| \ . \tag{4}$$

In the derivation, we assumed that α means the average amplitude of the squeezed light with the zero phase-factor ($e^{i\phi}=1$) in the q-p plane. The atomic quantities (a,b) are $(\alpha/\sqrt{N},\sqrt{1-\alpha^2/N})$ that are independent of the squeeze parameter R. Equation 3 reveals that the effect of the squeezing on the atomic quantum correlation can be characterized with the displacement parameter x.

The parameter x is an atomic quantity while the squeeze parameter R is an optical quantity. Those two different kinds of the parameters are related with

$$x = g(R)/N\,, \tag{5}$$

where

$$g(R) \cong 2R\cdot\left[\frac{1-2|R|(1-\sqrt{8|R|/3})+34R^2/3+\cdots}{1-2|R|+14R^2/3+\cdots}\right]\ . \tag{6}$$

The positive/negative squeeze parameter brings the phase/amplitude squeezed state. In the limitation of $|R|\to 0$, the parameter x also approaches to zero and the entangled state $|\Psi e\rangle$ becomes the product state $|\Psi_0\rangle$ that corresponds to the coherent state $|\alpha\rangle$ of light. As be described in the above, we can consider the pair-wise entangled state $|\Psi e\rangle$ to be a signature of the squeezed light stored in the atomic ensemble.

ACKNOWLEDGMENTS

The authors thank Dr. Y.Tokura for fruitful discussion.

REFERENCES

1. M.D. Lukin, *Rev. Mod. Phys* **75**, 457 (2003).
2. K.Honda, et al., *Phys. Rev. Lett.* **100**, 093601 (2008); J.Appel, et al., *Phys. Rev. Lett.* **100**, 093602 (2008) .
3. W.K. Wootters, *Phys. Rev. Lett.***80**, 2245 (1998).
4. M. Lewenstein and A. Sanpera, *Phys. Rev. Lett.***80**, 2261 (1998).
5. M. Fleischhauer and M.D. Lukin, *Phys. Rev. Lett.***84**, 5094 (2000).
6. K. Shimizu and D. Hashimoto, in preparation.

Simultaneous Slow Light Pulses With Matched Group Velocities via Double-EIT

A. MacRae, G. Campbell, A. Ordog and A.I. Lvovsky

Institute for Quantum Information Sciences, University of Calgary, Alberta T2N 1N4, Canada

Abstract. We demonstrate double electromagnetically induced transparency in rubidium vapor, in which two weak signal fields are simultaneously slowed or stored when a single pump field is applied. By manipulating the relative field strengths, we are able to control the group velocity of the pulses. In order to explore these effects, we develop a theoretical model of the atomic system used in the experiment.

Keywords: Electromagnetically Induced Transparency, double EIT, slow light, quantum memory
PACS: 01.30.Cc, 03.67.Hk, 42.50.Ex,

In recent years the electromagnetically induced transparency (EIT) effect [1] has become a workhorse for numerous methods of controlling optical fields, having applications ranging from precision interferometry to all-optical buffers in classical and quantum information networks. In double EIT (DEIT), a common pump creates EIT conditions for two independent signal fields, providing the possibility for coherent control [2, 3] and nonlinear interaction between weak optical fields [4, 5]. Allowing the propagation of two signal fields with minimal loss and increased interaction time makes DEIT a promising candidate for numerous applications in quantum computation and communication. A significant challenge in the experimental implementation of a DEIT system for the above applications is that the group velocities of the two signal pulses are generally not equal, significantly reducing their nonlinear interaction time.

In this paper we demonstrate a DEIT system and investigate the interdependency of the two signal fields in both continuous-wave (CW) and pulsed cases. In the CW case the effect of optical pumping due to the second signal field is investigated. We observe that it can be used to enhance the contrast of the transparency window. In the pulsed regime, we demonstrate that signal fields can be independantly slowed and stored. We demonstrate that the delay induced in the pulses can be adjusted and matched through preparation of the atomic states. In addition to quantum computational gates, this technique could find application in quantum communication protocols that use frequency multiplexed information channels, such as simultaneous quantum memory, simple qubit operations and correction of time delays [6].

The experimental setup is displayed in Fig. 1(a). We employed a magnetically shielded 12-cm vapor cell containing ^{87}Rb gas and 10 torr Ne buffer gas maintained at 45°C. The atomic level system was of tripod shape, formed by both hyperfine levels of the $5^2S_{1/2}$ shell as ground states and the $F = 2$ hyperfine state of the $5^2P_{1/2}$ shell as the excited state (Fig. 1(b)). The *pump* field was σ^+ polarized and coupled the $F = 2$ ground level to the $F' = 2$ excited level. The *Zeeman* signal field coupled the same levels as the pump, but was of σ^- polarization. The second signal field — the *hyperfine* field,

CP1110, *Quantum Communication, Measurement and Computing (QCMC), The Ninth International Conference,* edited by A. Lvovsky

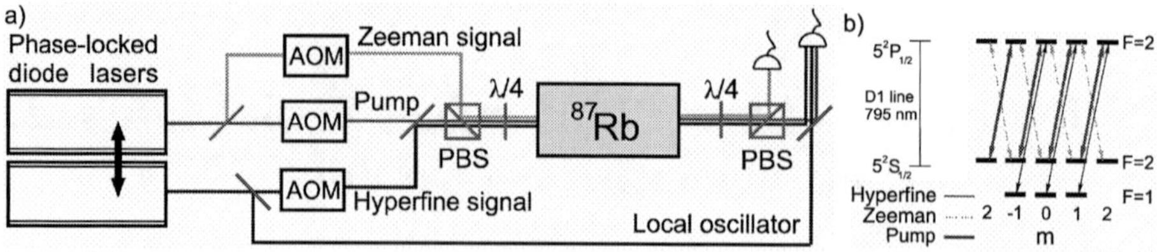

FIGURE 1. (a) Simplified experimental setup. (b) EIT dips for each field observed simultaneously.

was σ^+ polarized and coupled the $F = 1$ ground level to the $F' = 2$ excited level.

The fields were produced by two self-made external cavity diode lasers. In order that the two lasers acted coherently on the same set of atoms, they were phase locked so that their frequency difference remained fixed at 6.834 GHz. One of the lasers provided the pump and Zeeman fields; the other laser generated the hyperfine field and the local oscillator for its detection. The pump and each of the signal fields passed through acousto-optic modulators (AOMs) allowing them to be scanned in frequency, or switched on or off independently, which permits the creation of pulses. The beams were then spatially mode matched and passed through a quarter-wave plate to provide circular polarization. Typical laser beam power in this experiment amounted to 2.5 mW for the pump, and 1–150 μW for the signal fields. The beam diameters in the cell were about 750 μm. The Zeeman field was measured directly with a photodiode. Since the weak hyperfine field was in the same spatial and polarization mode as the strong pump, it could not be spatially separated. To measure this field, we employed heterodyne detection with a local oscillator differing in frequency by 200 MHz. The resultant beat frequency was observed with a spectrum analyzer in a time-resolved setting. In order to observe the EIT dip simultaneously for both signal fields in the CW regime, the pump field frequency was scanned, effectively varying over two-photon detuning for each signal field. A plot of double EIT is shown in Fig. 2(c).

In the CW case, there was a significant effect of one signal field on the other, even for moderate field strengths. Specifically, when the power of one signal field was increased, the transparency contrast in the other enhanced (Fig. 2). This can be explained by optical pumping from one ground state to the other. From the atomic level scheme (Fig. 1(b)), we see that when the hyperfine field is absent, the atomic population will collect in the $F = 1$ ground state. Turning on the hyperfine field will now serve to repopulate the $F = 2$ ground state, specifically the $m = 2$ sublevel, increasing the effective density of atoms experiencing EIT. Although the overall contrast of the EIT window improves in the presence of the repumping field, the transparency at the two-photon resonance also somewhat reduces [7].

In order to investigate the slow-light dynamics of the system, 1 μs pulses in the hyperfine and Zeeman fields were sent simultaneously through the atoms. While simultaneous group velocity reduction was noted, the group velocity of each pulse differed markedly. However, by preparing the atomic system with optical fields, we were able to manipulate the delay of the pulses. The initial preparation was performed by switching on one of the signal fields for some time prior to sending the pulses. Using a 500 μs preparation pulse in the hyperfine field and varying its power, we found that the group velocities were

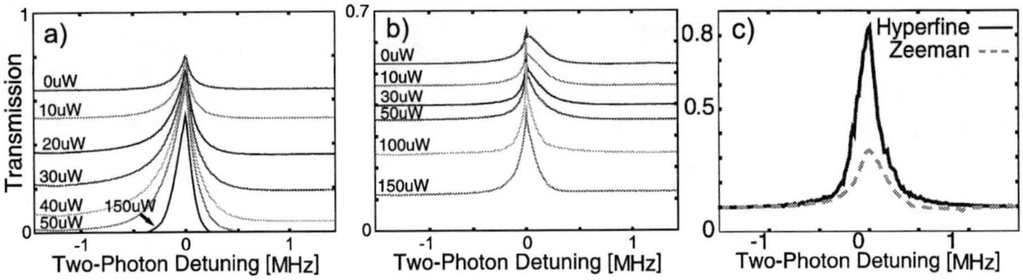

FIGURE 2. (Left) The effect of one signal field on the other in the DEIT system. In (a), the absorption spectrum of the hyperfine field, fixed at 50 μW, is measured for various powers of the Zeeman EIT signal. In (b), the Zeeman signal is of constant intensity and the hyperfine power is varied. The atomic level scheme is shown in (c)

matched at 135 km/s in the neighborhood of 50 μW preparation pulse power (Fig. 3). Conversely, the preparation pulse can be in the Zeeman field, having the effect of decreasing the group velocity of the hyperfine pulse. By using a strong (150 μW, 500 μs) preparation pulse, we were able to enhance the delay by an additional factor of 10, achieving a delay of several pulse lengths.

In order to gain understanding of our observations, we analyzed a model of a tripod atomic system consisting of a common excited state $|e\rangle$ and three ground states $|z\rangle$, $|h\rangle$, $|p\rangle$, coupled by two weak signal fields Ω_z, Ω_h, and a pump Ω_p respectively. The decoherence mechanisms included dephasing γ_{ij} (decay of the off-diagonal matrix elements) acting within each pair of ground states, and population exchange G between levels $|h\rangle$ and $|z\rangle$ (Fig. 3(a)). The Liouville equation was then solved in the steady state, noting that the time that the preparation pulse was off (10 μs) was much shorter than the relaxation time of the atomic system (measured to be on the order of several hundred microseconds). The group velocities of the pulses were then calculated using linear response theory with the initial state given by the steady state density matrix. The decoherence parameters of the system were set to fit the group velocity behavior to that observed experimentally. A good fit was obtained with a Doppler width of 500 MHz, $\Gamma_z = \Gamma_h = \Gamma_p = 6$ MHz, $\gamma_{hz} = \gamma_{hp} \sim 5$ kHz, $\gamma_{zp} \sim 40$ kHz, and $G = 50$ Hz. In the CW picture, our theoretical model also reproduces the enhancement of the transparency contrast for one signal field when increasing the strength of the other.

Of interest is the role of the population exchange rate G. As was found previously [7], the fraction of this mechanism in ground state decoherence is small compared to pure dephasing γ_{ij}. This finding is confirmed by the present experiment. Yet the population exchange mechanism cannot be completely neglected. If we set $G = 0$, the populations of states $|h\rangle$ and $|z\rangle$ would depend only on the ratio between Ω_h and Ω_z, but not on their absolute magnitude. The nonlinear effect of the signal fields on each other could then be observed at arbitrarily low field strength, which is not in agreement with our experimental observation. This discrepancy can be addressed by setting a small, but nonzero value of G, which governs the ground state populations at low signal intensities. Intuitively this can be understood by expressing the Zeeman and hyperfine ground states as linear combinations of "dark" and "bright" states: $|d\rangle = \Omega_h|z\rangle - \Omega_z|h\rangle$ and

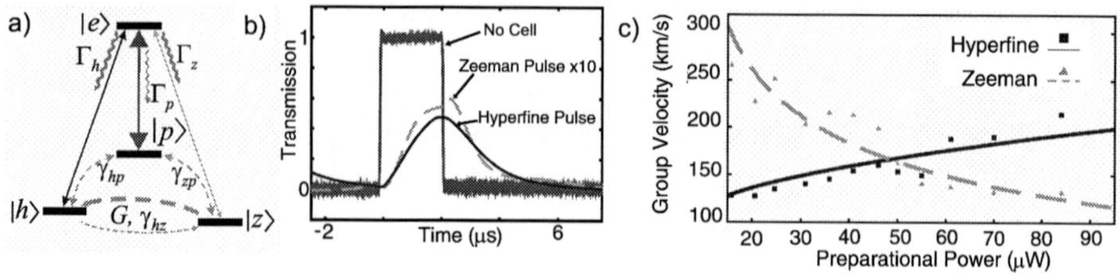

FIGURE 3. Group velocity matching. (a) The theoretical tripod scheme analyzed in the text. (b) Pulse waveforms with matched group velocities. The tail of the preparation hyperfine pulse (55 μW) is visible. (c) Group velocities of the co-propagating signal pulses as a function of the preparation power. Solid lines: theoretical model (see text).

$|b\rangle = \Omega_z|z\rangle + \Omega_h|h\rangle$, respectively. Of these only the bright state couples to $|e\rangle$. Hence, in the absence of population exchange between $|b\rangle$ and $|d\rangle$, the linear susceptibility is determined only by the composition of the bright state, i.e. the ratio between Ω_z and Ω_h.

By switching off the pump field during the propagation of the pulses, we were able to simultaneously store two pulses and retrieve them at a later time by switching the pump field back on. Storage times of more than 100 μs were observed, with decreasing pulse retrieval efficiency for longer times.

In summary, we have realized simultaneous group velocity reduction and storage for two signal fields using double EIT. Furthermore, we have demonstrated that the relative group velocities of the signal pulses can be varied and even matched. The observations are confirmed by a simple theoretical model. This development has potential applications in coherent control of multiple fields and increased interaction time for pulsed non-linear schemes. The observed simultaneous storage of two modes may find application in classical optical switching.

ACKNOWLEDGMENTS

We gratefully acknowledge K.-P.Marzlin and A. Gorshkov for fruitful discussions. This work was supported by NSERC, CIAR, AIF, CFI and Quantum*Works*.

REFERENCES

1. M. Fleishhauer, A. Imamoğlu, and J.P. Marangos, Rev. Mod. Phys. **77**, 633 (2005).
2. A. Imamoğlu M. D. Lukin, Nature **413**, 273–276 (2001).
3. F. Vewinger, J. Appel, E. Figueroa, and A. I. Lvovsky, Opt. Lett. **32**, 2771–2773 (2007).
4. K.-P. Marzlin, Z.-B. Wang, and B. Sanders, Phys. Rev. Lett. **97**, 063901 (2006).
5. X. Cao, C. Zang, C. Xie, H. Wang, S. Li, and X. Yang, Phys. Rev. Lett. **101**, 073602 (2008).
6. J. Khurgin, J. Opt. Soc. Am. B, **22**, 1062–1073 (2005).
7. E. Figueroa, F. Vewinger, J. Appel, and A.I. Lvovsky, Opt. Lett. **31**, 2625–2627 (2006).

CHAPTER 9

EXPERIMENTAL TOOLS

Detector imperfections in a recent scheme for Schrödinger-cat generation

Yuta Takahashi*,†, Shinichi Komatsu† and Shuichiro Inoue*

*Institute of Quantum Science, Nihon University, 1-8-14 Chiyoda-ku, Tokyo 101-8308, Japan
†Department of Applied Physics, Waseda University, 3-4-1 Shinjuku-ku, Tokyo 169-8555, Japan

Abstract. Nielsen and Mølmer have recently suggested a scheme transforming squeezed states into Schrödinger-cat-like states. Compared to previous schemes, this one is expected to generate cat-like states with larger amplitudes. Here, we analyze their scheme with consideration of experimental imperfections.

Keywords: Schrödinger-cat state, non-classical light
PACS: 42.50.Dv, 42.50.-p, 03.67.-a

INTRODUCTION

Schrödinger-cat states are well known to be useful in quantum information processing. The even (+) and odd (−) Schrödinger-cat states are defined as

$$|\psi_{\mathrm{cat}}^{\pm}\rangle \equiv \frac{1}{\sqrt{2(1\pm e^{-2|\alpha|^2})}}\left(|\alpha\rangle \pm |-\alpha\rangle\right). \tag{1}$$

Dakna *et al.* have shown that subtracting photons from a squeezed vacuum state can produce cat-like states [1], and some experimentalists have realized their scheme [2, 3, 4, 5]. In addition, Ourjoumtsev *et al.* have demonstrated a transformation of a two-photon Fock state into a squeezed Schrödinger-cat state [6]. Recently, Nielsen and Mølmer have suggested a new scheme to generate cat-like states with larger amplitudes [7]. In their scheme, the operations $\hat{O}_{2k} \equiv \prod_{i=1}^{k}(\hat{a}^2 - \beta_i^2)$ and $\hat{O}_{2k+1} \equiv \prod_{i=1}^{k}(\hat{a}^2 - \beta_i^2)\hat{a}$ applied to the squeezed vacuum generate the even and odd cat state, respectively. These operations can be experimentally realized by using auxiliary coherent states and beam splitters with high transmittivities $T_j = 1 - R_j$, as seen in the realization of $\hat{O}_3$ shown in Fig. 1. The cat state is then generated when all three photon-number resolving detectors (PNRDs) simultaneously detect exactly one photon.

Nielsen and Mølmer assumed the detectors to be ideal. In this work, we consider the effects of dark counts and non-unit quantum efficiency in order to evaluate the feasibility of their scheme. Our analysis indicates that the effects of dark counts are vanishingly small, but in the case of common single-photon detectors (SPDs), such as avalanche photodiodes, multiple photons cannot be resolved and drastically degrade the performance.

CP1110, *Quantum Communication, Measurement and Computing (QCMC), The Ninth International Conference*, edited by A. Lvovsky

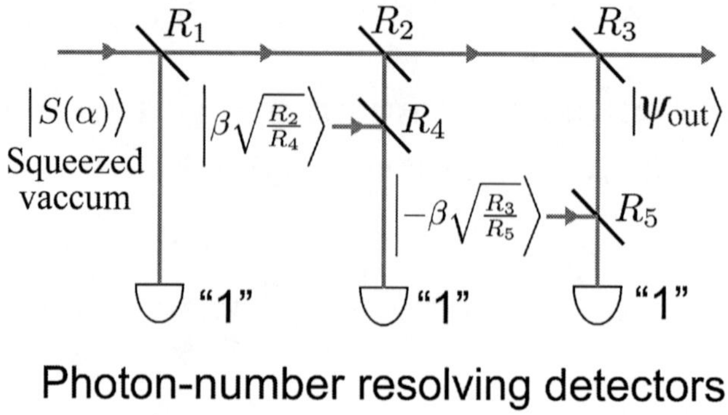

FIGURE 1. Experimental realization of the operation $\hat{O}_3$ employing high-transmittivity beam splitters and two coherent states.

CAT-STATE FIDELITY WHEN USING IMPERFECT DETECTORS

When the three PNRDs with perfect efficiency and without dark counts detect n_1, n_2 and n_3 photons, respectively, the output state can be expressed as

$$|\psi_{\text{out}}\rangle = e^{-\frac{\beta^2}{2}(R_2+R_3)} \sqrt{\frac{R_1^{n_1} R_2^{n_2} R_3^{n_3}}{T_1^{n_1} T_2^{n_1+n_2} T_3^{n_1+n_2+n_3} n_1! n_2! n_3!}} e^{\left(R_3 - \frac{R_2}{\sqrt{T_2}}\right)\frac{\beta^*}{\sqrt{T_3}}\hat{a}}$$

$$\times \left(\hat{a} - \sqrt{T_3}\beta\right)^{n_3} \left(\hat{a} + \sqrt{T_2 T_3}\beta\right)^{n_2} \hat{a}^{n_1} \sqrt{T_1 T_2 T_3}^{\hat{a}^\dagger \hat{a}} |S(\alpha)\rangle . \qquad (2)$$

In Eq. (2), we approximated the auxiliary coherent state $|\beta\sqrt{R_2/R_4}\rangle$ and the beam splitter with transmittivity T_4 by the displacement operator $\hat{D}(\sqrt{R_2}\beta) = \exp\left[\sqrt{R_2}\left(\beta\hat{a}^\dagger - \beta^*\hat{a}\right)\right]$. Therefore, the overlap between the generated state and the cat state when using PNRDs with perfect efficiency but non-negligible dark-count probability p_d can be written as

$$F_{\text{cat}}^{\text{PNRD}} = \frac{\sum_{n_1,n_2,n_3=0}^{1} p_d^{3-n_1-n_2-n_3} F_{n_1 n_2 n_3}}{\sum_{n_1,n_2,n_3=0}^{1} p_d^{3-n_1-n_2-n_3} P_{n_1 n_2 n_3}}, \qquad (3)$$

where

$$F_{n_1 n_2 n_3} \equiv |\langle \psi_{\text{cat}}^{-} | \psi_{\text{out}} \rangle|^2, \qquad (4)$$

$$P_{n_1 n_2 n_3} \equiv \langle \psi_{\text{out}} | \psi_{\text{out}} \rangle . \qquad (5)$$

Similarly, the cat-state fidelity when using SPDs with non-unit quantum efficiency and dark counts is

$$F_{\text{cat}}^{\text{SPD}} = \frac{\sum_{n_1,n_2,n_3=0}^{\infty} \gamma_{n_1} \gamma_{n_2} \gamma_{n_3} F_{n_1 n_2 n_3}}{\sum_{n_1,n_2,n_3=0}^{\infty} \gamma_{n_1} \gamma_{n_2} \gamma_{n_3} P_{n_1 n_2 n_3}} \qquad (6)$$

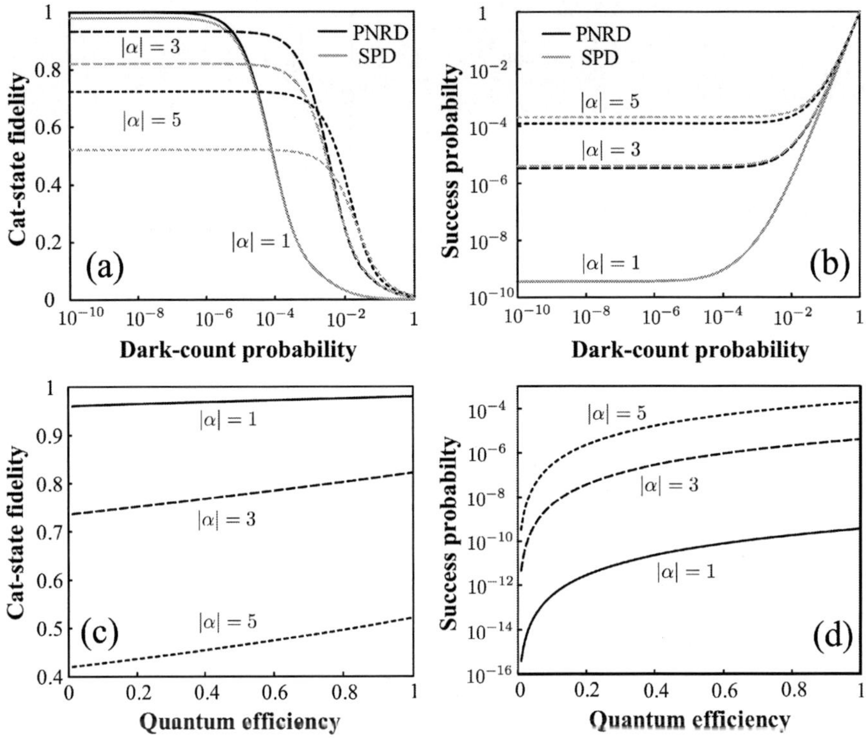

FIGURE 2. Dependence of (a) the fidelity and (b) the success probability on the dark-count probability for $|\alpha| = 1,3,5$. Effects of the quantum efficiency of the SPDs on (c) the fidelity and (d) the success probability for $|\alpha| = 1,3,5$ when the dark-count probability is 10^{-9}.

with

$$\gamma_n = 1-(1-\eta)^n + p_d(1-\eta)^n, \tag{7}$$

where η is the quantum efficiency of the SPDs.

ANALYSIS OF THE GENERATED STATE

In Figs. 2(a) and 2(b), we have plotted our exact results for PNRDs and SPDs with perfect efficiency, but non-negligible dark-count probabilities. For $|\alpha| = 3$, the probability that one of the three detector clicks is fake is $P_{011} + P_{101} + P_{110} \approx 9\times 10^{-4}$. For this reason, a fidelity of $F_{\text{cat}}^{\text{PNRD}} > 0.9$ is seen to require a dark-count probability of less than 10^{-3}. We note that the success probability, i.e., the probability for all detectors to indicate the presence of one photon, increases with the dark-count probability. The experimental parameters corresponding to $|\alpha| = 1,3,5$ are listed in Table 1. These values were calculated using Eqs. (6) and (7) in Ref. [7]. The fidelity of $F_{\text{cat}}^{\text{SPD}}$ becomes lower

TABLE 1. The experimental parameters

$\|\alpha\|$	**Squeezing Level (dB)**	β	$T_1\&T_2\&T_3$
1	1.1	0.47	0.99
3	6.7	1.42	0.99
5	10.9	2.37	0.99

than $F_{\rm cat}^{\rm PNRD}$ due to the detection of multiple photons. In order to prevent this degradation, we must use beam splitters with higher transmittivities.The dependence of the fidelity and success probability on the quantum efficiency of the SPDs are presented in Figs. 2(c) and 2(d). Figure 2(c) shows that for $|\alpha| = 1$ the experimental values required to obtain a fidelity of $F_{\rm cat}^{\rm SPD} > 0.9$ are easy to realize with detectors in the visible range.

CONCLUSION

We have analyzed the scheme proposed by Nielsen and Mølmer with consideration of the effects of dark counts and non-unit quantum efficiency. In conclusion, their scheme is lightly affected by dark counts. However, the effect of multiple photons coming to SPDs is considerably large, so PNRDs with close to perfect efficiency are required in order to produce cat-like states with larger amplitudes. In future work, we will consider the loss of the input squeezed states and the imperfections of PNRDs, i.e., dark counts, non-unit quantum efficiency and photon-number resolution.

ACKNOWLEDGMENTS

YT thanks Jonas Söderholm for his valuable comments and suggestions. This work was supported by the National Institute of Information and Communications Technology.

REFERENCES

1. M. Dakna, T. Anhut, T. Opatrný, L. Knöll, and D.-G. Welsch, Phys. Rev. A **55**, 3184 (1997).
2. A. Ourjoumtsev, R. Tualle-Brouri, J. Laurat, and P. Grangier, Science **312**, 83 (2006).
3. J. S. Neergaard-Nielsen, B. Melholt Nielsen, C. Hettich, K. Mølmer, and E. S. Polzik, Phys. Rev. Lett. **97**, 083604 (2006).
4. K. Wakui, H. Takahashi, A. Furusawa, and M. Sasaki, Opt. Express **15**, 3568 (2007).
5. H. Takahashi, K. Wakui, S. Suzuki, M. Takeoka, K. Hayasaka, A. Furusawa, and M. Sasaki, quant-ph/0806.2965 (2008).
6. A. Ourjoumtsev, H. Jeong, R. Tualle-Brouri and P. Grangier, Nature **448**, 784 (2007).
7. A. E. B. Nielsen and K. Mølmer, Phys. Rev. A. **76**, 043840 (2007).

Multi-channeled NbN superconducting single photon detectors (SSPDs) system with NbN meander nanowires

Mikio Fujiwara[a], Shigehito Miki[b], Masahide Sasaki[a], and Zhen Wang[b]

[a]*National Institute of Information and Communications Technology, 4-2-1 Nukui-Kitama, Koganei, Tokyo 184-8795, Japan..*
[b]*Kobe Advanced Research Center, National Institute of Information and Communications Technology, 588-2, Iwaoka, Nishi-ku, Kobe, Hyogo 651-2492, Japan*

Abstract. A superconducting single photon detector (SSPD) is promising candidate of the detector in a quantum key distribution (QKD) system, because of its low dark count and high speed repetition rate. We have developed the SSPD system cooled by a GM cryocooler. In this system, and the work surface can be cooled 2.95 K and up to 6 SSPDs can be installed. The active areas of SSPDs are 10×10 μm^2 or 20×20 μm^2, and the system detection efficiency at dark count rate of 100 Hz reached 2.6% at a wavelength of 1550 nm.

Keywords: NbN, superconducting single photon detector, GM cryocooler
PACS: 74.25.Gz

INTRODUCTION

The Quantum Key Distribution (QKD) allows two users, Alice and Bob, to communicate in absolute security. Unconditional security is guaranteed by the fundamental law of physics. BB 84[1] is the most famous protocol of QKD, and in such system, signals originated from single photon pulses emitted by Alice are guaranteed to be secure. Therefore, the key devices for QKD are a single photon emitter and a single photon detector at telecom wavelength. Especially, low dark count rate and high detection efficiencies of detectors are required for extending QKD at real field, because there is a powerful method to supplement of the imperfection in photon sources. The decoy state QKD[2] allows us to extend distribution length even using imperfect single photon source i.e. weak coherent pulses. We can achieve almost the same level of an ideal single photon source if we use decoy state. The dark count rate of a single photon detector limits transmission distance of QKD. Representative single photon detectors for telecom wavelengths are listed here. These semiconductor devices are commercial available. Especially the InGaAs APD[3] is usually used for QKD experiments. But this detector has large dark count and after pulse phenomenon and those phenomena limit the repetition rate of detection. Moreover, the InGaAs APD must be used in the gated Geiger mode. This mode demands precious synchronization between Alice and Bob. On the other hands, superconducting single

CP1110, *Quantum Communication, Measurement and Computing (QCMC), The Ninth International Conference,* edited by A. Lvovsky

photon detectors (SSPD)[4] have low dark count rate and no-gated operation can be done. Actually SSPDs have been applied for several QKD tests. In this paper, we report the performance of our NbN SSPD system installed in a GM cryocooler.

Experimental setup

We have developed a method for growing about 4 nm thick NbN epitaxial films directly on a single crystal MgO substrate by reactive dc magnetron sputtering at room temperature. These films show T_c ~12 K and a lower resistivity at 20 K (ρ_{20K}) ~ 120 μΩ-cm even for the thin films of 4.2 nm thickness. Using these high quality films we fabricated 100 nm wide meander lines covering a relatively large area of 10 × 10 and 20 × 20μm^2 with filling factors of 50%. NbN thin films are set in a mount block consisting of a pair of oxygen-free copper plates: the one is used to fix the device and the other has a capillary of a single mode fiber (SMF). An optical spot on the SSPD is aligned to the center of the active area preciously by checking the position from the backside of the thin film. The distance between a SMF and a device is less than 70μm at cryogenic temperature. Up to six SSPD mounts can be installed in a cryocooler, and brass semi-rigid coaxial cables are connected to the each package. The output port from the SSPD is connected to a parallel shunt resistance (50 Ω) and a bias tee at room temperature through a 50 Ω co-axial cable. The shunt resistance is placed outside of the cryocooler to allow back-reflections to be attenuated. The SSPD is current-biased via the dc arm of the bias tee. The output signal from the ac arm of the bias tee is amplified using a series of two low noise amplifiers (RF bay Inc., LNA 500 and LNA 1000). In this system, the detector mounts are cooled at 2.95 K. The picture of the GM cryocooler and the optical spot on the SSPD are shown in Fig. 1(a) and (b) respectively. In next section, we report the performance of this system. We measured single photon detection efficiencies (DEs) and dark count (DC) rates at telecom wavelength. The DEs included the efficiency of the optical coupling.

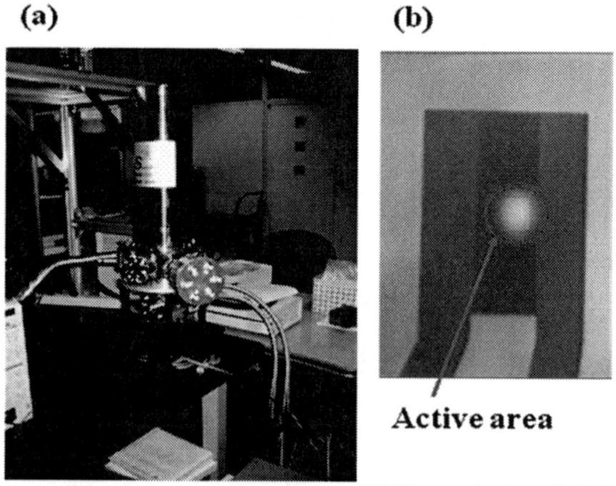

FIGURE 1. (a) Picture of the GM cryocooler. Up to 6 SSPDs can be installed. (b) The optical spot (1550 nm wavelength) on the SSPD through an MgO substrate. The optical spot is aligned on the center of the meander structure.

Results and discussion

The dc characteristics of the NbN nanowires are listed in TABLE. 1. Compare with our previous work (Ref [5]), the newly-formed devices have lower critical temperature (T_c) and critical current (I_c). Reasons include the thickness of newly-formed film is thinner than that of the previous one. Moreover, resistances at 20K and kinetic inductance (Lk) of newly-formed devices became higher. We also measured DEs and compared with that of previous ones. However, the counting rate of a 10 × 10μm^2 device is expected as 113 MHz.

TABLE 1. Nanowire design, dc characteristics, and measured device performance for two (plus one referred in [9]) typical SSPD designs. SSPD #A and #B have large and small active areas of 20 × 20 μm^2 and 10 × 10 μm^2.

	Ref [5]	#A	#B
Device size (μm)	20 × 20	20 × 20	10 × 10
Line width (nm)	100	100	100
Line pitch (nm)	200	200	200
films thickness (nm)	4.2	<4	<4
Tc	12	9.8	9.5
Ic	32	19	15
R20K	7	~10	~2.5
System DE (%)	0.4-3.5 at 2.5K	0.8-2.6 at 2.95 K	2.6 at 2.95 K
Lk, (μH)	1.0	2.0	0.44

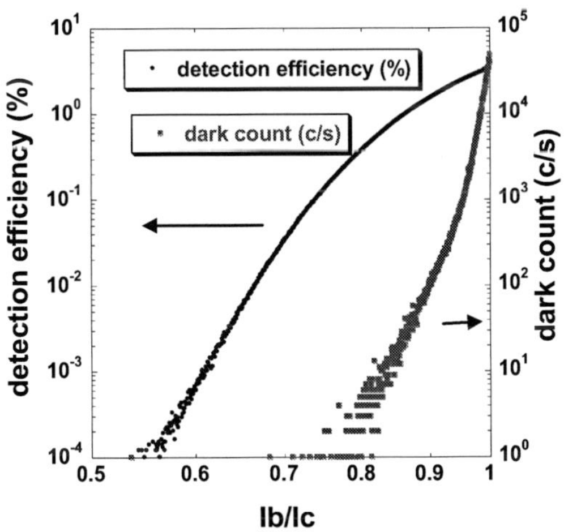

FIGURE 2. Detection efficiency (%) and dark count rate (c/s) of the typical device with active area of 10 × 10 μm^2 as function of bias current normalized by critical current.

Figure 2 shows the DE and DC rate of the typical sample with $10 \times 10\mu m^2$ active area as functions of a bias current. The bias current (Ib) is normalized by the critical current (Ic). The theoretical estimation predicts the coupling efficiency is more than 90% even if the active area is $10 \times 10\mu m^2$. The DE curve shows saturation in Fig 2. This result implies the uniformity of the nanowire. Histograms of DEs at DC of 100 c/s are summarized in Fig. 3(a) and (b). All $10 \times 10\mu m^2$ samples have DEs more than 1%. These results mean well reproducibility of the nanowire process, and good qualities of the NbN thin films. The thickness of films become thinner (< 4 nm), however, DEs are almost same compare with 4.2 nm devises.

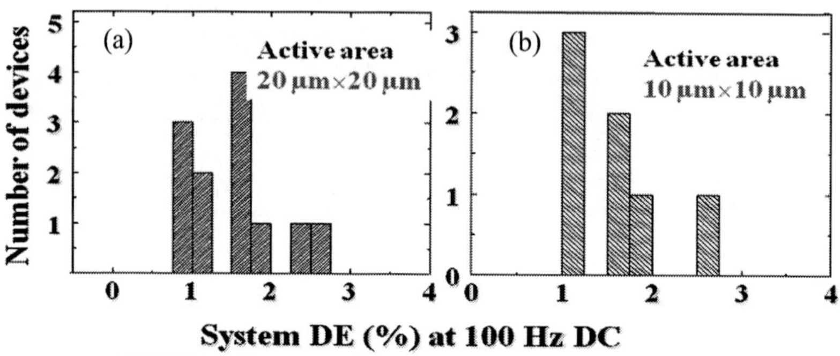

FIGURE 3. Histogram of system DE at DC of 100 c/s with active areas of (a) 20×20 μm^2, and (b) 10×10 μm^2.

SUMMARY

We report the system performance of the SSPD system cooled by a GM cryocooler. In most cases, we obtain DEs more than 1% at DC rate of 100 c/s at telecom wavelength. We will apply this system for QKD field tests and other quantum measurements.

ACKNOWLEDGMENTS

We would like to thank S. W. Nam, B. Baek at National Institute of Standards and Technology for helpful advices and discussions to develop the SSPD system.

REFERENCES

1. C. H. Bennett and G. Brassard, in *Proceedings of IEEE International Conference on Computers, Systems, and Signal Processing*, Bangalore, India, 1984 (IEEE, New York, 1984), pp. 175-179.
2. H. K. Lo, X. Ma, and K. Chen, Phys. Rev. Lett. **94**, pp. 230504 1-3 (2005).
3. A. Tomita and K. Nakamura, Opt. Lett. 27, pp. 1827-1829 (2002).
4. G. N. Gol'tsman, O. Okunev, G. Chulkova, A. Lipatov, A. Semenov, K. Smimov, B. Voronov, A. Dzardanov, C. Williams, and R. Sobolewski, Appl. Phys. Lett. 79, pp.705-707 (2001).
5. S. Miki, M. Fujiwara, M. Sasaki, B. Baek, A. J. Miller, R. H. Hadfield, S. W. Nam, and Z. Wang, Appl. Phys. Lett. vol. 92, pp.061116 1-3. (2008).

High Time Response and Low Dark Count Rate of Multipixel APD Operated at Liquid Nitrogen Temperature

Kenji Tsujino, Makoto Akiba and Masahide Sasaki

National Institute of Information and Communications Technology, New Generation Network Research Center, 4-2-1 Nukui-Kitamachi, Koganei, Tokyo 184-8795 Japan.

Abstract. Multipixel silicon avalanche photodiodes (Si APDs) are novel single-photon detection devices with a high repetition rate and the capability of determining the number of photons. A multipixel Si APD was operated in the Geiger-mode at liquid nitrogen temperature and it was observed to exhibit a time response of 2 ns and a dark count less than 0.2 counts/s. This device can be used in many experiments in optical quantum information science.

Keywords: Multipixel APD, Photon detector, Low temperature
PACS: 07.20.Mc

INTRODUCTION

Single-photon detection technology is widely used in optical quantum information science. Further advances in quantum information and communications technologies require devices with high photon detection efficiencies and low dark counts and capable of handling high count rates. In particular, the requirement relating to the count rate and dark count are important for fast quantum key distribution.

Recently, multipixel silicon avalanche photodiodes (Si APDs) (e.g., silicon photomultipliers (Si PM) and multipixel photon counters (MPPCs)) have been developed, and their photon detection capability has been investigated [1, 2]. Eraerds et al. demonstrated the capability of the photons to count at a speed as much as 430 MHz using a Si PM [3]. However, they observed the dark count to be large (more than 10 kHz). Dark count is generally low at liquid nitrogen temperature [4]. Hence, in this study, we cooled a multipixel Si APD to 78.5 K and then studied its dark count rate, afterpulse, photon detection efficiency, and photon counting ability.

EXPERIMENTAL SETUP AND TIME RESPONSE

We used a multipixel APD (detector) manufactured by Hamamatsu Photonics (Multi-Pixel Photon Counter, MPPC, S10362-11-100U). The pixel size and the number of pixels were 100 μm × 100 μm and 100, respectively, and the effective area of the APD was 1 mm^2. Figure 1 shows the experimental setup. The detector was

CP1110, *Quantum Communication, Measurement and Computing (QCMC), The Ninth International Conference,* edited by A. Lvovsky

placed in a liquid nitrogen cryostat. The temperature of the detector was 78.5 K. We used two different light sources in our experiments. One was a UV pulsed laser diode (Hamamatsu Photonics, PLP10-040C) with a wavelength of 407 nm and a pulse width of 75 ps, and the other was a light-emitting diode (LED) with a wavelength of 450 nm. Light from each source was focused on the APD through neutral density filters and a single-mode fiber. Output signals from the detector were amplified by using a voltage amplifier with a gain of 40 dB. The signals were then passed through a high-pass filter before being recorded by a high-speed oscilloscope (Lecroy SDA3000A).

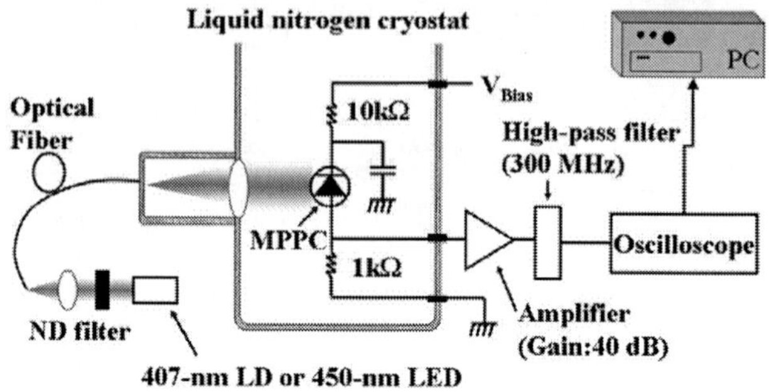

FIGURE 1. Schematic diagram of the apparatus used to measure the characteristics of the MPPC.

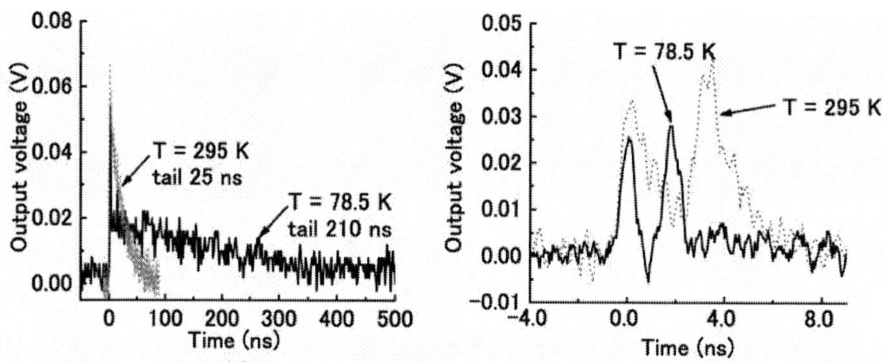

FIGURE 2. (a) Output signals from the MPPC when no high-pass filter is used. (b) Output signals from the high-pass filter when two photons happened to be incident successively.

Figure 2 shows the time response of our system. We first measured the time response without using the high-pass filter (Figure 2a). The time response was found to increase from 25 ns at 295 K to 210 ns at 78.5 K. However, it was observed that the cooled APD showed a rapid fall after the peak. Then, we passed the output signals through a high-pass filter and measured the time response (Figure 2b). As can be seen in the figure, two pulses separated by 2 ns can be distinguished at 78.5 K. All the following measurements were carried out using the high-pass filter.

DARK COUNT RATE, AFTERPULSE PROBABILITY, AND PHOTON DETECTION EFFICIENCY

To measure the dark count rates, the optical window of the cryostat was closed. Figure 3 shows the dark count rates that include afterpulses as a function of the bias voltage. The dark count rates increased rapidly when the bias voltage exceeded 60 V at 78.5 K. Hence, we fixed the variable range of the bias voltage to be between 59 V and 60 V.

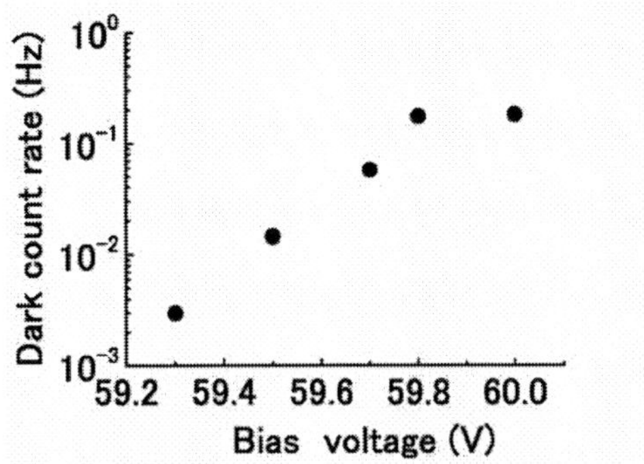

FIGURE 3. Dark count rate as a function of the bias voltage at 78.5 K.

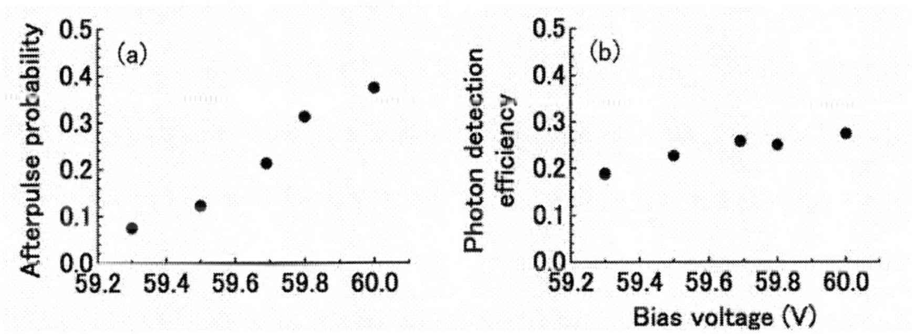

FIGURE 4. (a) Afterpulse probability and (b) photon detection efficiency as a function of the bias voltage at 78.5 K.

For the above-mentioned range of the bias voltage, we determined the afterpulse probability and the photon detection efficiency. The afterpulse probability was determined by using a correlation function. Figures 4(a) and (b) show the afterpulse probability and the photon detection efficiency as a function of the bias voltage. The change in afterpulse probability with the bias voltage is larger than that of photon detection efficiency. Thus, we determined the bias voltage for the minimum dark count rate and the minimum afterpulse probability.

COUNTING ABILITY

In order to investigate the counting ability of the detector, we used a CW light source. Figure 5 shows count rate as a function of the incident photon number. The maximum count rate was 30 MHz in the previous APD. This device has a high time response of 2 ns; however, the count rate is not very high. In our view, the count-rate saturation might have occurred because a stream of photons was continuously incident on all the pixels. The use of a large number of pixels in the device might lead to a high counting rate.

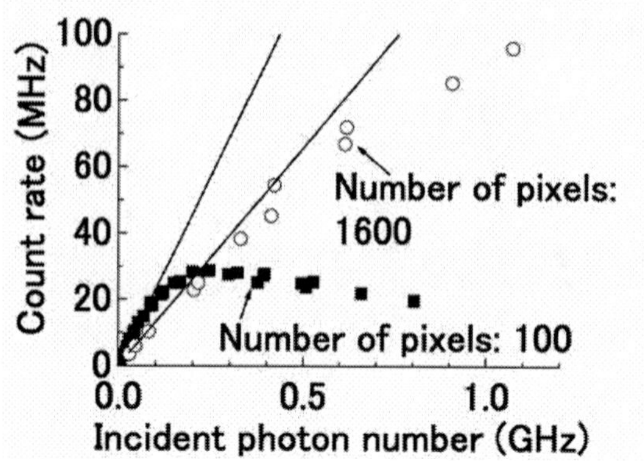

FIGURE 5. MPPC response to a CW source. The maximum count rate was up to 30 MHz for the previous APD. A device with a large number of pixels exhibited a high counting rate, more than 100 MHz.

CONCLUSION

We cooled a multipixel APD at 78.5 K and evaluated its characteristics as a photon detector. We observed that this device showed a dark count rate of less than 0.2 cps, detection efficiency of more than 18%, afterpulse probability of more than 0.08, time response of 2 ns, and count rate up to 30 MHz. Multipixel APDs are promising devices for use in experiments in quantum optics and optical quantum information science.

ACKNOWLEDGMENTS

We would like to thank E. Sasaki for technical support. We also thank K. Sato for useful discussions.

REFERENCES

1. P. Buzhan, B. Dolgoshein, L. Filatov, A. Ilyin, V. Kantserov, V. Kaplin, A. Karakash, F. Kayumov, S. Klemin, E. Popova, and S. Smirnov, "Silicon photomultiplier and its possible application," Nucl. Instr. and Meth. A **504**, 48–52 (2003).
2. S. Castelletto, I. P. Degiovanni, V. Schettini, and A. Migdall, "Reduced deadtime and higher rate photon counting detection using multiplexed detector array," J. Mod. Opt. **54**, 337–352 (2007).
3. P. Eraerds, M. Legre, A. Rochas, H. Zbinden, and N. Gisin, "SiPM for fast photon-counting and multiphoton detection," Opt. Express **15**, 14539–14549 (2007).
4. K. Tsujino, M. Akiba, and M. Sasaki, "Ultralow-noise readout circuit with an avalanche photodiode: Toward a photon-number-resolving detector," Appl. Opt. **48**, 1009–1014 (2007).

Telecom-band single-photon detection with a gigahertz repetition frequency using a sinusoidally gated avalanche photodiode

S. Adachi[1], N. Namekata[1], and S. Inoue[1]

[1]*Institute of Quantum Science, Nihon University, 1-8-14, Kanda Surugadai, Chiyoda-ku, Tokyo 101-8308, Japan.*

Abstract. Single-photon detection at 1550 nm with a high repetition rate was realized using an InGaAs/InP avalanche photodiode operated with a sinusoidal gating. Removing the AC noise due to the transferred gate signal by use of band elimination filters, it is realized to detect much smaller avalanche signals than that in the conventional gating scheme, which results in the suppression of the afterpulsing. Finally, we achieved the detection efficiency of 11.0 %, the dark count probability per gate of 7.7×10^{-7}, and the afterpulsing probability was 3.3 % at the repetition frequency of 1.2 GHz and the APD operating temperature of -50 degrees Celsius.

Keywords: single-photon detector, sinusoidal gating, telecom-band, afterpulsing.
PACS: 03.67.Hk

INTRODUCTION

A single-photon detector (SPD) is an essential component for the optical quantum information processing technologies. In the quantum communication, especially the quantum key distribution (QKD) based on the optical fiber links [1, 2], it is important to develop the SPD at the telecommunication wavelength (1550 nm) with a high quantum efficiency (*QE*) and low dark counts. Moreover, in practical use, the SPD is required to operate at a high repetition rate. Avalanche photodiodes (APDs) in geiger mode have been widely used for the SPD. A photo-excited carrier grows into a macroscopic current output via the carrier avalanche multiplication in the APD. Fractions of the many carriers trapped in the APD are subsequently emitted, and trigger additional avalanches. The phenomenon is well known as "afterpulsing" which disturbs single-photon detection. The candidate for the SPD at 1550 nm is an InGaAs/InP-APD. However, the APD in Geiger mode is very noisy, resulting from the afterpulsing and the dark counts caused by the thermal carriers. Therefore, the APD is generally cooled and operated with a gated mode. The gate duration (gate-on time) is generally set to a few nanoseconds in order to reduce the dark count probability per gate. Then the interval between a gate and the subsequent one is set to more than the lifetime of the trapped carriers so that the afterpulsing is suppressed. Using the cooled InGaAs/InP-APD operated with the short gated mode, the quantum efficiency at 1550 nm have reached 10 ~ 30% with the dark cont probability per gate (P_d) of $10^{-6} \sim 10^{-4}$ [3-7]. However, the repetition frequency of the gate, which allows the afterpulsing-

CP1110, *Quantum Communication, Measurement and Computing (QCMC), The Ninth International Conference,* edited by A. Lvovsky

free single-photon detection, was limited to only less than 1MHz because of a long lifetime of the trapped carriers.

The other potential SPD at telecommunication band are frequency-up-conversion-assisted Si-APD (up-conversion detector) [8] and superconducting-single-photon-detector (SSPD) [2, 9]. These detectors don't require the gating operation, and can be applied to the gigahertz clock systems. However, up-conversion detector is very noisy when *QE* set to high. As regards SSPD, it requires cryogenic environment (less than 4K).

Recently, high-speed SPDs based on InGaAs/InP-APD were reported. We developed the sinusoidally gated InGaAs/InP-APDs (SG-APD) [10] that achieved the gating repetition frequency in range of sub-gigahertz. Subsequently, the self-differential APD [11] achieved the gigahertz gating repetition frequency. These schemes enable us to detect extremely small avalanche signals, which results in suppression of afterpulsing. Then these schemes have possibility to operate the APD at a gating repetition frequency that was essentially determined by APD's response speed with low afterpulsing probability. However, as regards SG-APD, such intrinsic maximum gating repetition frequency has not been investigated.

In this paper, we show the recent progress of the SG-APD (by optimizing electronics), and demonstrate that the SG-APD can be operated with a gating frequency greater than 1 GHz. By use of a high-speed electronics and a discriminator that can discriminate a subnanosecond pulses, the undetected avalanche signal drastically decreased.

EXPERIMENT

The experimental setup for the measurements of quantum efficiency (*QE*), dark count probability (P_d), and afterpulsing probability (P_a) is shown in Fig. 1-(a). The tested InGaAs/InP-APD (Princeton-Lightwave, AGD-25-SE-1-T8) is mounted on the three stages Peltier cooler, and they are housed in TO-8 can package. The APD was cooled to maximally -50 degrees Celsius by the Peltier cooler. The APD module was used as the single-photon detection element after it was fiber-pigtailed. In order to

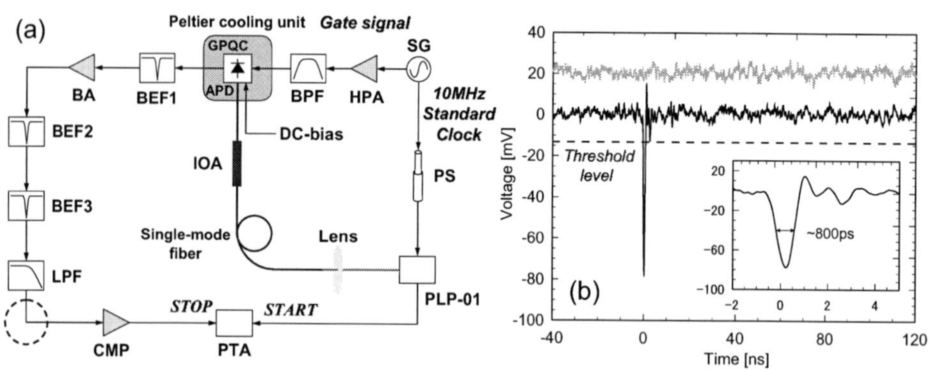

Figure 1: (a) The experimental setup for the measurements of quantum efficiency, dark count probability, and afterpulsing probability. (b) The oscilloscope trace of output signal from LPF (The dashed circle in (a)).

supply an AC voltage (gate voltage) superposed on the DC reverse bias voltage to the APD, we used the gated passive quenching circuit (GPQC) [10]. A sinusoidal voltage was generated by the signal generator (SG: Agilent N5181A). The sinusoidal voltage was used as the gate voltage after the amplification by the high power amplifier (HPA, ALDETEC, ALP-0520S430, BW: 0.5-2.0GHz). The band-pass-filter (BPF) rejected amplified sideband noises and the other redundant signal, which contributes to reduce the noise level of the GPQC output. The laser was triggered by the 10 MHz standard clock signal that was synchronized with sinusoidal gate voltage. Laser pulse delay was adjusted by the phase shifter (PS). The laser (HAMAMATSU, PLP-01) has a central wavelength of 1550 nm, and its pulse width was 50 ps. The optical pulse was led to the APD after it was attenuated to 0.1 photon / pulse by the inline-optical-attenuator (IOA). The GPQC output was amplified by the broadband amplifier (BA, R&K, AA040, BW: 3.0 GHz, gain: 40 dB), and transferred gate signal from GPQC was eliminated by three band-elimination-filters (BEF1-3). The total elimination ratio of the BEFs was 100 dB at a frequency ω_g of sinusoidal voltage. Then, BEFs distorted the avalanche pulse. The distortion degraded the time jitter of avalanche signal discrimination. Therefore, the distorted pulse form was fixed by low-pass-filter (LPF). The oscilloscope trace of output signal from LPF is shown in Fig. 1-(b). The black line shows the signal with avalanche signal. The gray line (20 mV offset is given) shows the signal without avalanche signal. The dashed line denotes threshold level. As shown in the figure, the transferred gate signal was eliminated completely. The inset in Fig. 1-(b) shows the avalanche signal in fast time scale. The width of avalanche signal (FWHM) is about 800 ps. The avalanche signal was discriminated by the ultra-high-speed comparator (CMP, MAXIM, MAX9600) circuit. Here, the noise level of output from LPF was less than 10 mV. Therefore, the threshold level is set to -16 mV. The comparator output is digital output in emitter-coupled-logic (ECL) level. The time histogram of detection events was measured by the picosecond time analyzer (PTA, ORTEC 9308).

Figure 2-(a) shows the relation between QE and P_d when ω_g was 1.0 and 1.2 GHz. The relation between QE and P_a was also plotted in the same figure. The APD operating temperature T_o was -50 degree Celsius. We employed a 400 ps time window to evaluate our SPD performances. When the ω_g was increased from 1.0 to 1.2 GHz,

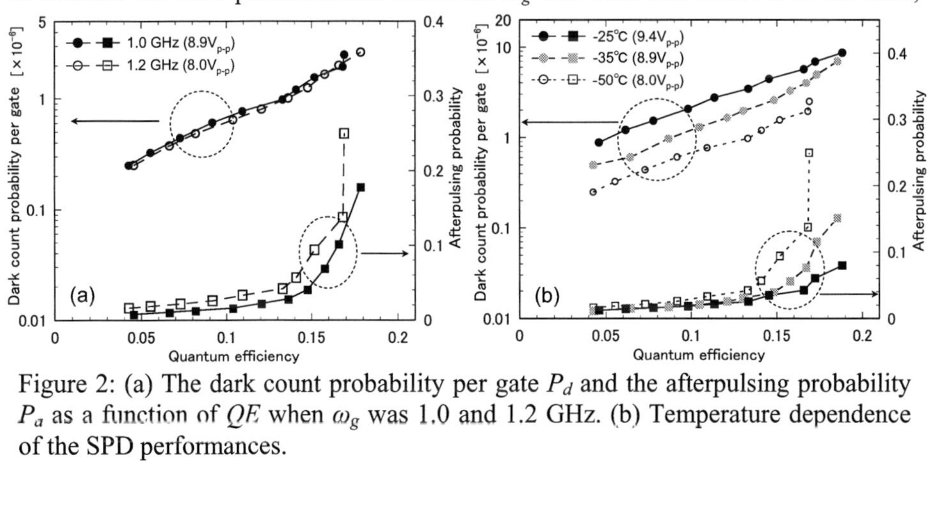

Figure 2: (a) The dark count probability per gate P_d and the afterpulsing probability P_a as a function of QE when ω_g was 1.0 and 1.2 GHz. (b) Temperature dependence of the SPD performances.

higher, the P_d slightly decreased with increasing ω_g. Temperature dependence of SPD performances is shown in Fig. 2-(b). Here, ω_g was fixd at 1.2 GHz. As the APD operating temperature was set to lower, the P_d suddenly decreased, while the P_a increased. Especially in the case that *QE* is less than 10%, P_d can be considerably decreased without surge of P_a.

The picosecond time analyzer we used has 50 ns dead time. The 50 ns dead time limits the maximum detection rate R_{max} of the SPD system to 20MHz. However, this limit is not disadvantageous to QKD, because the detection events weakly occur in almost all QKD systems.

Finally, the SPD performances are summarized in Table. 1. The SPD based on the SG-APD outperformed our previous result [10].

TABLE 1. Summary of SG-APD performance.

T_o	ω_g	*QE*	P_d	P_a	R_{max}
-50	1.2 GHz	11.0 %	7.7×10^{-7}	3.3 %	20 MHz

Conclusion

We developed the single-photon detector at 1550-nm using the sinusoidally gated InGaAs/InP APD. The gate repetition frequency of 1.2 GHz was achieved with the high quantum efficiency, low dark count probability, and low afterpulsing probability. The single-photon detector can be easily applied to gigahertz clocked QKD systems.

ACKNOWLEDGMENTS

This research was partially supported by the Grant-in-Aid for Scientific Research of Ministry of Education, Culture, Sports, Science and Technology, Japan (MEXT)

REFERENCES

1. H. Takesue, S. W. Nam, Q. Zhang, R. H. Hadfield, T. Honjo, K. Tamaki and Y. Yamamoto, Nature Photonics 1, 343 (2007).
2. N. Namekata, G. Hujii, T. Honjo, H. Takesue, and S. Inoue, Appl. Phys. Lett. **91**, 011112 (2007)
3. M. P. Brown and K. Austin, Appl. Phys. Lett. **85**, 2503-2504 (2004).
4. R. T. Wang, "Title of Chapter," in *Classic Physiques,* edited by R. B. Hamil, Publisher City: Publisher Name, 1999, pp. 212-213.
5. C. D. Smith and E. F. Jones, "Load-Cycling in Cubic Press" in *Shock Compression of Condensed Matter-2001*, edited by M. D. Furnish et al., AIP Conference Proceedings 620, American Institute of Physics, Melville, NY, 2002, pp. 651-654.
6. B. R. Jackson and T. Pitman, U.S. Patent No. 6,345,224 (8 July 2004)
7. D. L. Davids, "Recovery Effects in Binary Aluminum Alloys", Ph.D. Thesis, Harvard University, 1998.
8. E. Diamanti, H. Takesue, C. Langrock, M. M. Fejer, and Y. Yamamoto, Opt. Express **14**, 13073 (2006).
9. R. H. Hadfield, M. J. Stevens, S. S. Gruber, A. J. Miller, R. E. Schwall, R. P. Mirin, and S. W. Nam, Opt. Express **13**, 10846 (2005).
10. N. Namekata, S. Sasamori, and S. Inoue, Opt. Express 14, 10043 (2006).
11. Z. L. Yuan, B. E. Kardynal, A. W. Sharpe, and A. J. Shields, Appl. Pys. Lett. **91**, 041114 (2007).

Spatial and temporal characterization of a Bessel beam produced using a conical mirror

K. B. Kuntz,[1] B. Braverman,[2] S. H. Youn,[1] M. Lobino,[1] E. M. Pessina,[3] and A. I. Lvovsky[1]

[1] *Institute for Quantum Information Science, University of Calgary, Calgary, Alberta T2N 1N4, Canada*
[2] *Department of Physics, University of Toronto, Toronto, Ontario M5S 1A7, Canada*
[3] *Pirelli Labs, Viale Sarca 336, 20126 Milano, Italy*

Abstract. We experimentally analyze a Bessel beam produced with a conical mirror, paying particular attention to its superluminal and diffraction-free properties. We spatially characterized the beam in the radial and on-axis dimensions, and verified that the transverse profile is constant over a distance of 73 cm. In addition, we measured the superluminal phase and group velocities of the beam in free space. Both spatial and temporal measurements show good agreement with the theoretical predictions.

Keywords: Bessel beam, non-diffracting, superluminal
PACS: 42.25.Bs

A special class of non-diffracting light waves called Bessel beams [1] have been extensively studied over the last two decades. Bessel beams are characterized by a transverse field profile in the form of a zero-order first-kind Bessel function. They exhibit several intriguing properties, most notably the diffraction-free propagation of the central peak over a distance fixed only by the geometry of the source device, and superluminal phase and group velocities in free space. Due to their diffraction-free characteristics, they have applications in several fields of physics. Non-diffractive pump fields have been utilized in nonlinear optical processes like parametric down conversion [2], second [3], third [4], and higher-order harmonic generation [5]. Furthermore, the accuracy of optical tweezers [6], optical trapping [7] and the resolution of medical imaging [8] have been enhanced by the implementation of diffraction-free beams.

Here, we report complete characterization of spatiotemporal properties of an optical Bessel beam generated by reflection from a conical mirror [9]. We measured superluminal phase and group velocities in free space with greater accuracy than ever previously achieved, and we also spatially characterize the beam to examine its non-diffracting properties over its propagation range.

A non-diffractive Bessel beam is realized by a coherent superposition of equal-amplitude, equal-phase plane waves whose wavevectors form a constant angle θ, called the axicon angle, with the propagation axis z. The analytical solution of the Helmholtz equation for the electric field propagation is given by:

$$E(r,z,t) = A\exp\left[i(\beta z - \omega t)\right] J_0(\alpha r) \tag{1}$$

where A is constant, $\beta = k\cos\theta$, $\alpha = k\sin\theta$, $k = \omega/c$ is the wave-vector, ω is the

CP1110, *Quantum Communication, Measurement and Computing (QCMC), The Ninth International Conference,* edited by A. Lvovsky

angular frequency, r is the radial coordinate and J_0 is a zeroth-order Bessel function of the first kind. Equation (1) defines a non-diffracting beam since its transverse intensity distribution is independent of z and equal to $J_0^2(\alpha r)$. Furthermore, the field described in Eq. (1) propagates with a superluminal phase velocity given by $v_p = c/\cos\theta > c$. The independence of the phase velocity from the field frequency implies that in free space the group velocity is also superluminal and is equal to the phase velocity.

In practice the diffraction free region is limited by the finite extent of the beam. As evidenced by Fig. 1(a), the maximum diffraction free propagation distance depends on the radius R of the optical element used for producing the beam and it is given by $z_{max} = R/\tan\theta$. At propagation distances greater than z_{max}, the beam diffracts very quickly, spreading the energy over an annular region (Fig. 2(a)).

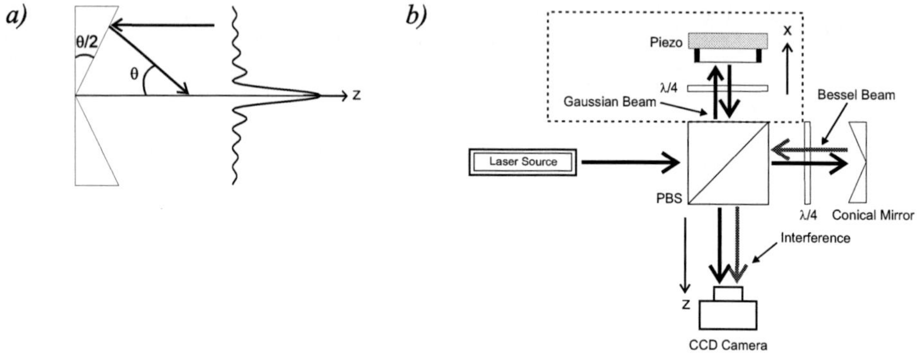

FIGURE 1. a) An illustration of the conical mirror used to produce the Bessel beam. b) Schematic of the experimental setup used for the spatial characterization of a Bessel beam. PBS, polarizing beamsplitter, $\lambda/4$, quarter waveplate. The components in the dashed box were used for the temporal measurements of the phase (continuous field) and group (pulsed field) velocities of Bessel beam.

In our experiment, the conical mirror has a radius of 1.27 cm, an apex angle of $\pi - \theta = 179°$ corresponding to a non-diffractive distance of 73 cm, and the phase and group velocities exceeding c by 0.015%. The non-diffractive propagation was verified by spatial characterization of the Bessel beam. The experimental setup used for these measurements is shown in Fig. 1(b). The light source is a continuous wave (CW) He:Ne laser operating at 543.5 nm. A photograph of the actual experimental setup is shown in Fig. 2(a). We compared the measured intensity pattern to a numerical solution, which is shown in Fig. 2(b). This theoretical curve was generated from the Rayleigh-Sommerfield diffraction integral [10] assuming the amplitude transmission function of the mirror is given by:

$$T(r) = e^{-2ikr\tan(\theta/2)} \tag{2}$$

The intensity of the central peak as a function of the propagation distance was compared with the numerical calculation in Fig. 2(c) and we observed good consistency between theory and experiment. The accuracy of the predicted 73 cm non-diffractive distance obtained from geometry consideration was verified as well [11].

The phase and group velocities in free space were measured in order to verify the superluminal nature of the Bessel beam. To measure the small velocity difference, we implemented an interferometric design using CW light for the phase velocity measure-

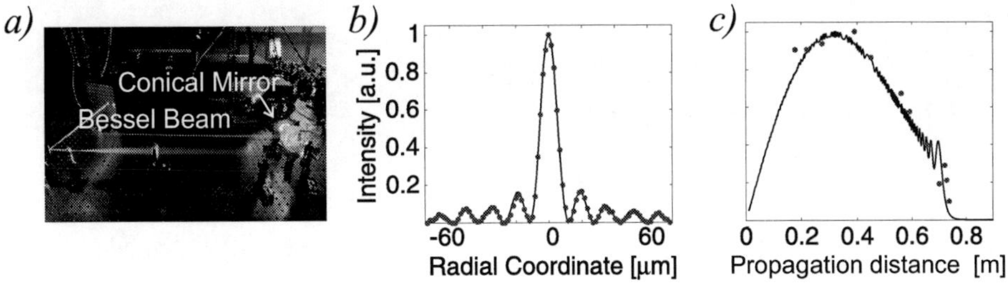

FIGURE 2. a) Experimental realization of setup used to produce a Bessel beam with a conical mirror. The Bessel beam is in the center propagating to the left. b) Measured intensity profile (dots) and numerical calculation (line). c) On-axis Intensity: experimental data (dots) and theoretical calculation (line). The intensity first grows approximately linearly because the light is emanating from a portion of the mirror that is equal to $2\pi R = 2\pi z \tan\theta$ and then decreases due to the Gaussian beam shape. The oscillations at the end of the curve are due to the Fresnel diffraction from the outer edge of the mirror.

ment, and femtosecond pulses for the group velocity measurement. The design, shown in Fig. 1(b), is a mixed Michelson interferometer, with a plane mirror in one arm and a conical mirror in the other. This arrangement allowed us to compare the propagation of the Bessel beam relative to the known group and phase velocities c of the Gaussian beam. Because the reference and Bessel waves have equal frequency but slightly different phase velocities along the z axis, they exhibit interference fringes spaced by $\Delta z = \lambda c / |\Delta v_p|$, where Δv_p is the difference between the z-components of the two group velocities [12].

Experimentally, we aligned the Gaussian beam along the optical rail and manipulated φ, the angle between the direction of propagation of the Gaussian and Bessel beam, by slightly tilting the conical mirror. We determined the value of φ by displacing the CCD along the optical rail and recording the transverse positions of the centres of the two beams at each location. The z-component of the reference field's phase velocity is then given by $c/\cos\varphi$. Figure 3(a) displays the measured interference period as a function of the relative angle φ together with a theoretical prediction. In particular, note that with increasing misalignment, Δz is increasing, indicating that the phase velocity of the Bessel beam is higher than that of the Gaussian beam. The data give a phase velocity of $v_p = (1.000155 \pm 0.000003)c$ which is within uncertainty of the expected value $c/\cos(\theta) = 1.000152c$, corresponding to an axicon angle of $1.009° \pm 0.01°$.

The group velocity measurement was performed using 200 fs pulses at $\lambda = 776.9$ nm from a Ti:Sapphire mode-locked laser pumped by a 532 nm solid-state laser. Using the CCD camera we recorded the intensity of the field of the interferometer. Interference fringes were detected if the pulses from both arms of the interferometer arrive at the camera's sensitive area simultaneously. The fringe visibility strongly depends on the temporal overlap of the pulses at the CCD. Because these pulses propagate at different group velocities, the z-position of the CCD at which the interference is optimized strongly depends on the path length difference in the two arms of the Michelson interferometer and thus on the position x of the plane interferometer mirror. We use this dependence to evaluate the group velocity of the Bessel pulse. We acquire

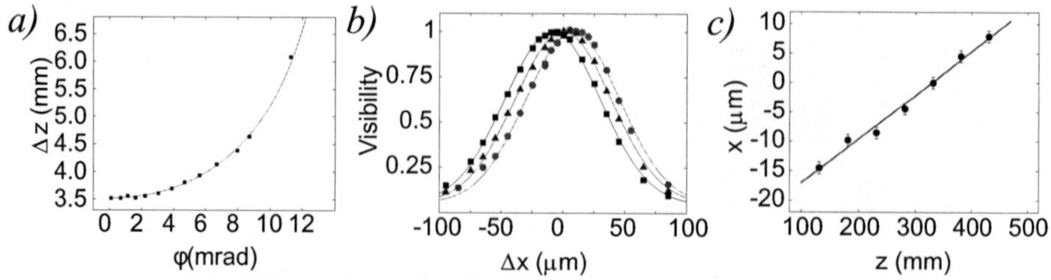

FIGURE 3. a) CW measurements: period of the interference fringes Δz as a function of the angle φ between the direction of propagation of the Gaussian and Bessel beams (dots) together with the theoretical fit (line). b) Pulsed measurements: fringe visibility as a function of the plane mirror translation Δx for three different CCD positions. The experimental data are fitted with a Gaussian curve. c) CCD camera positions z as a function of the displacement x of the Michelson interferometer plane mirror (dots) with a linear fit (line).

several pairs (x, z) for which the interference visibility is optimized (Fig. 3(b)). A linear relationship is evidenced by Fig. 3 (c), and is characterized by a linear regression slope $\partial x/\partial z = (-7.5 \pm 0.3) \times 10^{-5}$. The negativity of the slope indicates that the relative path length of the Gaussian beam had to be reduced as the overall travel distance was increased (i.e. larger z values), which corresponds to a superluminal group velocity, $v_g = [1 - 2(\partial x/\partial z)]c = (1.000150 \pm 0.000006)c$. This velocity corresponds to the axicon angle of $0.992° \pm 0.02°$.

REFERENCES

1. J. Durnin, *J. Opt. Soc. Am. A* **4**, 651 (1987).
2. S. Orlov, A. Stabinis, V. Smilgevicius, G. Valiulis, A. Piskarskas, *Opt. Lett.* **32**, 68 (2007)
3. O. Jedrkiewicz, *et al.*, *Phys. Rev. E* **68**, 026610 (2003)
4. V. E. Peet, R. V. Tsubin, *Phys. Rev. A* **56**, 1613 (1997)
5. A. Averchi, *et al.*, *Phys. Rev. A* **77**, 021802(R) (2008)
6. G. Milne, *et al.*, *Opt. Express* **15**, 13972 (2007)
7. I.V. Arakelyan, N. Chattrapiban, S. Mitra, W.T. Hill III, *Phys. Rev. A* **75**, 061604(R) (2007)
8. K. S. Lee, J. P. Rolland, *Opt. Lett.* **33**, 1696 (2008)
9. J.F. Fortin, G. Rousseau, N. McCarthy, M. Piche, *Proc. of SPIE* **4833**, 876 (2002)
10. J.W. Goodman, *Introduction to Fourier Optics 2nd edition* (McGraw-Hill, 1996).
11. K.B. Kuntz, B. Braverman, S.H. Youn, M. Lobino, E.M. Pessina, A.I. Lvovsky (to be published)
12. B. Braverman, K.B. Kuntz, M. Lobino, E.M. Pessina, A.I. Lvovsky, arXiv: 0811.4469

This work was supported by NSERC, CFI, AIF, Quantum*Works* and CIFAR (A.L.);

Laser Frequency Locking to an Optical Cavity using LQG Control

M. Heurs*, E. H. Huntington*, S. Z. Sayed Hassen*, I. R. Petersen* and M. R. James†

*School of Information Technology and Electrical Engineering, University College, The University of New South Wales, Canberra, ACT, 2600, Australia
†Department of Engineering, Faculty of Engineering and Information Technology, Australian National University, Canberra, ACT, 2600, Australia

Abstract. Systematic approaches to the design of quantum optical systems will become more and more important as their complexity grows. One such systematic technique is the Linear Quadratic Gaussian (LQG) methodology. Here we describe the design and implementation of a LQG feedback controller in a quantum optical experiment - frequency-locking an optical cavity to a laser. The successful implementation of the LQG design procedure to this particular problem lays the groundwork for the application of other modern control techniques to quantum optical systems. Our results are promising for future, more complex stabilisation problems in quantum optics, as the described approach is inherently multi-variable and naturally incorporates multiple sensors and actuators as well as nested loops.

Keywords: optimal control, quantum optics, linear-quadratic Gaussian
PACS: 42.60.Da, 42.60.Fc, 42.60.Mi, 42.62.Eh

INTRODUCTION

For the development of quantum information technologies systematic approaches to design and synthesis will be critical. Systematic methods such as Linear Quadratic Gaussian (LQG) optimal control have been studied in control engineering since the 1960s. We describe the application of systematic methods of LQG optimal control to the ubiquitous problem of cavity locking in quantum optics. To some extent, the purpose of this paper is to describe the approach required to implement optimal LQG control in a real-life quantum optical experiment. We describe practical techniques which aid in the use of LQG optimal control in such experiments. Fig. 1 schematically illustrates the experiment.

LINEAR QUADRATIC GAUSSIAN OPTIMAL CONTROL

The LQG optimal controller design methodology begins with a linear state-space model comprising a vector of system variables x, a vector input to the system u, (potentially) a vector measured output y, and w is a Gaussian white noise disturbance acting on the system. Also, the system matrix is represented by A, the input matrix is represented by B, C is the output matrix, $D_1 w$ represents process noise and $D_2 w$ represents measurement noise.

CP1110, *Quantum Communication, Measurement and Computing (QCMC), The Ninth International Conference,* edited by A. Lvovsky

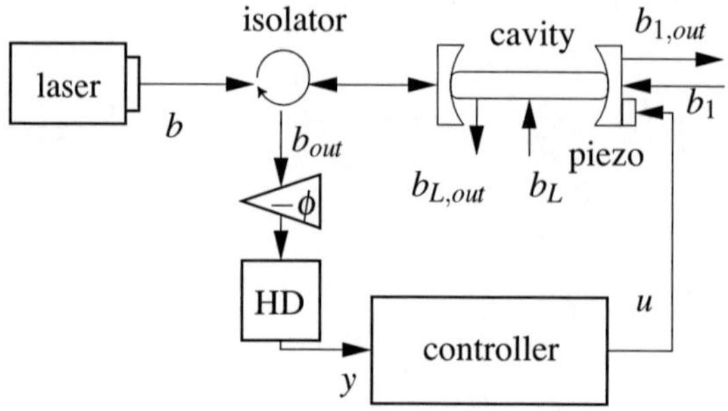

FIGURE 1. Cavity locking feedback control loop.

The goal of the LQG optimal control problem is to construct a feedback controller that minimises a quadratic cost functional. The cost functional codifies the design goal to minimise the system variables of interest (specifically, to keep the frequency detuning of the cavity small) and the requirement to limit the control energy.

The great advantage of the LQG optimal control approach to controller design is that it provides a tractable systematic way to construct output feedback controllers, even in the case of multi-input multi-output control systems. The solution to the LQG optimal control problem involves a Kalman filter which provides an optimal estimate $\hat{x}$ of the vector of system variables x based on the measured output y.

The standard LQG approach is not sufficient to generate a suitable controller as the system illustrated in Fig. 1 is subject to a large initial DC offset and slowly varying disturbances, which necessitates integral action. Here, an additional term is included in the cost function which involves the integral of a new quantity z. The new variable $\int z dt$ is also fed to the Kalman filter, which when combined with an optimal state-feedback control law leads to an integral LQG optimal controller; see [1].

Mathematically, the augmented system can be described in state-space form:

$$\dot{\tilde{x}} = \tilde{A}\tilde{x}+\tilde{B}w_1+\tilde{B}u; \quad \tilde{y}=\tilde{C}\tilde{x}+\begin{bmatrix} w_2 \\ w_3 \end{bmatrix}; \quad \tilde{x}=\begin{bmatrix} x \\ \int z d\tau \end{bmatrix} \text{ and } \tilde{y}=\begin{bmatrix} y_1 \\ y_2 \end{bmatrix}. \quad (1)$$

Here the matrices $\tilde{A},\tilde{B},\tilde{C}$ are constructed from the matrices A,B,C as follows:

$$\tilde{A}=\begin{bmatrix} A & 0 \\ C & 0 \end{bmatrix}, \quad \tilde{B}=\begin{bmatrix} B \\ 0 \end{bmatrix}, \text{ and } \tilde{C}=\begin{bmatrix} C & 0 \\ 0 & I \end{bmatrix}.$$

The integral LQG performance criterion can be written as:

$$\mathscr{J} = \lim_{T\to\infty} \mathbf{E}\left[\frac{1}{T}\int_0^T [x^T Qx+L(z)^T\bar{Q}L(z)+u^T Ru]dt\right] \quad (2)$$

where $L(z) = \int_0^t z(\tau)d\tau$ and we choose the matrices Q,R and $\bar{Q}$ such that $x^T Qx = |z|^2$, $u^T Ru = r|u|^2$, and $\bar{Q}=\bar{q}$.

The optimal LQG controller is given by the following set of simultaneous equations (e.g., see [2]):

$$u = -r^{-1}\tilde{B}^T X\hat{\tilde{x}} = F\hat{\tilde{x}}, \tag{3}$$

$$0 = X\tilde{A}+\tilde{A}^T X+\tilde{Q}-r^{-1}X\tilde{B}^T\tilde{B}X, \quad \tilde{Q}=\tilde{C}^T\begin{bmatrix} 1 & 0 \\ 0 & \bar{q}\end{bmatrix}\tilde{C}, \tag{4}$$

$$\dot{\hat{\tilde{x}}} = \tilde{A}\hat{\tilde{x}}+\tilde{B}u+K[\tilde{y}-\tilde{C}\hat{\tilde{x}}], \quad K = P\tilde{C}^T V_2^{-1}, \tag{5}$$

$$0 = \tilde{A}P+P\tilde{A}^T+V_1-P\tilde{C}^T V_2^{-1}\tilde{C}P. \tag{6}$$

LQG CONTROLLER FOR AN OPTICAL CAVITY

A complete state-space model of the system under control (or "plant"), the measurement and any noise sources is required in order to construct the optimal feedback controller. The cavity can be described in the Heisenberg picture by the standard cavity differential equations; e.g., see [3] and Section 9.2.4 of [4]. The linearised cavity dynamics and homodyne measurement can be expressed in state-space form in terms of the quadratures of the operators for the cavity mode and cavity inputs [5].

As it stands, that would not yet be a complete model of the system, as it does not include the dynamics for the detuning Δ. A reasonable hypothesis for these dynamics might be to treat them as arising from the dynamics of a cavity mirror, which is treated as a mass on a spring, subject to mechanical noise and laser frequency noise. We could then augment the optical state-space model with the equation $\Delta = Fx$, so that the matrix F relates the detuning to the vector of system variables, which includes those variables describing the mechanical dynamics of the cavity mirror as well as those for the optical mode.

Unfortunately, the true dynamics of the cavity mirror is much more complicated than a simple mass on a spring. Therefore, a more experimentally tractable approach must be adopted in order to arrive at the optimal integral LQG controller. In our modern control approach, the subspace identification method determines a state-space model from the input-output frequency response data and generates the system matrices A, B and C; see [6]. The frequency response data obtained is fitted to a 13th order model (including an 8th-order anti-aliasing filter with a corner frequency of 2.5 kHz) using subspace identification [7].

CONTROLLER AND RESULTS

The design values for the controller $\varepsilon_1 = 5\times10^{-2}$, $\varepsilon_2 = 500$ and $\varepsilon_3 = 3\times10^{-4}$ are the variances of the mechanical noise, detector noise and integrator noise. The control energy weighting in the controller cost function is $r = 1\times10^3$ and the integral output weighting is $\bar{q} = 1\times10^6$.

These parameter values lead to a 15th order LQG controller which is reduced to a 6th order controller using a frequency-weighted balanced controller reduction approach. The reduced controller is then discretised at a sampling rate of 50 kHz. This discretised controller provides good gain and phase margins of 16.2 dB and 63° respectively.

The discrete controller is implemented on a dSpace DS1103 Power PC DSP Board. This board is fully programmable from a Simulink block diagram and possesses 16-bit resolution. The controller successfully stabilises the frequency in the optical cavity, locking its resonance to that of the laser frequency, f_0; see [5]. This can be seen from the measured step response shown in Fig. 2, which is consistent with theory.

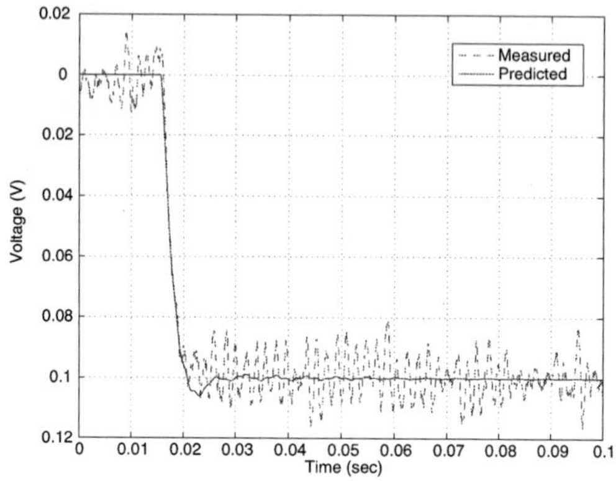

FIGURE 2. Step Response of the closed-loop system to an input of 0.1 V.

CONCLUSION

We have described the design and implementation of a LQG feedback controller in a quantum optical experiment - frequency-locking an optical cavity to a laser. Practical techniques are used to augment the basic LQG methodology, which facilitated the design and implementation of an optimal integral LQG controller. The controller was implemented digitally, and a step response measured. The performance of the closed-loop system is consistent with its state-space model.

REFERENCES

1. M. J. Grimble, Proceedings of the IEE, Part D **126**, 841 (1979).
2. H. Kwakernaak and R. Sivan, *Linear Optimal Control Systems* (Wiley, 1972).
3. H. Bachor and T. Ralph, *A Guide to Experiments in Quantum Optics* (Wiley-VCH, Weinheim, Germany, 2004), 2nd ed.
4. C. Gardiner and P. Zoller, *Quantum Noise* (Springer, Berlin, 2000).
5. S. Z. Sayed Hassen, M. Heurs E. H. Huntington, I. R. Petersen and M. R. James , submitted to J Phys. B (2008).
6. P. van Overschee and B. De Moor, *Subspace Identification for Linear Systems* (Kluwer Academic Publishers, 1996).
7. T. McKelvey, H. Akçay, and L. Ljung, IEEE Transactions on Automatic Control **41**, 960 (1996).

CHAPTER 10

QUANTUM INFORMATION PROCESSING IN SOLIDS AND LIQUIDS

Magnetic-field cycling triplet-DNP/NMR system for true quantum computation with hyperpolarized nuclear spins

A. Kagawa ,†, M. Negoro , K. Takeda and M. Kitagawa

Graduate School of Engineering Science, Osaka University, 560-8531 Toyonaka, Japan
† CREST, Japan Science and Technology Agency
Graduate School of Science, Kyoto University, 606-8502 Kyoto, Japan

Abstract. Toward NMR quantum computing with hyperpolarization beyond the entanglement threshold, a system has been developed enabling dynamic nuclear polarization (DNP) using photo-excited triplet electron spins with X-band microwave apparatus, followed by solid-state NMR experiments using the polarized nuclear spin system. In order to perform the triplet-DNP and NMR experiments in different magnetic fields, the triplet-DNP system and the NMR system are made to be spatially separated, between which the sample can be shuttled. It is shown that the system developed in this work fulfils the requirements for solid-state NMR quantum computing with hyperpolarized nuclear spin systems.

Keywords: NMR, dynamic nuclear polarization, triplet electron spins
PACS: 03.67.Lx, 76.70.Fz, 82.56.-b

A true demonstration of NMR quantum computing (NMRQC) demands hyperpolarized nuclei beyond the entanglement threshold on the order of 0.1, depending on the number of qubits involved in the spin system[1]. Since this threshold is 4-5 orders of magnitude larger than nuclear spin polarizations in typical experimental conditions, NMRQC experiments inevitably need some sort of polarization enhancement in such nuclear spin systems that quantum gate operation is applicable. Toward this goal, we have been exploring a strategy of exploiting dynamic nuclear polarization (DNP) using electron spins in the photo-excited triplet state (triplet-DNP), because of its following favorable properties in contrast to the DNP system using free radials. (i) In some of the photo-excitable molecules, the transient electron spin polarization in the metastable photo-excited triplet state is much larger than that in free radicals in thermal equilibrium irrespective of temperature and magnetic field, eliminating the technically challenging demands for performing DNP both at cryogenic temperature and in high magnetic fields requiring millimeter-wave apparatus. (ii) Owing to the finite lifetime of the triplet electron spins, the unwanted relaxation and decoherence effects in the nuclear spin system due to paramagnetic electrons are suppressed simply by stopping photo-excitation once the electron-to-nucleus polarization transfer has been completed.

By triplet-DNP, proton polarization of 0.7 has been attained at 100 K and in 0.3 T in naphthalene doped with small amount of pentacene as the triplet guest material[2]. Although such high proton polarization is quite attractive, the magnetic field of 0.3 T is rather low from the viewpoint of quantum gate operation, where the individual qubits have to be distinguishable by their chemical shifts. In this work, we have developed

CP1110, *Quantum Communication, Measurement and Computing (QCMC), The Ninth International Conference,* edited by A. Lvovsky

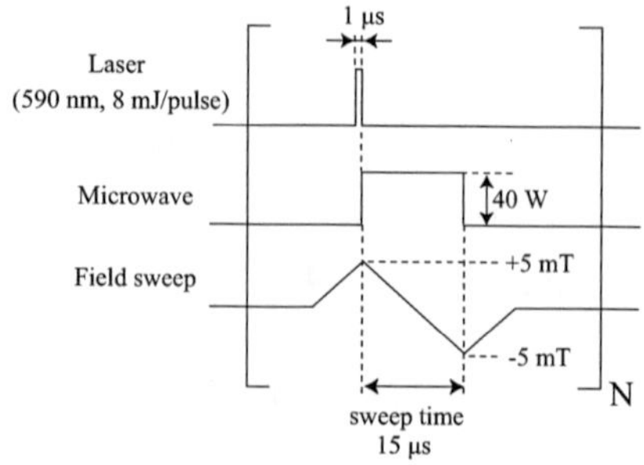

FIGURE 1. A pulse sequence for integrated cross polarization (ICP).

a field-cycling system by means of sample shuttling, enabling triplet-DNP and NMR experiments in different magnetic fields. Also, we have a strategy of co-doping another target material for NMRQC in pentacene-doped naphthalene, aiming at transferring polarization to the target molecules through ^{1}H spin diffusion. In order to attain the shorter triplet-DNP experimental time, studies on spin diffusion in partially deuterated naphthalene is also in progress. In this paper we focus on the hardware, leaving the issue of the material quest to our forthcoming paper elsewhere.

Among a number of electron-to-nucleus polarization transfer mechanism in triplet-DNP, Integrated Cross Polarization (ICP)[2, 3], is the most efficient polarization transfer mechanism reported so far. Figure 1 shows a pulse sequence of triplet-DNP by means of ICP, which requires, in addition to laser irradiation for photo-excitation and microwave irradiation at the electron spins in the photo-excited triplet state, adiabatic magnetic field sweep over the resonance line of the triplet electron spins. Also, it is critical to optimize crystal alignment with respect to the external magnetic field so as to maximize the orientation-dependent electron spin polarization in the photo-excited triplet state.

For NMR measurement and/or NMR quantum gate operation in the partially deuterated organic materials, we require that the NMR sample coil is triply-tuned at the resonance frequencies of the ^{1}H, ^{13}C, and ^{2}H spins. The ^{2}H channel can serve for improving spectral resolution through by ^{2}H decoupling. Also, another goniometer is necessary in the high field so as to manipulate the crystal orientation in the NMRQC experiments.

As schematically shown in Figure 2, a split-electromagnet and a solenoidal superconducting magnet are employed for triplet-DNP and NMR measurement, between which the sample is shuttled within ca. 5 seconds. In order to adjust the sample orientations in both the low and high fields in a reproducible fashion, the sample holder can be docked and undocked with both of the goniometers assembled in the triplet-DNP system and the NMR probe. Also, the sample is thermally insulated with a double-layered glass tube, and can be cooled with nitrogen gas flow to suppress spin-lattice relaxation of the nuclear spin system toward thermal equilibrium.

A single crystal sample of ca. 0.01 mol% pentacene-doped naphthalene was grown by the Bridgman method. The magnetic field in the triplet-DNP system was set to 0.2958 T, corresponding to the transition between the $|\quad 1\rangle$ and $|0\rangle$ sublevels in the photo-excited

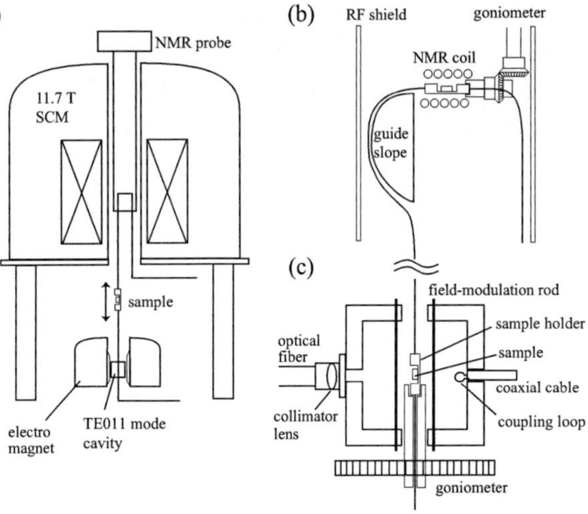

FIGURE 2. (a) Schematic diagram of a field-cycling triplet-DNP/NMR system. The sample is to be shuttled between the high magnetic field (11.7 T) in a superconducting magnet and the low field (~ 0 3 T) in an electromagnet. The system is thermally insulated with a vacuum double-layered tube (not shown in the figure), and the sample can be cooled with nitrogen gas flow. In the high field, the sample moves along a guide slope toward inside the NMR coil until the sample holder is docked with the goniometer, as described in (b). The NMR coil is triply tuned at 500 MHz, 125 MHz, and 76 MHz for ^{1}H, ^{13}C, and ^{2}H NMR experiments in 11.7 T. In the low field, a TE011 mode X-band cavity is employed as described in (c), and a laser beam and X-band microwave irradiations can be applied together with magnetic field sweep by means of a quadrupole coil which is composed of four field-modulation rods.

triplet state of the pentacene molecule when its long axis is parallel to the magnetic field. The ^{1}H polarization was built up by repeating the ICP sequence depicted in Fig. 1 at a rate of 50 Hz in the low-field triplet-DNP system. After the buildup process of the ^{1}H polarization, the sample was shuttled into the high-field NMR system. The goniometer axis normal to the magnetic field inside the superconducting magnet demands the sample to be shuttled along a curvy path. At the cost of this mechanical complexity, however, the solenoid coil could be used for solid-state NMR and the efficiency of rf irradiation and the detection sensitivity have become higher than that in the the previous shuttling systems along a straight path where a saddle coil is the natural choice. The attained rf intensities with 300 W power amplifiers were 130 kHz, 110 kHz, and 55 kHz for the ^{1}H, ^{13}C, and ^{2}H rf channels, respectively.

In order to examine the performance of the system, the following experiments were carried out. The first one is the buildup test of ^{1}H polarization by repeating the ICP sequence for 90 min. Figure 3(a) shows a ^{1}H spectrum measured in the high field. By comparing the signal intensity with that obtained in thermal equilibrium, the enhanced ^{1}H polarization was estimated to be 0.11, which corresponds to ~1900 times that in thermal equilibrium in 11.7 T at 203 K. The loss of the ^{1}H polarization due to sample shuttling was found to be negligible, since we found no discernible reduction in the signal intensity even the sample was shuttled between the high and low fields several

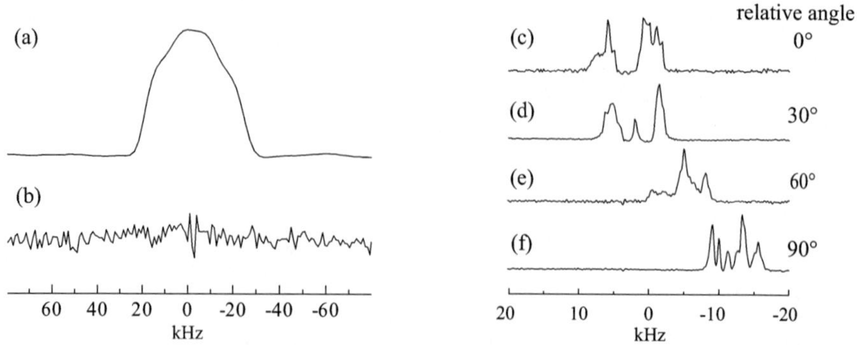

FIGURE 3. (Left) Single-scan ^{1}H NMR spectra in a single crystal sample of pentacene-doped naphthalene in 11.7 T and at 203 K obtained (a) after repeating the ICP sequences for 90 min, and (b) without the ICP sequence ($\times$50). (Right) Single-scan natural-abundance ^{13}C NMR spectra in a single crystal sample of pentacene-doped naphthalene. The buildup process of the ^{1}H polarization by repeating the ICP sequences for 10 min. ^{1}H-^{13}C cross polarization was performed before measuring the ^{13}C NMR signal under ^{1}H decoupling. Spectra in (c)-(f) were obtained with different crystal alignments by setting the goniometer angle.

times before NMR measurement.

The second NMR experiment is ^{1}H-^{13}C cross polarization, by which the enhanced polarization of the ^{1}H spins in naphthalene can further be transferred to the $\sim$1 % naturally-abundant ^{13}C spins. In a single crystal sample, the resonance frequency of the individual ^{13}C spin depends on the principal values and the orientation of the chemical shift tensor with respect to the external magnetic field. Thus, the profile of the ^{13}C NMR spectra varied as the goniometer was rotated, as demonstrated in Fig. 3(c)-(f). By the goniometer, we can choose an appropriate angle for operating quantum gates.

To summarize, we have developed a field cycling system for triplet-DNP and solid-state ^{1}H-^{13}C-^{2}H triple resonance NMR, which serves for NMRQC experiments with hyperpolarized nuclear spins. The experiments demonstrated above confirms that the hardware developed in this work fulfils the requirements for NMRQC.

ACKNOWLEDGMENTS

This work is supported by the CREST program of Japan Science and Technology Agency.

REFERENCES

1. S. L. Braunstein, C. M. Caves, R. Jozsa, N. Linden, S. Popescu and R. Schack, *Phys. Rev. Lett.* **83**, 1054–1057 (2000).
2. K. Takeda, K. Takegoshi and T. Terao, *J. Phys. Soc. Japan* **73**, 2313–2818 (2004).
3. A. Henstra, T.-S. Lin, J. Schmidt and W. Th. Wenckebach, *Chem. Phys. Lett.* **175**, 6–10 (1990).

Linear Preparation Method of Pseudo-Pure States while Correcting Errors and Enhancing Signals for NMR Quantum Computer

M. Kawamura[a], H. Sakai[b] and Y. Manmoto[a]

[a]*Department of Electrical and Electronic Engineering, Okayama University of Science,*
[b]*Graduate School of Engineering, Okayama University of Science,*
(1-1, Ridai-cho, Okayama,700-0005, Japan)

Abstract. An error cancellation method for producing pseudo-pure states in large number qubit systems is proposed that is aimed at creating initial states asymptotically within a specified error tolerance by using a spatial averaging method. In this work, it is shown that the present method enables almost complete error cancellation and even enhances the NMR signal compared with the initial state within the initialization process itself. An implementation of this method was successfully demonstrated using a five-qubit NMR quantum computer.

Keywords: Quantum Computer, NMR, Pseudo-Pure, Error Correction, Initialization
PACS: 03.67.Lx

INTRODUCTION

For NMR quantum computer, pseudo-pure states are generally used as initial states instead of pure states, which it is very hard to create. However, as the number of qubits increases, it becomes difficult to prepare even pseudo-pure states due to increase of the errors which are caused by imperfection of gate operations, decoherence and dissipation effects during the long process. In this work an efficient method of producing pseudo-pure states for large number qubit system is proposed, which is aimed at creating the states within a specified tolerance by using a few types of two-qubit gates, and it is shown that the present method enable to correct the errors almost completely and enhance the NMR signal intensities to a certain extent in the initialization process itself. The experimental demonstrations have been also performed successfully with a five-qubit NMR quantum computer.

There are predominantly two ways to prepare pseudo-pure state from the thermal equilibrium state. One is known as the temporal averaging method [1], which can be expressed as a sum of unitary transformations. The other is the spatial averaging method [2], which can be expressed as a series of unitary and non-unitary transformations. For the former method the total number of steps increases exponentially with the number of the qubits, so it is ineligible for large-qubit system. On the other hand, for the latter method it could be expected that the number of steps

CP1110, *Quantum Communication, Measurement and Computing (QCMC), The Ninth International Conference,* edited by A. Lvovsky

is proportional to the number of the qubits, abandoning analytic exact solution, which will not make sense practically for large-qubit system due to the enormous errors.

Preparation Method of Pseudo-Pure States with Error Correction

Let us next introduce a type of two-qubit gate, a *controlled-transfer* gate of angle θ, which is defined by the following transformation for diagonal density matrices.

$$diag(a,b,c,d) \Rightarrow diag\left(a,b,\frac{c+d+(c-d)\cos\theta}{2},\frac{c+d-(c-d)\cos\theta}{2}\right) \qquad (1)$$

This operation can be implemented as a controlled-rotation gate of angle θ followed by a magnetic gradient pulse. It can be shown that a sequence of controlled-transfer gates of angle $\pi/2$, $CT_{ij}(\pi/2)$, where i indicates the control-qubit and j indicates the target qubit, can create pseudo-pure states within a given tolerance, and the required number of gates and the signal intensity are proportional to the number of the qubits assuming that decoherence and dissipation effects can be ignored. Figure 1 shows the required number of the *controlled-transfer* gates required to achieve initialization within a tolerance of 1%, where the tolerance means the normalized standard deviation. From Figure 1, it can be seen that below six-qubit, the number of the gates increases as a second order polynomial, while it is proportional to the number of qubits above six-qubit. The linearity of the convergence can be understood from the fact that each of the gates averages out half of the diagonal elements at a time. The signal intensities increase with the number of qubits, n, as $\{n+n/(2n-1)\}/2$ compared to one of the initial states, because the energy of the ground state decreases proportionally with n.

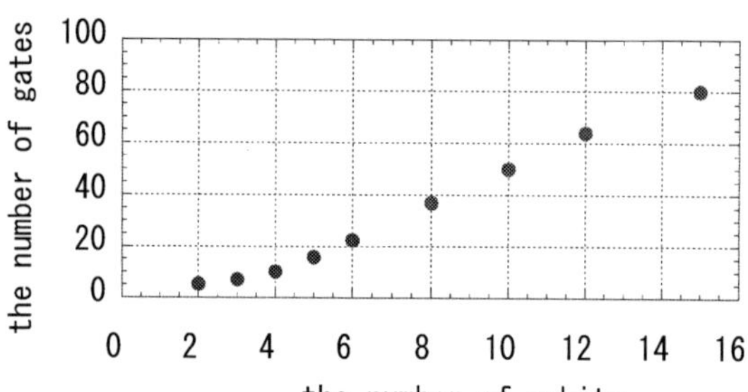

FIGURE 1. The required number of gates to obtain a tolerance of 1% with the number of qubits.

However, since the length of the sequence becomes much longer than in the temporal averaging method as the number of qubits is increased, it will fail to create pseudo-pure states even for a several-qubit system. Decoherence and dissipation during the gate operations affect the convergence and result in a lower limit of the convergence. The calculation results in a lower convergence limits shows that the limits increase proportionally with the number of qubits, where decoherence and dissipation effects are evaluated using the transverse and longitudinal relaxation rates, which were determined by experiments for a certain molecule. Although this situation seems to strand us, it would not be difficult to envision that a cooperative use of *controlled-transfer* gates of different angles would enable the convergence to accelerate or would enable the error correction. The calculation results when two types of *controlled-transfer* gates and *controlled-not* gates are used shows that the convergence speeds are considerably improved, and the lower limits of the convergence are improved by a factor of ten overall by using the three gate types.

Evaluation of Decoherence and Dissipation Effects

To embed an error correction function into the initialization sequences with a classical computer, the effects of decoherence, dissipation and imperfection of pulse operations on the density matrix have to be predicted as accurately as possible. The intensities of the NMR signals observed with a $\pi/2$ pulse after applying n units of *controlled-transfer* gate, $CT_{ij}(\pi)$, to the thermal equilibrium state can be obtained from expression (1) by using the Bloch equations. The results are given in Table 1, where R_1 and R_2 are the longitudinal and transverse relaxation rates. In Figure 2, the experimental results with a certain molecule and the theoretical results including expected decoherence and dissipation effects with R_1 and R_2 are shown. From the figure it can be seen that by using parameters, p_1 and p_2 in expression (2), the effects of decoherence and pulse imperfection are well described.

$$R_1' = e^{-p_1\frac{\Delta t}{T_1}}, \quad R_2' = e^{-\left(\frac{\Delta t}{T_2}\right)^m} - p_2 \tag{2}$$

TABLE 1. The signal intensities of each qubit after identical *controlled-transfer* gates of angle π are applied repeatedly, where the signal intensity in the thermal equilibrium state is assumed to be 1.

n	signal intensity of the control qubit		signal intensity of the target qubit	
	state of the target qubit		state of the control qubit	
	Up	Down	Up	down
odd	$2(R_1^n-1)/(\sqrt{R_1}+1)+1+R_2^n$	$2(R_1^n-1)/(\sqrt{R_1}+1)+1-R_2^n$	R_2^n	$-R_2^n$
even	$2(R_1^n-1)/(\sqrt{R_1}+1)+1$	$2(R_1^n-1)/(\sqrt{R_1}+1)+1$	R_2^n	R_2^n

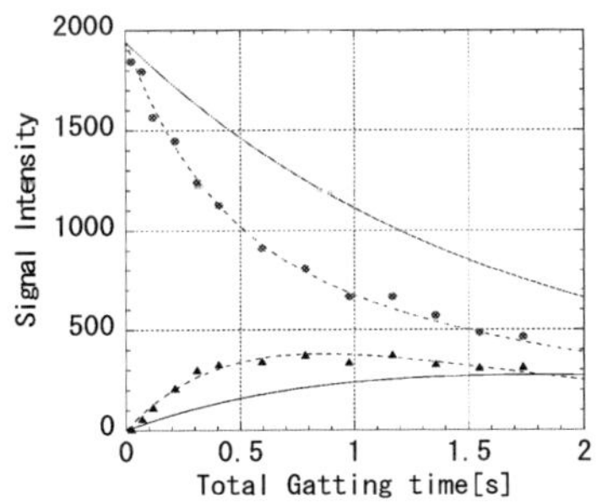

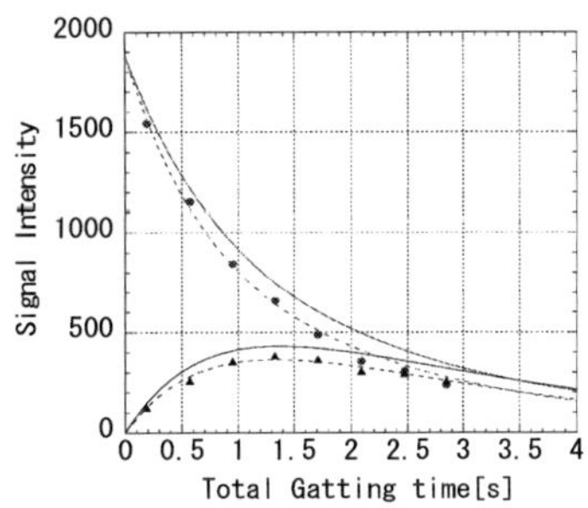

Figure 2. The signal intensities of a control-qubit after applying odd numbers of *controlled-transfer* gates of angle π to the thermal equilibrium state. (a) The experimental results when the shortest gates are applied, the duration time per gate is 29ms. (b) The experimental results of the longest gate, the duration time per gate is 190ms. The filled-circles indicate that the target-qubit state is |0>, and the filled-triangles indicate that the target-qubit state is |1>. The solid lines are the theoretical values obtained in the presence of T_1 and T_2 relaxation. The broken lines indicate the theoretical values when p_1 and p_2 are used as fitting parameters.

Producing of Initialization Sequence and Experimental Results

The sequence to prepare pseudo-pure state is produced by more than two processes. The optimization of the sequences of gates is performed with an annealing method. The first step is to create a basic sequence, which is prepared by appending one gate at a time in order that the standard deviation from the ideal state at this point becomes smallest including effects of the errors. The second step is to optimize the basic sequence, which is performed by examining and changing one gate at a time in order to minimize the standard deviation at the end of this sequence.

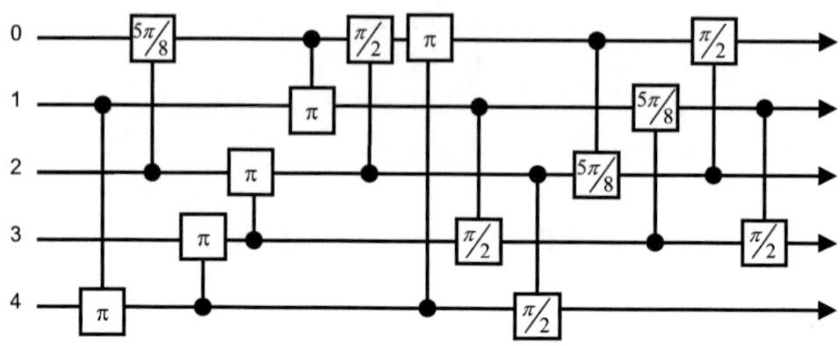

Figure 3. The quantum diagram of the optimized initialization sequence for a five-qubit NMR-QC, where three angles of controlled-transfer gates are used. The standard deviation from the ideal state is 1.6%.

To investigate the performance of the optimized sequence, initialization experiments are performed on a five-qubit NMR quantum computer with a 2,3,4,6-tetrafluoroaniline molecule, where four fluorine nuclear spins and a hydrogen nuclear spin were used as qubits. These experiments were taken under magnetic field of 9.4T and the sample was kept at a temperature of 308K. The diagonal elements before and after the initialization are shown in Figure 4. From the figure, it is seen that the experimental results agree well with the theoretically expected values. Although the population of the zero state would be reduced almost by half due to the decoherence effects, the signal intensity is increased approximately 1.25 times.

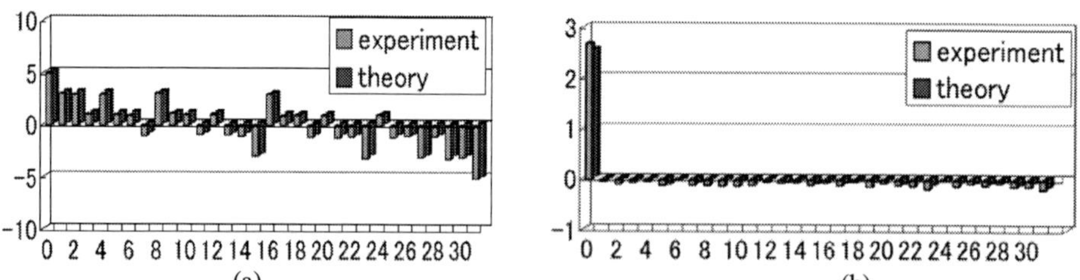

Figure 4. The theoretical and experimental results of the deviation density matrices' diagonal elements for both the thermal equilibrium state (a) and after initialization (b).

REFERENCES

1. N. A. Gershenfeld, I.L. Chuang, Science 275, 350(1997).
2. D. G. Cory, A. F. Fahmy and T. F. Havel, Proc. Natl. Acad. Sci. USA 94, 1634(1997).

Using nanoscale transistors to measure single donor spins in semiconductors

M. Sarovar*, K. C. Young†, T. Schenkel** and K. B. Whaley*

*Department of Chemistry, University of California, Berkeley, California 94720
†Department of Physics, University of California, Berkeley, California 94720
**Accelerator and Fusion Research Division, Lawrence Berkeley National Laboratory, Berkeley, California 94720

Abstract. We propose a technique for measuring the state of a single donor electron spin using a field-effect transistor induced two-dimensional electron gas and electrically detected magnetic resonance techniques. The scheme is facilitated by hyperfine coupling to the donor nucleus. We analyze the potential sensitivity and outline experimental requirements. Our measurement provides a single-shot, projective, and quantum non-demolition measurement of an electron-encoded qubit state.

Keywords: Quantum computation, Quantum measurement
PACS: 73.23.-b, 03.67.Lx, 76.30.-v, 84.37.+q

The spin of electrons bound to donors in silicon is considered a promising candidate qubit for quantum information processing [1]. An integral part of any quantum computation architecture is the capacity for high-fidelity qubit readout; however, the detection of spin states of single donor electrons and nuclei in silicon has remained elusive. In this paper we analyze spin dependent scattering between conduction electrons and neutral donors as a spin-to-charge-transport conversion technique, and show that quantum non-demolition (QND) measurements of single electron spin-encoded qubit states are realistically achievable when mediated via nuclear spin states. Such a measurement will also be of value to the developing field of spintronics where the electrical detection of spin states is valuable. Our readout takes advantage of two features: i) the ability to perform electron spin resonance spectroscopy using a two-dimensional electron gas (2DEG), and ii) the hyperfine shift induced on dopant electron Zeeman energies by the dopant nuclear spin state. In the following, we first describe the experimental apparatus and the techniques of 2DEG mediated spin resonance spectroscopy, and then we present our proposal for spin state measurement, analyze the sensitivity of the measurement scheme and establish the key factors that determine signal-to-noise.

The physical setting. Figure 1(a) shows a cross section of a 2DEG spin readout device with a single implanted donor. Prior studies have used similar devices with bulk-doping [2, 3] or a large number of implanted donors (10^6) [4] in the 2DEG channel. The 2DEG is operated in accumulation mode and thus scattering occurs between conduction electrons and electron(s) bound to the shallow donor(s). The basic principle exploited in these studies is the role of the exchange interaction in electron-electron scattering. At a scattering event between a conduction electron and a loosely bound donor impurity electron, the Pauli principle demands that the combined wave function of the two electrons be antisymmetric with respect to coordinate exchange. This constraint, together with the

CP1110, *Quantum Communication, Measurement and Computing (QCMC), The Ninth International Conference,* edited by A. Lvovsky

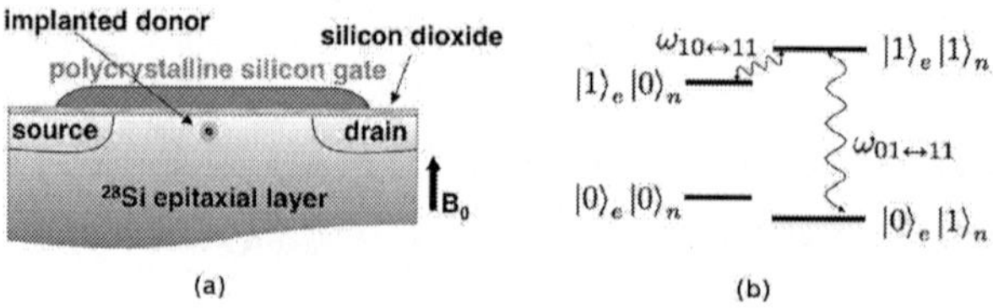

FIGURE 1. (a)A cross section of the field-effect transistor (FET) used to create the 2DEG. In order to reduce qubit decoherence, it is beneficial to implant into isotopically purified silicon. (b) Four-level system of electron-nuclear spin degrees of freedom in the secular approximation. The energy eigenstates are the eigenstates of σ_z^e and σ_z^n.

fact that the combined spin state can be symmetric (triplet) or antisymmetric (singlet), imposes a correlation between the spatial and spin parts of the wave function and results in an effective spin dependence of the scattering matrix, leading to a spin dependent conductance. Application of a static magnetic field will partially polarize conduction and impurity electrons leading to excess triplet scattering. A microwave drive will alter these equilibrium polarizations when on resonance with impurity (or conduction) electron Zeeman energies and hence alter the ratio of singlet versus triplet scattering events, registering as a change in the 2DEG current. Thus, the spin dependent 2DEG current can be used as a detector of spin resonance and accordingly this technique is commonly known as *electrically detected magnetic resonance* (EDMR).

From hereon we will consider the experimental setup described above in the particular situation where there is a single donor P nucleus present, the electron spin of which encodes the quantum information that we wish to measure. A crucial question in the context of quantum computing is whether the spin-dependent 2DEG current can be used to measure the state of an electron-spin qubit, as spin-dependent tunneling processes have been employed [5]. The fundamental concern here is whether the spin exchange scattering interaction at the core of the spin-dependent 2DEG current allows for a quantum state measurement of a single donor impurity electron spin. One might expect that because the exchange interaction is destructive (in the sense that it will change the state of the target (donor electron) spin with some probability), it will only produce a faithful measurement if the time over which it acts is extremely short. Indeed, one can show using a minimal model of the exchange scattering that the relaxation of a general electron spin state is rapid across virtually all reasonable scattering parameter ranges [6]. The estimated measurement induced population mixing time of $T_{mix} \sim 1-10ns$, is drastically smaller than the required measurement integration time: $\tau_m \sim 10^{-3}$s (see below for a calculation of τ_m). This makes it impossible to faithfully map the electron spin state onto the meter variable, and hence impossible to perform a single electron spin state measurement using the 2DEG current directly. However, as we will now show, it is possible to make use of the nuclear spin degree of freedom in order to utilize EDMR for a projective and QND measurement of single spin states. The key is that the state of the nuclear spin affects the Zeeman splitting of the electron spin (and thus its resonant frequency) via the mutual hyperfine coupling.

Nuclear spin mediated spin measurement. The low-energy, low-temperature Hamiltonian describing the electron and nuclear spins of a phosphorous dopant in a static magnetic field, $\mathbf{B} = B\hat{\mathbf{z}}$ is $H = \frac{1}{2}\left[g_e \mu_B B \sigma_z^e - g_n \mu_n B \sigma_z^n\right] + A\sigma^e \cdot \sigma^n$ where μ_B and μ_n are the Bohr and nuclear magnetons, g_e (g_n) is the electron (nuclear) g-factor, and A characterizes the strength of the hyperfine interaction between the two spins (we set $\hbar = 1$ throughout the paper). For moderate and large values of B, the σ_z terms dominate and we can make the *secular approximation*, to arrive at: $H \approx 1/2\left[g_e \mu_B B \sigma_z^e - g_n \mu_n B \sigma_z^n\right] + A\sigma_z^e \sigma_z^n$. The energy levels and eigenstates of this Hamiltonian are shown in Fig. 1(b). Note that we have ignored the coupling of both spins to uncontrolled degrees of freedom such as paramagnetic defects and phonons (coupling to lattice spins can be mitigated by the use of a ^{28}Si substrate). These environmental couplings will contribute to decoherence of the nuclear and electron spin states (e.g. [7]), and we will simply assume that this results in some effective relaxation and dephasing of the electron and nuclear spins. We see that the resonance frequency (Zeeman energy) of the electron is a function of the nuclear spin state. Therefore, our strategy will be to transfer the qubit state from the electron to the nucleus and then use EDMR to measure the nuclear spin. This is in effect a spin-to-resonance-to-charge conversion measurement. The state transfer can be performed by applying resonant microwave and radio frequency pulses to implement the qubit SWAP gate, or a variant of it (for more details, see Ref. [6]). Recently, coherent information transfer between electron and nuclear spins of P atoms in silicon was demonstrated in Ref. [8].

Measurement sensitivity and measurement-induced decoherence. Here we address the issue of the sensitivity of the differential EDMR current in the limit of single donor scattering. The on-resonant (I) and off-resonant (I_0) current differential (normalized) is [6]: $\frac{\Delta I}{I_0} \equiv \frac{I - I_0}{I_0} \approx -\alpha' s \langle \sigma_z \rangle_i^{ss} P_e^0 \frac{1/\tau_n}{1/\tau_t}$. Here $\alpha = \langle \Sigma_s - \Sigma_t \rangle|_{z=z_i} / \langle \Sigma_s + 3\Sigma_t \rangle|_{z=z_i}$, Σ_s and Σ_t are singlet and triplet scattering cross sections, respectively, and $\langle \cdot \rangle|_{z=z_i}$ denotes an average over the scattering region with the donor location in z (see Fig. 1(a)) held fixed [2, 9]. $\langle \sigma_z \rangle_i^{ss}$ is the equilibrium "polarization" of the single donor, an equilibrium spin projection value that is achieved in $\sim 10ns$, much faster than the observable times scales of the measurement [6]. s is a saturation parameters that characterizes how much microwave power is absorbed by the impurity and conduction electrons. The final term in the expression for the current differential represents the ratio between impurity scattering ($1/\tau_n$) and total scattering ($1/\tau_t$) rates. We assume $1/\tau_t = 1/\tau_0 + 1/\tau_n$, where $1/\tau_0$ is the scattering rate due to all other processes (such as surface roughness scattering and Coulomb scattering by charged defects).

To estimate the expected magnitude of this current differential, we begin by considering present state of the art 2DEG mediated EDMR experiments where this current differential is $\sim 10^{-7}$ (with T $\sim 5K, B \sim 0.3T$, a 2DEG channel area of $160 \times 20\mu m^2$, probe current 1μA, and a donor density of 2×10^{11}donors/cm^2) [4]. We assume that α' will be similar for the single donor device as in current experiments. Then in scaling down to a single donor, the first aspect to consider is the scattering rate ratio: $\rho \equiv \frac{1/\tau_n}{1/\tau_t}$. To first order this ratio can be kept constant if we scale the 2DEG area concomitantly with the donor number. From the channel area and density of current experiments, we extrapolate that a 2DEG area of $\sim 30 \times 30nm^2$ – well within the realm of current technology [10] – would keep ρ unchanged. Optimization of donor depth might relax this size requirement [9]. A higher order analysis would require detailed investigation of the device specific interface and intrinsic contributions to the other scattering processes, and hence to the channel mobility and τ_0. Related to this concern, the mobility of the 2DEG channel can be improved – e.g., by using hydrogen passivation to mitigate surface roughness at the oxide interface [11] – to increase ρ. We conservatively estimate a factor of 10 increase in $\Delta I/I_0$ from such improvements. The saturation parameter $s \sim 1$ for large enough microwave powers in the recent measurements [4] and so does not present an area for improvement. Finally, an avenue for significant improvement in signal is to increase the conduction electron polarization, P_c^0, which is currently $\sim 0.1 - 1\%$. This polarization is roughly proportional to the applied static magnetic field, and therefore a factor of 10 improvement is possible by operating at $B = 3T$. Additionally, spin injection techniques can be employed to achieve $P_c^0 > 10\%$ (e.g. [12, 13]), resulting in a 100-fold improvement in $\Delta I/I_0$. Hence, by improvements in device scaling and channel mobility, and by incorporating spin injection, we estimate a realistic, improved current differential of $\Delta I/I_0 \sim 10^{-4}$. Given this $\Delta I/I_0$ and a probe current of $I_0 \sim 1\mu A$, to achieve an SNR of 10 through shot-noise limited detection we require a τ_m satisfying $\left(\frac{\Delta I}{I_0}\right)\frac{I_0\tau_m}{e} > 10\sqrt{\frac{I_0\tau_m}{e}}$, where the left hand side is the signal, the right hand side is the accumulated shot-noise multiplied by the SNR, and e is the fundamental unit of electric charge. Solving this yields a measurement integration time of $\tau_m \sim 10^{-3}\ s$.

The final step in completing the measurement analysis is to confirm that the state of the nucleus does not flip within the measurement time, especially in the presence of the EDMR electron driving. We have shown that in moderate to large magnetic fields this is indeed the case, and the nuclear T_1 in the presence of electron driving will be comfortably larger than τ_m [6]. Also, we note that this has very recently been confirmed by explicit measurements of nuclear T_1 under the conditions of electron driving [14]. This analysis implies that once the measurement collapses onto a nuclear basis state, the nuclear spin state effectively remains there, and therefore the EDMR measurement satisfies the QND requirement on the qubit state. This QND aspect also makes the measurement technique an effective method for initializing the state of the nuclear spin.

REFERENCES

1. R. Vrijen, *et al.* Phys. Rev. A **62**, 012306, 2000; C. D. Hill, *et al.* Phys. Rev. B **72**, 045350, 2005; T. Schenkel, *et al.* App. Phys. Lett. **88**, 112101, 2006.
2. R. N. Ghosh and R. H. Silsbee, Phys. Rev. B **46**, 12508 (1992).
3. L. H. Willems van Beveren, *et al.* App. Phys. Lett. **93**, 072102 (2008).
4. C. C. Lo, *et al.* App. Phys. Lett. **91**, 242106 (2007).
5. J. M. Elzerman, *et al.* Nature **430**, 431 (2004); M. Xiao, *et al.* Nature **430**, 435 (2004).
6. M. Sarovar, *et al.* arXiv:0711.2343 [cond-mat] (2007).
7. R. de Sousa and S. Das Sarma, Phys. Rev. B **67**, 033301 (2003).
8. J. J. L. Morton, *et al.* Nature (in press). arXiv:0803.2021 [quant-ph] (2008).
9. R. de Sousa, C. C. Lo, and J. Bokor (2008), arXiv:0806.4638 [cond-mat.mes-hall].
10. S. J. Park, *et al.* J. Vac. Sci. Tech. B **22**, 3115 (2004).
11. K. Eng, R. N. McFarland, and B. E. Kane, Phys. Rev. Lett. **99**, 016801 (2007).
12. I. Appelbaum, B. Huang, and D. J. Monsma, Nature **447**, 295 (2007).
13. B. T. Jonker, *et al.* Nature Physics **3**, 542 (2007).
14. S. A. Lyon and A. M. Tyryshkin, *Private communication* (2008).

Realization of quantum non-demolition measurement of nuclear spin 1/2 of cold ytterbium atom

T. Takano*, R. Namiki* and Y. Takahashi†,*

*Department of Physics, Graduate School of Science, Kyoto University, Kyoto 606-8502, Japan
†CREST, JST, 4-1-8 Honcho Kawaguchi, Saitama 332-0012, Japan

Abstract. We have demonstrated a quantum non-demolition (QND) measurement with a collective spin of cold ytterbium atoms (^{171}Yb) via Faraday rotation interaction, and have observed $1.8^{+2.4}_{-1.5}$ dB spin squeezing. Since ^{171}Yb atoms have only a nuclear spin of one-half in the ground state, the system constitutes the simplest spin ensemble and is thus robust against decoherence. Furthermore, we have considered the atomic quantum swapping gate as a quantum information device using multiple Faraday rotation interactions, and have found that we can realize the quantum-domain performance for a realistic experimental condition.

Keywords: Quantum non-demolition measurement, Spin squeezing, Swapping
PACS: 03.67.-a, 42.50.Ct, 42.50Dv

Light has been a central player in quantum communication as a carrier of quantum information [1, 2]. On the other hand, atoms are suitable for the storage and processing of the quantum information, so the construction of the quantum interface between light and atoms is very important and directly linked to the construction of the quantum network [3, 4]. Especially, spin ensemble and light polarization is a good system for continuous variable quantum information processing with spin quantum non-demolition measurement (spin-QNDM) through Faraday rotation (FR) interaction as a quantum interface due to the robustness against decoherence and availability of relatively large interaction [5, 6, 7]. In fact, spin entanglement and spin squeezing by spin-QNDM was achieved by several groups [8, 9] and the atomic quantum memory was achieved [10].

In this paper, we report our successful experimental realization of spin-QNDM with laser-cooled ytterbium (^{171}Yb) atoms and short light pulses [11]. The important difference between our experiment and that based on alkali atoms [8, 9] is the fact that ^{171}Yb atoms have the simplest ground state with a nuclear spin of one-half and have no electron spin. Therefore, the system is robust against stray magnetic fields since the magnetic moment of nuclear spin is a thousandth of that of electron spin. By using short light pulses with a width of 100 ns, more than a hundred time operations are expected to be performed within the coherence time. Furthermore, we report our theoretical consideration of a quantum swapping gate between the light polarization and spin ensemble by three successive Faraday rotation interactions [12].

The interaction Hamiltonian between the atomic system composed of spin $\frac{1}{2}$ particles and the near-resonant light travelling in the z-direction is described by $H_{int} = \alpha S_z J_z$, and its time evolution operator is $U_{int}(t) = \exp(-i\alpha t S_z J_z)$, where α is a real constant, $\mathbf{S} = (S_x, S_y, S_z)$ is a Stokes operator in the Poincarè sphere representing the state of the

CP1110, *Quantum Communication, Measurement and Computing (QCMC), The Ninth International Conference,* edited by A. Lvovsky

light polarization, and $\mathbf{J}=(J_x,J_y,J_z)$ is a total spin operator, which is written as $\mathbf{J}=\sum\mathbf{j}$ ($\mathbf{j}$ is a spin operator of a single particle)[5, 6]. This interaction causes the rotation of the polarization proportional to J_z. This phenomenon is called paramagnetic Faraday rotation. From now, let us consider the situation that the light and the spins are initially polarized in the z-direction, i.e., $\langle S_x\rangle \simeq S$, and $\langle J_x\rangle \simeq J$. For simplicity, we use the normalized operators $\tilde{S}_y=S_y/\sqrt{S}$, $\tilde{S}_z=S_z/\sqrt{S}$, $\tilde{J}_y=J_y/\sqrt{J}$, and $\tilde{J}_z=J_z/\sqrt{J}$. Then, the time evolution operator is written as $U_{int}(t) = \exp(-i\kappa\tilde{S}_z\tilde{J}_z)$, where the interaction strength κ is given by $\kappa=\alpha t\sqrt{SJ}$. Then, the time evolution of these operators is $\tilde{S}_y(t) = \tilde{S}_y(0)+\kappa\tilde{J}_z(0)$, $\tilde{S}_z(t)=\tilde{S}_z(0)$, $\tilde{J}_y(t)=\tilde{J}_y(0)+\kappa\tilde{S}_z(0)$, and $\tilde{J}_z(t)=\tilde{J}_z(0)$. In the case of our experiment with ^{171}Yb atoms, κ is written as $\kappa = \frac{\Gamma\sigma_0\sqrt{SJ}}{3\pi w_0^2}\left(\frac{\delta-\delta_0}{(\delta-\delta_0)^2+(\Gamma/2)^2} - \frac{\delta}{\delta^2+(\Gamma/2)^2}\right)$, where $J\simeq 3.4\times 10^5$ is the total spin [13], $\Gamma=2\pi\times 29$ MHz is the natural full line width, σ_0=7.6×10^{-14} m^2 is the photon-absorption cross section of ^{171}Yb atom, $w_0=58$ μm is the beam waist, $\delta=2\pi\times 160$ MHz is the detuning from the $^1S_0 \leftrightarrow {}^1P_1$ ($F'=3/2$) states, and δ_0=$2\pi\times 320$ MHz is the frequency difference between the $F'=1/2$ and $F'=3/2$ states in the $^1P_1$ state [7, 14]. In our experimental conditions, the maximum value of κ is 0.62. At this value, the loss parameter $\varepsilon\equiv rt/2$ is 9.3×10^{-2} for S of 1.6×10^6, where r is the absorption rate [7].

The FR interaction satisfies a back-action evading condition $[H_{int},\tilde{J}_z]$=0, and makes a quantum correlation between $\tilde{J}_z$ and $\tilde{S}_y$. Thereby, the measurement of $\tilde{S}_y$ will essentially project the spin state into an eigenstate of $\tilde{J}_z$ and the variance of $\tilde{J}_z$ will be squeezed. The measurement of $\tilde{S}_y$ is said to be the QND measurement of $\tilde{J}_z$, and induces spin squeezing in the z direction [5, 6].

In contrast with the QND measurement in optics [15, 16], it is difficult to directly measure the spin states, and hence, we experimentally investigate the relations of two successive spin-QNDM. Suppose that two successive pulses interact with the atoms. Then, the Stokes operator of the first pulse $(\tilde{S}_{1,y},\tilde{S}_{1,z})$ and that of the second pulse $(\tilde{S}_{2,y},\tilde{S}_{2,z})$ are transformed as $\tilde{S}_{1,y}^{(f)}=\tilde{S}_{1,y}^{(i)}+\kappa\tilde{J}_{1,z}$, $\tilde{S}_{1,z}^{(f)}=\tilde{S}_{1,z}^{(i)}$, $\tilde{S}_{2,y}^{(f)}=\tilde{S}_{2,y}^{(i)}+\kappa\tilde{J}_{2,z}$, $\tilde{S}_{2,z}^{(f)}=\tilde{S}_{2,z}^{(i)}$, where $\tilde{J}_{1,z}$ and $\tilde{J}_{2,z}$ are the collective spin operators for the first pulse and second pulse, respectively. If both of the initial states are coherent states and the QND condition $\tilde{J}_{1,z}=\tilde{J}_{2,z}$ is satisfied, we have $\sigma_{1(2)}\equiv\langle\Delta(\tilde{S}_{1(2),y}^{(f)})^2\rangle=(1+\kappa^2)/2, \sigma_\pm\equiv\langle\Delta(\tilde{S}_{1,y}^{(f)}\pm\tilde{S}_{2,y}^{(f)})^2\rangle/2=(1+\kappa^2\pm\kappa^2)/2$. This implies that the FR interaction increases the individual variances $\sigma_{1,2}$ by the same factor of $\kappa^2/2$ and the positive correlation σ_+ by a factor of κ^2, while the negative correlation σ_- does not change.

To confirm these relations, we measured about 2600 pairs of the Stokes operators of the pulses for each arrangement of the experimental parameters in the following method. At first, typically 10^6 atoms are loaded in a magneto-optical trap (MOT) in 1 s and are released in the next 250 μs. Secondly, the atoms are polarized by the circularly-polarized resonant pumping pulse whose width is 14 μs. Then, two linearly-polarized probe pulses pass through the atoms and go into the polarization detector. The pulses have the same width of 100 ns and the interval between them is 15 μs. The detection of the second pulse completes the single measurement process of our experiment. Then, the atoms are recaptured by MOT and reused for the next cycle. For the single loading of atoms, we repeated 10 release-and-recapture cycles [17].

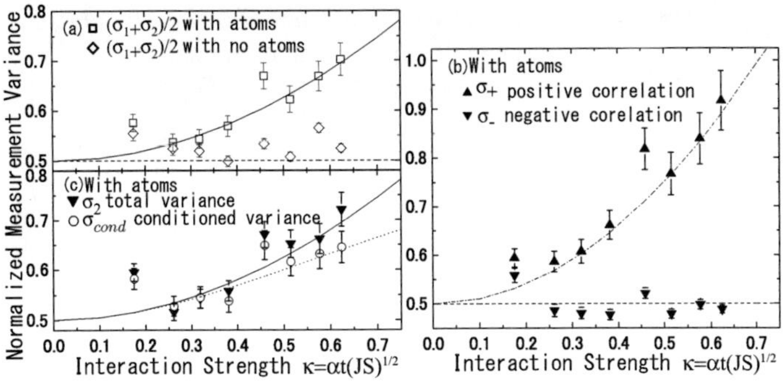

FIGURE 1. Measured variance as a function of the interaction strength κ. The solid curve, dash-dotted curve, dashed curve, and dotted curve are given by $(1+\kappa^2)/2$, $(1+2\kappa^2)/2$, $1/2$, and $\{1+\kappa^2/(1+\kappa^2)\}/2$ respectively. (a)Average of the individual variances $\sigma_{1,2}$ with and without atoms. At a large κ, it is apparent that $\sigma_{1,2}$ with atoms became larger than $\sigma_{1,2}$ without atoms. (b)$\sigma_\pm$ with atoms. We successfully observed larger value of σ_+ than $\sigma_{1,2}$ with atoms in Fig. 1(a), while σ_- remains the value of 1/2 corresponding to the light-shot noise. (c)Conditioned variance σ_{cond} and σ_2 with atoms. At a large κ, it is apparent that σ_{cond} became smaller than σ_2.

We have performed this measurement for a various interaction strength κ by changing the incident photon number $2S$ and observed the behavior of the variances and correlations. Figure 1(a) shows the average of the individual variances $\sigma_{1,2}$ with and without atoms. At a large κ, it is apparent that $\sigma_{1,2}$ with atoms became larger than $\sigma_{1,2}$ without atoms. Figure 1(b) shows the values of $\sigma_\pm$ with atoms. We successfully observed larger value of σ_+ than $\sigma_{1,2}$ with atoms in Fig. 1(a), while σ_- remained the value of 1/2 corresponding to the light-shot noise. Finally, we show that the z-component of the spin J_z is conditionally squeezed. In order to observe this spin squeezing, one may count the data of $\tilde{S}_{2,y}^{(f)}$ only when $\tilde{S}_{1,y}^{(f)}$ takes a specific value. Consequently, we divided the data around the center into 21 bins of $\tilde{S}_{1,y}^{(f)}$ and took the average of the variance of $\tilde{S}_{2,y}^{(f)}$ of each bin. In Fig. 1(c), we showed the conditioned variance σ_{cond} and σ_2 for various interaction strength κ. Note that the error bars are calculated from the statistical error of the variance measurement and the atomic number fluctuation. At a large κ, it is apparent that σ_{cond} became smaller than σ_2. Here, the solid curve, dash-dotted curve, dashed curve, and dotted curve are given by $(1+\kappa^2)/2$, $(1+2\kappa^2)/2$, $1/2$, and $\{1+\kappa^2/(1+\kappa^2)\}/2$. As one can see, the experimental results have the values near to the above theoretical estimation and spin squeezing was achieved with the degree of $1.8^{+2.4}_{-1.5}$ dB when $\kappa = 0.62$.

We also consider the atomic quantum swapping gate as a quantum information device [12]. Figure 2(a) shows the setup. It is realized by three successive Faraday rotation interactions with the $\lambda/4$ wave plates and delay lines. In an ideal condition, the quantum states of the light and atoms are exchanged. We have clarified the criterion of the quantum-domain (QD) performance of the swapping gate [12, 18, 19]. The colored plane in Fig. 2(b) shows the average fidelity $\bar{F}$ of the swapping gate with an input set of coherent states, which has the two-mode Gaussian distribution whose width is characterized by $1/\lambda$, and the clear plane means the classical bound $\bar{F}$. If $\bar{F} > \bar{F}_C$ is satisfied, the swapping gate is in QD. We have found that the QD performance can be achieved with a relatively wide range of experimental parameters.

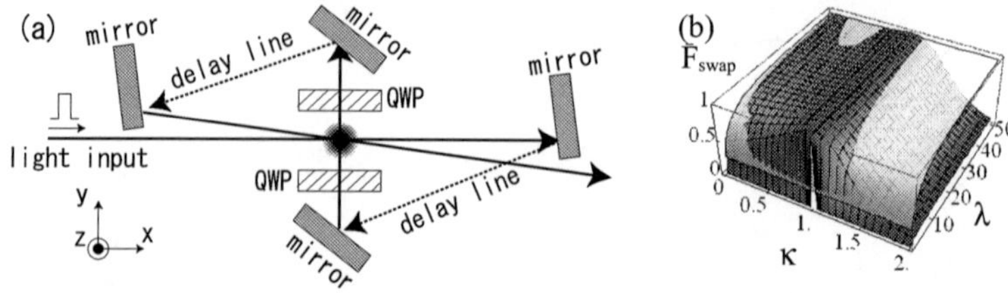

FIGURE 2. (a)Setup for the swapping gate. A z-polarized light pulse propagates through a z-polarized atomic spin ensemble three times. To evade the overlap in the atomic area, delay lines are set in each paths. A QWP represents a $\lambda/4$ wave plates. First $\lambda/4$ wave plate rotates S_x-S_y space by -90 degrees and second one rotates it by 90 degrees. After the third interaction, the information of light polarization is swapped for the one of the atomic spin ensemble. (b)Average fidelity of our swapping gate $\bar{F}_{swap}$. The clear plane shows the classical bound $\bar{F}_C$. The area in which $\bar{F}_{swap}$ surpasses $\bar{F}_C$ is QD.

This work was supported by the Grant-in-Aid for Scientific Research of JSPS (No. 18204035) and GCOE Program "The Next Generation of Physics, Spun from Universality and Emergence" from MEXT of Japan. T. T. and R. N. are supported by JSPS Research Fellowships.

REFERENCES

1. D. Bouwmeester, *et al.*, Nature (London) **390**, 595 (1997).
2. F. Grosshans, *et al.*, Nature (London) **421**, 238 (2003).
3. L.-M. Duan, M. D. Lukin, J. I. Cirac , and P. Zoller, Nature (London) **414**, 413 (2001).
4. J. I. Cirac, P. Zoller, H. J. Kimble, and H. Mabuchi, Phys. Rev. Lett. **78**, 3221 (1997).
5. Y. Takahashi *et al.*, Phys. Rev. A **60**, 4974 (1999).
6. A. Kuzmich, N. P. Bigelow and L. Mandel, Europhys. Lett. **42**, 481 (1998)
7. L.-M. Duan, *et al.*, Phys. Rev. Lett. **85**, 5643 (2000).
8. B. Julsgaard, A. Kozhekin, and E. S. Polzik, Nature (London) **413**, 400 (2001).
9. A. Kuzmich, L. Mandel, and N. P. Bigelow, Phys. Rev. Lett. **85**, 1594 (2000).
10. B. Julsgaard, *et al.*, Nature **432**, 482 (2004).
11. T. Takano, *et al.*, ArXiv:0808.2353[quant-ph].
12. T. Takano, *et al.*, Phys. Rev. A **78**, 010307(R) (2008).
13. Note that our atomic number is much smaller than previous experiments[8, 9]. By this character, we needed a magetic field control by only a small amount.
14. M. Takeuchi, *et al.*, Appl.Phys.B **83**, 107 (2006).
15. N. Imoto, H. A. Haus, and Y. Yamamoto, Phys. Rev. A **32**, 2287 (1985).
16. R. Bruckmeier, H. Hansen, and S. Schiller, Phys. Rev. Lett. **79**, 1463 (1997).
17. Note that we have compensated a zero-vias of the measured data depending on which one of the 10 cycles the data belongs to, but the behavior of the data is essentially the same when there is no compensation.
18. K. Hammerer, M. M. Wolf, E. S. Polzik, and J. I. Cirac, Phys. Rev. Lett. **94**, 150503 (2005).
19. R. Namiki, M. Koashi, and N. Imoto, Phys. Rev. Lett. **101**, 100502 (2008).

CHAPTER 11

QUANTUM KEY DISTRIBUTION

Squashing Models for Optical Measurements in Quantum Communication

Normand J. Beaudry*, Tobias Moroder*,† and Norbert Lütkenhaus*,†

*Institute for Quantum Computing, University of Waterloo, Waterloo, Canada
†Quantum Information Theory Group, Institute of Theoretical Physics I, and Max-Planck Research Group, Institute of Optics, Photonics and Information, University Erlangen-Nuremberg, Erlangen, Germany

Abstract. Measurements with photodetectors are naturally described in the infinite dimensional Fock space of one or several modes. For some measurements a model has been postulated which describes the full mode measurement as a composition of a mapping (squashing) of the signal into a small dimensional Hilbert space followed by a specified target measurement. We present a formalism to investigate whether a given measurement pair of full and target measurements can be connected by a squashing model. We show that some measurements used in the BB84 protocol do allow a squashing description, although the corresponding six-state protocol measurement does not. As a result, security proofs for detectors with squashing models can be based on the assumption that the eavesdropper forwards at most one photon, while the same does not hold for measurements without squashing models.

Keywords: Quantum Communication, Quantum Optics
PACS: 42.50.Ex, 03.67Dd, 03.67.Hk, 42.79.Sz

In quantum communication, theories of protocols are often created under the assumption that only single photons (or qubits) are used. However, in the realization of these protocols, the photons are described in the Fock space of spatio-temporal modes. The goal of this work is to reduce the large-dimensional measurements of these systems to the small-dimensional ones required for many theories, such as security proofs for quantum key distribution (QKD) protocols. Here we present a powerful tool that can be used to ease the analysis of optical implementations of quantum communication protocols, which has been outlined previously in [1].

For some QKD protocols, it has been assumed that there exists a model that equates a measurement on multiple photons to one that is first preceded by a map that reduces (squashes) the incoming signal to a single photon or the vacuum, which is forwarded to the same measurement but now on a low-dimensional Hilbert space. Our result specifies how to determine whether such a squashing map exists. In the context of QKD, one typically assumes the *calibrated device scenario* in which the detection device is trusted and known. Then if a squash model exists, the corresponding squashing map can become part of the eavesdropper's (Eve's) attack: without loss of generality we can assume that Eve sends a signal in a qubit or the vacuum to the receiver, Bob. However, our analysis is more general than described here; the small-dimensional measurement need not be on a qubit and the vacuum, it can be on any small-dimensional system.

CP1110, *Quantum Communication, Measurement and Computing (QCMC), The Ninth International Conference,* edited by A. Lvovsky

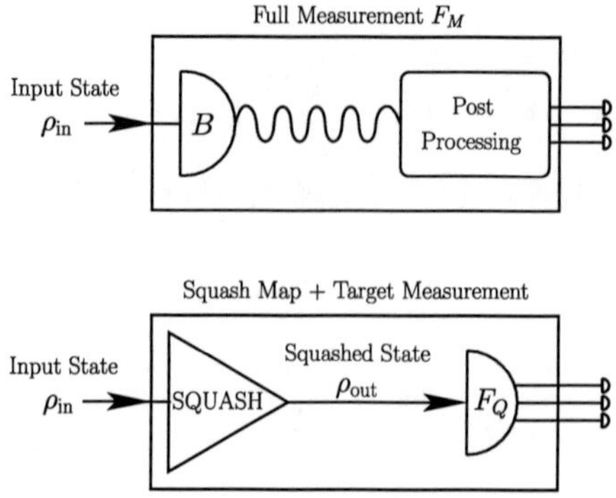

FIGURE 1. The full measurement F_M (above) has a general optical input ρ_{in}, which is first measured by a receiver's physical detector B, followed by classical post-processing. The squashed measurement (below) has the same general optical input ρ_{in}, which is then squashed by a map Λ to a smaller Hilbert space, followed by a fixed physical measurement F_Q. It is required that both of these measurements produce the same output statistics for all ρ_{in}.

DEFINITION OF A SQUASHING MODEL

A squashing model is the statistical equivalence of a high-dimensional full measurement, F_M, and a squashing map, Λ, followed by a low-dimensional target measurement, F_Q. In other words, the probability of any outcome of the measurement F_M for any input is the same as the probability of any outcome of the measurement F_Q when it is preceded by a specific squashing map (see Fig. 1). If a squashing map exists such that this condition holds, then we say that the full measurement F_M has a squashing model with respect to the target measurement F_Q.

The existence of a squashing model connecting a full measurement and target measurement depends on the existence of a squashing map that precedes the target measurement. A physical map that does this must be a trace-preserving completely positive map, Λ, whose adjoint can be described by a positive semidefinite operator τ via the Choi-Jamiołkowski isomorphism [2]. The operator τ is given on a Hilbert space QQ' by applying the map to half of a maximally entangled state $|\psi^+\rangle = 1/\sqrt{d}\sum_{i=1}^{d}|i\rangle_Q|i\rangle_{Q'}$, where $d = \dim(QQ')$ as $\tau = \Lambda^\dagger \otimes \mathrm{id}(|\psi^+\rangle\langle\psi^+|)$. The conditions for the existence of a squashing map can be phrased formally as:

$$\tau^R|F_Q^{(i)}\rangle\rangle = |F_M^{(i)}\rangle\rangle, \tag{1a}$$

$$\langle k,k'|\tau^R|l,l'\rangle = \langle k,l|\tau|k',l'\rangle, \tag{1b}$$

$$\tau^\dagger = \tau \geq 0, \tag{1c}$$

where $F_Q^{(i)}, F_M^{(i)}$ are the POVM elements for the target and full detectors respectively. Here we introduce the vector notation of an operator, $O = \sum_{i,j} o_{i,j}|i\rangle\langle j|$, as $|O\rangle\rangle = \sum_{i,j} o_{i,j}|i\rangle|j\rangle$. These conditions can be solved analytically, and can also be efficiently solved using convex optimization; searching for completely positive maps using these techniques has been used, for example, in [3, 4].

Now we will discuss two examples of detectors that detect the polarization of photons: the measurement that is performed in the BB84 protocol [5], and the measurement performed in the six-state protocol [6, 7]. We investigate whether there exists a squashing model for these measurements from their detectors which accept any number of photons in the Hilbert space of two orthogonal polarization modes and the vacuum, to one which accepts only single photons, also in two orthogonal polarization modes and the vacuum.

BB84 Protocol: Active Basis Choice

The BB84 measurement is composed of a polarizing beam splitter, which can be set to separate in one of two bases: either the horizontal/vertical basis or the +45/-45 degree basis. At the end of each arm of the polarizing beam splitter there is a threshold photodetector which cannot resolve the number of incoming photons. In the event that multiple photons are input to this detector, then double clicks (clicks between both detectors) can occur, and so a post processing of these events is neccesary (see Fig. 2). For this measurement there exists a squashing model to the target detector described above. The squashing map is made up of first a QND measurement of the number of incoming photons, followed by a projection onto a four-dimensional subspace, and its orthogonal complement, so that a simple solution to Eqns. (1) can be found. This squashing map has also been found independently by Tsurumaru and Tamaki [8]. The squashing model is also useful in other quantum communication contexts that use this same detector, since the model depends on the detector and not on the method of its use. For example, in security proofs of entanglement-based QKD that use the same polarization measurement as the BB84 protocol can also be generalized via the squashing model.

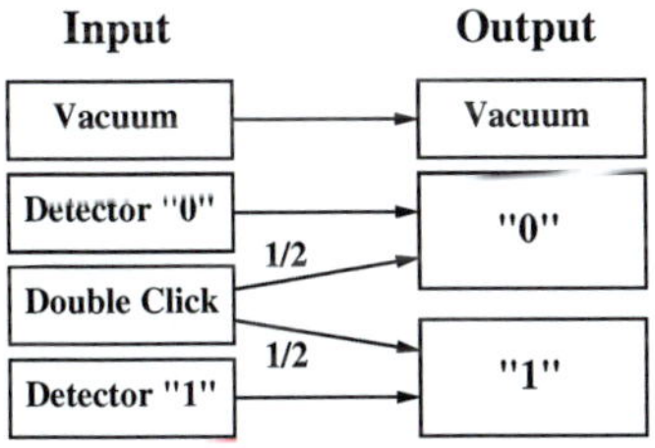

FIGURE 2. The standard classical post processing of the basic BB84 and six-state protocol detector events in which double clicks are randomly assigned to either a 0 or 1. This post processing has been evaluated previously in the analysis of QKD [9].

Six-state Protocol: Active Basis Choice

The six-state measurement is the same as the BB84 measurement, except that there is a third setting to the polarizing beam splitter which splits photons according to a circular basis. The post processing of double click events is again randomly assigned to either single detection events (see Fig. 2). In this case, the operator τ is completely determined by the linear constraints Eqn. (1a), since the measurement operators F_Q form a complete basis for their Hilbert space. In this case, we find that τ is not a positive map, and therefore there does not exist a squashing model for this detection setup.

Passive Basis Choice

The above BB84 and six-state measurements can be passive, meaning that the basis choice is made by a 1:1 ratioed beamsplitter (or 1:1:1 for the six-state measurement) followed by copies of the a polarizing beam splitter oriented in their respective bases, followed by threshold detectors. Here the measurement is the same as the active case, with the exception of double-clicks between different arms of the ratioed beamsplitter, which are post processed to a random bit and basis. As a result, the squashing map can be thought of as $\tau_{passive} = p\tau_{active} + (1-p)\mathbb{1}$, where p is a fixed probability, and the identity map, $\mathbb{1}$, outputs a mixed qubit $\mathbb{1}_Q/2$. Clearly for the BB84 measurment this map exists and is positive semidefinite. As for the six-state measurement, there is enough of the identity map added to the negative active squashing map to make the whole map positive, and therefore the squashing map exists for the passive six-state measurement.

CONCLUSION

To conclude, we have provided necessary and sufficient linear conditions to solve for a squashing map that connects a high-dimensional full measurement to a low-dimensional target measurement. In QKD, this generalizes a qubit-based security proof to one that applies to the general optical implementation. Such is the case in the BB84 detector with active or passive basis choice, and the six-state detector with passive basis choice. If there does not exist a squashing map for a given full and target measurement, then this method does not help to generalize the security proof, as is the case with the six-state measurement with an active basis choice. Our results will help to simplify quantum communication in other contexts, such as entanglement verification of optical modes with threshold detectors.

ACKNOWLEDGMENTS

The authors acknowledge financial support via the European Projects SECOQC and QAP, by the NSERC Discovery Grant, QuantumWorks and CSEC. We also thank Marco Piani and Marcos Curty for helpful discussions and feedback.

REFERENCES

1. N. J. Beaudry, T. Moroder, and N. Lütkenhaus, *Phys. Rev. Lett.* **101**, 093601 (2008).
2. (see for example) K. Życzkowski, and I. Bengtsson, *Open Syst. Inf. Dyn.* **11**, 3–42 (2004).
3. M. Reimpell, and R. F. Werner, *Phys. Rev. Lett.* **94**, 080501 (2005).
4. A. S. Fletcher, P. S. Shor, and M. Z. Win, *Phys. Rev. A* **75**, 7 (2007).
5. C. H. Bennett, and G. Brassard, "Quantum Cryptography: Public key distribution and coin tossing.," in *Proceedings of IEEE International Conference on Computers, Systems, and Signal Processing, Bangalore, India*, IEEE, New York, 1984, pp. 175–179.
6. D. Bruß, *Phys. Rev. Lett.* **81**, 3018–3021 (1998).
7. H. Bechmann-Pasquinucci, and N. Gisin, *Phys. Rev. A* **59**, 4238–4248 (1999).
8. T. Tsurumaru, and K. Tamaki, *Phys. Rev. A* **78**, 032302 (2008).
9. N. Lütkenhaus, *Appl. Phys. B* **69**, 395 (1999).

SECOQC: Major results, the QKD-Network Prototype in Vienna

Momtchil Peev

Austrian Research Centers GmbH – ARC, TechGate Vienna, Donau-City-Straße 1, 1220 Vienna, Austria.
on behalf of the European Integrated Project SECOQC

Abstract. The main results of the integrated EC project SECOQC are presented with a focus on the SECOQC prototype – a Quantum Key Distribution (QKD) back-bone network, formed by eight QKD-Link devices.

Keywords: Quantum Key Distribution, Trusted Repeater QKD Network.
PACS: 03.67.Dd, 42.50.Ex

INTRODUCTION

The rapid progress in theory and experiment of Quantum Key Distribution (QKD) techniques has been reflected by a number of successful demonstrations in the past years. Many groups all over the world have put forward QKD-Links, operating in the standard point-to-point modus.

Although the basic enabling technology has yet to mature in terms of performance (i.e. maximal distance and key generation rate) as well as in terms of stability of operation of QKD devices, several research spin-offs have already entered the market (IdQuantique, MagiQ, Smart Quantum). Additionally a number of internationally operating enterprises (e.g. Toshiba, Thales, HP, NEC) currently develop and demonstrate prototypes. The secure communication solutions pursued are typically based on dedicated high-end symmetric encryptors with frequent key change, fresh key being constantly generated by QKD devices. The resulting QKD-Link-Encryptors offer comparable functionality to existing products while featuring higher levels of security. However, broad proliferation of QKD systems is hindered by a number of road blocks revolving typically around: the point-to-point paradigm and correspondingly the quadratic scaling of the initial secrets with the number of users, the question of integrability in existing networks, the price of QKD devices, but also around issues like missing standards. As a result, QKD appears to be restricted to a relatively narrow niche market.

The EC funded FP6 Integrated Project SECOQC (2004-2008) [1] focused on overcoming these major limitations by implementing the novel concept of QKD-networks [2]. Another major goal of the project was further development of a of QKD technology in general, and decisively increasing the technological maturity of a

CP1110, *Quantum Communication, Measurement and Computing (QCMC), The Ninth International Conference,* edited by A. Lvovsky

number of QKD device-types in particular. The results of this development, and specifically the functionality of the operating SECOQC prototype, have been demonstrated on October 8, 2008 in Vienna.

SECOQC QUANTUM-BACK-BONE (QBB) NETWORK

The main SECOQC goal was to design, implement, take into operation and test an integrated secrets' distribution network that connects, for test purposes, five subsidiaries of SIEMENS in the city of Vienna (Fig.1.). A wide range of heterogeneous QKD-systems engineered by SECOQC partners have been connected via Node Modules, which provide standard interfaces and overtake the novel network-wide secret distribution protocols. The QBB, deployed on a typical telecom metropolitan area network, is a crucial step towards demonstrating the feasibility of networked, multi-point to multi-point QKD. A target market of quantum based communication solutions for organizations with distributed subsidiaries/facilities such as governmental institutions, companies and banks is envisaged. To this end a number of supporting but future oriented efforts such as a SECOQC/ETSI standardization initiative, launched in summer 2008, are underway.

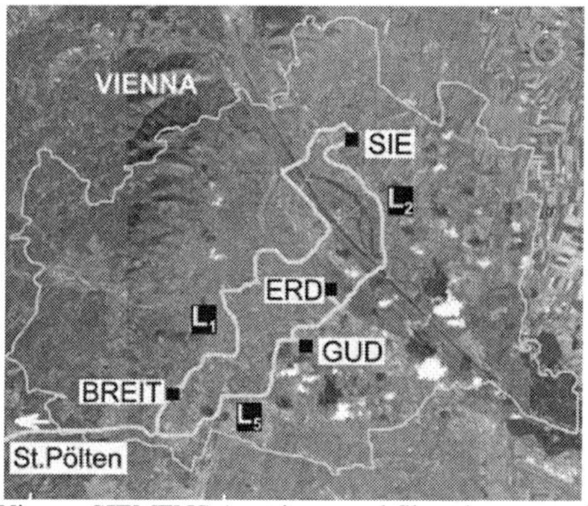

FIGURE 1. City map of Vienna. SIEMENS Austria-owned fiber ring network with stations SIE, ERD, GUD, BREIT and "St. Pölten", connected by QBB-links.

A QKD-network offers advantages similar to those that a telecommunication network provides in comparison with direct point-to-point links: longer communication distances and higher rates; reduced number of new point-to-point links needed to register new users (linear instead of quadratic scaling); improved reliability of communication, etc.

In general, there are different ways to extend QKD from a point-to-point secure key generation service to a network-wide, information-theoretically secure key agreement [3]. SECOQC follows the only wide-range solution compatible with current technology – the trusted repeater paradigm. In the latter it is assumed that the network nodes, situated in secure locations are connected by QKD-Links and can be trusted. The trusted-repeater QKD network put forward by SECOQC features a specific highly integrated design:

- The key provided on a QKD-Link basis is collected in dedicated key-stores in the nodes.
- The point-to-point secure links allow distribution of secrets along the network on a node-to-node basis.
- The protocols for secrets' routing and transport ensure information-theoretic security while enabling optimal utilization of secret transport capacities.

Finally it should be underlined that the pay-load secure communication of data is logically completely decoupled from the distribution of secrets. It takes place on an independent application layer for which the functionality of the QKD network is encapsulated and reduces to provision of key material on demand.

QKD-LINK DEVICES FOR THE SECOQC-QBB

The dedicated effort of SECOQC for advancing the QKD technology has resulted in the development of highly-mature QKD-Link devices. For the stage of prototype deployment, all devices had to comply with stringent performance criteria: continuous, stand-alone key-generation rate of more than 1 kbit/s at 25 km over standard telecom-fiber. The following QKD devices form the QBB-network [1]:

Attenuated Laser Pulses: The Swiss company IdQuantique delivers upgraded versions three commercially available "Clavis" device-pairs, usually referred to as "Plug & Play" [4]. *Coherent-One-Way:* The team of N. Gisin at GAP (University of Geneva) provides one QKD system in which the position of non-empty time slots of attenuated light pulses carries the transmitted bit value [5]. *One-way, decoy states:* The group of A. Shields (Toshiba UK) brings a phase-encoding QKD system with two interferometers, stabilized by pulses that are time-multiplexed with the quantum signals [6]. *Entangled photons:* The group of A. Zeilinger (University of Vienna) provides a QKD-system in which correlations of entangled photon-pairs are used to generate the key [7]. *Continuous Variables:* The consortium CNRS-Thales-ULB lead by P. Grangier develops a continuous variable QKD system, in which homodyne detection of coherent light pulses replaces single photon detection [8].

The "last mile" (80 m), access *Free Space Link* has been developed by an LMU (Munich, Germany) team led by H. Weinfurter [9]. It offers a high rate (>10kbit/s), stable connectivity to the QBB in a 24/7 operation regime.

The security of the prototype QKD devices has been thoroughly analyzed by a dedicated quantum information theory group lead by N. Lütkenhaus. This analysis catalyzed a comprehensive review of security of practical QKD systems [10]. The work of this group resulted also in a number of fundamental advances in QKD theory such as novel security proof techniques based on information-theoretical notions [11-12], finite key-analysis [13], and recently, squashing detector models [14].

An important aspect of the SECOQC effort towards the advancement of QKD technology was a dedicated initiative on developing novel and/or significantly increasing the performance of single photon sources and detectors. A detailed survey of this activity surpasses the scope of this presentation. We would only highlight the advent of self-differencing circuits, allowing clock rates that exceed 1GHz with InGaAs APDs [15].

QKD NETWORK PROTOCOLS

SECOQC has designed and implemented a novel interface between the different QKD-Links and the node module, which hosts all higher level protocols. This interface interacts with a Quantum Point-to-Point Protocol (Q3P) providing the QKD-Link devices with a classical authentic communication channel needed for key distillation. Q3P simultaneously offers crypto-services to the upper layers in the network. Most importantly, Q3P serves as Point-to-Point Protocol between a pair of QBB nodes and manages the key-stores.

The QKD Routing Layer (QKD-RL) Protocol [16] manages the routing information within the QBB nodes. QKD-RL follows the pattern of OSPF but includes essential modifications to address the specific requirements of secure key material routing.

The QKD Transport Layer (QKD-TL) Protocol [16] is inspired by TCP/IP but introduces new approaches dealing with highly congested networks based on QKD-generated key resources. QKD-TL enables users to exchange confidential and authentic information across the network on an end-to-end basis.

All three protocols outlined above are independently designed and each presents a standard-like interface, which can easily be introduced in current telecom networks. Upper layers need not be modified to use unconditionally secure network connectivity.

CONCLUSIONS

SECOQC has successfully demonstrated the functionality of an integrated QKD network, and pushed forward the development of QKD technology. It is an important step towards practical QKD, achieved with the efforts of all 41 project partners.

ACKNOWLEDGMENTS

This paper has been written within the EC Project SECOQC (IST-2002-506813).

REFERENCES

1. http://www.secoqc.net.
2. A. Poppe, M. Peev, and O. Maurhart, *IJQI* **6**; 209-218 (2008).
3. R. Alléaume et al, *quant-ph*/0701168.
4. G. Ribordy et al., *Electronics Letters* **34**, 2116 – 2117 (1998); http://www.idquantique.com
5. D. Stucki, N. Brunner, N. Gisin, V. Scarani, and H. Zbinden, *Appl. Phys. Lett.* **87**, 194108 (2005).
6. J. F. Dynes, Z. L. Yuan, A. W. Sharpe, and A. J. Shields, *Opt. Expr.* **15**, 8465-8471 (2007).
7. H. Hübel et al., *Opt. Expr.* **15**, 7853-7862 (2007).
8. J. Lodewyck et al. *Phys. Rev. A* **76**, 042305 (2007).
9. H. Weier et al., *Fortschr. Phys.* **54**, 840-845 (2006).
10. V. Scarani et al., *Rev. Mod. Phys.* – accepted for publication; *quant-ph*/0806.0120
11. B. Kraus, N. Gisin, R. Renner, *Phys. Rev. Lett.* **95**, 080501 (2005).
12. R. Renner, N. Gisin and B. Kraus, *Phys. Rev. A* **72**, 012332 (2005).
13. V. Scarani and R. Renner, *Phys. Rev. Lett.* **100**, 200501 (2008).
14. N. J. Beaudry, T. Moroder and N. Lütkenhaus, *Phys. Rev. Lett.* **101**, 093601 (2008).
15. Z. L. Yuan, B. E. Kardynal, A. W. Sharpe, and A. J. Shields., *Appl. Phys. Lett.* **91**, 041114 (2007).
16. M. Dianati, R. Alléaume, M. Gagnaire, und X. Shen, *J. Sec. Com. Net.* **1**, 57-74 (2008).

Scattering Effects on QKD Employing Simultaneous Classical and Quantum Channels in Telecom Optical Fibers in the C-band

G. B. Xavier[a,*], G. Vilela de Faria[a], G. P. Temporão[a] and J. P. von der Weid[a]

[a]*Center for Telecommunication Studies, Pontifical Catholic University of Rio de Janeiro, Rua Marquês de São Vicente, 225 Gávea, Rio de Janeiro, Brazil.*
[*]*Corresponding author: guix@opto.cetuc.puc-rio.br*

Abstract. Almost all Quantum Key Distribution experiments in optical fibers performed so far have used the so called "dark fibers", which are exclusively used as the quantum channel. As quantum communications make its way towards commercial applications, there is a clear advantage in using fibers with simultaneous classical signals, via wavelength multiplexing, for that purpose. We show what is the noise impact of transmitting classical optical signals alongside a quantum channel for a fixed fiber length.

Keywords: Quantum communications, Optical networks, Quantum Key Distribution
PACS: 03.67.Hk, 03.67.Dd

Quantum Key Distribution (QKD) [1] has evolved into a mature technology. It has reached commercial status, and as such it is attracting increasing attention from the engineering community. There is a growing trend in moving QKD from laboratory experiments into real world systems. Since QKD employs single photons as the means of transmission, optical fibers are extensively used for this purpose.

In order to implement QKD into existing optical networks, the ability to perform a quantum key exchange on a fiber already used by a telecommunication carrier will be needed. This is also important in terms of lowering costs for system implementations. Most QKD experiments have been carried out employing so called dark fibers [2], which is a fiber only used for the quantum channel. We discuss here if the noise generated from classical channels can allow QKD to take place in a lit fiber.

We focus on the impact of Raman and Rayleigh scattering [3] caused by a standard laser at a C-band wavelength on nearby quantum channels in the same fiber. Previous works have recently studied this problem [4-5]; however, the focus was on employing the QKD channel at 1310 nm, far from the telecom C-band around 1550 nm. Broadband Raman and Rayleigh scattering are generated inside an optical fiber from any classical channel. At low powers this noise has no impact on telecom systems, but it may be overwhelming for a QKD setup.

Our experiment consists of observing the noise collected by an InGaAs commercial single photon avalanche detector (SPAD), generated from a tunable laser operating at typical telecom level optical power. We initially performed the noise measurements with a counter-propagating setup, as shown in Fig. 1. We used a double Dense

CP1110, *Quantum Communication, Measurement and Computing (QCMC), The Ninth International Conference,* edited by A. Lvovsky

Wavelength Division Multiplexer (DWDM) optical filter before the SPAD centered at 1546.12 nm, each filter with over 50 dB extinction ratio. A circulator was used to couple the light from the tunable laser into the optical fiber, in a counter propagating way. A notch filter (Bragg grating) centered at 1546.12 nm was connected in series with the tunable laser, to prevent Amplified Spontaneous Emission (ASE) noise generated from the laser to reach the SPAD through Rayleigh backscattering and connector reflections. We verified that the notch filter removed all noise contributions from Rayleigh backscattering by using a strong reflection (-8dB) replacing the fiber. We also verified that the noise contributions come strictly from the fiber, by performing a measurement with the same setup without the spool.

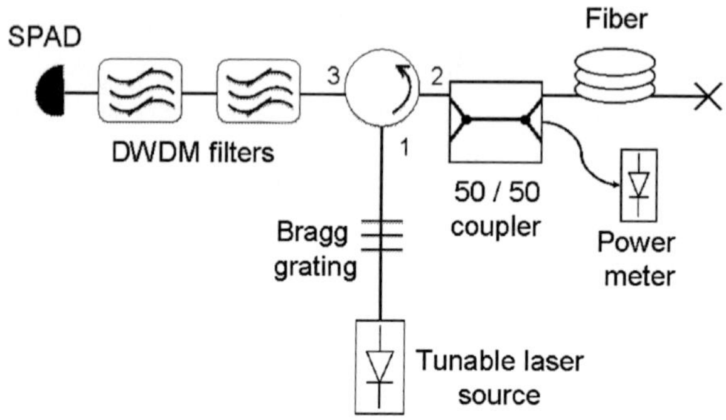

FIGURE 1. Counter-propagating setup. The power meter is used to normalize the measurement results due to the tunable laser source power fluctuations over the wavelength range.

The measured results are shown in Fig. 2, for similar lengths of Dispersion Shifted (DS) and Standard (SMF-28) fibers.

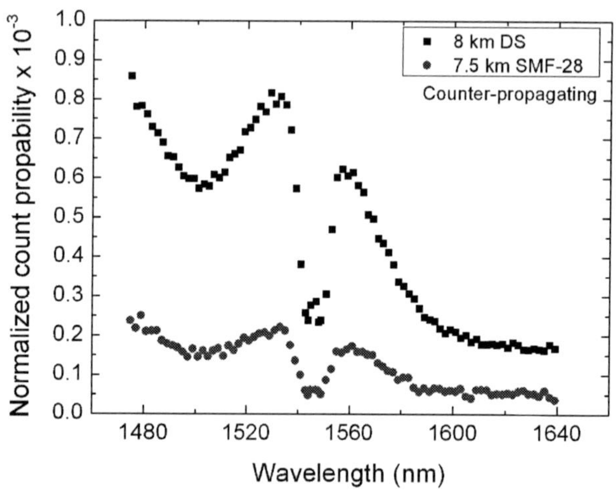

FIGURE 2. (Color online) Count probability per 1ns gate for 1mW (0dBm) of input CW optical power as a function of the tunable laser wavelength.

Next we measured the noise dependency on the input CW optical power:

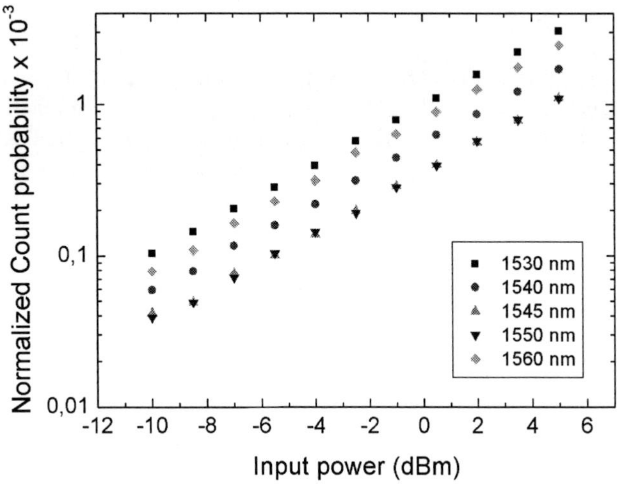

FIGURE 3. (Color online) Count probability for a 1 ns gate in a counter-propagating setup, for 8 km of DS fiber as a function of input power and wavelength.

It is clear that the count probability varies linearly with input optical power. We can be certain that the noise we are measuring is not Rayleigh backscattering (since we removed it by proper filtering). In Fig. 4 we show the experimental setup to measure the co-propagating noise.

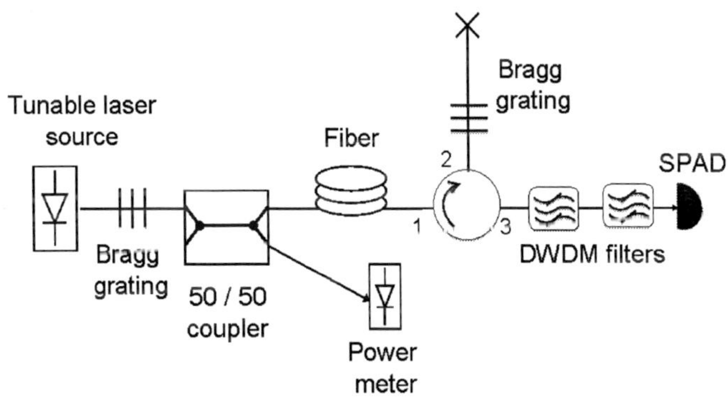

FIGURE 4. Co-propagating setup.

The results obtained from the co-propagating setup are presented in Fig. 5. The results obtained show that the noise contribution from a 0 dBm input power classical channel (a typical value in classical lightwave systems) at a different wavelength than the quantum one is around two orders of magnitude higher than the dark count probability of commercial InGaAs SPADs. This fact cannot be overlooked when attempting to perform a QKD transmission over a lit fiber. As expected, the noise contribution is higher from a DS fiber than a standard one, due to the differences in fiber core radii.

From the results we can conclude that if one wishes to perform QKD in the C-band with wavelength multiplexed classical optical traffic, the best approach is to place the classical channels as close as possible to the quantum channel, as both Stokes and anti-Stokes Raman noise increases with the wavelength separation in the C-band. In order to completely remove the noise contribution, a solution would be to synchronously shut down the classical channel for the single photon to be transmitted. Also, since Raman scattering is a linear process, we find that even if the power is considerably reduced, the noise is still present contributing to the quantum bit error rate.

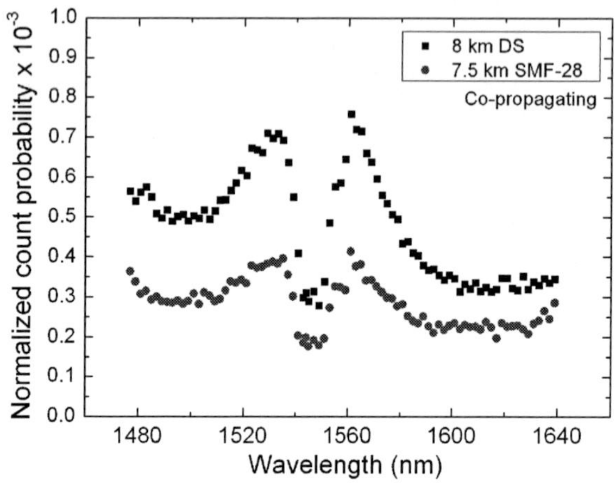

FIGURE 5. (Color online) Count probability per 1ns gate for 1mW (0dBm) of input CW optical power as a function of the tunable laser wavelength.

ACKNOWLEDGMENTS

The authors wish to thank H. Zbinden and N. Gisin for helpful discussions and for the single photon detector lent, and FAPERJ and CNPq for the financial support.

REFERENCES

1. N. Gisin, G. Ribordy, W. Tittel and H. Zbinden, *Rev. of Mod. Phys.* **74,** 145 (2002).
2. C. Gobby, Z. L. Yuan and A. J. Shields, *Appl. Phys. Lett.*, **84**, 3762 (2004).
3. G. P. Agrawal, "*Nonlinear Fiber Optics Third Edition*" Academic Press, USA 2001.
4. D. Subacius et al, *Appl. Phys. Lett.* 86, 011103, December 2004.
5. N. Iweke et al, Applied Physics Letters, Vol. 87, 174103, October 2005.

192 km relay transmission and HDTV transmission experiments by Quantum Yuen-2000 transceiver

S.Akutsu*, Y.Doi*, T.Hosoi*, M.Honda*, K.Harasawa*, O.Hirota† and T.Katayama**

*Hitachi Information and Communication Engineering Ltd, Yokohama, Japan
†Research Institute of Tamagawa University, Tokyo, Japan
**SoftBank Telecommunication,Tokyo, Japan

Abstract. We are concerned with the realization of high speed secure communication system which is an urgent desire from the data center system. It has been proved that a quantum stream cipher by Yuen 2000 protocol (so called Y-00 or $\alpha\eta$) has remarkable security under quantum individual attack. So it is very attractive in implementing $1 \sim 10$ G bit/sec optical data center communication system. This paper introduces a system with some randomizations to reinforcement of the security in future implementation. Furthermore, we report the data transmission of 2.5 Gbit/sec in the real commercial network of 192 Km in Tokyo, and also a demonstration of High Definition Television (HDTV: 1.4 Gbit/sec) transmission of 40 Km by 2.5 Gbit/sec transceiver based on the basic model, which is immediately applicable to medical and broadcasting system.

Keywords: Quantum stream cipher, Yuen protocol, HDTV
PACS: 03.67.Dd, 42.50.Lc

INTRODUCTION

Recently, so many data center systems have been constructed by using dark fiber service of communication companies. However, all system have no security guarantee. Even if the data itself is encrypted by the conventional cipher, it is not secure, and also the encryption for each channel of each user in multi-user fiber transmission is inconvenient. Yuen-2000 protocol (Y-00) can be an attractive new quantum cryptography which can realize the ultra high speed data encryption system with the provable security for optical networks. In fact, so far, many remarkable experiments have been demonstrated by using phase modulation [1] and intensity (or amplitude) modulation [2,3] based on the basic model. To realize provable practical security which means unbreakable in the real world, in general we will need some modifications of Y-00, which consists of additional randomizations or a general encryption box as the driver. The principle and basic features of the security have been explained in [4], and a design method of Y-00 with provable security under some assumptions has been discussed in [5,6,7]. In this paper, we summarize methods which improve the security of Y-00, and show how to apply those to a real system. Furthermore,we realize the transceiver of 2.5 Gbit/sec which practices Y-00 protocol by intensity modulation, in which the transceiver has some randomizations. And also, we demonstrate the data transmission in the real commercial network in Tokyo city, and test for the HDTV transmission in laboratory.

CP1110, *Quantum Communication, Measurement and Computing (QCMC), The Ninth International Conference,* edited by A. Lvovsky

BASIC MODEL OF Y-00 AND RANDOMIZATIONS

The transmitter of a basic model of Y-00 consists of a LFSR with a secret key K_s and M-ary signal modulator, and the receiver consists of the same LFSR with the same key as the transmitter, but the receiver employs a binary detection. The key length is $|K_s| = 100 \sim 4000$ bits. The key is extended to a running key by a LFSR. The output bit sequence of the LFSR, i.e., the running key K_R is divided into blocks of $logM$ bits, and each $logM$ bits is regarded as the running key: $K_R = \{1,2,\ldots,K_i,\ldots,\}$, and $K_i \in \{1,2,\ldots,M\}$. A running key sequence is called keystream. The function taking the running key to the signal amplitude α_i of coherent state $|\alpha_i\rangle$ emitted from a laser is called mapping or mapper. In general, the mapping is defined as regular mapping which is a mapper with regular ordering. That is, the mapping $K_i \to \theta_i$ means that $K_i \to \{\alpha_i, \alpha_i + \alpha_M/2\}$, $\alpha_1 < \alpha_2 < \ldots < \alpha_M$, and $i,j,k \in \mathscr{M} = (1 \sim 2M)$. Quantum state sequences emitted from the transmitter can be described as follows:

$$\begin{aligned} |\Psi\rangle &= |\alpha(K_R,X)\rangle_1 |\alpha(K_R,X)\rangle_2 |\alpha(K_R,X)\rangle_3 \ldots \\ &= |\alpha_i\rangle_1 |\alpha_j\rangle_2 |\alpha_k\rangle_3 \ldots, \end{aligned} \tag{1}$$

where $|\alpha_i\rangle$ is one of $2M$ coherent states.

However, the regular mapping produces unequal distribution for the bit error position in M-ary signal detection. In order to decrease such a correlation, the following irregular mapping[6] can be used effectively.

Definition, Irregular mapping is a mapper with irregular ordering. The following is an example.

$$L = \begin{pmatrix} K_i \\ \alpha_i \\ x \end{pmatrix} = \begin{pmatrix} 53 & 26 & 1 & M & \ldots & 9 \\ \alpha_1 & \alpha_2 & \alpha_3 & \alpha_4 & \ldots & \alpha_M \\ 0 & 1 & 0 & 1 & \ldots & 0 \end{pmatrix} \tag{2}$$

where x is the data bit. Thus, the mapping relation between running key and physical signal set as the basis of M-ary modulation is designed by irregular ordering, which is not stochastically random. Once the mapping is set, such a mapping is fixed. In such a scheme, one of regulations of the security is evaluated by an assisted exhaustive search complexity:

$$Q = \Gamma^{|K_s|/logM} \tag{3}$$

where Γ is the undistinguishable basis number by quantum noise.

In general, it is very difficult to realize the ideal theoretical model in physical encryption. So we need additional ideas for approaching to the theoretically designed performance.

An overlapping selection keying :OSK is a method to change the polarity of bit sequence by the additional LFSR. By this, one can obtain the same effect of quantum noise in known plaintext attack as that of ciphertext only attack.

The quantum noise effect in the quantum measurement by heterodyne et al is not large in experiments. Since the probability distribution which describes the quantum noise at the measurement of the attacker is Gaussian, the effect of the edge of the distribution may degrade the security. To improve such an effect, we can employ keyed deliberate signal randomization:DSR. This is to randomize the position of a true basis or corresponding signal by another LFSR or another quantum noise. By this scheme, one can design such that the conditional probability for the attacker becomes uniform. That is, it is equivalent to making the wedge approximation in the theoretical model.

REALIZATION OF YUEN-2000 TRANSCEIVER

In this section, we describe an implemention of the transceiver which realizes Yuen-2000 protocol based on intensity modulation scheme. In order to provide SONET/SDH scheme which is the commercial network interface, we chose 2.48832 Gbit/sec as the data rate. Our transmitter and receiver employ the basic model of Y-00. The number of basis is 512, and seed key is 128 bits. At transmitter, OSK and keyed DSR are employed and they are driven by additional LFSRs. The irregular mapping will be installed in the next version. The launching power is 0 dBm, and the wave length is 1550 nm. The receiver does not need the LFSR information of keyed DSR, and the receiver's threshold dynamically follows the M-ary driver output which is driven by the LFSR of the same key as that of the transmitter.

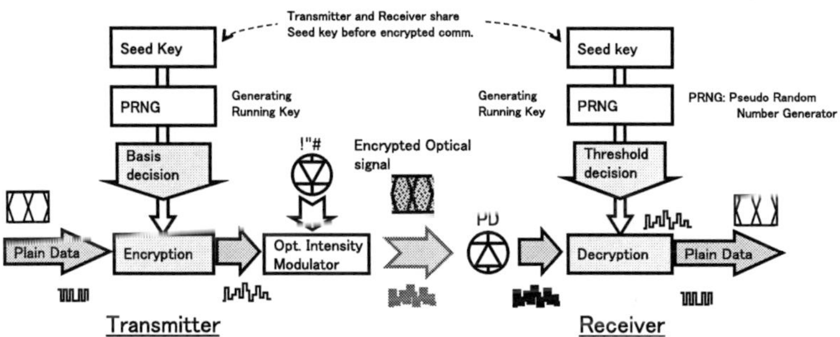

FIGURE 1. Configuration of Y-00 transceiver

TRANSMISSIONS OF DOCUMENT DATA AND HDTV

To realize the practical use, we have to take all disturbance from outside of the fiber cable into account. In general, the fiber communication systems are built in bad circumstances. So it is very important to demonstrate the system in the real commercial line. We used the commercial line of SoftBank Telecom from the Tokyo station to Ikebukuro station along Japan Rail line. To get long line, we used turnover scheme and the total length was 192 km. The interface is SONET/SDH of 2.5 Gbit/sec. Even though the cable circumstance was very bad, for instance, the vibration by train and the temperature variation by sunshine, our system worked during one week without trouble for data transmission. In addition, we demonstrated two wavelength division multiplex. On the other hand, there is a strong demand of HDTV transmission with high security and no transmission delay from the medical application. We also demonstrated the transmission of 40 km for HDTV without compression (1.4 Gbit/sec) by our 2.5 Gbit/sec Y-00 transceiver. This provides a remote medical operation system by secure communication.

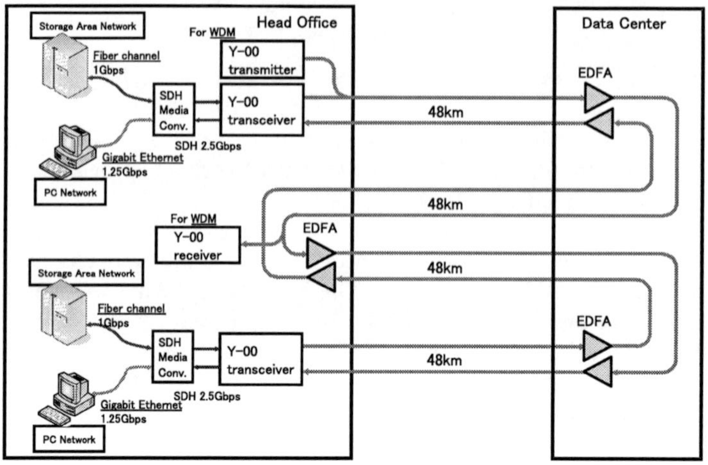

FIGURE 2. Demonstration of data transmission in the commercial network

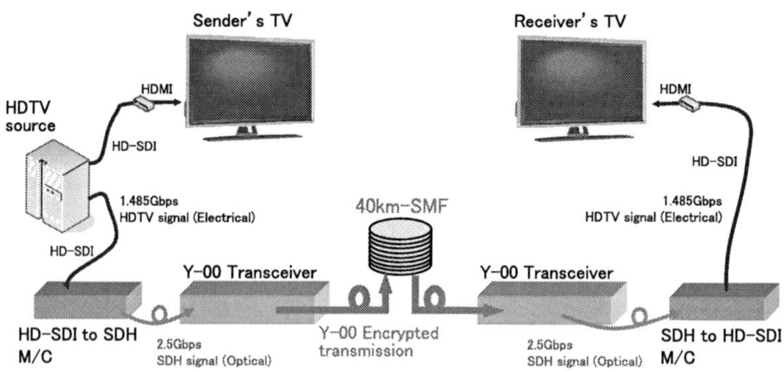

FIGURE 3. Demonstration of HDTV transmission

REFERENCES

1. G.A.Barbosa, et al *Physical Review Letters*, vol-90, 227901-4, 2003.
2. O.Hirota,et al, *Physical Review A*, vol 72, 022335, 2005.
3. K.Harasawa, et al, *Trans. of IEICE of Japan*, vol J91-C, no-8, p399, 2008.
4. H.P.Yuen, et al, *Quantum Information and Computing*, vol-8, p561, 2006.
5. O.Hirota, *Physical Review A*, vol 76, 032307, 2007.
6. T.Shimizu, et al, *Physical Review A*, vol 77, 034305,2008.
7. R.Nair,et al, *Physical Review A*, vol 74, 052309, 2006.

InGaAs Avalanche Photodiodes for Gigahertz Quantum Key Distribution

Z. L. Yuan*, A. R. Dixon*,†, J. F. Dynes*, A. W. Sharpe* and A. J. Shields*

*Toshiba Research Europe Ltd, 208 Cambridge Science Park, Cambridge CB4 0GZ, UK
†Cavendish Laboratory, University of Cambridge, J.J. Thomson Avenue, Cambridge CB3 0HE, UK

Abstract. We report the use of compact InGaAs avalanche photodiodes in the first gigahertz decoy-protocol quantum key distribution to achieve a record secure key rate exceeding 1 Mbit/s for a fibre distance of 20 km.

Keywords: avalanche photodiodes, quantum cryptography, optical fibre communication
PACS: 03.67.Dd; 85.60.Dw; 42.79.Sz

INTRODUCTION

Quantum key distribution (QKD) uniquely allows two remote parties to share a verifiably secret key [1]. Although promising, the secure key rate, limited to 10 kbit/s for a fibre distance of 20 km [2] and ~10 bit/s for 100 km [3] for an unconditionally secure protocol, is insufficient for practical use. The limitation arises from the difficulty in counting frequently arriving single photons at a telecom wavelength.

InGaAs avalanche photodiodes (APDs) have been the most widely used detectors for QKD, mainly due to their compactness and cryogenic-free operation. However there is a drawback; these detectors suffer significant noise due to afterpulsing. When an avalanche occurs, carriers can become trapped. If these carriers are subsequently released they can cause another avalanche, known as an afterpulse, which adds to the detector noise. The greater the charge of an avalanche the more carriers become trapped, leading to an increased probability for afterpulsing. In order to suppress the afterpulse noise the avalanche charge must be limited, which is typically done by gating an APD on for a duration of a few nanoseconds when a photon is expected. This has greatly reduced the avalanche charge to a few pC, allowing an InGaAs APD to operate at a gating frequency of 10 MHz. However, even with such a modest charge, a dead-time of a few microsecond must be applied to allowed trapped carriers to decay after a photon detection, resulting a maximum count rate of ~100 kHz. These considerations have rendered that the perforamnce of InGaAs APDs must be drastically improved for high speed QKD.

SELF-DIFFERENCING InGaAs AVALANCHE PHOTODIODES

To increase the QKD key rate it is necessary to increase the gating speed of the detectors, up to GHz frequencies. For this to be possible while limiting the afterpulsing noise described above it is necessary to detect weak avalanches. Unfortunately these signals are

CP1110, *Quantum Communication, Measurement and Computing (QCMC), The Ninth International Conference,* edited by A. Lvovsky

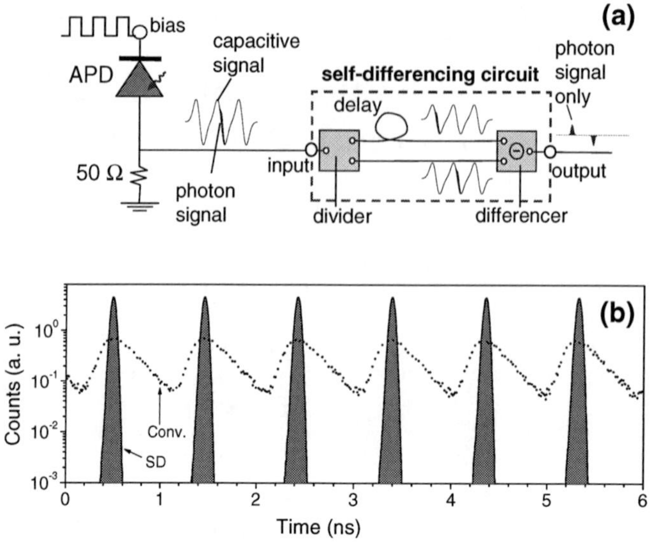

FIGURE 1. (a) Schematic for an APD in SD mode; (b) A comparison of histograms of photon arrival times obtained with the same APD under conventional gated Geiger mode operation (dotted line) or SD mode. The APD was illuminated with a 1550-nm pulsed laser operating at 1.036 GHz clock frequency.

usually hidden by the detectors capacitive response, especially at high gating frequencies. Recently, we demonstrated a simple self-differencing (SD) circuit [4] that enables weak avalanches to be clearly seen in the output of an APD operated with a GHz gating signal. The circuit is shown in Fig. 1(a) and consists of a 50:50 pulse splitter to divide the APD output into two equal components, which are then combined by a differencer. One of the coaxial cables connecting the splitter and differencer is longer than the other, introducing a delay corresponding to one gate period. This causes the output signal from the previous gate period to be subtracted away from the current periods signal, thus cancelling out the repeated capacitive response and leaving only avalanche signals. With a SD-circuit, an avalanche as small as 0.1 pC can be detected.

Under SD mode, InGaAs APDs can gated in a frequency exceeding 1 GHz. They tyically have a dark count probability of $1 - 10 \times 10^{-6}$ per gate and an afterpulse probability of $< 6\%$ for a detection efficiency of 10% under –30°C cooling. The detector deadtime is significantly reduced. The smallest separation between avalanches is measured to be two clock periods, as expected for a perfect SD, suggesting an ultra short detector dead time. Such a short dead time, three orders of magnitude less than typical values for conventional gated mode InGaAs APDs, is attributed to the reduced avalanche charges. It helps to increase the maximum count rate to 100 MHz.

The SD APD has excellent timing performance, as shown in Fig. 1(b). The detector here is biased active during a 0.48-ns gate in each cycle. We found a sharp distribution of photon arrivals at each gate, which can be fitted well with a Gaussian lineshape with a full width at half maximum (FWHM) of 50 ps. A clean gap between adjacent peaks prevents any ambiguity in assigning the receiving clock cycle and hence the bit value.

In comparison, InGaAs APDs under conventional mode display much poorer timing performance (FWHM: 370 ps). Moreover, adjacent photon peaks overlap. Consequently, ambiguity arises in assigning the clock cycle to photons received in the overlapping area. Typically, such a problem is solved by applying a small time window to reject the most ambiguous photons, but such a solution not only reduces the detector efficiency but also adds to the complexity of the system. In light of such observation, we believe that SD-APDs are preferable for applications, such as QKD, where photons arrive at regular intervals, even if improvements in InGaAs APD technology allows continuous operation in future.

GIGAHERTZ QKD WITH RECORD SECURE KEY RATES

We used SD-APDs, operating at 1.036 GHz gating frequency, to enable the first gigahertz QKD using decoy pulses [5, 6]. Use of decoy pulses allow generation of unconditionally secure keys even when using a laser, instead of a single photon source. Photons are generated by a 1.55 μm pulsed laser. The three intensity levels required for signal, decoy and near vacuum pulses are created using a fibre-optic intensity modulator. The quantum information is transmitted in the phase of the photons, with the phase encoding/decoding optics based on an asymmetric fibre Mach-Zender interferometer (AMZI) [7]. The optical pulses are strongly attenuated to the desired levels before leaving the sender's apparatus, with the intensity level of the pulses monitored using a beam-splitter and optical power meter. Dispersion shifted single mode fibre is used as the quantum channel, with an attenuation loss of 0.20 dB/km. A polarization controller at Bob's side recovers the polarization alignment before his AMZI. Each photon detection event is recorded with a unique time-stamp by time tagging electronics, and these events are used by Alice and Bob to sift a key.

We first performed QKD across a fibre distance of 20 km. A high clock rate allows accumulation of sufficient photon events from a short session to account for statistical fluctuations. During an uninterrupted session that lasts 2.3 s, a total of 7.91×10^6 sifted signal bits are accumulated with a quantum bit error rate (QBER) of 2.53%. Based on parameters obtained experimentally, we are able to calculate single photon pulse characteristics and thus the secure key rate using the results of Ma et al. [8]. In order to account for the finite key size we use statistical upper and lower bounds calculated at 10 standard deviations for experimental parameters, to give a conservative bound on key rate. We have achieved a secure key rate of 1.02 Mbit/s, being two orders of magnitude higher than the previous record of 10 kbit/s [2] for an unconditionally secure QKD protocol. We believe that this is the first time the secure key rate has exceeded 1 Mbit/s for any QKD protocol for this fibre length or longer.

Figure 2 shows the fibre distance dependence of the raw and secure key rates, and the QBER. The same parameters as for the 20 km experiment are used for all distances studied. Theoretical simulations are also shown in Fig. 2. The measured QBER is dominated by the detector afterpulse noise for fibre distances less than 60 km. The raw key as well as the secure key rates, are in good agreement with the simulation. They decrease exponentially with fibre attenuation at 0.2 dB/km, apart from for short fibres (<20 km). At 5.6 km, the secure bit rate is 1.65 Mbit/s, noticeably less than

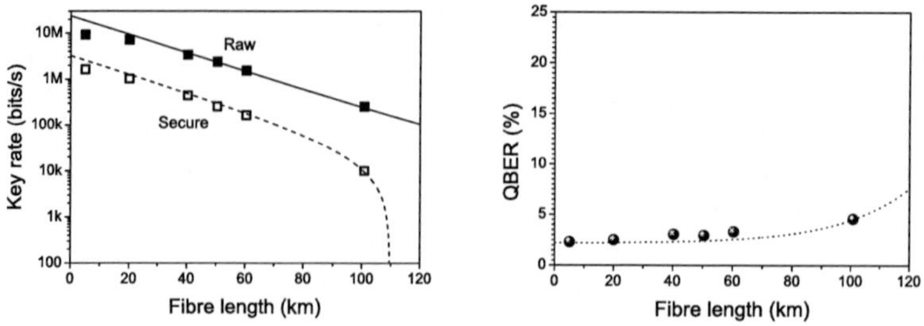

FIGURE 2. Decoy QKD experimental results. Raw (filled squares), secure (open squares) key rate, and the QBER (solid circles). Three intensity levels were used with nominal intensities of 0.55, 0.10, 7.5×10^{-4} photons/pulse and probabilities of 0.80, 0.16 and 0.04 for signal, weak decoy and near-vacuum decoy pulses respectively. Theoretical simulations are also shown for raw (solid lines), secure (dashed lines) key rates and the QBER (dotted line). In simulations, the receiver's detection efficiency is set to 5% (corresponding 10% of detector efficiency), detector dark count probability of 6.8×10^{-6} per gate and afterpulse rate 4.7%, and a QBER of 0.3% due to optical misalignment. All simulation parameters used are consistent with experimental results.

expected, due to the count rate limitation imposed by the time-tagging electronics. For fibre lengths beyond 60 km, the detector dark count noise deteriorates the QKD performance, causing the secure key rate to fall faster than the fibre loss and eventually preventing key formation at a fibre distance of 110 km, as simulated in Fig. 2.

CONCLUSION

In conclusion, we have successfully applied compact and cryogenic-free InGaAs APDs into gigahertz decoy-protocol QKD. Our results open the prospect of low cost and practical high bandwidth QKD systems for future applications.

ACKNOWLEDGMENTS

Partial support from the EU integrated project FP6 SECOQC is acknowledged.

REFERENCES

1. C. H. Bennett and G. Brassard, *Proceedings of the IEEE International Conference on Computers, Systems and Signal Processing*, Bangalore, India, 1984, pp. 175-179.
2. J. F. Dynes, Z. L. Yuan, A. W. Sharpe and A. J. Shields, *Opt. Express* **15**, 8465 (2007).
3. D. Rosenberg, J. W. Harrington, P. R. Rice, P. A. Hiskett, C. G. Peterson, R. J. Hughes, A. E. Lita, S. W. Nam and J. E. Nordholt, *Phys. Rev. Lett.* **98**, 010503 (2007).
4. Z. L. Yuan, B. E. Kardynal, A. W. Sharpe and A. J. Shields, *Appl. Phys. Lett.* **91**, 041114 (2007).
5. H. K. Lo, X. Ma and K. Chen, *Phys. Rev. Lett.* **94**, 230504 (2005).
6. X. B. Wang, *Phys. Rev. Lett.* **94**, 230503 (2005).
7. C. Gobby, Z. L. Yuan and A. J. Shields, *Appl. Phys. Lett.* **84**, 3762–3764 (2002).
8. X. Ma, B. Qi, Y. Zhao and H. K. Lo, *Phys. Rev. A* **72**, 012326 (2005).

Effects of antisqueezing for ASK-type Y00 cryptosystem

Masaki Sohma*, Yuki Nagasako† and Osamu Hirota*

*Research Center for Quantum Information Science, Tamagawa University, 6-1-1, Tamagawa-gakuen, Machida, Tokyo, 194-8610, Japan
†Internet Research Institute, Inc., Shinjuku-Nomura Building 11F 1-26-2 NishiSinjuku, Shinjuku-ku, Tokyo, 163-0511, Japan

Abstract. We estimate effects of antisqueezing in the Y00 (or $\alpha\eta$) cryptographic system with ASK signal states. In addition we consider attenuation channel with phase sensitive amplifiers and evaluate the maximum optical fiber length for the system.

Keywords: Y00 cryptosystem, squeezing ,antisqueezing
PACS: 03.67.Hk,03.67.Dd,42.50.Lc

INTRODUCTION

The Y00 system is a kind of stream cipher randomized by inherent quantum noise [1]. Several modulation scheme have been used for it: phase shift keying (PSK) [1, 2], amplitude shift keying (ASK) [3], and intensity modulation(IM). PSK signal states are given as $2M$ coherent states, $|\alpha_\ell\rangle = |\alpha_0 e^{i\theta_\ell}\rangle$, $\theta_\ell = \pi\ell/M$, $\ell = 0,1,2,.....,2M-1$, for a real α_0, forming M pairs $B_\ell = \{|\alpha_\ell\rangle, |\alpha_{\ell+M}\rangle\}$, $\ell = 0,1,...,M-1$. For each data bit Alice and Bob communicate using a different basis B_ℓ, where the value of ℓ is determined by the running key generated from a shared seed key K. The eavesdropper Eve does not know the key and hence her error probability P_E inevitably takes a large value when the fidelity between neighboring states is nearly equal to 1. On the other hand, ASK signal states are given as $|\tilde{\alpha}_\ell\rangle$, $\ell = 0,1,2,.....,2M-1$, with $\tilde{\alpha}_\ell = \alpha_0 + \ell R/M$. These form M bases, $\tilde{B}_\ell = \{|\tilde{\alpha}_\ell\rangle, |\tilde{\alpha}_{\ell+M}\rangle\}$,$\ell = 0,1,2,.....,M-1$.

Our purpose is to find a better way to implement the ASK-type Y00 cryptosystem. Tomaru and Ban found that using antisqueezed light in the PSK modulation increases the error probability of Eve [4]. Applying their idea to the case of ASK, we consider $2M$ squeezed states $|\tilde{\alpha}_\ell; -\gamma\rangle$, $\ell = 0,1,...,2M-1$. Then we can find antisqueezing increases the value of P_E like in the case of PSK. Unfortunately, it also increases the error probability P_B of the legitimate receiver Bob, but the increase of P_B can be neglected for not so large value of γ.

GAUSSIAN STATE

Throughout the present paper, a physical carrier of information is described by a Gaussian state, which is a generalization of coherent and squeezed states. The Gaussian state is characterized by a mean vector and a correlation matrix through the quantum charac-

CP1110, *Quantum Communication, Measurement and Computing (QCMC), The Ninth International Conference,* edited by A. Lvovsky

teristic function[5]. For example let us consider a squeezed state $|\alpha;\gamma\rangle$, that is, a pure Gaussian state $D(\alpha)S(\gamma)|0\rangle$, where $S(\gamma)=\exp[(\gamma/2)(a^2-(a^\dagger)^2)]$ is a squeezing operator with a real parameter γ, and $D(\alpha)=\exp[\alpha a^\dagger-\bar{\alpha}a]$ is a displacement operator. Then its mean vector and correlation matrix are given by $\sqrt{2\hbar}(\mathrm{Re}\alpha,\mathrm{Im}\alpha)^T$ and $\mathscr{A}(\gamma)=\frac{\hbar}{2}\mathrm{diag}[e^{-2\gamma},e^{2\gamma}]$ respectively, where $\alpha=\mathrm{Re}\alpha+i\mathrm{Im}\alpha$, and $\mathrm{diag}[x,y]$ denotes a diagonal matrix with diagonal elements x and y.

The attenuation channel with a transmittance $k<1$ is described by the transformation in the Heisenberg picture, $a'=k^{1/2}a+(1-k)^{1/2}a_0$, where $a=(2\hbar)^{-1/2}(q+ip)$ and $a'=(2\hbar)^{-1/2}(q'+ip')$ are input and output modal photon annihilation operators, $a_0^\dagger$ is a creation operator. The change in Gaussian states caused by attenuation can be described in terms of a mean vector and a correlation matrix as follows. As an input state let us consider a Gaussian states with a mean $(m^q,m^p)^T$ and a correlation matrix

$$\mathscr{A}=\begin{bmatrix} A^{qq} & A^{qp} \\ A^{qp} & A^{pp}\end{bmatrix}.$$

Then the output state from the attenuation channel with a transmittance $k(<1)$ has the mean vector $(k^{1/2}m^q,k^{1/2}m^p)^T$ and the correlation matrix

$$\phi_k(\mathscr{A})=k\mathscr{A}+\hbar\frac{1-k}{2}I_2, \tag{1}$$

with the identity matrix I_2.

Next we compute two important quantities for Gaussian states. Let us consider quantum Gaussian states ρ_i with mean m_i and the common correlation matrix $\mathscr{A}$ with $i=1,2$. Then the fidelity between quantum Gaussian states ρ_1 and ρ_2 is given by

$$F=\exp\left[-\frac{1}{4}(m_1-m_2)^T\mathscr{A}^{-1}(m_1-m_2)\right]. \tag{2}$$

On the other hand, the error probability of homodyne detection for quantum Gaussian states ρ_1 and ρ_2 is given by

$$P_B=\frac{1}{2}-\frac{1}{2}\mathrm{erf}\left(\frac{|m_1^q-m_2^q|}{2\sqrt{2A^{qq}}}\right), \tag{3}$$

where $\mathrm{erf}(x)=(2/\sqrt{\pi})\int_0^x e^{-t^2}dt$ is the Gauss error function.

DESIGN OF PARAMETERS OF Y00 SYSTEM

Let us estimate effects of antisqueezing in the proposed system. We consider the attenuation channel with a transmittance k $(0<k\le 1)$ and assume that Bob employs the homodyne detection. Then, from Eqs. (2) and (3), the error probability P_B of Bob and the fidelity F between neighboring states, $|\tilde{\alpha}_\ell;-\gamma\rangle$and $|\tilde{\alpha}_{\ell+1};-\gamma\rangle$ are given respectively as

$$P_B=(1/2)(1-\mathrm{erf}(R/\sqrt{2(e^{2\gamma}+1/k-1)})), \tag{4}$$

and

$$F = \exp(-R^2/M^2 e^{2\gamma}). \tag{5}$$

We have to design values of R and M satisfying $P_B \sim 0$ and $F \sim 1$. As a realistic setting, we put $k = 0.01$, $P_B = 10^{-12}$ and $F = 0.99$, which means that the error probability P_E for the optimal quantum measurement of neighboring states is equal to 0.45. Then R and M are considered as functions of squeezing parameter γ by Eqs. (4) and (5). Figure 1 gives graphs of these functions and shows that antisqueezing can decrease the number M of bases with R kept almost the same value.

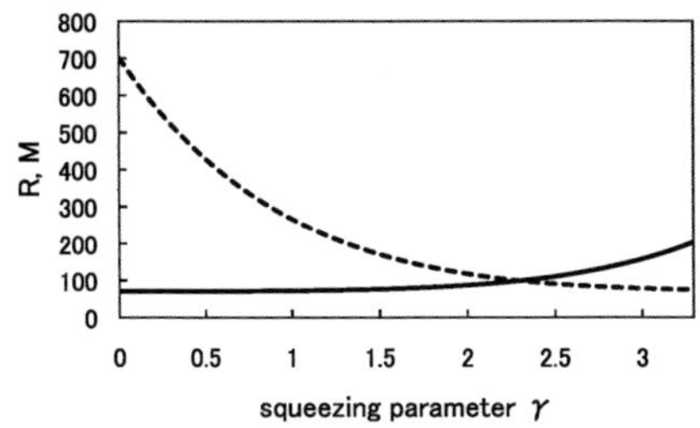

FIGURE 1. Signal distance R (solid line) between states in a basis $\tilde{B}_\ell$ and the number M (dot line) of bases with respect to squeezing parameter γ.

ATTENUATION CHANNEL ASSISTED BY PHASE SENSITIVE AMPLIFIERS

In this section we estimate effect of amplification. In particular we consider phase sensitive amplifier (PSA). The PSA is an ideal parametric amplifier represented by $q' = k^{1/2}q, \quad p' = k^{-1/2}p$, where the quadrature q is maximally amplified while the quadrature p is correspondingly maximally attenuated[6]. The change in Gaussian states caused by phase sensitive amplifier can be described as follows. As an input state we consider a Gaussian states with a mean (m^q, m^p) and a correlation matrix $\mathscr{A} = \mathrm{diag}[A^{qq}, A^{pp}]$. Then the output state from the phase sensitive amplifier with a power gain $k(>1)$ has the mean vector $(k^{1/2}m^q, k^{-1/2}m^p)$ and the correlation matrix

$$\tilde{\phi}_k(\mathscr{A}) = \mathrm{diag}[kA^{qq}, k^{-1}A^{pp}]. \tag{6}$$

In our optical communication system[7], n amplifiers are placed equally spaced on the attenuation channel. Here the channel is divided into $n+1$ parts, whose transmittances take the same value $k(<1)$, by n amplifiers with the power gain $G = 1/k$. An amplifier at the receiver is not necessary, because the capacity is a quantity obtained for the optimum measurement, which is never improved with an additional device, e.g. pre-amplifier[8]. We assume that a transmitter transmits a pure Gaussian states ρ_m with a mean vector (m^q, m^p) corresponding to classical information and the fixed correlation matrix $\mathscr{A}(-\gamma)$. Output states from the channels can be obtained by using Eq. (6) as follows. The output state $\Phi^{PSA}(\rho_m)$ from the attenuation channel assisted by PSAs is the Gaussian state with the mean vector $(k^{1/2}m^q, k^{(2n+1)/2}m^p)$ and the correlation matrix

$$\begin{aligned}\phi^{PSA}(\mathscr{A}(-\gamma)) &= \phi_k \circ \tilde{\phi}_{1/k} \circ \phi_k \circ \cdots \circ \tilde{\phi}_{1/k} \circ \phi_k(\mathscr{A}(-\gamma)) \\ &= \frac{\hbar}{2}\mathrm{diag}[ke^{2\gamma} + \varphi_{qq}, k^{2n+1}e^{-2\gamma} + \varphi_{pp}]\end{aligned} \tag{7}$$

where $\varphi_{qq} = \varphi_{qq}(k,n) = (1-k)(1+n)$ and $\varphi_{pp} = \varphi_{pp}(k,n) = (1-k^{2(n+1)})/(1+k)$ are functions of k and n.

Now we are ready to estimate the error probability of Bob. Using Eqs. (3) and (7), we obtain

$$P_B = \frac{1}{2} - \frac{1}{2}\mathrm{erf}\left(R/\sqrt{2(\mathrm{e}^{2\gamma} + (1/k-1)(1+n))}\right) \tag{8}$$

Putting $\gamma = 2$, $R = 100$ in Eq. (8) and assuming loss of optical fiber is 0.2dB/km, we can obtain the maximum optical fiber length $L = -50(n+1)\log_{10} k$ satisfying $P_B = 10^{-12}$ or 10^{-9}. Figure 2 gives graphs of L with respect to number n of amplifiers.

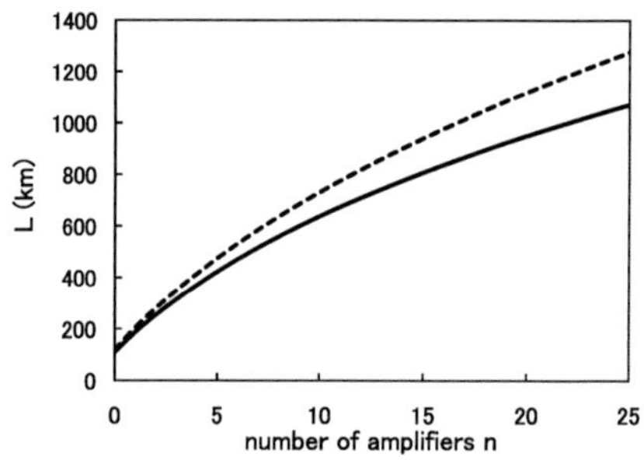

FIGURE 2. Maximum optical fiber length L (km) satisfying $P_B = 10^{-12}$ (solid line) or $P_B = 10^{-9}$ (dot line) with respect to number n of phase sensitive amplifiers.

ACKNOWLEDGMENTS

This work was supported by National Institute of Information and Communications Technology under Grant No. NICT-11901.

REFERENCES

1. H. P. Yuen, *Quantum Communication, Computing, and Measurement 3, KA/PP* pp. 285–293 (2001).
2. G. A. Barbosa, E. Corndorf, P. Kumar, and H. P. Yuen, *Phys. Rev. Lett* **90**, 227901 (2003).
3. O. Hirota, M. Sohma, M. Fuse, and K. Kato, *Phys. Rev. A* **72**, 022335 (2005).
4. T.Tomaru, and M. Ban, *Phys. Rev. A* **74**, 032312 (2006).
5. A. S. Holevo, M. Sohma, and O. Hirota, *Phys. Rev. A* **59**, 1820–1828 (1999).
6. H. P. Yuen, *Optics Letters* **17**, 73–75 (1992).
7. M. Sohma, and O. Hirota, *Physical Review A* **76**, 024303 (2007).
8. H. P. Yuen, *Quantum Squeezing(Springer)* pp. 227–258 (2003), LANLquant-ph/019054v1.

Exact Minimum and Maximum of Yield with a Finite Number of Decoy Light Intensities

Toyohiro Tsurumaru[a], Alexandre Soujaeff[b] and Shigeki Takeuchi[b]

[a]Mitsubishi Electric Corporation, Information Technology R&D Center, 5-1-1 Ofuna, Kamakura-shi, Kanagawa, 247-8501, Japan
[b]JST-CREST / Research Institute for Electronic Sciences, Hokkaido University, Kita 12 Nishi 6, Kita ku, Sapporo, 060-0812, Japan

Abstract. In this presentation, for the decoy state method using a finite number of decoy light intensities, we present an improved upper and lower bounds for the asymptotic yield y_n for n-photon states. In particular if all the light intensities are less than or equal to 1, they are not only a lower or upper bound, but in fact are the exact minimum or maximum.

Keywords: Quantum Cryptography, Decoy States.
PACS: 03.67.Dd.

INTRODUCTION

The decoy state method is a technique used in quantum key distribution (QKD) for determining the possible range of the yield y_1 and the error rate e_1, by a statistical test using several different light intensities [1-3]. Here the yield y_n is the probability that an n-photon state emitted by Alice is detected in Bob's apparatus, and e_n is the error rate caused solely by the n-photon states. In this method, Alice first chooses the average photon number of each of her coherent light pulses randomly out of $\mu_1,\ldots,\mu_M$, and Bob records every detection events. After quantum communications are completed, Alice reveals the average photon number of each pulse over an authenticated public channel. Then referring to these data, Bob calculates the detection rates corresponding to each μ_i, and estimates a lower bound or the minimum of y_1 that is consistent with them. Similarly, he can also estimate an upper bound or the maximum of e_1.

These values are then used to calculate the key generation rate R by plugging them into well-known formulas, *e.g.*, $R = Q(\mu)f(E(\mu))H_2(E(\mu)) + Q_0(\mu) + Q_1(\mu)[1 - H_2(e_1)]$ for the Bennett-Brassard 1984 (BB84) protocol [4]. Here $Q(\mu)$ is the overall detection rate in Bob's detector, and $Q_0(\mu)$, $Q_1(\mu)$ are the contributions to it from the pulses containing zero and one photon respectively. $E(\mu)$ is the overall error rate, and $H_2(e)$ is the binary entropy function $H_2(e) = -e\log_2 e - (1-e)\log(1-e)$, hence $H_2(E(\mu))$ corresponds to the length of a syndrome consumed to correct bit errors. The factor

CP1110, *Quantum Communication, Measurement and Computing (QCMC), The Ninth International Conference,* edited by A. Lvovsky

$f(E(\mu))$ is inserted to take into account the information rate of practical error correcting codes which is usually below the Shannon limit. Lower bounds on y_1 with a finite number of decoy intensities have been discussed in many papers (see, *e.g.*, [1-3] and references therein), and the best and the most general one is due to Hayashi [7]. Adding to these results, in this presentation, we present improved upper and lower bounds X_n, Z_n for the asymptotic yield y_n [8]. In particular, if all the light intensities μ_i are less than or equal to 1, X_n and Z_n are not only a lower or an upper bound, but in fact are the exact minimum or maximum.

The main difference between preceding approaches and ours is as follows. The original decoy problem is an optimization problem involving an infinite number of variables y_n. In order to reduce the number of variables to finite, Wang devised a decomposition of a phase-randomized coherent state [3], which was later generalized to the case of an arbitrary number of decoy light intensities by Hayashi [7]; what they did was to decompose the state ρ sent by Alice as a sum $\rho = \sum_{n=1}^{N} a_n \rho_n$ of mixed states ρ_n. Then by regarding a_n as independent variables and using a linear-programming-like approach, Hayashi presented a general algorithm for obtaining the minimum of y_1, which is linear in a_n. At first this method may seem general enough and capable of giving the exact minimum of y_n. So what needs to be improved further? The answer is that a_n cannot be considered as independent in reality, since ρ_n are not completely distinguishable to each other. Thus, by regarding that way they give Eve more power than she actually has, and there is no guarantee that the obtained minimum is also that of the original problem involving an infinite number of y_n.

In contrast, in this presentation, we present a method for finding the minimum of y_1 without reducing the variables concerned, *i.e.*, we treat all y_n independently as in the original decoy method problem. The key observation here is that when regarding variables y_n as an infinite-dimensional vector y, any difference Eve can make to y without being noticed by Alice or Bob can be expanded by a set of basis vectors $w^{(m)}$, each of which is written in a simple form with the Schur polynomials (see, *e.g.*, [9,10]).

What is remarkable about our result is that the configuration of y_n leading to the smallest y_1 varies depending on whether the number M of decoy light intensities (including the signal) is even or odd. The analysis is especially simple for M even and $\mu_i \leq 1$; because of the positivity of the Schur polynomials, it is readily seen that y_1 is minimized when $y_n = 0$ for $n > M$ and that the problem is automatically reduced to that involving only a finite number of variables $y_1, \ldots, y_M$. Thus by simply inverting a matrix, the minimum of y_1 is expressed in an explicit and simple form. On the other hand, for M odd, the analysis turns out to be somewhat more complicated; however, we can still specify the configuration that corresponds to the smallest y_1 and write down an explicit algorithm for finding it out within a finite number of steps.

SETUPS AND THE MAIN RESULTS

For the sake of simplicity, here we consider the case where the yield for vacuum states y_0 is already known precisely by using vacuum decoy states, and discuss the minimum and maximum of y_1 under the condition that

$$Q_+(\mu_i) := \exp(-\mu_i)\sum_{n=1}^{\infty}\frac{\mu_i^n}{n!}y_n$$

is satisfied for $i = 1,\dots,M$. Note, however, that our analysis is equally valid even without vacuum decoy states. $Q_+(\mu_i)$ appearing in the above equation denotes the contribution from the non-zero photon number state to the detection rate in Bob's detector, *i.e.*, $Q_+(\mu) := Q(\mu) - e^{-\mu}y_0$. Being a probability, each y_n is of course constrained as

$$0 \le y_n \le 1 \qquad \text{for all } n \ge 1$$

The explicit form of the detection rate $Q_+(\mu)$ depends on the physical model that one employs for describing the quantum channel. In this presentation, we assume that in the absence of Eve, the yield takes the value $y_n = q_n$ with

$$q_n := A\eta_n + B,$$
$$\eta_n := 1 - (1-\eta)^n,$$

and that each parameter is conditioned as

$$0 \le A \le 1, \quad 0 \le B \le 1/10. \tag{1}$$

Here η is the channel transmission rate including the quantum efficiency of Bob's detector, and B is roughly the dark count rate p_{dark}.

Note that for practical QKD systems, Eq. 1 is not really a restriction; η is already around 0.1 at 0km due to the detector efficiency. Additionally, for sufficiently small η, we have $B \le \frac{1}{2}Q(\mu_i)E(\mu_i) \cong \frac{1}{2}\eta\mu_i E(\mu_i) \le \frac{1}{2}\eta\mu_i$. Thus, with the signal light intensity (say, μ_1) normally being around 0.5 or less, $B \le 1$ is automatically satisfied.

According to Lo et al. [2] and Hayashi [7], these parameters are $A = 1$, $B = p_{\text{dark}}$, from which we have

$$Q(\mu_i) = 1 - \exp(-\eta\mu_i) + p_{\text{dark}}$$

whereas in some other references (*e.g.*, [4]), slightly different models such as $A = 1 - p_{\text{dark}}$ and $B = p_{\text{dark}}$ are used.

Under these settings, we present upper and lower bounds on y_n in terms of quantities X_n and Z_n; for any M and $n < M$,

$$X_n \le y_n \le Z_n \quad \text{if} \quad M - n \text{ is odd},$$
$$Z_n \le y_n \le X_n \quad \text{if} \quad M - n \text{ is even},$$

where X_n are expressed in a simple form (for details, see [8]). For instance, X_1 takes the form

$$X_1 = \sum_{n=1}^{M} \frac{\exp(\mu_i) Q_+(\mu_i)}{\mu_i} \prod_{j=1, j\neq i}^{M} \frac{\mu_j}{\mu_j - \mu_i}$$

On the other hand Z_n cannot be written in a simple form as X_n, however, they can always be obtained by a numerical calculation within a finite number of steps [8].

In addition, it can be shown that at least when $\mu_i \leq 1$, Eve can actually attain $y_n = X_n$ and $y_n = Z_n$. Hence they are not only a lower (resp. upper) bound, but in fact are the minimum (resp. maximum) of y_n.

In order to demonstrate how effective our approach is, take a typical set of experimental parameters, *e.g.*, $A = 1$, $\eta = 10^{-2}$, $B = p_{\text{dark}} = 10^{-5}$, $M = 3$, and $(\mu_1, \mu_2, \mu_3) = (0.07, 0.2, 0.5)$. It turns out that $Z_1 = 0.993 \times 10^{-2} \leq y_1 \leq 1.003 \times 10^{-2} = X_1$, where the yield in the absense of Eve is $y_1 = q_1 = 1.001 \times 10^{-2}$. Hence by using only four decoy light intensities including vacuum, we can determine y_1 within accuracy of less than one percent.

ACKNOWLEDGMENTS

This work was supported by the project "Research and Development on Quantum Cryptography" of the National Institute of Information and Communications Technology, as part of Ministry of Internal Affairs and Communications of Japan's program "R&D on Quantum Communication Technology."

REFERENCES

1. W. -Y. Hwang, Phys. Rev. Lett. 91, 057901 (2003).
2. H. -K. Lo, X. Ma, and K. Chen, Phys. Rev. Lett. 94, 230504 (2005).
3. X. -B. Wang, Phys. Rev. Lett., 94, 230503 (2005).
4. H. Inamori et al., Euro. Phys. J. D 41, 599 (2007); D. Gottesman et al., Quant. Inf. Comput. 5, 325 (2004); R. Renner et al., Phys. Rev. A 72, 012332 (2005); M. Hayashi, Phys. Rev. A 76, 012329 (2007); M. Koashi, arXiv:quant-ph/0609180, (2006).
5. X. -B. Wang, Phys. Rev. A 72, 012322 (2005).
6. X. Ma et al., Phys. Rev. A 72, 012326 (2005).
7. M. Hayashi, New J. Phys, 9, 284 (2007).
8. T. Tsurumaru, A. Soujaeff, and S. Takeuchi, Phys. Rev. A 77, 022319 (2008).
9. W. Fulton and J. Harris, *Representation Theory*, (Springer Verlag, Berlin, 1991).
10. W. Fulton, *Young Tableaux*, (Cambridge Univ. Press, 1997).

Quantum Key Distribution with Multi Letter Continuous Variable Alphabets

Denis Sych and Gerd Leuchs

Institut für Optik, Information und Photonik, Max–Planck Forschungsgruppe, Universität Erlangen–Nürnberg, Günther–Scharowsky–Straße 1, 91058 Erlangen, Germany

Abstract. We present a new protocol for continuous variable quantum key distribution. The novelty of the protocol is a multi letter alphabet represented by coherent states of light with a fixed amplitude and variable phase. Information is encoded in the phase of a coherent state which can be chosen from a regular discrete set consisting, however, of an arbitrary number of letters. We evaluate the security of the protocol against the beam splitting attack. As a result we show the proposed protocol has advantages over the standard two letter coherent state protocol, especially in the case when losses in the communication channel are low.

Keywords: quantum cryptography, quantum key distribution, qkd, continuous variable, coherent states, multi letter alphabet
PACS: 03.67.Dd, 42.50.Ex, 89.70.Cf

Quantum key distribution (QKD) is a procedure of information exchange between two parties, the sender Alice and the receiver Bob, which allows to distribute the absolutely secure data between them [1]. The distinctive part of QKD with respect to classical communication is use of a quantum information channel. In this case secrecy is guaranteed by laws of quantum mechanics [2, 3, 4, 5]. Mathematically, carriers of informational can be classified as discrete variable (DV) and continuous variable (CV) [6], although physically they can be of various kind: single photons [4], weak coherent pulses [7], squeezed states [8, 9] and other systems where a signal possesses essentially quantum properties.

The universal characteristics that can be compared for different QKD protocols (apart from their experimental realizations) are the secret key generation rate (shortly, *key rate*) and the maximum possible length of the channel at which key rate is still reasonable. In the case of DV QKD, it has been shown that extensions of the standard four–letter BB84 protocol [4] to higher number of letters in the alphabet can improve performance in terms of higher critical error rate or longer communication lines [10, 11, 12, 13]. CV QKD protocols are known to be not very robust to the high channel losses. In the case of low losses, however, they can establish significantly higher key rate than the key rate provided by DV QKD protocols.

In this paper we introduce a new CV QKD protocol and address a question how one can increase key rate of CV QKD by varying the number of letters and geometry of the quantum alphabet. The protocol based on two coherent states [14, 15] is a CV analogy of the discrete B92 protocol [7]. Based on the idea of improving the properties of DV QKD by use of higher number of letters in a quantum alphabet [13], we perform a generalization of the two letter CV QKD protocol to a higher number of states. Namely, instead of two coherent states $|\pm a\rangle$ which have relative phases 0 and π, we use N

CP1110, *Quantum Communication, Measurement and Computing (QCMC), The Ninth International Conference,* edited by A. Lvovsky

coherent states $|\alpha_k\rangle = \left|ae^{i\frac{2\pi}{N}k}\right\rangle$ which have different phases $\frac{2\pi}{N}k$ and a fixed amplitude a. We perform a security analysis of the proposed protocol and show how the number of letters affects the key rate.

An elementary information transmission in the multi letter protocol is as follows:

- The sender (Alice) chooses a random equiprobable number $k = 1 \ldots N$ and sends the respective coherent state $|\alpha_k\rangle = \left|ae^{i\frac{2\pi}{N}k}\right\rangle$.
- The receiver (Bob) uses a double homodyne, or heterodyne, measurement [16], i.e. measures orthogonal quadratures x and p, and obtains a state $|\beta\rangle = |\beta_x + i\beta_p\rangle$.
- Then he assigns a classical number l to the measured state $|\beta\rangle$ by finding a state $|\alpha_l\rangle$ which is the closest alphabet's state to the state $|\beta\rangle$: $|\langle\alpha_l|\,\beta\rangle|^2 = \max_n |\langle\alpha_n|\,\beta\rangle|^2$. Generally speaking, this classical value l decoded by Bob can be different from the initial value k sent by Alice even in the absence of eavesdropping or any excess noise. In the phase space representation the alphabet looks like a regular constellation, and the region, where a state $|\alpha_l\rangle$ is assigned to the result of Bob's measurement $|\beta\rangle$, looks like a sector with an angle $2\pi/N$.

The elementary information transmission from Alice to Bob can be schematically shown as a "$k \to l$" channel:

$$(encoding: k) \to (transmission: |\alpha_k\rangle) \to (measurement: |\beta\rangle) \to (decoding: l). \quad (1)$$

The amplitude a and number of letters N can be flexibly adjusted for a given information channel. These parameters are supposed to be publicly known, particularly for the possible eavesdropper. We note that the number of states can be arbitrary large. In the infinite limit $N \to \infty$ we have a continuous phase modulation. Thus our protocol can be viewed as a smooth transition between discrete [14, 15] and continuous [16, 17] modulation of CV. After the measurement, Bob can use the postselection idea [14], when he decides whether to keep the transmission depending on the value β. The elementary transmission (1) followed by the postselection is repeated until Alice and Bob collect enough data to perform classical error correction and privacy amplification procedures [1]. The resulting data is a secret key.

We investigate the security of our protocol against the beam splitting attack [15]. We assume Eve might replace the real channel with a lossless one, so all losses in the channel can be in her hands. The beam splitting transformation is $|\alpha_k\rangle_A \to |\beta_k\rangle_B \otimes |\varepsilon_k\rangle_E$, where Alice's initial state $|\alpha_k\rangle$ is split to Bob's state $|\beta_k\rangle = \left|\sqrt{\eta}\alpha_k\right\rangle$ and Eve's state $|\varepsilon_k\rangle = \left|\sqrt{1-\eta}\alpha_k\right\rangle$, and η is the channel transmittance. In this scenario Eve's states are not entangled with Bob's states. This can be seen as absence of excess noise on Bob's side. Although this is not the most general and optimal strategy of eavesdropping, it has perfectly agreed with experimental observations [18].

The amount of classical mutual Shannon information I_{AB} transmitted from Alice to Bob via the channel (1) is equal to the difference of *a priori* (before measurement) and *a posteriori* (after measurement) entropies [19]: $I_{AB} = H_{Bob}^{prior} - H_{Bob}^{post}$. Before any measurement all possibilities are equal for Bob, so his *a priori* entropy is the unconditional "pure guess" entropy $H_{Bob}^{prior} = \log_2 N$. After the heterodyne measurement Bob obtains a state $|\beta\rangle = |\beta_x + i\beta_p\rangle$. The conditional probability density to measure the state $|\beta\rangle$ when a state $|\alpha_k\rangle$ has been sent is $p(\beta|k) = |\langle\alpha_k|\,\beta\rangle|^2 = \frac{1}{\pi}e^{-|\beta-\beta_k|^2}$. The total unconditional probability density to measure a state $|\beta\rangle$ is $p(\beta) = \frac{1}{N}\sum_{k=1}^{N} p(\beta|k)$. After the measurement of $|\beta\rangle$, the probability $p_k(\beta)$ that the state $|\alpha_k\rangle$ was initially sent is the normalized conditional probability $p_k(\beta) = \frac{p(\beta|k)}{Np(\beta)} = \frac{1}{\pi Np(\beta)}e^{-|\beta-\beta_k|^2}$.

As far as the quantum alphabet we consider is a regular constellation, and all states have equal probability to be sent, all the channel inputs and outputs (1) are equiprobable. Thus Bob's *a posteriori* entropy H_{Bob}^{post} is the Shannon entropy of the total probability distribution $P(\beta) = \{p_1(\beta), p_2(\beta), \ldots, p_N(\beta)\}$ conditioned on the measured state $|\beta\rangle$:

$$H_{Bob}^{post}[P(\beta)] = -\sum_{k=1}^{N} p_k(\beta)\log_2 p_k(\beta). \tag{2}$$

To calculate Eve's information we remind that we consider a beam splitting attack. Thus in each transmission (1) Eve has a state $|\varepsilon_k\rangle = |\sqrt{1-\eta}\,\alpha_k\rangle$ conditioned only on a sent state $|\alpha_k\rangle$. Eve is not restricted to any practical way of information extraction from this state, thus her potential knowledge is bounded by the Holevo information [20]. In our case Eve's conditional states are pure, so Eve's information I_{AE} is equal to the von Neumann entropy S of her unconditional state: $I_{AE} = S(\hat{\rho}_E)$, where $\hat{\rho}_E = \frac{1}{N}\sum_{k=1}^{N}\hat{\rho}_k$.

The key rate, i.e. the amount of secret information per transmission (1), is equal to

$$G = \int p(\beta)\,(I_{AB} - I_{AE})\,d\beta, \tag{3}$$

where we integrate over the whole phase space. In this case the key rate can be positive only for transmittance higher than 50%, i.e. losses are less than 3dB [17]. To overcome this limit we use the postselection idea [14], which means we select only that part of transmissions which give us positive terms in the total integral (3).

The postselection procedure can be qualitatively described as follows: Eve's information I_{AE} depends only on the transmitted state $|\alpha_k\rangle$ and the channel transmittance η, whereas Bob's information I_{AB} additionally depends on the measured value β. The key rate can be increased if Bob accepts only those transmissions where β is such that he has higher information than Eve: $I_{AB}(\beta) > I_{AE}$. Then instead of integration over the whole phase space (3) we have an integration over the postselected area (PSA). To find this PSA we numerically solve an equation $I_{AB}(\beta) > I_{AE}$. To find the key rate G_{PS} after postselection as a function of transmittance η we optimize the amplitude a in such a way that for each given channel transmittance η the optimal amplitude $a_0 = a_0(\eta)$ leads to the maximal key rate $G_{PS}(\eta) = \max_a G_{PS}(\eta, a)$.

The calculated secret key rate $G_{PS}(\eta)$ and the optimal amplitude $a_0(\eta)$ for several alphabets are shown in Fig.1. An interesting result we can see in Fig. 1 is that lines $G(\eta)$ for various numbers of letters are intersecting in different points. This means that for different values of transmittance η there are different optimal numbers of letters N_0. We derive the following values of N_0 among the alphabets which are shown

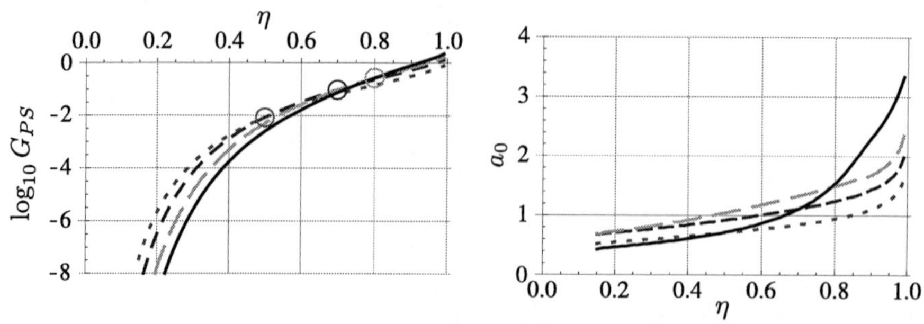

FIGURE 1. The postselected secret key rate G_{PS} in logarithmic scale (the left figure) and the optimal amplitude of the signal (the right figure) for 2, 3, 4, and 8 letter alphabets (dot, short hyphen, long hyphen and solid lines respectively). Intersection points in the left figure are marked by circles.

in Fig. 1: $N_0(0 < \eta < 0.5) = 2$, $N_0(0.5 < \eta < 0.7) = 3$, $N_0(0.7 < \eta < 0.8) = 4$, and $N_0(\eta > 0.8) = 8$. The higher the transmittance, the higher the optimal number of letters, and the higher key rate. In the limit $\eta \to 1$ we have continuous phase modulation ($N_0 \to \infty$) and, ideally, infinitely large key rate. Thus the most significant advantage of the multi letter protocol over the two letter protocol one can get in the case of high transmittance.

Authors thank Norbert Lütkenhaus for helpful discussions. D.S. acknowledges the Alexander von Humboldt Foundation for a fellowship.

REFERENCES

1. N. Gisin, G. Ribordy, W. Tittel, and H. Zbinden, *Rev. Mod. Phys.* **74**, 145 (2002)
2. W.K. Wootters and W.H. Zurek, *Nature* **299**, 802 (1982)
3. D. Dieks, *Phys. Lett. A* **92** 271 (1982)
4. Ch. Bennett and G. Brassard, in Proceedings IEEE Int. Conf. on Computers, Systems and Signal Processing, Bangalore, India; IEEE, New York, (1984), 175
5. A.K. Ekert, *Phys. Rev. Lett.* **67**, 661 (1991).
6. V. Scarani, H. Bechmann–Pasquinucci, N. Cerf, M. Dušek, N. Lütkenhaus, M. Peev, quant-ph/08024155
7. Ch. Bennett, *Phys. Rev. Lett.* **68**, 3121 (1992)
8. T.C. Ralph, *Phys. Rev. A* **61**, 010303 (1999)
9. M. Hillery, *Phys. Rev. A* **61**, 022309 (2000)
10. D. Bruss, *Phys. Rev. Lett.* **81**, 3018 (1998)
11. H. Bechmann–Pasquinucci and N. Gisin, *Phys. Rev. A* **59**, 4238 (1999)
12. D. Sych, B. Grishanin, and V. Zadkov, *Phys. Rev. A* **70**, 052331 (2004)
13. D. Sych, B. Grishanin, and V. Zadkov, *Sov. J. Quant. Electron*. **35**, 80 (2005)
14. Ch. Silberhorn, T. C. Ralph, N. Lütkenhaus, and G. Leuchs, *Phys. Rev. Lett.* **89**, 167901 (2002)
15. M. Heid and N. Lütkenhaus, *Phys. Rev. A* **73**, 052316 (2006)
16. C. Weedbrook, A. M. Lance, W. P. Bowen, T. Symul, T. C. Ralph, and P. K. Lam, *Phys. Rev. Lett.* **93**, 170504 (2004)
17. F. Grosshans, and P. Grangier, *Phys. Rev. Lett.* **88**, 057902 (2002)
18. S. K. Lorenz, N. Korolkova, and G. Leuchs, *Appl. Phys. B: Lasers Opt.* **79**, 273 (2004).
19. C. Shannon, *Bell Syst. Tech. J.* **27**, 379 (1948)
20. A. Holevo, *Probl. Inf. Trans.* **9**, 177 (1973)

Upper bounds for the security of differential-phase-shift quantum key distribution with weak coherent states

Marcos Curty*, Kiyoshi Tamaki†,**, Tobias Moroder‡,§ and Hipólito Gómez-Sousa*

*ETSI Telecomunicación, Department of Signal Theory and Communications, University of Vigo, Campus Universitario, E-36310 Vigo, Spain
†NTT Basic Research Laboratories, NTT Corporation, 3-1 Morinosato Wakamiya Atsugi-Shi, Kanagawa, 243-0198, Japan
**CREST, JST Agency, 4-1-8 Honcho, Kawaguchi, Saitana, 332-0012, Japan
‡Institute for Quantum Computing, University of Waterloo, Waterloo, ON, N2L 3G1, Canada
§Quantum Information Theory Group, Institut für Theoretische Physik I, and Max-Planck Research Group, Institute of Optics, Information and Photonics, University of Erlangen-Nürnberg, 91058 Erlangen, Germany

Abstract. In this paper we present limitations imposed by sequential attacks on the maximal distance achievable by a differential-phase-shift (DPS) quantum key distribution (QKD) protocol with weak coherent pulses. Specifically, we compare the performance of two possible sequential attacks against DPS QKD where Eve realizes, respectively, optimal unambiguous state discrimination of Alice's signal states, and optimal unambiguous discrimination of the relative phases between consecutive signal states. We show that the second eavesdropping strategy provides tighter upper bounds for the security of a DPS QKD scheme than the former one.

Keywords: quantum cryptography, differential-phase-shift quantum key distribution, unambiguous state discrimination, sequential attack, intercept-resend attack, security
PACS: 03.67.-a, 03.67.Dd, 03.67.Hk

INTRODUCTION

The main figure of merit when comparing different quantum key distribution (QKD) schemes is the achievable secret key rate over a given distance. For example, it is well known that the standard BB84 protocol with weak coherent pulses (WCP) can deliver a key generation rate of order $O(\eta^2)$, where η denotes the total transmission efficiency of the quantum channel.

In recent years, several schemes improving the performance of QKD setups with practical signals have been proposed. One of these schemes is the so-called differential-phase-shift (DPS) QKD [1]. Its basic setup is illustrated in Fig. 1. Alice sends to Bob a train of WCP $|\alpha\rangle$ whose phases are randomly modulated by 0 or π. On the receiving side, Bob measures out each incoming signal by means of an interferometer whose path-length difference is set equal to the time difference Δt between two consecutive pulses. Whenever the relative phase between two consecutive signals is 0 ($\pm\pi$) only the photon detector $D0$ ($D1$) may produce a "click". Once the quantum communication phase of the protocol concludes, Bob uses an authenticated classical channel to announce the

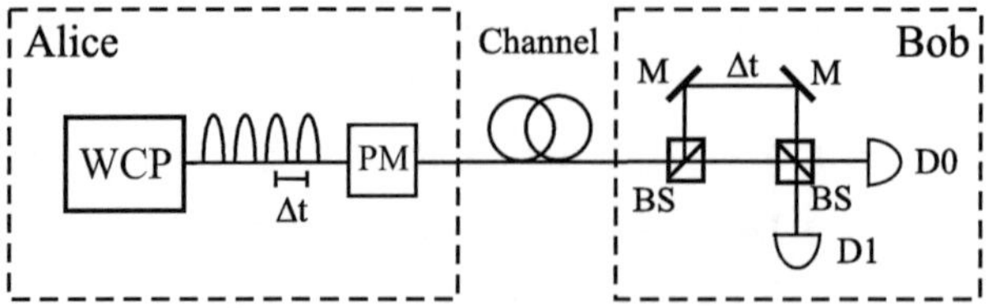

FIGURE 1. Basic setup of a DPS QKD scheme. PM denotes a phase modulator, BS, a 50 : 50 beam splitter, M, a mirror, D0 and D1 are two photon detectors and Δt represents the time difference between two consecutive pulses.

time slots where he obtained a click, but he does not reveal which detector fired each time. From this information, Alice can infer which photon detector had clicked at Bob's side each given time. Then, Alice and Bob agree, for instance, to select a bit value "0" whenever the photon detector $D0$ fired, and a bit value "1" if the detector $D1$ clicked. In this scenario, however, a secure key rate of order $O(\eta)$ has only been proven so far against a special type of individual attacks [1], and also against a particular class of collective attacks [2]. Whether DPS QKD is secure against the most general attack remains an important open question in the field [3].

The main goal of this paper is to present limitations imposed by sequential attacks [4] on the performance of DPS QKD protocols. For that, we consider two possible attacks. In the first one, Eve realizes unambiguous state discrimination (USD) of Alice's signal states [5]. In the second one, Eve identifies unambiguously the relative phase between consecutive signal states [6]. As a result, we obtain upper bounds on the maximal distance achievable by a DPS QKD scheme as a function of the error rate in the sifted key, the mean photon-number of Alice's signals, and the detection efficiency, the dark count probability, and the dead-time of Bob's detectors. According to our results, the long-distance implementations of DPS QKD reported so far in the literature would be insecure against a sequential attack for the case of uncalibrated devices.

SEQUENTIAL ATTACKS AGAINST DPS QKD

A sequential attack [4] can be seen as a special type of intercept-resend attack. First, Eve measures out every signal state emitted by Alice with a detection apparatus located very close to the sender. Afterwards, she transmits each measurement result through a lossless classical channel to a source close to Bob. Whenever Eve obtains a predetermined number of consecutive *successful* measurement outcomes, this source prepares a new train of non-vacuum signal states that is forwarded to Bob. Otherwise, Eve typically sends vacuum signals to Bob to avoid errors. Sequential attacks transform the original quantum channel between Alice and Bob into an entanglement breaking channel and, therefore, they do not allow the distribution of quantum correlations needed to establish a secret key.

As mentioned above, we shall consider two possible sequential attacks. First, Eve realizes USD of Alice's signal states, since she can always access a local oscillator that is phase-locked to the coherent light source employed by the sender [5]. Whenever Eve identifies unambiguously a signal state sent by Alice, *i.e.*, she determines without error whether it is $|\alpha\rangle$ or $|-\alpha\rangle$, then she considers this result as successful. Otherwise, she considers it a failure. We define the integer parameter M_{min} as the minimum number of consecutive USD successful results that Eve needs to obtain to regard the sequence of measurement outcomes as successful events. More precisely, if $k \geq 0$ denotes the total number of consecutive USD successful outcomes obtained by Eve before she obtains an inconclusive result, then, whenever $k > M_{min}$, Eve prepares a new train of signal states

$$|\psi_{\mathrm{e}}^{k}\rangle = \sum_{n=1}^{k} A_n^{(k)} \exp(i\theta_n)\hat{a}_n^{\dagger}|vac\rangle, \tag{1}$$

followed by $1+d$ vacuum states for the inconclusive result, and she sends these signals to Bob. The parameter d is defined as $d = \lceil t_{\mathrm{d}} f_{\mathrm{c}} \rceil$ with t_{d} and f_{c} denoting, respectively, the dead-time of Bob's detectors and the clock frequency of the system [5]. The angles θ_n in Eq. (1) fulfill $\theta_n = 0$ if the signal state identified by Eve's USD measurement at the time instance n is $|\alpha\rangle$ and $\theta_n = \pi$ if the signal state identified by Eve is $|-\alpha\rangle$, the operator $\hat{a}_n^{\dagger}$ represents a creation operator for one photon in temporal mode n, and the state $|vac\rangle$ refers to the vacuum state.

On the other hand, if $k < M_{min}$ Eve sends to Bob $k+1$ vacuum states, where the last vacuum state corresponds to Eve's inconclusive result. When $k = M_{min}$ Eve employs a probabilistic strategy that combines the two previous ones. In particular, Eve sends to Bob the signal state $|\psi_{\mathrm{e}}^{M_{min}}\rangle$, together with $1+d$ vacuum states, with probability q and, with probability $1-q$, she sends to Bob $M_{min}+1$ vacuum states. That is, the parameter q allows Eve to smoothly fit her eavesdropping strategy to the observed data. Moreover, to simplify our calculations, we define the integer parameter $M_{max} > M_{min}$ as the maximum number of consecutive USD successful results that Eve can obtain in order to send to Bob a train of signal states. Whenever Eve obtains M_{max} consecutive USD successful outcomes then she discards the next measurement outcome and directly sends to Bob the quantum state $|\psi_{\mathrm{e}}^{M_{max}}\rangle$ together with $1+d$ vacuum states.

In the second eavesdropping strategy, the goal of Eve is to identify unambiguously a predetermined number of consecutive relative phases sent by Alice [6]. Again, if M_{min} denotes now the minimum number of consecutive relative phases that has to be correctly identified by Eve to regard the sequence of measurement outcomes as successful, then whenever $k > M_{min}$ Eve forwards to Bob the state $|\psi_{\mathrm{e}}^{k+1}\rangle$ followed by $1+d$ vacuum states. On the other hand, if $k < M_{min}$ Eve sends to Bob $k+2$ vacuum states. The case $k = M_{min}$ is completely analogous to the one discussed above, *i.e.*, Eve follows a probabilistic strategy. Note, however, that the angles θ_n are now selected such that they reproduce the relative phases identified by Eve. That is, $\theta_n - \theta_{n-1}$, with $1 < n \leq k+1$, is equal to the relative phase between pulse n and pulse $n-1$ sent by Alice.

Figure 2 shows a graphical representation of the gain versus the quantum bit error rate (QBER) in both sequential attacks presented above for a given value of the mean photon number $\mu_{\alpha} = |\alpha|^2$ of Alice's signal states, and the parameter d [7]. It states that no key distillation protocol can provide a secret key from the correlations established by the legitimate users above the curves, *i.e.*, the secret key rate in that region is zero. In these examples we have considered the optimal distribution for the state coefficients $A_n^{(k)}$ [5]. That is, the one which provides the lowest QBER for a given value of the gain.

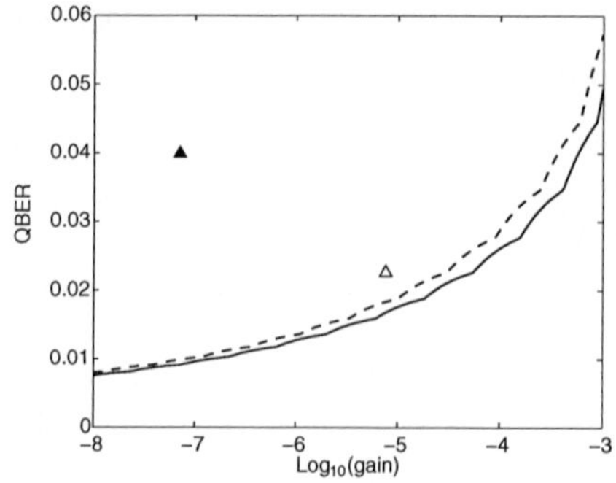

FIGURE 2. Gain versus QBER in a sequential attack that follows the first eavesdropping strategy (dashed line), and the second eavesdropping strategy (solid line). The mean photon number of Alice's signals is $\mu_\alpha = 0.2$, and the parameter $d = 500$. The triangles represent experimental data from [7].

CONCLUSION

In this paper we have analyzed limitations imposed by sequential attacks on the performance of a differential-phase-shift (DPS) quantum key distribution (QKD) protocol based on weak coherent pulses. Specifically, we have compared the behaviour of two possible sequential attacks, and we have derived ultimate upper bounds on the maximal distance achievable by a DPS QKD scheme. It states that there exists no improved classical communication protocol or improved security analysis which can turn the correlations established by the users into a secret key.

ACKNOWLEDGMENTS

This work was supported by the European Projects SECOQC and QAP, by the NSERC Discovery Grant, Quantum Works, CSEC, and by Xunta de Galicia (Spain, Grant No. INCITE08PXIB322257PR).

REFERENCES

1. K. Inoue, E. Waks and Y. Yamamoto, *Phys. Rev. Lett.* **89**, 037902 (2002); K. Inoue, E. Waks and Y. Yamamoto, *Phys. Rev. A* **68**, 022317 (2003).
2. C. Branciard, N. Gisin and V. Scarani, *New J. Phys.* **10**, 013031 (2008).
3. Y.-B. Zhao, C.-H. F. Fung, Z.-F. Han and G.-C. Guo, *Phys. Rev. A* **78**, 042330 (2008); K. Wen, K. Tamaki and Y. Yamamoto, arXiv:0806.2684.
4. E. Waks, H. Takesue and Y. Yamamoto, *Phys. Rev. A* **73**, 012344 (2006); M. Curty, L. L. Zhang, H.-K. Lo and N. Lütkenhaus, *Quant. Inf. Comp.* **7**, 665–688 (2007); T. Tsurumaru, *Phys. Rev. A* **75**, 062319 (2007).
5. M. Curty, K. Tamaki and T. Moroder, *Phys. Rev. A* **77**, 052321 (2008).
6. H. Gómez-Sousa and M. Curty, *Quant. Inf. Comp.* **9**, 62–80 (2009).
7. H. Takesue, S. W. Nam, Q. Zhang, R. H. Hadfield, T. Honjo, K. Tamaki and Y. Yamamoto, *Nature Photonics* **1**, 343–348 (2007).

Proposal of an eavesdropping experiment for BB84 QKD protocol with 1→3 phase-covariant quantum cloner

Yuta Okubo[*,†], Francesco Buscemi[†] and Akihisa Tomita[†]

[*]*Department of Frontier Science, University of Tsukuba*
[†]*ERATO-SORST Quantum Computation and Information Project, Japan Science and Technology Agency*

Abstract. We propose an experimental implementation of a particular attack against the BB84 quantum key distribution protocol, utilizing a 1→3 asymmetric phase-covariant quantum cloning. Our eavesdropping strategy is numerically optimized to extract the information about the input state. The proposed cloning circuit is easily implementable with linear optics and its performances turn out to be well close to the optimum attack.

Keywords: Quantum Key Distribution, Quantum Cloning, Linear optics
PACS: 03.67.Dd

INTRODUCTION

Quantum key distribution(QKD), which ensures the secure communication between two parties, is one of the most promising technologies in the field of quantum information processing. A number of theoretical researches have been done on the security of QKD[1, 2]. They consider various kinds of attacks by the eavesdropper[3, 4], Eve, and show the security criteria as the permissive error rate not only in the ideal condition, but in practical conditions. However, only a few experimental researches have been done on the eavesdropping attacks to QKD. Therefore, experimental verification on the security of QKD has not been fully demonstrated.

On the cloning attack, Eve generates the multiple clones of the input state, and sends the one of the clones to Bob. Finally Eve extracts the information about the input state from the remains. Since it is impossible to prepare the perfect clone of quantum states (No-cloning theorem), these clones are imperfect. Buzek and Hillery[5] proposed the universal quantum cloner, which duplicates the unknown qubits. However, the universal cloning is superfluous for eavesdropping BB84 protocol [1], since the input state is restricted to a part of Hilbert space. Phase-covariant cloning[6] offers a better quality of output clones, at the cost of the restriction for the input state to the equator of the Bloch sphere. So far a 1→2 symmetric / asymmetric[2] phase-covariant cloner has been implemented with linear optics[7, 8] and NMR [9]. A 1→3 symmetric phase-

[1] The universal cloner is still effective for eavesdropping six-state protocol

[2] As for the cloning attack, asymmetric cloner, on which the information from the input is asymmetrically distributed to the outputs, is required.

CP1110, *Quantum Communication, Measurement and Computing (QCMC), The Ninth International Conference,* edited by A. Lvovsky

covariant cloner was realized by combining the linear and the nonlinear optics[10, 11, 12]. In this paper, we propose a linear optical implementation of 1→3 asymmetric phase-covariant cloner, and show how to utilize it for cloning attack.

A PHASE-COVARIANT QUANTUM CLONER

Experimental setup

An experimental setup to implement linear optical 1→3 asymmetric phase-covariant quantum cloner is shown in Fig.1(Left). This setup consists of a tritter[13], which is realized with a beam splitter array, one input port is for Alice's qubit, and the others are for Eve's ancillae. Also, one output port is for Bob's qubit, and the others are for Eve's. In order to obtain an asymmetric cloning, we use one beam splitter which has variable reflectance of r, while the other two are 50:50 beam splitters. We assume classical bit "0", "1" of Alice is encoded to $\left\{|\pm x\rangle = \frac{|0\rangle\pm|1\rangle}{\sqrt{2}}\right\}$ or $\left\{|\pm y\rangle = \frac{|0\rangle\pm i|1\rangle}{\sqrt{2}}\right\}$, each of which is physically represented either by the photonic polarization or the relative phase between two coherent pulses (time-bin). After combining the three inputs at the tritter, the tripartite entangled state $|\xi(\phi,r)\rangle$ is obtained, where r is the reflectance of the variable beam splitter (VBS). Eve sends the first qubit to Bob and keeps the remaining two qubits. The success probability of this cloning operation is $\frac{1-3r^2+6r^3}{4}$[3].

EAVESDROPPING

In order to evaluate the performance of the proposed cloner as an eavesdropper, we derive the relation between disturbance and the amount of information leaked to Eve. The disturbance-leaked information trade off yields the amount of permissive error in which finally Alice and Bob can distill the secure key[14].

Disturbance

The mutual information between Alice and Bob is given by

$$I(A:B) = 1 - H(1-D) \tag{1}$$

where $D \equiv 1 - \mathrm{Tr}_{\mathrm{Eve}}|\xi(\phi,\mathrm{r})\rangle\langle\xi(\phi,\mathrm{r})|$ is the disturbance parameter that characterizes the error of communication channel from Alice to Bob, and Eve can control it by changing r (reflectance).

---

[3] We *post-selected* the events where one and only one photon is detected on one detector.

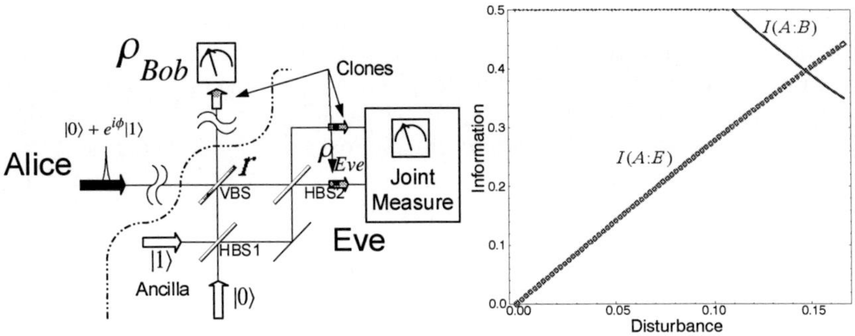

FIGURE 1. (Left)Experimental Setup
(Right)Disturbance v.s. Information

Eve's information

Let us calculate the amount of the information leakage for Eve $I(A:E)$[4] as a function of disturbance. We assume that Eve can keep her qubits in quantum memory until Alice reveals the basis. This means that Eve knows the basis on which Alice encoded her classical bit, therefore she only has to discriminate $\rho_0,\ \rho_\pi\ \left(\rho_{\pi/2},\ \rho_{-\pi/2}\right)$ to obtain the information about the input state, where ρ_ϕ denotes Eve's bipartite density matrix defined as $\rho_\phi \equiv \mathrm{Tr}_{\mathrm{Bob}}|\xi(\phi,\mathrm{r})\rangle\langle\xi(\phi,\mathrm{r})|$. The information leakage $I(A:E)$ is written by[15]

$$I(A:E)=\sum_k \mathrm{Tr}\hat{M}_k \sum_\phi \rho_\phi \log \frac{\mathrm{Tr}(\rho_\phi \hat{\mathrm{M}}_\mathrm{k})}{\Sigma_l \mathrm{Tr}\rho_\phi \hat{\mathrm{M}}_\mathrm{l}\ \Sigma_{\phi'} \mathrm{Tr}(\rho_{\phi'}\hat{\mathrm{M}}_\mathrm{k})}. \tag{2}$$

where $\hat{M}_k$ is Eve's measurement operators, satisfying $\hat{M}_k \geq 0\ \forall k$, and $\sum_k \hat{M}_k = \mathbf{1}$. The amount of information refers to how well Eve can discriminate the state by choosing appropriate measurement $\{\hat{M}_k\}$. It should be noted that such optimized measurement may be in general a joint-measurement on the two qubits of Eve's.

Optimal POVM

We have optimized the measurement $\{\hat{M}_k\}$ according to a steepest-ascent method[15]. The optimized POVM is given by

$$\begin{cases} \hat{M}_1 = |\chi_{-\theta},\chi_\theta\rangle\langle\chi_{-\theta},\chi_\theta| & \hat{M}_2 = |\chi_{-\theta},\chi_{\pi+\theta}\rangle\langle\chi_{-\theta},\chi_{\pi+\theta}| \\ \hat{M}_3 = |\chi_{\pi-\theta},\chi_\theta\rangle\langle\chi_{\pi-\theta},\chi_\theta| & \hat{M}_4 = |\chi_{\pi-\theta},\chi_{\pi+\theta}\rangle\langle\chi_{\pi-\theta},\chi_{\pi+\theta}|. \end{cases} \tag{3}$$

$$|\chi_\alpha,\chi_\beta\rangle \equiv \frac{|0\rangle+e^{i\alpha}|1\rangle}{\sqrt{2}} \otimes \frac{|0\rangle+e^{i\beta}|1\rangle}{\sqrt{2}}, \quad \cos\theta = \frac{2\sqrt{D}}{\sqrt{1-2D}}$$

---

[4] The mutual information between Alice and Eve

The phase factor θ is determined by the disturbance parameter. A crucial point now is that the optimum measurement is separable and easy to be implemented with linear optics: the measurement operators are simply realized by adjusting the measurement basis (i.e. the angle of the polarizers) for each output port. In Fig.1(Right), we plot the mutual information $I(A:B)$ and information extracted by Eve ($I(A:E)$) using the optimum measurement(3), which corresponds to the accessible information[16]. We obtained a permissive error rate as 14.74%, as shown in Fig.1(Right). With the error rate below that, Alice and Bob can distill the secure key by the privacy amplification[17]. Since the permissive error rate for the optimal individual attack is 14.64%[5], our method compares favorably with the optimal one.

CONCLUSION

We have proposed an eavesdropping experiment for BB84 protocol, and show the relation between the disturbance parameter D and the information leakage $I(A:E)$ We have designed not only the optical cloning circuit but the measurement scheme in order that Eve obtains the accessible information.

ACKNOWLEDGEMENT

We would like to thank Xiang-Bin Wang and Yun-Kun Jiang for helpful discussion.

REFERENCES

1. P. W. Shor, J. Preskill, Phys. Rev. Lett. 84, 441(2000)
2. D. Gottesman, H-K. Lo, N. Lutkenhaus, J. Preskill, Quant. Inf. Comput. 5,325(2004)
3. N. Gisin, G. Ribordy, W. Tittel, H. Zbinden, Rev. Mod. Phys. 74. 145(2002)
4. V. Makarov, D. R. Hjelme, J. Mod. Opt. 52, 691(2005)
5. V. Buzek and M. Hillery, Phys. Rev. A 54, 1844(1996)
6. D. Bru, M. Cinchetti, G. M. D'Ariano, C. Macchiavello, Phys. Rev. A 62,012301(2000)
7. J. Fiurasek, Phys. Rev. A 67, 052314(2003)
8. L. Bartuskova, M. Dusek, A. Cernoch, J. Soubusta, J. Fiurasek, Phys. Rev. Lett 99, 120505(2007)
9. H. Chen, X Zhou, D. Suter, J. Du, Phys. Rev. A 75, 012317(2007)
10. X. Zou and W. Mathis, Phys. Rev. A 72, 022306(2005)
11. F. Sciarrino, and F. De Martini, Phys. Rev. A 76,012330(2007)
12. F. Sciarrino, and F. De Martini, Phys. Rev. A 72,062313(2005)
13. M. Zukowski, A. Zeilinger, M. A. Horne, Phys. Rev. A 55, 55(1997); M. Reck, A. Zeilinger, H. J. Bernstein, P. Bertani, Phys. Rev. Lett. 73,58(1994)
14. C. A. Fuchs, N. Gisin, R. B. Griffith, C-S Niu, A. Peres, Phys. Rev. A 56, 1163(1997)
15. J. Rehacek, B-G. Englert, D. Kaszlikowski, Phys. Rev. A 71, 054303(2003)
16. G. Chen, H-L. Kauffman, S. J. Lomonaco(EDT), "Mathematics of Quantum Computation and Quantum Technology", CRC Press(2007)
17. C.H. Bennett, G. Brassard, C. Crepeau, and U. Maurer, IEEE Transaction on Information Theory , 1995. Volume 41, Number 6, 1915(1995)

---

[5] It is attained by 1-2 asymmetric phase covariant cloning attack.

Symmetric extension and its application in QKD

Geir Ove Myhr*,†, Norbert Lütkenhaus*,†, Andrew C. Doherty** and Joseph M. Renes‡

*Institute for Quantum Computing, University of Waterloo, 200 University Ave. W., Waterloo ON N2L 3G1, Canada
†Universität Erlangen-Nürnberg, Staudtstr. 7/B2, 91058 Erlangen, Germany
**School of Physical Sciences, The University of Queensland, Queensland 4072, Australia
‡Institut für Angewandte Physik, Technische Universität Darmstadt, Hochschulstr. 4a, 64289 Darmstadt, Germany

Abstract. We investigate the distillation of secret key from classical data when these data are derived from quantum states by measurements. Underlying quantum states with a so-called symmetric extension can never lead to a secret key via one-way communication but must be processed further by LOCC operations. We show that if a symmetric extension can be broken by LOCC operations, it is also possible to break it with a single filter. For Bell-diagonal and some other two-qubit states we characterize the states with symmetric extension, and give a conjectured formula for the general two-qubit case. We apply these results to the the 4-state and 6-state QKD protocols to show that the two-way procedure proposed by Chau fails for error rates beyond 20% and 27.6% precisely because at that point it fails to break symmetric extensions.

Keywords: quantum key distribution, symmetric extension
PACS: 03.67.Dd, 03.67.Mn

QKD AND SYMMETRIC EXTENSION

Any implementation of quantum key distribution consists of two parts. First the hardware distributes and measures quantum particles, producing correlated data. Then the software subjects the generated data to various tests and, if it passes, creates a secret key from it. Whether the implementation is entanglement-based or an equivalent prepare and measure scheme, the data must show evidence of effective entanglement for the output to be a secret key [1]. However, the converse is not known, that is, whether the data can be fashioned into a secret key given proof of effective entanglement.

Certainly this is not always possible if the key distillation protocol is restricted to one way communication. In this case there exist entangled states with a so-called symmetric extension [2], meaning that an eavesdropper will be in at least as good a position as one party to guess the other party's data. Formally, a bipartite state ρ_{AB} has a symmetric extension if there exists a state $\rho_{ABB'}$ on an extended system ABB', which satisfies the two properties

$$\mathrm{tr}_{B'}(\rho_{ABB'}) = \rho_{AB} \qquad \text{and} \qquad V_{BB'}\rho_{ABB'}V_{BB'}^{\dagger} = \rho_{ABB'}.$$

Here $V_{BB'}$ is the unitary swap operator, swapping systems B and B'. The system B' is then a copy of B and, for QKD purposes, is assumed to be in the hands of an eavesdropper along with the rest of the purifying system. The symmetry requirement

CP1110, *Quantum Communication, Measurement and Computing (QCMC), The Ninth International Conference,* edited by A. Lvovsky

on $\rho_{ABB'}$ may seem to be more restrictive than necessary on first sight. The seemingly weaker requirement $\mathrm{tr}_{B'}(\rho_{ABB'}) = \mathrm{tr}_B(\rho_{ABB'}) = \rho_{AB}$ gives an equivalent condition, since if such a state exists, the state $\frac{1}{2}(\rho_{ABB'} + V_{BB'}\rho_{ABB'}V_{BB'}^{\dagger})$ also exists and has the required symmetry.

We argue that symmetric extensions are also relevant in the general context of QKD with two-way postprocessing. Any such protocol consists of a finite number of rounds of one-way communication in alternating directions, and since the protocol must eventually terminate, the effective state shared prior to the final round must have no symmetric extension in order for the protocol to succeed. If the effective state prior to the postprocessing had a symmetric extension, it must be broken at some point along the way.

REDUCTION TO TWO COMMUNICATION ROUNDS

By concerning ourselves only with the question of whether a key can be distilled or not, and not with the rate of distillation, the problem can be simplified further. Assume that each round of one-way postprocessing is done on blocks with a finite number of systems. We may then limit ourselves to filtering operations, which are general quantum operations which can succeed or fail. If more than one outcome of Bob's generalized filtering measurement succeeds, we could as well preselect one of them and only consider this a successful outcome.

Moreover, the number of rounds in the protocol can be reduced to two, as follows. Assume that the last one-way round is from Alice to Bob. For each block that Bob operates on in his last round, he can guess Alice's announcements relating to the quantum systems in that block ahead of time and start immediately with his last round. Alice will notice if this guess was wrong and tell Bob to discard that block. For the tiny fraction of blocks where Bob was right, Alice can proceed with her last round. This means that if a QKD protocol succeeds in distilling secret key from an underlying quantum state with symmetric extension, it must be possible to break the symmetric extension only with a filter on a block of copies of the state. [3]

WHICH STATES HAVE SYMMETRIC EXTENSION?

To identify strategies for breaking a symmetric extension, we need to characterize states with symmetric extension. We can show that these states can be written as convex combinations of states with a *pure* symmetric extension. For a pure state on ABB', the spectra on both sides of a splitting must be the same because of the Schmidt decomposition, so $\vec{\lambda}(\rho_{AB}) = \vec{\lambda}(\rho'_B)$. If a pure state is a symmetric extension of ρ_{AB}, it must satisfy $\rho_{AB} = \rho_{AB'}$ and therefore also $\rho_B = \rho_{B'}$. Any state with a pure symmetric extension therefore has a spectrum which is equal to the spectrum of its reduced density operator on B,

$$\vec{\lambda}(\rho_{AB}) = \vec{\lambda}(\rho_B).$$

We can prove that in the special case when A and B are both qubits—but never otherwise—this condition is also sufficient for having a pure symmetric extension [4].

Using this condition and other techniques we have found necessary and sufficient conditions for special classes of two-qubit states to have a general symmetric extension. These classes are Bell-diagonal states, rank-2 states and states where only one of the off-diagonal terms in the density matrix is non-zero (in the latter case the state is either separable or the non-zero element can be chosen to be $\rho_{00,11} = \rho_{11,00}$ by a local change of basis). In all these cases the conditions can be shown to be equivalent to [4]

$$\mathrm{tr}(\rho_B^2) \geq \mathrm{tr}(\rho_{AB}^2) - 4\sqrt{\det(\rho_{AB})}.$$

We conjecture that this also is a necessary and sufficient condition for all other two-qubit states. The conjecture is also supported by numerical evidence.

BB84 AND 6-STATE PROTOCOLS

In the BB84 and 6-state protocols with standard sifting, the states from which the correlations arise can be taken to be Bell-diagonal without loss of generality. Moreover, most postprocessing schemes preserve this property throughout the postprocessing. It is therefore possible to check for which states a given postprocessing breaks a symmetric extension. The currently best known procedure in terms of distilling key from more states is to first perform so-called B-steps on two qubit pairs and then finish off with one-way error correction and privacy amplification [5]. The B-steps is the only part where two-way communication is required, and it therefore has to break the symmetric extension if there is one to begin with. In a B-step Alice and Bob both announce the parity of two bits and if they are equal they keep the first bit. If the parity bits are different they know there was an error on one of the bits and discard both. If this is done n times it requires n rounds of communication in each direction, since both parties need to know which pairs to discard before proceeding to the next round. As mentioned above, this can be reduced to two one-way rounds without changing the threshold at the cost of reducing the key rate drastically. Alice and Bob would then operate on blocks of $N = 2^n$ qubits, and they both announce the parity of one bit with all the other bits and keep this bit only if all the parities agree. This procedure is called advantage distillation [6] and can of course be done on any $N \neq 2^n$ as well.

Chau [7] showed that this procedure can distill key for up to a quantum bit error rate of 20.0 % for BB84 and 27.6 % for the 6-state protocol. More precisely, if the initial Bell-diagonal state is characterized by the eigenvalues $p_i, i \in \{0, x, y, z\}$ for eigenvectors $(\mathbb{1} \otimes \sigma_i)(|00\rangle + |11\rangle)/2$, it will distill key whenever $(p_0 - p_z)^2 > (p_0 + p_z)(p_x + p_y)$. We define

$$D_C := \log_2 \left(\frac{(p_0 - p_z)^2}{(p_0 + p_z)(p_x + p_y)} \right)$$

so that the condition for secret key generation becomes $D_C > 0$. One B-step doubles D_C so if $D_C \leq 0$ it will always remain so, while if $D_C > 0$ it can be increased to an arbitrary positive value by repeating the B-steps. Similarly, advantage distillation on blocks of N copies gives $D_C^{\mathrm{out}} = ND_C$.

The values of D_C can also be connected to symmetric extension. It turns out [3] that all states that have $D_C \leq 0$ have a symmetric extension. Since B-steps and advantage distillation cannot change the sign of D_C, they will never break the symmetric extension in this case. Similarly, no states that have $D_C \geq 2$ have a symmetric extension. Therefore, $n \geq 1 - \log_2(D_C)$ rounds of B-steps, or advantage distillation on block size $N \geq \frac{2}{D_C}$ will break the symmetric extension.

Acín et al. [8] have previously shown that if $D_C \leq 0$, there is an attack by Eve that will prohibit Alice and Bob from generating a secret key by advantage distillation followed by one-way processing. From our analysis it follows that this impossibility can be understood as a failure of the two-way part of the protocol to break a symmetric extension. By using our criterion for symmetric extension of Bell-diagonal states, no actions by the eavesdropper need to be taken into account explicitly.

ACKNOWLEDGMENTS

The authors would like to thank Debbie Leung, Matthias Christandl, Barbara Kraus and especially Marco Piani and Tobias Moroder for stimulating discussions. This work was funded by the Research Council of Norway, project No. 166842/V30, the European Union through the IST Integrated Project SECOQC and the IST-FET Integrated Project QAP, Ontario Centres of Excellence, the NSERC Innovation Platform Quantum Works and by the NSERC Discovery Grant. JMR acknowledges the support of the Alexander von Humboldt Foundation and the Visiting Scholars Program at the University of Queensland, where part of this work was carried out.

REFERENCES

1. M. Curty, M. Lewenstein, and N. Lütkenhaus, *Phys. Rev. Lett.* **92**, 217903 (2004).
2. A. C. Doherty, P. A. Parrilo, and F. M. Spedalieri, *Phys. Rev. Lett.* **88**, 187904 (2002).
3. G. O. Myhr, J. M. Renes, A. C. Doherty, and N. Lütkenhaus, Symmetric extension in two-way qauntum key distribution (2008), in preparation.
4. G. O. Myhr, and N. Lütkenhaus, Spectrum conditions for symmetric extendible states (2008), in preparation.
5. D. Gottesman, and H.-K. Lo, *IEEE Trans. Inf. Theory* **49**, 457 (2003).
6. U. M. Maurer, *IEEE Trans. Inf. Theory* **39**, 733–742 (1993).
7. H. Chau, *Phys. Rev. A* **66**, 60302 (2002).
8. A. Acín, J. Bae, E. Bagan, M. Baig, L. Masanes, and R. Muñoz-Tapia, *Phys. Rev. A* **73**, 012327 (2006).

Sub optimum quantum measurement for M-ary ASK applicable to Y-00 cryptosystem

Y. Nagasako*, H.Suemitu†, R.Sugawara†, M.Sohma** and O.Hirota**

*Internet Research Institute, Tokyo, Japan
†Faculty of Enginnering, Tamagawa University, Tokyo, Japan
**Research Institute of Tamagawa University, Tokyo, Japan

Abstract. Recently quantum stream cipher based on quantum detection theory so called Yuen-2000 protocol has been developed in both theory and experiment. There are two types of implementation of Y-00 scheme. One is by phase shift keying (PSK), and another is by amplitude or intensity shift keying (ASK). One of the evaluation methods for such stream cipher is a success probability by quantum optimum receiver. In the case of PSK, the signal states have a property of symmetric group, and the quantum limitation has been calculated by our group. In that case, the optimum detection becomes so called square root measurement. Since we are developing Y-00 quantum stream cipher by ASK, we want to clarify the limitation of the attacker in the case of ASK. But in this case, it has no property as group theory, so it is at present difficult to obtain the optimum solution.

In this letter, we show the performance of a square root measurement for ASK as sub-optimum receiver and discuss an evaluation of the Y-00 protocol by ASK system.

Keywords: Quantum stream cipher, Yuen protocol, ASK system
PACS: 03.67.Dd, 42.50.Lc

INTRODUCTION

So far two schemes have been proposed in order to realize secure communication based on physical phenomena. One of methods is one time pads supported by the quantum key distribution :QKD. The other is a kind of randomized stream cipher. However, in the case of QKD, there is a big gap in the performance between the experimental realization and the real communication network requirement. On the other hand, a stream cipher invented by Yuen may provide a randomized stream cipher with provable security by randomization based on quantum noise[1]. This scheme is called quantum stream cipher by Yuen 2000 protocol (Y-00), or $\alpha\eta$ scheme. It has been discussed by Yuen group [2,3,4] and Hirota [5] that Y-00 protocol has sufficient security in comparison with the conventional stream cipher. In addition, in order to clarify an intrinsic difference between them in the sense of information theoretic security, Kato and Hirota [6] calculated the exact success probability for secret key recovery in phase shif keying (PSK) scheme of Y-00 under small samples. Consequently, the success probability in the case of Y-00 is very small, while the success probability of the conventional cipher is always 1. Thus, the quantum stream cipher by Y-00 is, in principle and in practice, always secure than the conventional cipher such as AES(Advanced encryption standard) and so on.

We are developing the different type, so called amplitude (ASK) or intensity (IMDD) modulation scheme for the realization of Y-00 [7]. Unfortunately, the quantum optimum measurement for amplitude or intensity modulation scheme are not yet clarified, while

CP1110, *Quantum Communication, Measurement and Computing (QCMC), The Ninth International Conference,* edited by A. Lvovsky

the quantum optimum measurement for PSK scheme has been given. In this letter, we employ so called square root measurement which may give sub-optimum measurement for many modulation schemes, in order to make clear the difference between Y-00 by ASK (IMDD) and the conventional cipher. As a result, we show the fact that Y-00 by ASK or IMDD is also secure than the conventional cipher such as AES and so on.

SUCCESS PROBABILITY ANALYSIS FOR Y-00 BY ASK

Square root measurement

The model of signal in the quantum stream cipher by Y-00 is described in the following. One sequence of coherent states emitted from the transmitter is regarded as a code word state of the length $|K_s|$ or as one quantum state on the extended space $H_s^{\otimes m}$. So a number of possible code words is $J = 2^{|K_s|}$. Eve's problem reduces to discrimination for such $2^{|K_s|}$ code words with minimum error probability. Quantum states for the code words are

$$
\begin{aligned}
|\Psi_1> &= |\alpha_1>_1 |\alpha_1>_2 |\alpha_1>_3 \ldots |\alpha_1>_m \\
|\Psi_2> &= |\alpha_2>_1 |\alpha_1>_2 |\alpha_1>_3 \ldots |\alpha_1>_m \\
|\Psi_3> &= |\alpha_1>_1 |\alpha_2>_2 |\alpha_1>_3 \ldots |\alpha_1>_m \\
&\vdots \\
|\Psi_J> &= |\alpha_M>_1 |\alpha_M>_2 \ldots |\alpha_M>_m
\end{aligned}
\tag{1}
$$

where $m = |K_s|/logM$.

Our target is to solve the following formula from known Helstrom-Holevo-Yuen formalism for quantum detection theory [8]

$$
\begin{aligned}
\tilde{P}_e &= min\frac{1}{J}\sum(1- <\Psi_i|\bar{\Pi}_i|\Psi_i>), or \\
\tilde{P}_d &= 1-\tilde{P}_e
\end{aligned}
\tag{2}
$$

where $\{\bar{\Pi}_i\}$ is detection operator or POVM (positive operator valued measure) which represents general quantum detection procedures, $|\Psi_i> \in H_s^{\otimes m}$ means a state of m length of M coherent states. This gives an evaluation of the success probability of key recovery.

In this letter, we only discuss a physical decryption by the quantum individual measurement. Even so, it is difficult to find the optimum measurement. Here we employ so called square root measurement for each slot as follows [9,10]:

$$
\Pi_i = |\mu_i><\mu_i|, \quad |\mu_i> = \hat{G}^{-1/2}|\alpha_i>, i = 1,2,\ldots,M \tag{3}
$$

where

$$
\hat{G} = \sum_i |\alpha_i><\alpha_i| \tag{4}
$$

and where $\hat{G}$ is Gram operator. Let λ_i be eigenvalue of the Gram matrix corresponding to the Gram operator. In the case of PSK, the detection probability is, in general, given by [10]

$$P_d = \frac{1}{M^2}|\sum \sqrt{\lambda_i}|^2 \tag{5}$$

The quantum gain comes from the cross term of eigenvalues in the above formula, for example, $\sqrt{\lambda_i \lambda_j}$. In the case of ASK, there is no guarntee for the optimality. We will show how the square root measurement and the above formula provide sub optimum performance for ASK system by numerical simulation.

Properties of square root measurment for ASK signals

Let us define the ASK signals in Y-00 as follows:

$$L = \begin{pmatrix} K_i \\ |\alpha_i> \\ x \end{pmatrix} = \begin{pmatrix} 1 & 2 & 3 & 4 & \dots & M \\ |\alpha_1> & |\alpha_2> & |\alpha_3> & |\alpha_4> & \dots & |\alpha_M> \\ 0 & 1 & 0 & 1 & \dots & 0 \end{pmatrix}, \tag{6}$$

where K_i is running key, x is data bit, $|\alpha_i>$ is coherent state, respectively. Here, $\alpha_1 = \alpha_{min}$ and $\alpha_M = \alpha_{max}$ are minimum and maximum amplitude which are ,in general, fixed. The number of basis M is designed to give enough error to a receiver of Eve taking the following relation into account:

$$\frac{|\alpha_{max}| - |\alpha_{min}|}{M} \quad vs \quad noise \tag{7}$$

Here we assume that Eve tries directly to recover the seed key of LFSR by quantum detection for $m = |K_s|/logM$ slots. That is, it attempts directly to determine the running keystream by square root measurement scheme for M-ary signals. We calculate the error probability or detection probability per slot and show the numerical example in Fig.1.

Finally, the success probability is given by

$$P_{suc}(Key) = P_d{}^{|K_s|/logM} \tag{8}$$

Consequently, the success probability of seed key of LFSR is very small. This is one of evidences that Y-00 by ASK is also secure than any conventional cipher such as AES and so on.

CONCLUSION

To find the quantum optimum measurement for ASK system is in general difficult. In this latter, we have clarified a performance of the square root measurement as a sub optimum quantum measurement for ASK. This result can be used to show the superiority of the quantum stream cipher by ASK Y-00 against any conventional data encryption such as AES,and so on. Indeed when the number of signal M is very large, the success probability is very small while the success probability is unity for any conventional data encryption.

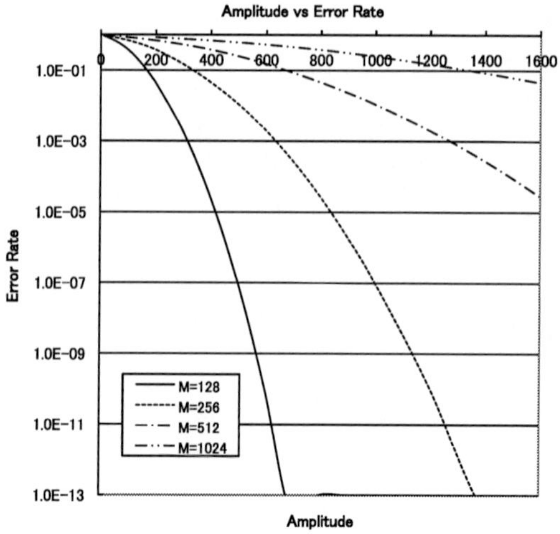

FIGURE 1. Error probability for M-ary ASK system. Ampitude corresponds to the difference between maximum and minimum amplitude.

REFERENCES

1. H.P.Yuen, R.Nair, E.Corndorf, G.K.Kanter, and P.Kumar, *Quantum Information and Computing*, vol-8, p561, 2006.
2. H.P.Yuen and R.Nair, *Physics Letters A*, vol-364, p112, 2007.
3. R.Nair, H.P.Yuen, E.Corndorf, T.Eguchi,P.Kumar, *Physical Review A* vol 74, 052309,2006.
4. R.Nair and H.P.Yuen, *arXiv* quant-ph/0712.1092v5,2008, to be appeared in Physics Letters A.
5. K.Kato,and O.Hirota, *Physical Review A*, to be submitted in 2008.
6. O.Hirota, *Physical Review A*, vol 76, 032307, 2007.
7. O.Hirota, M.Sohma, M.Fuse, and K.Kato, *Physical Review A*, vol 72, 022335, 2005.
8. C.Helstrom, *Quantum detection and estimation theory*, Academic Press, 1976.
9. M.Ban, K.Kurokawa, R.Momose, and O.Hirota, *International Journal of Theoretical Physics*, vol 35, pp1269-1288, 1997.
10. K.Kato, M.Osaki, M.Sasaki, and O.Hirota, *IEEE Trans. on Communications*, vol 47, pp248-254, 1999.

Security proof of differential phase shift quantum key distribution in the noiseless case[1]

Yi-Bo Zhao[2]

Key Lab of Quantum Information, University of Science and Technology of China, (CAS), Hefei, Anhui 230026, China

Abstract. Differential phase shift quantum key distribution systems have a high potential for achieving high speed key generation. However, its unconditional security proof is still missing, even though it has been proposed for many years. Here, we prove its security against collective attacks with a weak coherent light source in the noiseless case (i.e. no bit error). The only assumptions are that quantum theory is correct, the devices are perfect and trusted and the key size is infinite. Our proof works on threshold detectors. We compute the lower bound of the secret key generation rate using the information-theoretical security proof method. Our final result shows that the lower bound of the secret key generation rate per pulse is linearly proportional to the channel transmission probability if Bob's detection counts obey the binomial distribution.

Keywords: Differential phase shift, quantum key distribution, QKD, security proof.
PACS: 03.67.Dd, 03.67.Hk

In this paper, we prove the security of DPSQKD against collective attacks with a weak coherent light source in the noiseless case. The security in this case follows from a key result that Eve's state is independent of the positions of Bob's detected signals. This result makes sense since the fact that no bit error restricts what Eve can do to Bob's signals. In particular, she has to ensure that Bob receives all signals with equal intensities, since signals with different intensities will result in bit errors with non-zero probability. Our final result on the lower bound on the key generation rate is a function of the estimated parameters of the channel. In order to understand this result further, we compute this bound by considering a channel that produces a binomial distribution in Bob's detection statistics. Specifically, we show that the lower bound of the secret key generation rate per pulse is linearly proportional to the channel transmission probability. The only assumptions used in the proof are that quantum theory is correct, the key size is infinite and the devices are perfect and trusted. Our proof works on threshold detectors, which are the detectors commonly used in practice. In the following analysis, we map the traditional DPS protocol into a big-state protocol and give the security proof for this big-state protocol.

Protocol 1 - original protocol: Alice sends a sequence of coherent states, each with amplitude α, but with a randomly selected phase, $|\alpha\rangle$ or $|-\alpha\rangle$, to Bob. Then she records each state with a binary variable x_i, by setting $x_i = 0$ if the i-th state is $|-\alpha\rangle$ and $x_i = 1$ if the i-th state is $|\alpha\rangle$. Using the Mach-Zehnder (M-Z) interferometers, Bob measures

[1] A full version of this conference paper has been appeared as Phys. Rev. A 78, 042330 (2008).
[2] Email: zhaoyibo@mail.ustc.edu.cn

CP1110, *Quantum Communication, Measurement and Computing (QCMC), The Ninth International Conference,* edited by A. Lvovsky

the phase difference between every two adjacent states. Bob stores his measurement result into binary variables y_i and z_i. While measuring the phase difference of the i-th and $(i-1)$-th state, if Bob gets a photon count, he sets $z_i = 1$ and if not, he sets $z_i = 0$. Also, if $i = kN + 1$ ($k = 0, 1, 2, 3, ...$), Bob sets $z_i = 0$. If $z_i = 1$ and the measurement results indicates that the phase difference is zero, he sets $y_i = 0$; otherwise, if it indicates a non-zero phase difference, he sets $y_i = 1$. If $z_i = 0$, Bob sets $y_i = 0$. After Bob receives each set of N states, he announces all z_i's for these N states. Alice generates another variable $l_i = x_i \oplus x_{i-1}$ for $i > 1$. After many rounds of such communications, Alice and Bob randomly publish some of l_i and y_i corresponding to $z_i = 1$ to test the bit error rate (BER) between them. Alice and Bob generate new binary variables u_i^A and u_i^B respectively. Alice sets $u_i^A = l_i \cdot z_i$. Bob sets $u_i^B = y_i \cdot z_i$. Alice and Bob estimate the mutual information between the binary strings $\vec{u}^A$ and $\vec{u}^B$ conditioned on Bob's announcement. Alice announces the error correction information of binary string $\vec{u}^A$ and Bob uses it to reconcile his string $\vec{u}^B$ to the corresponding string $\vec{u}^A$. Alice and Bob perform privacy amplification on their common binary string $\vec{u}^A$ to generate the final secret key.

Instead of above classical data process, Alice and Bob can simply discard the l_i and y_i that correspond to $z_i = 0$, and perform error correction and privacy amplification on the remaining sifted bits. This protocol is then equivalent to the original DPSQKD protocol [1]. In this protocol, Bob measures the phase difference between every two adjacent pulses. Thus, it may not be a good idea to try to map them into single bits and then discuss the security. Here, our basic idea is to regard N states as one big state and to discuss the security of this big state. In the following, we will introduce a protocol that maps the above protocol into a big state protocol, in which Alice sends Bob a big state and Bob measures it with N equipments. The security of this big state protocol is weaker than the above one. Thus, if the security of this inferior protocol is proved, then the security of the above protocol is proved.

Protocol 2: In the above protocol Alice generates a state $|\Psi_{\vec{x}}^N\rangle = \bigotimes_{i=kN+1}^{(k+1)N} |(-1)^{x_i+1}\alpha\rangle$ according to the binary string $\vec{x} = (x_{kN+1}, x_{kN+2}, ..., x_{(k+1)N})$, and sends it to Bob through N fibers. While receiving this big state, Bob measures the phase difference between each two adjacent states at the same time with the equipment shown in Fig. 1. Other Steps remains the same as in Protocol 1. It can be proved that Protocol 2 is inferior to Protocol 1. In the following we will prove the security of Protocol 2 first. Then the security of Protocol 1 follows.

In the DPS protocol, what Alice sends to Bob are two weak coherent states. Since these two states are non-orthogonal Eve in principle cannot always distinguish them and cannot always know the phase difference between two adjacent pulses. The mutual information per pulse between Alice and Eve is thus less than one. Bob uses the M-Z system to measure the phase difference between two adjacent pulses. This system can let Bob definitely know the deterministic phase difference between the two pulses with certain probability (Bob can definitely know the phase difference if he gets a count.). Finally Alice and Bob only keep the measurement results of the phase differences that Bob has good knowledge of. The point is that either Bob knows that Eve has obtained good information on a pulse pair (in which case he has not gotten a photon count and this result is discarded) or Bob knows more about this result than Eve does (in which case

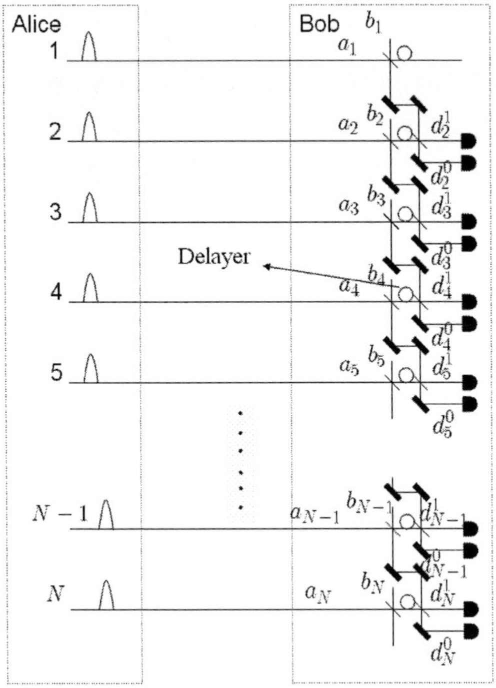

FIGURE 1. Illustration of Protocol 2, where the delayer generates the time delay that makes the two pulses meet at the beam splitter before the detector.

Alice and Bob proceed to distill a secret key by applying error correction and privacy amplification to many of such result).

Here we only discuss the security against the collective attack under the infinite-key length case. Let $\vec{X}$, $\vec{Y}$, $\vec{Z}$, $\vec{L}$, $\vec{U}_A$ and $\vec{U}_B$ denote the variables that can take the values $\vec{x}$, $\vec{y}$, $\vec{z}$, $\vec{l}$, $\vec{u}^A$ and $\vec{u}^B$, respectively. According to the information-theoretical security proof, the secret key rate of above protocol can be given by

$$G \geq I(\vec{U}_A : \vec{U}_B|\vec{Z}) - S(\vec{U}_A : E|\vec{Z}) \tag{1}$$

where I and S denote the Shannon mutual information and the Von Neuemann mutual information respectively. We can prove that if there is no bit error, Eve cannot control Bob's count position and let Bob get a count from the state she know well. Then after the postselection, Bob knows Alice's selected data better than Eve. More precisely, this condition actually shows that Eve's information about Alice satisfies

$$S(\vec{U}_A : E|\vec{Z}) \leq \sum_{w=1}^{N-1} P(w)w - \frac{1}{N-1}\sum_{w=1}^{N-1} P(w)wS(\vec{L}|E, W = w). \tag{2}$$

where w denotes the number of Bob's total counts. The upper bound of Eve's information can be obtained from Eq. (2) by applying the fact that the mutual information between Alice and Eve plus W system is no larger than the Holevo quantity of Alice's sent out state.

We finally can prove that the final lower bound on the secret key rate per N pulses is given by

$$G \geq \sum_{w=1}^{w_0} P(w)w - \Delta \tag{3}$$

where the w_0 is the solution to the equations

$$\begin{cases} \sum_{w=1}^{w_0} P(w)(N-1) \leq K \\ \sum_{w=1}^{w_0+1} P(w)(N-1) \geq K \end{cases} \tag{4}$$

Δ is given given by $\Delta = \sum_w P(w)w - I(\vec{U}_A : \vec{U}_B|\vec{Z})$ and K is defined by

$$\sum_{w=1}^{N-1} P(w)S(\vec{L}|E, W = w) \geq (N-1)[1 - P(w=0)] - NS(A) =: K. \tag{5}$$

$S(A)$ is the Holevo quantity of Alice's sent out state.

Consider the specific case where Bob's total count obeys the binomial distribution, $\Delta = 0$ and $N \rightarrow \infty$. Under this case the secret key rate per pulse can be given by

$$g \geq \eta|\alpha|^2\{1 - h[\frac{1}{2}(1 - e^{-4|\alpha|^2})]\} \tag{6}$$

where α is the amplitude of Alice's sent out coherent state. It can be calculated that when $\alpha = 0.338$, the right part in the Eq. (6) is maximized and is given by 0.0357η.

We proved the security for DPSQKD with a weak-coherent light source against collective attacks in the noiseless case. The only assumption we employ are that the quantum theory is true, the device is trusted and the key size is infinite. Although we have only proved the security of DPSQKD for the noiseless case, we hope that our work can offer some insights into the security of DPSQKD and may serve as a stepping stone for proving the security for the noisy case.

ACKNOWLEDGMENTS

This work is supported in part by Innovation Fund of the University of Science and Technology of China under Grant No. KD2006005, the Research Grants Council of Hong Kong under Grant No. HKU 701007P, and the Natural Sciences and Engineering Research Council of Canada.

REFERENCES

1. K. Inoue, E. Waks and Y. Yamamoto, Phys. Rev. A, 68, 022317 (2003).
2. Y.-B. Zhao, C.-H. F. Fung, Z.-F. Han, G.-C. Guo, Phys. Rev. A 78, 042330 (2008).
3. R. Renner, arXiv: quant-ph/0512258 (2005).

CHAPTER 12

QUANTUM COMMUNICATION PROTOCOLS
(OTHER THAN CRYPTOGRAPHY)

Quantum Energy Teleportation

Masahiro Hotta

Graduate School of Science, Tohoku University, Sendai, 980-8578, Japan

Abstract.
We report a protocol of quantum energy teleportation that transports energy by LOCC respecting causality and local energy conservation.

Keywords: quantum teleportation, entanglement, spin chains
PACS: 03.67.-a

INTRODUCTION

Quantum teleportation (QT) [1] teleports an unknown quantum state to a distant place by local operations and classical communication (LOCC). However, QT cannot transport the excitation energy of the transported states. For example, let us imagine that Alice (A) sends to Bob (B) a spin-up state of a qubit in an external uniform magnetic field parallel to the z axis. The Hamiltonian is given by

$$H_s = g\sigma_z$$

with a positive constant g. The ground state of the qubit is the spin-down state with energy $-g$ and the excited energy of the spin-up state is $+g$. We note that a contracted state at A or B of a Bell pair shared by A and B for QT in the magnetic field is the maximal entropy state and has zero energy on average. Therefore, both A and B must first supply energy +g on average to create the Bell pair from two spin-down qubits in the ground state. B has to supply an additional $+g$ energy to his spin to receive the spin-up state from A's site by QT. Hence, the total amount of energy-input at B's site for teleportation is $+2g$, which should be locally prepared by B.

Recently, a protocol called quantum energy teleportation (QET) was proposed for qubit chains, which transports energy by LOCC respecting causality and local energy conservation [2]. Without any fine-tuning of the Hamiltonian H of the qubit chain, entanglement is generated in the ground state $|g\rangle$. H is given by a sum of the energy density operators T_n. The origin of T_n can be defined such that $\langle g|T_n|g\rangle = 0$. Here, it is interesting to note that a negative energy density appears because of the entanglement in $|g\rangle$. For example, let us consider a case in which a correlation function $\langle g|T_nO_m|g\rangle$ is *not* decomposed into $\langle g|T_n|g\rangle\langle g|O_m|g\rangle$ because of the entanglement for a local operator O_m at site m far from n. In this case, $|g\rangle$ cannot be an eigenstate of T_n, because if $T_n|g\rangle = \varepsilon|g\rangle$ is satisfied, the correlation function must be written as $\langle g|T_nO_m|g\rangle = \varepsilon\langle g|O_m|g\rangle = \langle g|T_n|g\rangle\langle g|O_m|g\rangle$, which clearly contradicts the above assumption. Next, we spectrally decompose T_n as $T_n = \sum_{\nu,k_\nu} \varepsilon_\nu|\varepsilon_\nu,k_\nu\rangle\langle\varepsilon_\nu,k_\nu|$, where ε_ν are eigenvalues of T_n, $|\varepsilon_\nu,k_\nu\rangle$ are the corresponding eigenstates and the index k_ν denotes the degeneracy freedom of

CP1110, *Quantum Communication, Measurement and Computing (QCMC), The Ninth International Conference,* edited by A. Lvovsky

the eigenvalue ε_ν. The ground state can then be expanded as $|g\rangle = \sum_{\nu,k_\nu} g_{\nu,k_\nu} |\varepsilon_\nu, k_\nu\rangle$. By this decomposition, the above definition of the energy density origin gives the following equation: $\langle g|T_n|g\rangle = \sum_{\nu,k_\nu} \varepsilon_\nu \left|g_{\nu,k_\nu}\right|^2 = 0$. Clearly, this equation has no solution when the lowest eigenvalue ε_- of T_n is positive. The case of $\varepsilon_- = 0$ is also prohibited because $|g\rangle$ would then become an eigenstate of T_n corresponding to the eigenvalue ε_-. This means that ε_- must be negative. It is thereby verified that there exist quantum states $|\varepsilon_-, k_-\rangle$ with negative energy density due to entanglement. It should be stressed that even if a state has negative energy density over a certain region, there exists positive energy density at other regions and the total energy is not negative.

PROTOCOL

By using negative energy, a protocol of QET can be constructed as follows. (i) For the ground state $|g\rangle$, A (energy sender) performs a local measurement of $\sigma_A = \vec{u}_A \cdot \vec{\sigma}_A$, where $\vec{u}_A$ is a three-dimensional real unit vector and $\vec{\sigma}_A$ is the Pauli spin operator acting on A's qubit. The eigenvalues of the observable are $(-1)^\mu$ with $\mu = 0, 1$. We write the spectral decomposition of σ_A as $\sigma_A = \sum_{\mu=0,1} (-1)^\mu P_A(\mu)$, where $P_A(\mu)$ are projective operators onto the eigenspaces. In this measurement process, A must input a positive amount of energy $E_A = \sum_{\mu=0,1} \langle g|P_A(\mu) H P_A(\mu)|g\rangle$ to the qubit chain. (ii) A announces μ to B (energy receiver) by a classical channel. (iii) B performs a local unitary operation depending on the value of μ. The unitary operator is defined by $V_B(\mu) = I\cos\theta + i(-1)^\mu \sigma_B \sin\theta$, where $\sigma_B = \vec{u}_B \cdot \vec{\sigma}_B$, $\vec{u}_B$ is a three-dimensional real unit vector, $\vec{\sigma}_B$ is the Pauli spin operator acting on B's qubit and the angle parameter θ is adequately chosen. The quantum state after (iii) is written as follows: $\rho = \sum_{\mu=0,1} V_B(\mu) P_A(\mu) |g\rangle\langle g| P_A(\mu) V_B^\dagger(\mu)$. Using this state, it can be shown that B extracts positive energy $+E_B$ on average from the qubit chain, accompanied by excitations with negative energy $-E_B$ in the qubit chain around B's site in (iii). The maximum value of E_B is written explicitly [2] as $E_B = \frac{1}{2}\left[\sqrt{\xi^2+\eta^2} - \xi\right]$, where ξ and η are given by $\xi = \langle g|\sigma_B H \sigma_B|g\rangle \geq 0$ and $\eta = i\langle g|\sigma_A [H, \sigma_B]|g\rangle$. η is real and a two-point correlation function of (semi-)local operators of A and B. It is a notable point that η does not vanish because of entanglement in general and yields a positive value of E_B. In the above analysis, we assume that dynamical evolution of the system induced by H is negligible during a short time interval for the protocol: $\exp[-itH] \sim I$. It is also assumed that LOCC for the qubits can be repeated many times even over the short time period we consider by taking the non-relativistic limit: the light velocity $c \sim \infty$. These assumptions are not essential to QET, and can be omitted if we do not mind that the protocol becomes comlicated. Explicit examples of QET have been proposed for near-critical Ising chains with large correlation lengths [3] and for massless field theory [4]. Transportation of energy to distant places can be achieved in both these examples. For the field QET case [4], time evolution of the system between A's measurement and B's operation is taken account of, and not projective but POVM measurements are adopted in step (i).

It is pertinent to note here that even after the last step (iii), there remains energy E_A around A's site in the qubit chain which A had to first deposit to the qubit chain for her measurement. Let us imagine that A attempts to completely withdraw E_A by local operations after (iii). If possible, the energy gain of B from the qubit chain might require no cost. Then, the total energy of the qubit chain would become equal to $-E_B$ and negative. Consequently, due to the non-negativity of H, A cannot withdraw energy larger than $E_A - E_B$ by A's local operations because her local measurement breaks entanglement between her qubit and the other qubits. If A wants to recover the original state of her qubit with zero energy density, she must recreate the entanglement. However entanglement generation needs nonlocal operations. Because of the locality of her operations, A cannot withdraw all the energy. We note also that B cannot get any energy by the protocol if $E_A = 0$. Thus it is concluded that B has borrowed E_B in advance from a local ground state of the qubit chain, based on security of the energy deposited by A. The measurement information from A can be regarded as a password for B to withdraw a part of the deposited energy.

How then might we intuitively understand how energy is teleported without any physical agent in the protocol? The reason why the protocol works may be briefly summarized as follows. As mentioned above, quantum mechanics allows negative energy density regions by controlling quantum fluctuation. Hence, even if B has no energy on average around him, the value of the energy density around him can decrease by local operations of B with the measurement results of A. Therefore, the ground state can be locally regarded as "an excited state", compared with the state with negative energy around B. In the protocol, B extracts this local excitation energy. It can be said that the energy, which B will obtain, existed around B *before* the start of the protocol. Therefore, we do not need any transfer of energy from A to B in the protocol. Of course, the energy hidden behind the regions with zero energy density is not always available. In the protocol, classical information about the measurement result of A becomes a key for B to access this energy. Because the qubit of A is entangled with qubits around B, the measurement result includes information about quantum fluctuation around B. B infers from the measurement information how the quantum fluctuation behaves around him and chooses his unitary operation acting on his qubit to extract energy from the spin chain and not to give energy. Consequently, the protocol is able to transport energy by LOCC without breaking local energy conservation and without any excited physical entity.

ACKNOWLEDGMENTS

This research was partially supported by the SCOPE project of the MIC.

REFERENCES

1. C.H. Bennett, G. Brassard, C. Crépeau, R. Jozsa, A. Peres, and W.K. Wootters: Phys. Rev. Lett. 70 (1993) 1895.
2. M. Hotta, arXiv: 0803.0348.
3. M. Hotta, Phys. Lett. A 372 (2008) 5671.
4. M. Hotta, Phys. Rev. D 78 (2008) 045006.

Variants of Controlled Quantum Teleportation: Use of W-like States

Robabeh Rahimi*, Akira SaiToh* and Mikio Nakahara†

*Research Center for Quantum Computing, Interdisciplinary Graduate School of Science and Engineering, Kinki University, Higashi-Osaka, Osaka 577-8502, Japan
†Department of Physics, Kinki University, Higashi-Osaka, Osaka 577-8502, Japan

Abstract. A standard quantum controlled teleportation uses a GHZ state shared by the two players and controllers. We introduce and evaluate a similar scheme using a W state. We secondly introduce a scheme to build in a majority vote of controllers for making their decision.

Keywords: Quantum controlled teleportation
PACS: 03.67.Hk, 03.67.Mn

INTRODUCTION

Quantum controlled teleportation is a variant of the quantum teleportation [1], in which teleportation channels are under the supervision of controllers. We pick up one of standard quantum controlled teleportation schemes [2], in which a single EPR pair among N EPR pairs is supervised by controllers. The protocol involves Alice, Bob, and M controllers $\mathrm{C}_1,\ldots,\mathrm{C}_M$, as illustrated in Fig. 1. Alice tries to teleport the N-qubit state $|\mathrm{A}'\rangle = \sum_{x_1\ldots x_N=0\ldots0}^{1\ldots1} p_{x_1\ldots x_N}|x_1\ldots x_N\rangle_{\mathrm{A}'_1\ldots\mathrm{A}'_N}$ (here, $\sum_{x_1\ldots x_N}|p_{x_1\ldots x_N}|^2=1$) to Bob. The protocol works as follows.

[i] First Alice ($\mathrm{A}=\mathrm{A}_1\ldots\mathrm{A}_N$), Bob ($\mathrm{B}=\mathrm{B}_1\ldots\mathrm{B}_N$), and controllers ($\mathrm{C}=\mathrm{C}_1\ldots\mathrm{C}_M$) share the state $|\mathrm{ABC}\rangle = \left(\bigotimes_{s=1}^{N-1}\frac{1}{\sqrt{2}}\sum_{y_s=0}^{1}|y_s\rangle_{\mathrm{A}_s}|y_s\rangle_{\mathrm{B}_s}\right)\otimes|\xi\rangle_{\mathrm{A}_N\mathrm{B}_N\mathrm{C}_1\ldots\mathrm{C}_M}$ with $|\xi\rangle_{\mathrm{A}_N\mathrm{B}_N\mathrm{C}_1\ldots\mathrm{C}_M} = \frac{1}{\sqrt{2}}\sum_{y=0}^{1}|y\rangle_{\mathrm{A}_N}|y\rangle_{\mathrm{B}_N}|y\ldots y\rangle_{\mathrm{C}_1\ldots\mathrm{C}_M}$.

[ii] For the first step, Alice makes a Bell measurement on every $\mathrm{A}'_\mathrm{k}\mathrm{A}_\mathrm{k}$ pair for all $1\le k\le N$. To clarify the process, we utilize the well-known relation:

$$|x,y\rangle = \frac{1}{\sqrt{2}}\sum_{i=0}^{1}(-1)^{x\cdot i}|B_{i,x\oplus y}\rangle, \tag{1}$$

where $|x,y\rangle$ is a computational basis vector and $|B_{i,j}\rangle$ is the (i,j)th Bell basis vector (here, $j=x\oplus y$, i.e., $y=j\oplus x$). We can rewrite the initial state involving A'ABC as

$$|\mathrm{A}'\mathrm{ABC}\rangle = \frac{1}{2^N}\sum_{x_1\ldots x_N} p_{x_1\ldots x_N}\left(\bigotimes_{s=1}^{N-1}\sum_{i_s=0}^{1}\sum_{j_s=0}^{1}(-1)^{x_s\cdot i_s}|B_{i_s,j_s}\rangle_{\mathrm{A}'_s\mathrm{A}_s}|j_s\oplus x_s\rangle_{\mathrm{B}_s}\right)$$
$$\otimes\sum_{i_N=0}^{1}\sum_{j_N=0}^{1}(-1)^{x_N\cdot i_N}|B_{i_N,j_N}\rangle_{\mathrm{A}'_N\mathrm{A}_N}|j_N\oplus x_N\rangle_{\mathrm{B}_N}\otimes|(j_N\oplus x_N)\ldots(j_N\oplus x_N)\rangle_{\mathrm{C}_1\ldots\mathrm{C}_M}.$$

CP1110, *Quantum Communication, Measurement and Computing (QCMC), The Ninth International Conference,* edited by A. Lvovsky

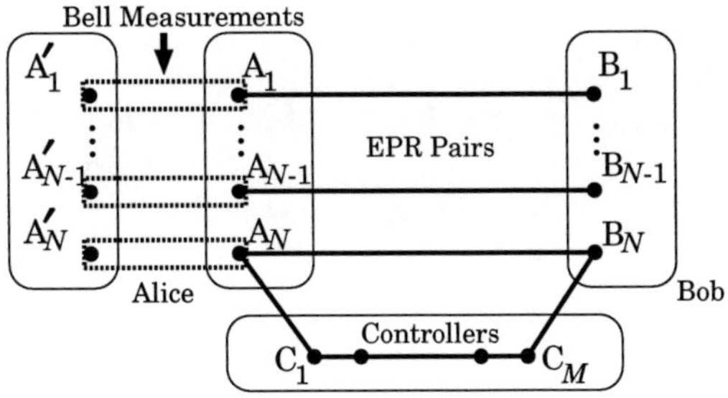

FIGURE 1. Illustration of a controlled quantum teleportation explained in the text.

It is now easy to follow the effect of Alice's Bell measurements with the outcomes $\{i_s, j_s\}_{s=1}^N$.

[iii] Each controller individually measures her/his qubit in the X basis. Note that a computational basis state $|c\rangle$ of a qubit can be written as $|c\rangle = \frac{1}{\sqrt{2}} \sum_{h=+1,-1} (-1)^{c \cdot (h==-1)} |h\rangle$ using the X-basis states $|h\rangle$ (here, the operation "==" returns one if its two arguments are equal; zero otherwise). A single component $|j_N \oplus x_N\rangle_{C_t}$ (here, $t \in \{1, ..., M\}$) is written as $\frac{1}{\sqrt{2}} \sum_{h_t} (-1)^{(j_N \oplus x_N) \cdot (h_t == -1)} |h_t\rangle$. Thus measuring C_t in the X basis results in the phase factor $(-1)^{(j_N \oplus x_N) \cdot (h_t == -1)}$ depending on the outcome h_t.

After all the controllers make measurements, the resultant state of the system B becomes

$$\sum_{x_1 \ldots x_N} p_{x_1 \ldots x_N} \left(\bigotimes_{s=1}^{N-1} (-1)^{x_s \cdot i_s} |j_s \oplus x_s\rangle_{B_s} \right) (-1)^{x_N \cdot i_N} \left[\prod_t (-1)^{(j_N \oplus x_N) \cdot (h_t == -1)} \right] |j_N \oplus x_N\rangle_{B_N}.$$

[iv] Bob receives the results (i_s, j_s) $(s = 1, ..., N)$ from Alice and applies $\bigotimes_{s=1}^N Z^{i_s} X^{j_s}$ to his qubits. In addition, Bob receives the results "+1" or "−1" from individual controllers. If the total number of "−1" is odd, then he applies Z^{j_N} to his Nth qubit. The final state of Bob after these operations becomes $\sum_{x_1 \ldots x_N} p_{x_1 \ldots x_N} \bigotimes_{s=1}^N |x_s\rangle$. The teleportation is successful in this way.

In this contribution, we aim to introduce schemes in which the shared state $|\xi\rangle_{A_N B_N C_1 \ldots C_M}$ is engineered to be different from the GHZ state.

USE OF A SHARED W STATE

First we consider the scheme where we use the same setup as illustrated in Fig. 1 but with a single W state $\frac{1}{\sqrt{M+2}} \sum_{l=1}^{M+2} |0 \cdots 0_{l-1} 1_l 0_{l+1} \cdots 0\rangle_{A_N B_N C_1 \ldots C_M}$ for the shared state $|\xi\rangle_{A_N B_N C_1 \ldots C_M}$ instead of the GHZ state. Let us denote the value of the bit C_l as c_l, etc. Let us also denote $\diamondsuit = b_N, c_1, ..., c_M$. Without confusion we also denote $B_N C_1 \ldots C_M$ as $\diamondsuit$. One can rewrite the W state as

$$\frac{1}{\sqrt{M+2}} |y(\diamondsuit)\rangle_{A_N} (|0 \cdots 0\rangle + \sum_{l=1}^{M+1} |l\text{th}\rangle)_{\diamondsuit}$$

with $|l\text{th}\rangle = |0\cdots 0_{l-1}1_l0_{l+1}\cdots 0\rangle$ and $|y(\diamondsuit)\rangle_{A_N} = |\overline{b_N + c_1 + \cdots + c_M}\rangle_{A_N}$. (Namely that $y(\diamondsuit)$ is one iff $B_N, C_1, ..., C_M$ are 0's; otherwise it is zero.)

Using Eq. (1), it is easy to find that the initial state is written as

$$|\phi_0\rangle = \frac{1}{\sqrt{2M+4}}\sum_{x=0}^{1}|\alpha_x\rangle_\clubsuit \sum_{i_N=0}^{1}(-1)^{x\cdot i_N}|B_{i_N,x\oplus y(\diamondsuit)}\rangle_{A'_N A_N}(|0\cdots 0\rangle + \sum_{l=1}^{M+1}|l\text{th}\rangle)_\diamondsuit,$$

where $|\alpha_x\rangle_\clubsuit$ is the state of the subpart independent of the controllers, i.e.,

$$|\alpha_x\rangle_\clubsuit = {}_{A'_N}\langle x|\sum_k p_k|k\rangle_{A'_1...A'_N} \otimes (|\text{EPR}\rangle^{\otimes(N-1)})_{A_1...A_{N-1}B_1...B_{N-1}}.$$

Let us omit describing the teleportation happening in $\clubsuit$ since this is a basic quantum teleportation. Thus we set the state of $\clubsuit$ after a successful teleportation to $|\tilde{\alpha}_x\rangle_\clubsuit$.

We start from the Bell measurement acting on $A'_N A_N$. Suppose that Alice obtains a measurement outcome (i_N, j_N). Then, this implies $x \oplus y(\diamondsuit) = j_N$, and $x = j_N \oplus y(\diamondsuit)$. The state after the measurement is

$$|\phi_1\rangle = \frac{1}{\sqrt{M+2}}\sum_{j_N\oplus y(\diamondsuit)=0}^{1}|\tilde{\alpha}_{j_N\oplus y(\diamondsuit)}\rangle_\clubsuit(-1)^{i_N\cdot[j_N\oplus y(\diamondsuit)]}|B_{i_N,j_N}\rangle_{A'_N A_N}(|0\cdots 0\rangle + \sum_{l=1}^{M+1}|l\text{th}\rangle)_\diamondsuit.$$

Results (i_s, j_s) $(s = 1, ..., N)$ are sent to Bob. Recall the relation $|y(b_N, c_1, ..., c_N)\rangle_{A_N} = |\overline{b_N}\cdot\overline{c_1}\cdot\cdots\cdot\overline{c_M}\rangle_{A_N}$. In order to recover the initial state, we need the following condition: $y(\diamondsuit) = 1$ if and only if $b_N = 0$ and $y(\diamondsuit) = 0$ if and only if $b_N = 1$. This is because of the following reason. Suppose that $y(\diamondsuit) = 1$ may occur for both cases of $b_N = 0$ and $b_N = 1$. Then $|x\rangle_{A'_N}$ and $|b_N\rangle_{B_N}$ have no one-to-one correspondence. The condition is equivalent to the condition of having M zeros as a result of measurement on $C_1...C_M$. The probability of achieving this result is $2/(M+2)$.

When the measurement outcome of $C_1...C_M$ is $0\cdots 0$, we have the state

$$|\phi_2\rangle = \sum_{j_N\oplus y=0}^{1}|\tilde{\alpha}_{j_N\oplus y}\rangle_\clubsuit(-1)^{i_N\cdot(j_N\oplus y)}|B_{i_N,j_N}\rangle_{A'_N A_N}|\bar{y}\rangle_{B_N}|0\cdots 0\rangle_{C_1\cdots C_M}.$$

We need to recover the state $\sum_{j_N\oplus y=0}^{1}|\widetilde{\tilde{\alpha}_{j_N\oplus y}}\rangle_{B_1...B_{N-1}}|j_N\oplus y\rangle_{B_N}$ (here, $|\widetilde{\tilde{\alpha}_{j_N\oplus y}}\rangle_{B_1...B_{N-1}}$ is the state after tracing out $A'_1...A'_{N-1}A_1...A_{N-1}$ from $\clubsuit$). Thus, Bob has to apply $U_{B_N} = Z^{i_N}X^{j_N+1} = Z^{i_N}X^{\bar{j}_N}$ to B_N to reproduce the original state, in addition to applying $\bigotimes_{s=1}^{N-1}Z^{i_s}X^{j_s}$ to $B_1...B_{N-1}$. (Bob's operations for the qubits $B_1...B_{N-1}$ are, of course, the same as those used in a basic quantum teleportation.)

A drawback is the success probability, which decreases as $2/(M+2)$. To improve it, one can use a biased W state

$$|\widetilde{W}\rangle_{A_N B_N C_1...C_M} = A(|100\cdots 0\rangle + |010\cdots 0\rangle)_{A_N B_N C_1...C_M} + \sqrt{1-2|A|^2}|00\rangle_{A_N B_N} \otimes \frac{1}{\sqrt{M}}\sum_{l=1}^{M}|0\cdots 01_l0\cdots 0\rangle_{C_1...C_M}$$

with $|A|^2 \leq 1/2$ for the shared state $|\xi\rangle_{A_N B_N C_1 \ldots C_M}$. Then the success probability becomes $2|A|^2$. Nevertheless, there is a clear trade-off between the success probability and the fidelity that can be achieved by ignoring controllers (i.e., by cheating). In case $N=1$, Bob recovers the state $p_0|0\rangle + p_1|1\rangle$ with the probability $2|A|^2$ but he obtains the state $|0\rangle$ otherwise. Therefore, the fidelity of the state when Bob ignores controllers with respect to the original state is $F_{\text{cheat}} = \sqrt{2|A|^2 + (1-2|A|^2)|p_0|^2}$.

SHARED ENTANGLED STATE FOR MAJORITY VOTE

It is possible to add a special function in a way to control a teleportation by engineering the shared state $|\xi\rangle_{A_N B_N C_1 \ldots C_M}$. Here we present a majority vote of three controllers. Let us set

$$|\xi\rangle_{A_N B_N C_1 C_2 C_3} = \frac{1}{2\sqrt{2}} \sum_{y=0}^{1} |yy\rangle_{A_N B_N} (e^{iy\theta_1}|001\rangle + e^{iy\theta_2}|010\rangle + e^{iy\theta_3}|100\rangle + e^{iy\theta_4}|111\rangle)_{C_1 C_2 C_3}$$

with $\theta_i \neq \theta_j \pmod{2\pi}$ for $i \neq j$. Then, Bob can determine the phase factor to be corrected if two of the three controllers make measurements in the computational basis and send results to him.

Similarly, a majority vote of four controllers is possible with the shared state

$$|\xi\rangle_{A_N B_N C_1 C_2 C_3 C_4} = \frac{1}{2} \sum_{y=0}^{1} |yy\rangle_{A_N B_N} [|W(y)\rangle + |\overline{W(y)}\rangle^*]_{C_1 C_2 C_3 C_4}$$

with $|W(y)\rangle = \frac{1}{2}(e^{iy\theta_1}|0001\rangle + e^{iy\theta_2}|0010\rangle + e^{iy\theta_3}|0100\rangle + e^{iy\theta_4}|1000\rangle)$ and $|\overline{W(y)}\rangle^*$ is the complex conjugate of the bitwise inverse of $|W(y)\rangle$; θ_i's are non-zero $\pmod{2\pi}$ and $\theta_i \neq \pm\theta_j \pmod{2\pi}$ for $i \neq j$.

A general majority vote scheme should involve a systematic way, such as a polynomial coding [3] common for secret sharing. An alternative way of coding will be reported elsewhere together with a different application of engineering in the shared state.

ACKNOWLEDGMENTS

RR is supported by the Grant-in-Aid from JSPS (Grant No. 1907329). AS and MN are supported by "Open Research Center" Project for Private Universities: matching fund subsidy from MEXT. MN would like to thank a partial support of the Grant-in-Aid for Scientific Research from JSPS (Grant No. 19540422).

REFERENCES

1. C. H. Bennett et al., *Phys. Rev. Lett.* **70**, 1895 (1993).
2. J. Zhou et al., quant-ph/0006030; N. B. An, *Phys. Rev. A* **68**, 022321 (2003); N. B. An and J. Kim, quant-ph/0303149; F. G. Deung et al., *Phys. Rev. A* **72**, 022338 (2005); Z.-X. Man et al., *J. Phys. B* **40**, 1767 (2007).
3. A. Shamir, *Commun. ACM* **22**, 612 (1979); R. Cleve et al., *Phys. Rev. Lett.* **83**, 648 (1999).

Composability of biased quantum coin flipping

Satoshi Ishizaka

Nano Electronics Res. Labs., NEC Corporation, 34 Miyukigaoka, Tsukuba 305-8501, Japan
INQIE, the University of Tokyo, 4-6-1 Komaba, Meguro-ku, Tokyo 153-8505, Japan

Abstract. A biased quantum coin flip (QCF) does not realize the functionality of a classical black-boxed coin flip, if it satisfies some fidelity condition. We consider a quantum black-boxed coin flip and show that the fidelity condition is well-explained as the functionality of the quantum black-boxed coin flip. Moreover, we show the results of the security analysis for a QCF like protocol.

INTRODUCTION

A coin flip [1] is an important cryptographic primitive on a communication network. Although a perfectly fair quantum coin flip (QCF) is impossible [2], biased quantum coin flipping is possible [3]. In the biased QCF, a dishonest party can cheat to bias the probability to $1/2+\varepsilon$, but it is ensured that the bias satisfies $|\varepsilon|<1/2$; so a dishonest party cannot fully control the outcome. On the other hand, let us imagine ideal biased coin flipping such that a (classical) black box outputs a common random bit to both parties. A dishonest party can bias the probability of the outcome but can do nothing else because the party cannot touch the inside of the box at all. A biased QCF, at first glance, realizes the black-boxed coin flip, because it is ensured by the laws of physics that the bias range is limited to $|\varepsilon|<1/2$ against all possible operations for cheating. However, a biased QCF is distinguishable from a (classical) black-boxed coin flip, if we consider the situation such that Alice and Bob scramble for a maximally entangled state according to the outcome of coin flipping [4]. Let us consider the following protocol:

Protocol 1 (sharing and maintaining entanglement).
Stage 1: Alice prepares $|\phi^+\rangle_{AB}=(|00\rangle+|11\rangle)/\sqrt{2}$ and sends the B qubit to Bob.
Stage 2: Alice and Bob execute a coin flipping subprotocol.
Stage 3: The winner of the coin flip obtains both A and B qubits and checks whether or not the state of the AB qubits is $|\phi^+\rangle_{AB}$. If the state is not $|\phi^+\rangle_{AB}$, the party detects the cheating of the other party with nonzero probability.

It is then found that a QCF even allows the cheating that is completely prohibited by a black-boxed coin flip, if it satisfies

$$\Delta a \equiv \max_{\varepsilon',\varepsilon'',F=1} (\varepsilon'-\varepsilon'')/[2(1+\varepsilon'+\varepsilon'')] > 0. \tag{1}$$

Here, the maximization is taken over all the pairs of the two biasing operations subject to $F(\rho'_c,\rho''_c)=1$, F is the fidelity, and ρ'_c (ρ''_c) is the reduced state of a honest party corresponding to the bias of ε' (ε''), when the outcome of the QCF is c.

CP1110, *Quantum Communication, Measurement and Computing (QCMC), The Ninth International Conference,* edited by A. Lvovsky

QUANTUM BLACK-BOXED COIN FLIPPING

A biased QCF does not realize the ideal functionality of a (classical) black-boxed coin flip, if the QCF satisfied Eq. (1). How should we regard the functionality of such a QCF? To partially answer this question, let us imagine a quantum black box which outputs a state of $|\psi\rangle = \sqrt{1/2+\varepsilon}|00\rangle_{CD} + \sqrt{1/2-\varepsilon}|11\rangle_{CD}$. The C (D) qubit is sent to Alice (Bob). Each party then measures the qubit in the $\{|0\rangle, |1\rangle\}$ basis, and the result corresponds to the outcome of coin flipping. Therefore, this quantum black-boxed coin flip outputs the outcome 0 with probability $1/2+\varepsilon$. A dishonest party can control the bias ε by inputing ε to the quantum box. An important difference from a classical black-boxed coin flip is that the quantum box accepts a quantum state as the input of ε, and as a result a superposed input is allowed. Let us denote the input state by X. If a dishonest party inputs $|\varepsilon'\rangle_X + |\varepsilon''\rangle_X$, the quantum box outputs

$$|\psi\rangle \propto \left(\sqrt{\tfrac{1}{2}+\varepsilon'}|00\rangle_{CD} + \sqrt{\tfrac{1}{2}-\varepsilon'}|11\rangle_{CD}\right) + \left(\sqrt{\tfrac{1}{2}+\varepsilon''}|00\rangle_{CD} + \sqrt{\tfrac{1}{2}-\varepsilon''}|11\rangle_{CD}\right).$$

Now, let us consider the protocol 1. Suppose that Alice is dishonest and sends a partially entangled state $|\Phi\rangle = \sqrt{1/2+\Delta a}|00\rangle_{AB} + \sqrt{1/2-\Delta a}|11\rangle_{AB}$ instead of $|\phi^+\rangle_{AB}$ in the stage 1. In the stage 2, the above quantum black-boxed coin flip is used as a coin flipping subprotocol. Let us consider Alice's cheating strategy such that she inputs the state of $\sqrt{1/2+\Delta a}|\varepsilon''\rangle_X|00\rangle_{AB} + \sqrt{1/2-\Delta a}|\varepsilon'\rangle_X|11\rangle_{AB}$ to the quantum box. The point is that, since the quantum box accepts a quantum input, the box also accepts the above state which is entangled with the state of the AB qubits. The quantum box then outputs

$$\begin{aligned} |\psi\rangle &\propto \sqrt{1/2+\Delta a}\left(\sqrt{1/2+\varepsilon''}|00\rangle_{CD} + \sqrt{1/2-\varepsilon''}|11\rangle_{CD}\right)|00\rangle_{AB} \\ &+ \sqrt{1/2-\Delta a}\left(\sqrt{1/2+\varepsilon'}|00\rangle_{CD} + \sqrt{1/2-\varepsilon'}|11\rangle_{CD}\right)|11\rangle_{AB}. \end{aligned}$$

Suppose that the outcome of the coin flip is 0. Alice then loses the coin flip, and the state of the AB qubits will be checked by Bob in the stage 3. However, the whole state after the quantum black-boxed coin flip is

$$|\psi'\rangle \propto \left[\sqrt{(\tfrac{1}{2}+\Delta a)(\tfrac{1}{2}+\varepsilon'')}|00\rangle_{AB} + \sqrt{(\tfrac{1}{2}-\Delta a)(\tfrac{1}{2}+\varepsilon')}|11\rangle_{AB}\right] \otimes |00\rangle_{CD}. \qquad (2)$$

Note that Bob's states for the biases ε' and ε'' are both $|0\rangle_D$, and thus $F(\rho_0', \rho_0'') = 1$. This plays a crucial role so that the AB qubits are disentangled from the CD qubits as in Eq. (2). It is then found from Eq. (2) that the state of the AB qubits becomes $|\phi^+\rangle_{AB}$ if $\Delta a = (\varepsilon' - \varepsilon'')/[2(1+\varepsilon'+\varepsilon'')]$. Therefore, whenever dishonest Alice loses the coin flip, she successfully disguises the partially entangled state $|\Phi\rangle_{AB}$ as $|\phi^+\rangle_{AB}$, if the quantum black-boxed coin flip allows the biases ε' and ε'' such that $(\varepsilon' - \varepsilon'')/[2(1+\varepsilon'+\varepsilon'')] > 0$ and $F(\rho_0', \rho_0'') = 1$. This condition for the successful cheating is nothing but Eq. (1). In this way, the condition of Eq. (1) is well-understood if we regard the functionality of a biased QCF as that of the quantum black-boxed coin flip discussed above.

EXAMPLE OF SECURITY ANALYSIS

Let us consider the following QCF like protocol to be used in the protocol 1:

Protocol 2 (QCF like).

Alice prepares $|\phi^+\rangle_{CD}$ and sends the D qubit to Bob. He optionally checks $|\phi^+\rangle_{CD}$ getting the C qubit in his hand. If Bob uses the option, he automatically loses the coin flip. Otherwise, he measures the D qubit in the $\{|0\rangle, |1\rangle\}$ basis, sends the result to Alice, and she confirms the validity by measuring the C qubit. Alice (Bob) loses the coin flip if the measurement result is 0 (1).

Suppose that Bob is dishonest and his dishonest action in the stage 1 is to partially destroy the entanglement of $|\phi^+\rangle_{AB}$ by a generalized measurement as discussed in [4]. Suppose further that the post-measured state is $|\Phi_i\rangle_{ABa}$, where Ba denotes Bob's large system including his ancilla qubits, i denotes the outcome of the measurement, and p_i is the probability of obtaining the i-th outcome. In the stage 2 where the protocol 2 is used as a coin flipping subprotocol, Bob has three choices after he receives the D qubit: using the option or declaring either 0 or 1 as the measurement result for the D qubit. The general action is described by another generalized measurement with three outcomes, which is described by a positive operator valued measure (POVM). Let L^i_y $(y=0,1,2)$ be the POVM elements acting on BaD qubits and let $\tilde{L}^i_y = L^{i\dagger}_y L^i_y$.

Suppose that Bob obtains the outcome $y=2$. He then declares that he uses the option, automatically loses the coin flip, and the state of the AB qubits will be checked by Alice in the stage 3. The probability that Bob obtains $y=2$ and passes Alice's check is $P_2 = \mathrm{tr}\left[\Lambda(L^i_2(|\Phi_i\rangle\langle\Phi_i| \otimes P^+_{CD})L^{i\dagger}_2)P^+_{AB}\right]$, where $P^+ = |\phi^+\rangle\langle\phi^+|$ and Λ is a trace-preserving operation acting on the $BaCD$ qubits (Bob can access the C qubit for $y=2$). The singlet fraction $F(\sigma) = \mathrm{tr}\left[\sigma P^+_{AB}\right]$ is bounded as $F(\sigma) \leq [\mathrm{tr}\sigma + N_A(\sigma)]/2$. Here, $N_A(\sigma) = \mathrm{tr}|\sigma^{\Gamma_A}| - \mathrm{tr}\sigma$ is negativity [5], and $N_A(\Lambda(\sigma)) \leq N_A(\sigma)$ holds because N_A is an entanglement monotone. Since σ is a pure state, $N_A(\sigma) = 2\sqrt{\lambda^+\lambda^-}$, where $\lambda^\pm$ are two eigenvalues of $\sigma_A = \mathrm{tr}_{BaCD}\sigma$. In this way, we have $P_2 \leq (m^+_i + m^-_i)/2 + \sqrt{m^+_i m^-_i}$, where $m^\pm_i$ are eigenvalues of $\mathrm{tr}_{BaD}\tilde{L}^i_2(|\Phi_i\rangle\langle\Phi_i| \otimes \frac{I_D}{2})$. This bound is tight for the best cheating strategy because the entanglement of the pure state σ is two-qubit entanglement, and $F(\Lambda(\sigma)) = [\mathrm{tr}\sigma + N_A(\sigma)]/2$ is achieved when Λ is an appropriate local unitary transformation on the $BaCD$ qubits. Likewise, the probability P_1 that Bob obtains the outcome $y=1$ (and hence declares 1 as the measurement result for the D qubit) and passes Alice's check, is tightly bounded as $P_1 \leq (l^+_i + l^-_i)/2 + \sqrt{l^+_i l^-_i}$, where $l^\pm_i$ are eigenvalues of $\mathrm{tr}_{BaD}\tilde{L}^i_1(|\Phi_i\rangle\langle\Phi_i| \otimes \frac{(|1\rangle\langle 1|)_D}{2})$. The probability of detecting Bob's cheating is then

$$P_B \geq 1 - \sum_i p_i \Big\{\mathrm{tr}\big[\tilde{L}^i_0(|\Phi_i\rangle\langle\Phi_i| \otimes \tfrac{(|0\rangle\langle 0|)_D}{2})\big] + \big[\tfrac{l^+_i + l^-_i}{2} + \sqrt{l^+_i l^-_i} + \tfrac{m^+_i + m^-_i}{2} + \sqrt{m^+_i m^-_i}\big]\Big\}$$

$$\geq \sum_{k=0,1} \sum_i p_i \mathrm{tr}\big[\tilde{L}^i_k(|\Phi_i\rangle\langle\Phi_i| \otimes \tfrac{(|1-k\rangle\langle 1-k|)_D}{2})\big] + \sum_i p_i^2 \big[\tfrac{(l^+_i - l^-_i)^2}{4} + \tfrac{(m^+_i - m^-_i)^2}{4}\big], \qquad (3)$$

where $\sqrt{\alpha\beta} \leq (\alpha+\beta)/2 - (\alpha-\beta)^2/4$ $(0 \leq \alpha, \beta \leq 1)$ was used. Since all terms in Eq. (3) are non-negative, all these terms must be zero so that $P_B = 0$. By expressing $|\Phi_i\rangle$ in the Schmidt decomposed form as $|\Phi_i\rangle_{ABa} = a_i|\phi_0^i\rangle_A|\psi_0^i\rangle_{Ba} + b_i|\phi_1^i\rangle_A|\psi_1^i\rangle_{Ba}$ $(a_i \geq b_i)$, we have the necessary conditions for $P_B = 0$ as

$$a_i^2\langle\psi_0^i 1|\tilde{L}_0^i|\psi_0^i 1\rangle = b_i^2\langle\psi_1^i 1|\tilde{L}_0^i|\psi_1^i 1\rangle = 0, \quad l_i^+ = l_i^-,$$
$$a_i^2\langle\psi_0^i 0|\tilde{L}_1^i|\psi_0^i 0\rangle = b_i^2\langle\psi_1^i 0|\tilde{L}_1^i|\psi_1^i 0\rangle = 0, \quad m_i^+ = m_i^-, \tag{4}$$

where $|\psi_0^i\rangle, |\psi_1^i\rangle$ are the states of the Ba qubits and $|0\rangle, |1\rangle$ are those of the D qubit. The conditions for $l_i^+ = l_i^-$ are $a_i^2\langle\psi_0^i 1|\tilde{L}_1^i|\psi_0^i 1\rangle = b_i^2\langle\psi_1^i 1|\tilde{L}_1^i|\psi_1^i 1\rangle$ and $a_i b_i\langle\psi_0^i 1|\tilde{L}_1^i|\psi_1^i 1\rangle = 0$, and likewise for $m_i^+ = m_i^-$. Since $\tilde{L}_0^i + \tilde{L}_1^i + \tilde{L}_2^i = I$, it is then found that L_y^i satisfying Eq. (4) exists if and only if $a_i^2/b_i^2 \leq 2$. Moreover, $a_i^2/b_i^2 \leq 2$ is sufficient for $P_B = 0$ because the second inequality in Eq. (3) becomes an equality for $l_i^+ = l_i^-$ and $m_i^+ = m_i^-$, and the first inequality is tight for the best cheating strategy (the upper bounds of P_1 and P_2 are tight). Namely, if and only if $a_i^2/b_i^2 \leq 2$, Bob can successfully disguise the collapsed pure state of $|\Phi_i\rangle_{AB_a}$ as $|\phi^+\rangle_{AB}$, whenever he loses the coin flip.

This best cheating strategy of Bob is well-understood by using Eq. (1). In fact, it is found in the protocol 2 that $F(\rho_1', \rho_1'') = 1$ for Bob's two biasing operations of (i) he always uses the option $(\varepsilon' = 1/2)$, and (ii) he measures the D qubit, and if the result is 1, he uses the option $(\varepsilon'' = 0)$. For these two biasing operations, we have from Eq. (1) that $\Delta a = 1/6$ and $a = 2/3$, which agrees with the the best bound of $a_i^2/b_i^2 = 2$ shown above.

SUMMARY

The biased QCF satisfying Eq. (1) does not realize the functionality of a (classical) black-boxed coin flip. Therefore, *if we define the ideal functionality of a biased coin flip as that of the black-boxed coin flip*, it can be concluded that the QCF does not fulfill the conditions for universally composable security [6]. This definition of the ideal functionality seems to be natural, because an indeed *unexpected* security hole occurs, if the QCF is combined with another cryptographic protocol such as the protocol 1. How should we regard the functionality of the QCF to avoid the unexpected security hole? We considered a quantum black-boxed coin flip and showed that the uncomposability condition of Eq. (1) is well-explained as the functionality of it. Moreover, we showed the results of the rigorous security analysis for a QCF like protocol as an example.

REFERENCES

1. M. Blum, *Proc. of the IEEE Spring Computer Conference* (IEEE, New York, 1982), p. 133.
2. H.-K. Lo and H. F. Chau, Physica (Amsterdam) **120D**, 177 (1998); D. Mayers *et al.*, arXiv:quant-ph/9904078 (1999); A. Ambainis *et al.*, *Proc. of the 19th IEEE Conference on Computational Complexity* (IEEE Computer Society, Washington, DC, 2004), p. 250.
3. D. Aharonov, A. Ta-Shma, U. V. Vazirani, and A. C. Yao, *Proc. of the 32nd ACM Symposium on Theory of Computing* (ACM, New York, 2000), p. 705; R. W. Spekkens and T. Rudolph, Phys. Rev. Lett. **89**, 227901 (2002); A. Ambainis, J. Comput. Syst. Sci. **68**, 398 (2004); C. Mochon, *Proc. of the 45th IEEE Symposium on Foundation of Computer Science* (IEEE Computer Society, Washington, DC, 2004), p. 2; R. Colbeck, Phys. Lett. A **362**, 390 (2007).
4. S. Ishizaka, Phys. Rev. Lett. **100**, 070501 (2008).
5. G. Vidal and R. F. Werner, Phys. Rev. A **65**, 032314 (2002).
6. M. Ben-Or and D. Mayers, quant-ph/0409062.

A Fair Loss-Tolerant Quantum Coin Flipping Protocol
(extended abstract)

Guido Berlín*, Gilles Brassard*, Félix Bussières† and Nicolas Godbout†

** Département d'informatique et de recherche opérationnelle, Université de Montréal*
CP 6128, Succursale centre-ville, Montréal (Québec), H3C 3J7 Canada
email: {berlingu,brassard}@iro.umontreal.ca
† Laboratoire des fibres optiques, Centre d'optique, photonique et laser
Département de génie physique, École Polytechnique de Montréal
CP 6079, Succursale centre-ville, Montréal (Québec), H3C 3A7 Canada
email: {felix.bussieres,nicolas.godbout}@polymtl.ca

Abstract.
Coin flipping is a cryptographic primitive in which two spatially separated players, who in principle do not trust each other, wish to agree on a random bit. Classical and quantum coin flipping protocols have been studied extensively for more than twenty-five years. However, until recently, quantum coin flipping protocols were designed without taking into consideration the losses of quantum information that would be unavoidable in any realistic implementation. We introduced in 2008 a novel protocol and proved its security even when losses are taken into account: no cheating player could obtain a desired outcome with a probability greater than $(6+\sqrt{2})/8 \approx 93\%$. Here, we refine our earlier protocol by making it *fair* in the sense that the optimal cheating strategies allow either player to bias the outcome by the same amount. Specifically, either player can cheat to obtain a desired outcome with probability exactly 90%, but no more. An implementation is underway.

Keywords: Coin flipping, Loss tolerance, Quantum cryptography, Unconditional security.
PACS: 03.67.Hk, 03.67.Dd, 42.50.Dv.

INTRODUCTION

Coin Flipping is a cryptographic primitive in which two spatially separated players, who in principle do not trust each other, wish to agree on a random bit [4]. This is useful whenever the two players need to pick a random bit even though it could be to the advantage of one of them (or perhaps both) to determine, or at least bias, the outcome of the protocol. If we only allow for classical communication, it is always the case that one player, given sufficient computing power, can force the outcome of the protocol to be the bit of his or her choosing, unless we consider *relativistic* protocols in which messages are sent with sufficient simultaneity by both players to force their complete independence [5]. In this paper, we shall consider only *asynchronous* protocols, which consist of a sequence of rounds in which the two players alternate in sending messages to each other.

If we allow for quantum communication, a number of protocols have been introduced in which the bias on the final outcome any player can introduce is limited [1, 2]: either player can influence the outcome but neither can determine it. Nevertheless, until recently, quantum coin flipping protocols did not take account of the losses unavoidable

CP1110, *Quantum Communication, Measurement and Computing (QCMC), The Ninth International Conference,* edited by A. Lvovsky

in any real implementation (losses that occur during the transmission of quantum information between the payers, during the storage of quantum information required in many protocols or when the information is measured by one player). This issue was addressed in two recent papers [3, 6]. In particular, our earlier paper [3] introduced the first quantum coin flipping protocol capable of resisting losses and guarantee as outcome a shared bit that is not under the complete control of either player (unless one player cheats and the other detects it). However, our earlier protocol was unfair in the sense that one of the two players could bias the outcome to her choosing with probability $(6+\sqrt{2})/8 \approx 93\%$ of success [1], whereas the other player could obtain his desired outcome with probability $(\sqrt{2}+2)/4 \approx 85\%$ only. (These probabilities assume that one player is honestly following the protocol whereas the other is doing her or his best to bias the outcome.) Here, we introduce a *fair* loss-tolerant quantum coin flipping protocol in which either player can determine the outcome with a 90% probability of success, but no more.

After a brief overview of the general template followed by most quantum coin flipping protocols and an explanation of the problem caused by unavoidable losses, we present our new fair protocol. We state theorems about its security, but we leave formal proofs to our forthcoming full paper on the subject. An experimental realization of our protocol is currently underway at the University of Calgary (and will no doubt have been completed by the time these Proceedings appear!), which will be the subject of yet another paper.

THE NEED FOR LOSS-TOLERANCE

A typical quantum coin flipping protocol [1, 2] starts with a commitment phase, in which one of the payers, called Alice, commits to a bit b by sending the other player, called Bob, a quantum state $|\psi_{b,x}\rangle$ for a randomly chosen x (in general, x is not limited to being a single bit). This state is kept unmeasured by Bob in a quantum memory. Bob then chooses his own random bit g and sends it to Alice. To unveil the commitment, Alice reveals b and x, which allows Bob to measure the stored quantum state and verify that it is consistent with what Alice claims she had sent. If Bob is satisfied with this verification, the outcome of the protocol is the exclusive-or (sum modulo 2) of Alice's bit b and Bob's bit g. If not, the protocol aborts because Alice was caught cheating by Bob.

If we take into account the losses unavoidable in any real implementation (in Alice's quantum state generation apparatus, through the quantum channel between Alice and Bob, or in Bob's quantum storage or measurement apparatus), Bob must be allowed to request Alice to send another quantum state when his measurement apparatus does not register anything. However, this opens the door to a variety of cheating strategies in any realistic attempt at implementing the above template, as pointed out in [3, 6]. For instance, Bob could pretend he did not receive anything if he is not happy with Alice's announced value of b because the exclusive-or of b and g is not to his liking! In fact, this problem is sufficiently serious to completely undermine the security of all previously published quantum coin flipping protocols, except [3, 6].

---

[1] It is proved in [3] that the probability of successful cheating cannot exceed $(2\sqrt{2}+1)/4 \approx 96\%$, which is indeed correct, but in fact it is exactly $(6+\sqrt{2})/8$ with the optimal cheating strategy.

In [3], we introduced a novel protocol in which these vulnerabilities were eradicated. For this, Bob is allowed to ask Alice for a new quantum state in case he fails to record a measurement result. However, he is requested to measure the quantum state sent by Alice immediately upon reception, which prevents him from requesting retransmission after he has learned from Alice that the forthcoming coin flip outcome would not suit him. More subtly, it is essential that a mixture of $|\psi_{0,x}\rangle$ cannot be *unambiguously discriminated* from a mixture of $|\psi_{1,x}\rangle$ (for all possible x), for otherwise Bob could ask for retransmission precisely when he was unable to learn that Alice's b is to his liking. However, our protocol was unfair in the sense explained above in the Introduction.

A *FAIR* PROTOCOL

Consider the following states:

$$\left.\begin{array}{l} |\psi_{0,0}\rangle = |0\rangle \\ |\psi_{0,1}\rangle = |\phi^+\rangle \end{array}\right\} \quad b=0$$

$$\left.\begin{array}{l} |\psi_{1,0}\rangle = |1\rangle \\ |\psi_{1,1}\rangle = |\phi^-\rangle \end{array}\right\} \quad b=1\,,$$

where $|\phi^+\rangle = (\cos\phi)|0\rangle + (\sin\phi)|1\rangle$ and $|\phi^-\rangle = (\sin\phi)|0\rangle - (\cos\phi)|1\rangle$ for some angle ϕ to be determined later.

Protocol.

1. Alice chooses random bits b and x; she prepares and transmits state $|\psi_{b,x}\rangle$ to Bob.
2. Bob chooses a random bit $\hat{x}$ and measures the received qubit in the basis formed of $|\psi_{0,\hat{x}}\rangle$ and $|\psi_{1,\hat{x}}\rangle$. If Bob fails to register a measurement outcome (either because he did not receive Alice's qubit or because his measurement apparatus did not react), he asks Alice to restart at step 1. Otherwise, let the outcome be $\hat{b}$.
3. Bob sends a randomly chosen classical bit g to Alice.
4. Alice reveals her original b and x to Bob.
5. If $x=\hat{x}$ but $b\neq\hat{b}$, Bob aborts the protocol, *calling Alice a cheater*; if $x\neq\hat{x}$, Bob accepts Alice's honesty on faith.
6. If Bob did not abort the protocol, the outcome of the coin flip is the exclusive-or (sum modulo 2) of b and g.

Theorem 1 *With the above protocol, Alice can bias the protocol and obtain her desired outcome with probability* $(3+\sin\phi)/4$. *No cheating strategy on Alice's part can allow her to enjoy a larger bias. (We assume that Bob follows the protocol exactly)*

Theorem 2 *With the above protocol, Bob can bias the protocol and obtain his desired outcome with probability* $(1+\cos\phi)/2$. *No cheating strategy on Bob's part can allow him to enjoy a larger bias. (We assume that Alice follows the protocol exactly)*

Proofs. In the forthcoming full paper. □

The protocol introduced in [3] used $\phi = 45°$, so that Alice or Bob (albeit not both simultaneously, of course) could obtain her or his desired outcome with probability $(6+\sqrt{2})/8 \approx 93\%$ or $(\sqrt{2}+2)/4 \approx 85\%$, respectively, as mentioned in the Introduction. In order to achieve fairness, we must balance Alice's and Bob's bias, which requires choosing angle ϕ so that

$$(3+\sin\phi)/4 = (1+\cos\phi)/2\,.$$

In other words, we need

$$2\cos\phi - \sin\phi = 1\,.$$

This is achieved when

$$\sin\phi = {}^{3}/_{5} \text{ and } \cos\phi = {}^{4}/_{5}$$

(i.e. $\phi \approx 36.87°$), in which case either Alice or Bob can bias the protocol to obtain a desired outcome with probability exactly 90%, but no more.

CONCLUSION AND OPEN PROBLEMS

We have presented an improvement on our earlier loss-tolerant quantum coin flipping protocol [3], in which neither player has an unfair advantage over the other. As a result, our new protocol is more resilient to cheating than the former one. An implementation of our new protocol is currently underway at the University of Calgary.

It is still an open question to determine if this probability of 90% is optimal among all possible fair loss-tolerant quantum coin flipping protocols. Furthermore, loss-tolerant *weak* coin flipping protocols should be studied, with the possibility that such protocols might exist in which neither player can introduce more than an arbitrarily negligible bias into the outcome. It is also an open question to determine what can be done, if anything, if we take account not only of losses but also or errors in the creation, transmission and measurement of quantum states, when we insist (as we have done) that the outcome of the protocol should either be one bit on which both players agree or the correct decision by one player that the other has cheated. An original approach to error tolerance was introduced in [6], but in the context that the two players are willing to accept that the protocol may produce no outcome even when both players are honest.

1. D. Aharonov, A. Ta-Shma, U. Vazirani and A. C.-C. Yao, "Quantum bit escrow", *Proceedings of 32nd Annual ACM Symposium on Theory of Computing*, Portland, Oregon, USA, pp. 705–714, 2000.
2. A. Ambainis, "A new protocol and lower bounds for quantum coin flipping", *Journal of Computer and System Sciences* **68**(2):398–416, 2004.
3. G. Berlín, G, Brassard, F. Bussières and N. Godbout, "Loss-tolerant quantum coin flipping", *Proceedings of Second International Conference on Quantum, Nano and Micro Technologies (ICQNM)*, Sainte-Luce, Martinique, pp. 1–9, 2008.
4. M. Blum, "Coin flipping by telephone: A protocol for solving impossible problems", *Advances in Cryptology: A Report on CRYPTO'81*, Santa Barbara, California, USA, pp. 11–15, 1981.
5. A. Kent, "Coin tossing is strictly weaker than bit commitment", *Physical Review Letters* **83**(25): 5382–5384, 1999.
6. A. T. Nguyen, J. Frison, K. Phan Huy and S. Massar, "Experimental quantum tossing of a single coin", *New Journal of Physics* **10**, 083037, 2008.

A Protocol of Quantum Authentication with Secure Quantum Passwords

Masahiro Hotta* and Masanao Ozawa†

*Graduate School of Science, Tohoku University, Sendai, 980-8578, Japan
†Graduate School of Information Science, Nagoya University, Chikusa-ku, Nagoya, 464-8601, Japan

Abstract. A protocol is proposed robust against the theft of smart cards for authentication quantum keys. The protocol also prevents the leakage of secret information of a certification agent.

Keywords: quantum authentication, entanglement, quantum key
PACS: 03.67.-a

INTRODUCTION

Recently there is a surge in crimes by skimming and stealing of passwords in smart cards. Because of computer viruses and criminals, authentication systems also suffer from leakage of customer's passwords from certification agents which should keep secrets under normal circumstances. This is due to fragility of classical authentication based on clonable classical information for cross-checking. Quantum authentication is a possible way to overcome the above problems. Several quantum protocols have been already proposed. Those protocols are basically classified into two categories. In one of them, common secret keys are generated and stored as classical information. After that the secret information is converted into quantum information and sent from a user A to a certification agent B. For example, Guo et al [1] proposed a protocol in which a classical password is encoded by use of four orthogonal Bell states. B performs a Bell measurement in order to decode the password and checks whether the obtained result is consistent with B's classical password or not. In another category, common secret keys are shared as quantum information from the start, without sharing any classical information. A stores quantum common secret keys in a specific portable device called a quantum smart card. For example, Barnum proposed quantum protocols in this category [2]. A and B share a Bell state or a catalyst state. A keeps a half of qubits in those states in her smart card. Though those protocols have advantages which cannot be achieved in any classical protocols, several unsatisfactory aspects remain. For example, clone leakage of classical secret keys from B cannot be prohibited in principle in the first category. For the second category, impersonation may be allowed when quantum smart cards are stolen.

In this report, we explain our secure quantum authentication protocol proposed in our paper [3]. In the protocol, those weak points can be resolved. A and B share Bell pairs as cross-checking information. A can lock those Bell pairs by use of a classical password and local operations. By virtue of this locking, an adversary Eve (E) cannot impersonate

CP1110, *Quantum Communication, Measurement and Computing (QCMC), The Ninth International Conference,* edited by A. Lvovsky

A without knowing A's password even if the smart card of A is stolen by E. In the stolen card no available information is contained at all. It is also a remarkable property of our protocol that there is no need for B to keep A's clonable passwords. This completely eliminates risks of clone leakage of A's password from B. When quantum passwords are transferred, E cannot make perfect copies of passwords by eavesdropping. Moreover hiding quantum passwords behind many decoy qubits makes possible to detect E's eavesdropping with a high rate. It is also attainable that A's quantum password becomes disposable as a one-time-pad key in each round of the protocol.

PROTOCOL

In this section, we explain a basic protocol in order to outline the essence of the idea behind our full. More secure protocol is given in our paper [3]. We now give a detailed explanation using a password example. The protocol is composed of nine steps, as follows.

(1) A and B meet and generate N qubit pairs in a Bell state $|+\rangle$. For example, let us consider a case with $N=4$. The state of the system is then given by $|+\rangle|+\rangle|+\rangle|+\rangle$. A stores half of the qubits Q_A of the Bell pairs in a quantum smart card. B keeps the other half Q_B of the Bell pairs in a quantum storage device. (2) A generates an N-bit classical password K composed of 0s and 1s, keeping it secret from B and others. For instance, let us consider $K=(0101)$. A performs a unitary transformation R on the qubits of Q_A corresponding to the bit values 1 of K. The action of R changes $|+\rangle$ into another Bell state $|\xi\rangle$. In the example, the state of Q_A and Q_B is transformed into $|+\rangle|\xi\rangle|+\rangle|\xi\rangle$. (3) In authentication, A generates a quantum password Q_K. The bit values 0 of K is encoded into a qubit state $|0\rangle$ and 1 of K is encoded into a state $|\alpha\rangle$ non-orthogonal to $|0\rangle$. In the example, the quantum password is generated as $Q_K=|0\rangle|\alpha\rangle|0\rangle|\alpha\rangle$. (4) A sends Q_A and Q_K to B through a quantum channel. (5) B unlocks Q_A and Q_B using a quantum device UL with the quantum password Q_K. The device UL operates so as to perform a unitary transformation U on the composite system of Q_K, Q_A and Q_B. U does not change the input state $|0\rangle|+\rangle$ with bit values 0 of K and transforms the input state $|\alpha\rangle|\xi\rangle$ with bit values 1 of K into $|c\rangle|+\rangle$, where $|c\rangle$ is a quantum state of a qubit of Q_K. (6) B checks whether the output state of Q_A and Q_B is $|+\rangle^{\otimes N}$. This is done by a Bell measurement of $|+\rangle$. If a positive result is obtained, B recognizes A. If not, B stops the process. (7) B locks Q_A and Q_B by a quantum device L with Q_K. The action of L is the inverse transformation of U. In the example, the output state of L for Q_A and Q_B is given by $|+\rangle|\xi\rangle|+\rangle|\xi\rangle$. The output state of L for Q_K is given by $|0\rangle|\alpha\rangle|0\rangle|\alpha\rangle$. (8) B breaks off Q_K and erases the information so it cannot be stolen by others. (9) B returns Q_A to A and A restores Q_A to the smart card.

We note that Q_A and Q_B cannot be decoded correctly by B without the information of K or Q_K. Therefore, even if the smart card is stolen by E, E cannot impersonate A without K. Moreover, the contracted states of the qubits of Q_A become the maximal entropy state:

$$\rho_{\max}=\frac{I}{2}.$$

Similarly, the contracted states of Q_B also become the maximal entropy state. Therefore, the storage device of B does not contain any information of K. Hence the information about K is not leaked from B's storage device. In step (4), the information of K is encoded by two non-orthogonal states. Thus, E cannot perfectly obtain Q_K by eavesdropping. By using an extension in [3], rapid detection of eavesdropping also becomes possible. It is notable that there exists no entanglement between A's quantum password Q_K and the composite system of Q_A and Q_B. Therefore, Q_K is disposable in each round of the protocol. B cannot get perfect information of K even in steps (5) and (6) because the quantum qubits accessible by B are all non-orthogonal to each other.

In the following, we give explicit forms of the quantum states. Let us define four Bell states orthogonal to each other as follows:

$$|\pm\rangle = \frac{1}{\sqrt{2}}\left[|0\rangle|0\rangle \pm |1\rangle|1\rangle\right],$$
$$|B_{\pm}\rangle = \frac{1}{\sqrt{2}}\left[|0\rangle|1\rangle \pm |1\rangle|0\rangle\right],$$

where $|0\rangle$ and $|1\rangle$ are two orthonormal states of a qubit. The first qubit is stored by A and the second by B. The state $|\alpha\rangle$, which is used when A's quantum password is generated, is given by

$$|\alpha\rangle = \alpha|0\rangle + \beta|1\rangle,$$

where α and β are real constants such that $0 \leq \alpha < 1$ and $\beta = \sqrt{1-\alpha^2}$. The unitary transformation R is defined as follows:

$$R|0\rangle = e^{i\delta}|0\rangle,$$
$$R|1\rangle = e^{-i\delta}|1\rangle,$$

where δ is a real parameter. For later convenience, we introduce two real parameters ξ and η such that $e^{i\delta} = \xi + i\eta$, $0 \leq \xi < 1$ and $\eta = \sqrt{1-\xi^2}$. Acting on A's qubit in $|+\rangle$ with R yields a new Bell state $|\xi\rangle$:

$$R \otimes I|+\rangle = |\xi\rangle, \tag{1}$$

where $|\xi\rangle$ is given by

$$|\xi\rangle = \xi|+\rangle + i\eta|-\rangle.$$

It is easy to check explicitly that $|\xi\rangle$ is a Bell state because the following relations hold:

$$\rho_A = \operatorname*{Tr}_B\left[|\xi\rangle\langle\xi|\right] = \frac{I}{2},$$
$$\rho_B = \operatorname*{Tr}_A\left[|\xi\rangle\langle\xi|\right] = \frac{I}{2}.$$

Because $|+\rangle$ is also a Bell state, the following relations are also satisfied:

$$\rho_A = \operatorname*{Tr}_B [|+\rangle\langle+|] = \frac{I}{2},$$
$$\rho_B = \operatorname*{Tr}_A [|+\rangle\langle+] = \frac{I}{2}.$$

Consequently no information of K can be extracted from only Q_A or Q_B. Next, we consider a case where E succeeds in stealing A's card. However, we assume that E does not know K. In this case, E must make a forged quantum password. The optimal state for each qubit is denoted by ρ_E. Assuming K is randomly generated, the appearance probability of each bit value of K is $1/2$. For bit values 0 of K, the state of $Q_A + Q_B$ is $|+\rangle$. Thus, the detection probability of E per qubit is given by

$$p_{E0} = \mathrm{Tr}\left[U\left(\rho_E \otimes |+\rangle\langle+|\right)U^\dagger\left(I - |+\rangle\langle+|\right)\right].$$

For bit values 1, the state is $|\xi\rangle$. Hence, the detection probability is written as

$$p_{E1} = \mathrm{Tr}\left[U\left(\rho_E \otimes |\xi\rangle\langle\xi|\right)U^\dagger\left(I - |+\rangle\langle+|\right)\right].$$

The average detection probability per qubit is given by

$$\Delta_E = \frac{1}{2}p_{E0} + \frac{1}{2}p_{E1}.$$

The minimum value of Δ_E can be obtained as

$$p_n = \min \Delta_E = \frac{1}{2}\frac{\left(1-\xi^2\right)}{1-\alpha^2\xi^2}\left(1-\alpha\right). \tag{2}$$

For instance, taking typical values of α and ξ as $\alpha = \xi = 1/2$, the value of p_n is evaluated as $p_n = 1/5$. It is noted that the total detection probability of E is given by

$$p_n^{(N)} = 1 - (1 - p_n)^N.$$

Consequently, E can be detected with a high rate for a large N.

ACKNOWLEDGMENTS

This research was partially supported by the SCOPE project of the MIC.

REFERENCES

1. C.-P. Guo, et al, Phys.Lett.A286(2001)401.
2. H. N. Barnum, arXiv:quant-ph/9910072.
3. M. Hotta and M. Ozawa, arXiv:0806.1076.

Quantum Metagame Extensions of Noncooperative Bimatrix Games

Akira SaiToh*, Robabeh Rahimi* and Mikio Nakahara†

*Research Center for Quantum Computing, Interdisciplinary Graduate School of Science and Engineering, Kinki University, Higashi-Osaka, Osaka 577-8502, Japan
†Department of Physics, Kinki University, Higashi-Osaka, Osaka 577-8502, Japan

Abstract. A quantum metagame approach is introduced to extend classical noncooperative bimatrix games to their quantum counterparts. This approach is free from a certain unnaturalness found in conventional quantum games.

Keywords: Quantum game, Metagame
PACS: 02.50.Le, 03.67.AC

INTRODUCTION

Formulations of quantum games so far follow the original proposal by Meyer [1] or that of Eisert et al. [2] except for a small number of different formulations [3]. The formulations in Eisert et al.'s type allow players to share an initially correlated quantum system to which players apply local operations as their strategies. A quantum game is said natural [4] when it reduces to a classical counterpart for some set of quantum strategies given by local unitary operations available for players. Nevertheless, an undeniable unnaturalness exists: there is no validity to introduce an initial correlation in the quantum states distributed to players because classical noncooperative games usually do not involve an initial correlation beneficial to players. A referee should avoid, in principle, any action if it introduces a benefit for players. A natural extension of a game should be made by introducing cleverness of players rather than introducing a player-friendly referee.

Much attention should have been paid to the fact that dilemmas in some classical noncooperative games, such as that in Prisoner's Dilemma (PD), were resolved by (classical) metagame extensions in old literatures (*e.g.*, Refs. [5, 6]). The metagame extension is not to change the rule of a game but to allow players to use predictions on the other players' strategies so that they may choose cleverer strategies. Suppose that players' original strategies are limited to s_1 and s_2. A metalevel-one strategy is denoted as a pair of one's strategies chosen by one's prediction of the opponent's strategy. It looks like, say, "s_2/s_1", which indicates that the player chooses the strategy s_2 (s_1) in case the player predicts that the opponent chooses s_1 (s_2). Possible metalevel-one strategies are thus $m_1 = s_1/s_1$, $m_2 = s_1/s_2$, $m_3 = s_2/s_1$, and $m_4 = s_2/s_2$. A metalevel-two strategy is a further extension of a metalevel-one strategy. It looks like, say, "$s_2/s_1/s_2/s_2$", indicating that a player chooses the strategy s_2 (s_1, s_2, s_2) in case the opponent is predicted to choose m_1 (m_2, m_3, m_4). There are 16 possible metalevel-two strategies for a player. It was shown [5] that the dilemma in PD is resolved in a level-two (or full-full) metagame extension but not in a level-one (or full) metagame extension.

CP1110, *Quantum Communication, Measurement and Computing (QCMC), The Ninth International Conference,* edited by A. Lvovsky

In this contribution, we investigate level-one quantum metagame extensions of classical games. These extensions employ a clever player who changes a probability amplitude of her/his quantum state according to that of the opponent player. It is found that a quantum level-one metagame extension of the Samaritan's Dilemma (SD) game resolves the dilemma and that of the PD game changes the type of the dilemma.

QUANTUM METAGAME EXTENSION

Samaritan's Dilemma. The classical SD game is characterized by the payoff matrix which is a bimatrix illustrated in Table 1. The set of Alice's rational outcomes (namely, the set of outcomes with Alice's best responses) is $R_{\mathrm{A}} = \{<\mathrm{A},\mathrm{W}>,<\mathrm{NA},\mathrm{L}>\}$ and that of Bob's rational outcomes is $R_{\mathrm{B}} = \{<\mathrm{A},\mathrm{L}>,<\mathrm{NA},\mathrm{W}>\}$. Thus the set of equilibria is $R_{\mathrm{A}} \cap R_{\mathrm{B}} = \emptyset$, indicating that there is no equilibrium; this is the dilemma of the game.

TABLE 1. Payoff matrix of the classical SD. ♠: Alice's rational outcome, ♣: Bob's rational outcome.

		Bob	
		W	L
Alice	A	♠(3,2)	(−1,3)♣
	NA	(−1,1)♣	♠(0,0)

Let us firstly reformulate SD using quantum states of single qubits as strategies of players. Let us also set strategies of Alice and Bob to normalized quantum states $|s_a\rangle = \alpha_0|0\rangle + \alpha_1|1\rangle$ and $|s_b\rangle = \beta_0|0\rangle + \beta_1|1\rangle$, respectively with complex probability amplitudes $\alpha_0, \alpha_1, \beta_0, \beta_1$. Let $A = |\alpha_0|^2$ and $B = |\beta_0|^2$. A referee determines Alice's payoff and Bob's payoff $\mathrm{Tr}\rho W_a$ and $\mathrm{Tr}\rho W_b$, respectively, with $\rho = |s_a\rangle\langle s_a| \otimes |s_b\rangle\langle s_b|$; $W_a = \mathrm{diag}(3,-1,-1,0)$ and $W_b = \mathrm{diag}(2,3,1,0)$. Then, Alice's payoff and Bob's payoff are given by $p_a = A(5B-1) - B$ and $p_b = B(1-2A) + 3A$, respectively. This leads to the rational outcomes of Alice and Bob, $R_a = \{<A=0, B \leq 1/5>,<A=1, B \geq 1/5>\}$ and $R_b = \{<A \leq 1/2, B=1>,<A \geq 1/2, B=0>\}$. Thus $R_a \cap R_b = \emptyset$ and the dilemma remains.

Secondly, we extend the above quantum SD by introducing a clever Alice. In this metalevel-one game, Alice uses a predicted value B' of Bob's B to choose her strategy, namely a choice of the value of A. Let her strategy be $A \in \{f_1, f_2\}$ with the functions $f_1 = B'$ and $f_2 = 1 - B'$. A referee follows the previous process to determine payoffs. Then, the payoff matrix is described by Table 2. This leads to the rational outcomes of Alice and Bob, $R_a = \{<f_1, (B' \leq 1/5,\ 1/2 \leq B')>,<f_2, (1/5 \leq B' \leq 1/2)>\}$ and $R_b = \{<f_1, B'=1>,<f_2, B'=0>\}$, respectively. Consequently, an equilibrium $R_a \cap R_b =<f_1, B'=1>$ appears with the payoff $(3,2)$ which is a Pareto optimum. Thus the level-one quantum metagame extension resolves the dilemma of SD.

Prisoner's Dilemma. The classical PD game is characterized by the payoff matrix described in Table 3. Alice and Bob's rational outcomes are $R_a = \{<\mathrm{D},\mathrm{C}>,<\mathrm{D},\mathrm{D}>\}$

TABLE 2. Payoff matrix of the metalevel-one quantum SD game with a clever Alice.

Alice \ Bob	B' (continuous)
f_1	$(5B'^2-2B',-2B'^2+4B')$
f_2	$(-5B'^2+5B'-1,2B'^2-4B'+3)$

and $R_b=\{<\mathrm{C,D}>,<\mathrm{D,D}>\}$; the equilibrium is thus $R_a\cap R_b=<\mathrm{D,D}>$, which is not a Pareto optimum.

TABLE 3. Classical payoff matrix of the PD game. ♠: Alice's rational outcome, ♣: Bob's rational outcome. The set of payoffs for the equilibrium is denoted in a bold face.

		Bob	
		C	D
Alice	C	(3,3)	(0,4)♣
	D	♠(4,0)	♠**(2,2)**♣

We follow the same method as the previous example, so as to introduce a metalevel-one extension of a quantum PD game. Players choose their states $|s_a\rangle$ and $|s_b\rangle$, and a referee uses observables $W_a=\mathrm{diag}(3,0,4,2)$ and $W_b=\mathrm{diag}(3,4,0,2)$. Then payoffs for Alice and Bob are given by $p_a=-2A+AB+2B+2$ and $p_b=-2B+AB+2A+2$. A clever Alice chooses $A\in\{f_1,f_2\}$ with the functions $f_1=B'$ and $f_2=1-B'$. The payoff matrix under these conditions becomes one described by Table 4. The rational outcomes

TABLE 4. Payoff matrix of the metalevel-one quantum PD game with a clever Alice.

Alice \ Bob	B' (continuous)
f_1	(B'^2+2,B'^2+2)
f_2	$(-(B'-\frac{5}{2})^2+\frac{25}{4},-(B'+\frac{3}{2})^2+\frac{25}{4})$

of players are $R_a=\{<f_1,(B'\leq 1/2)>,<f_2,(1/2\leq B')>\}$ and $R_b=\{<f_1,B'=1>,<f_2,B'=0>\}$. As $R_a\cap R_b=\emptyset$, there is no equilibrium point; the original PD dilemma has been changed to a dilemma similar to the classical SD.

REMARKS

We have introduced quantum level-one metagame extensions of the SD and PD games. It has been shown that the dilemma of SD is resolved and that of PD is changed to an SD-like one, in the extensions. As for the SD game, it has been known [7] that classical mixed-strategy extension does not resolve the dilemma while an Eisert-type quantum extension using a certain entangling operation applied by a referee resolves the dilemma. We have newly shown that resolving the dilemma is possible without using the entangling operation (hence, without introducing a player-friendly referee), in the case of a quantum metagame extension.

Nevertheless, this fact does not guarantee an advantage of introducing quantumness in games. These metalevel-one quantum games can be simulated by metalevel-one classical mixed strategy games because, as we have seen, they involve product states whose probability amplitudes are selected by players as their strategies. An advantage of quantumness is thus unclear, so far, for noncooperative bimatrix games; a clear discussion should wait for further investigations of different ways of metagame extensions.

ACKNOWLEDGMENTS

AS is thankful to Şahin Kaya Özdemir for helpful comments and discussions. AS and MN are supported by "Open Research Center" Project for Private Universities: matching fund subsidy from MEXT. RR is supported by the Grant-in-Aid from JSPS (Grant No. 1907329). MN would like to thank a partial support of the Grant-in-Aid for Scientific Research from JSPS (Grant No. 19540422).

REFERENCES

1. D. A. Meyer, *Phys. Rev. Lett.* **82**, 1052 (1999).
2. J. Eisert, M. Wilkens, and M. Lewenstein, *Phys. Rev. Lett.* **83**, 3077 (1999).
3. J. Wu, quant-ph/0405032.
4. Ş. K. Özdemir, J. Shimamura, and N. Imoto, *New J. Phys.* **9**, 43 (2007).
5. N. Howard, *General Systems* **11**, 167 (1966); *ibid.* 187 (1966).
6. N. Howard, *Behavioral Science* **21**, 524 (1976).
7. Ş. K. Özdemir, J. Shimamura, and N. Imoto, *Playing games in quantum mechanical settings: features of quantum games*, in *Mathematical Aspects of Quantum Computing 2007* (World Scientific, Singapore, 2008).

CHAPTER 13

QUANTUM COMPUTATION AND ALGORITHMS

Optical CNOT gates with Quantum Interrogation

Juan Carlos García-Escartín and Pedro Chamorro-Posada

Departamento de Teoría de la Señal y Comunicaciones e Ingeniería Telemática. Universidad de Valladolid. ETSI Telecomunicación. Camino del Cementerio s/n, 47011, Valladolid, Spain.

Abstract. We present three families of optical CNOT gates where the photon interaction is mediated by ancillary particle qubits. The gates are based on a matter to light CZ gate, in which the possibility of a particle absorbing a photon is enough for the coupling. The employed quantum interrogation scheme allows to have a negligible probability of an actual absorption taking place.

Keywords: Quantum interrogation, optical CNOT gates, entanglement creation.
PACS: 42.50.Ex, 03.65.Ta, 03.67.Bg.

1. INTRODUCTION

Linear optics quantum computation is a promising candidate for simple quantum computers [1]. Single qubit gates are easy to build with optical elements, but the implementation of entangling operations, such as the CNOT gate, requires of advanced experimental equipment. Existing CNOT gates are probabilistic or rely on strong nonlinearities or matter to light coupling inside cavities.

We put forward an alternative approach for the implementation of CNOT gates based on the Quantum Interrogation of quantum particles. This approach can relax some of the experimental demands on optical gates. Our gates are composed of optical single qubit gates and a CZ gate between light and matter. Table 1 presents all the quantum gates used in the discussion.

TABLE 1. Elementary quantum gates used in the design.

Input	$\alpha\|0\rangle+\beta\|1\rangle$		$\alpha\|00\rangle+\beta\|01\rangle+\gamma\|10\rangle+\delta\|11\rangle$
Gate	**Output**	**Gate**	**Output**
NOT (X)	$\alpha\|1\rangle+\beta\|0\rangle$	CNOT	$\alpha\|00\rangle+\beta\|01\rangle+\gamma\|11\rangle+\delta\|10\rangle$
Z	$\alpha\|0\rangle-\beta\|1\rangle$	CZ	$\alpha\|00\rangle+\beta\|01\rangle+\gamma\|10\rangle-\delta\|11\rangle$
H	$\alpha\|+\rangle+\beta\|-\rangle$*		

* $|+\rangle=\frac{|0\rangle+|1\rangle}{\sqrt{2}}$ and $|-\rangle=\frac{|0\rangle-|1\rangle}{\sqrt{2}}$.

CP1110, *Quantum Communication, Measurement and Computing (QCMC), The Ninth International Conference,* edited by A. Lvovsky

2. OPTICAL TO MATTER CZ GATE

The core element of our proposal is the Quantum Interrogation (QI) setup of Kwiat [2]. With it, we want to ascertain the presence of a bomb so sensitive that a single photon can make it explode. It is possible to detect the bomb with a negligible probability of explosion by taking advantage of the continuous projective measurement the possibility of explosion provides.

Figure 1 shows the QI setup that performs the CZ operation. The light qubit is encoded into the photon's path ($|0\rangle$ for the upper path, $|1\rangle$ for the lower) and the matter qubit is encoded in the position of a particle ($|0\rangle$ if it blocks a potential path of the photon, $|1\rangle$ if it stands out of the photon's way). The particle will be represented as a bomb that absorbs the photon ("explodes") if they meet.

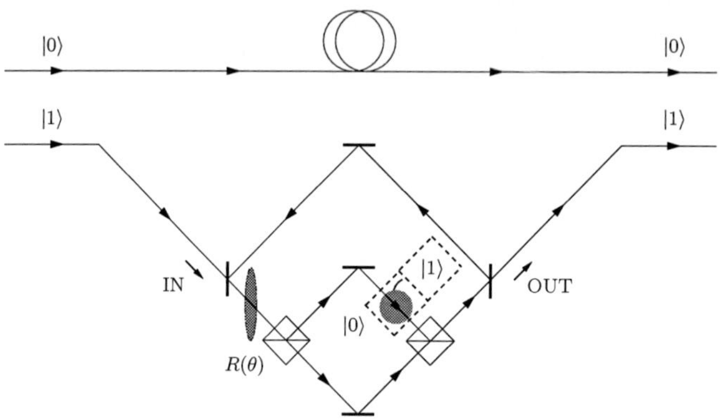

FIGURE 1. CZ gate for a quantum bomb (shaded) and a dual-rail qubit.

The setup is completed by two polarizing beamsplitters, mirrors to direct the photons and a polarization rotator (represented as an oval) that rotates the photon's polarization by an angle $\theta = \frac{\pi}{N}$. The points marked as IN and OUT indicate the places where the photon is introduced and released from the inner loop, where it will complete N round cycles.

If there is no bomb in the path, the polarization of the photon is rotated an angle $N\theta = \pi$ and a horizontally polarized input photon, in state $|H\rangle$, will go out with a sign shift. If there is a bomb (particle state $|0\rangle$), the photon, which, after the first rotation, is in the state $\cos\theta|H\rangle + \sin\theta|V\rangle$, takes the path of the bomb with probability $\sin^2\theta$, triggering an explosion. However, if the explosion does not happen, the photon state is projected back to $|H\rangle$. The same evolution is repeated at each loop and, after the N cycles, the photon remains in $|H\rangle$ with probability $\cos^{2N}\theta$. For a large N, this probability is, approximately, $1 - \frac{\pi^2}{N}$, which can be made arbitrarily close to 1 if we increase N.

This evolution, when applied on an arbitrary input state, corresponds to that of CZ gate. There will be a sign shift if, and only if, the photon goes through the $|1\rangle$ path and there is no bomb there (the global state is $|11\rangle$).

This gate has to its advantage that it is nondestructive and valid for arbitrary input states, unlike some previous proposals [3]. The gate is a simplification of our previous QICNOT gate [4] and is equivalent to some recent QI gate proposals [5].

3. THE OPTICAL GATES

We can derive three different families of optical CNOT gates from the QI CZ gate between particles and photons. All of them will have ancillary particle qubits, which will start in a $|+\rangle$ state and will be measured in the $\{|+\rangle, |-\rangle\}$ basis.

Classically controlled gates, cX and cZ, (represented with the customary double lines) will play an essential role in the design. Whenever possible, we have substituted quantum gates by the easier to realize classically controlled or purely classical gates.

3.1. Memory CNOT gates

Figure 2 shows two possible implementations of an optical CNOT gate based on gate teleportation [6]. In the following, $|C\rangle$ will represent the control qubit state and $|T\rangle$ the state of the target qubit. The input can be any superposition of different $|C\rangle|T\rangle$ states and will result in the corresponding superposition at the output.

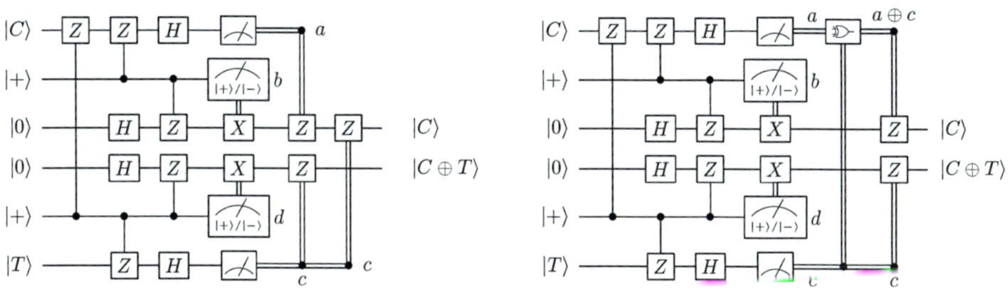

FIGURE 2. Memory CNOT gates, with two cX and three cZ gates (left) and with a XOR classical gate that permits to merge two of the previous cZ gates (right).

Figure 2 shows the basic configuration and an interesting simplification in which a classical XOR gate allows to save one cZ gate. Both gates can be easily interpreted as a memory read and write circuit. Two CZ gates map $|0\rangle$ and $|1\rangle$ into $|+\rangle$ and $|-\rangle$. If the control qubit is $|1\rangle$, the state of the second memory is flipped to $|-\rangle$ before the memory write. Then we have an inverting memory, which stores the negated value of the target qubit. This is the origin of the CNOT operation.

3.2. Half-memory CNOT gates

In the memory CNOT gate, only the second of the two particle qubits takes part in the CNOT operation. The first particle qubit only contributes to teleport the control. We can save two ancillary qubits, one photon and one particle, by avoiding this teleportation. Figure 3 shows two half-memory gates that result from this reduction. In the first one, only the target is teleported. If the control is $|1\rangle$, we have, again, an inverting memory. In the second gate, it is the control qubit which is teleported. Now, the stored control qubit, in the $\{|+\rangle, |-\rangle\}$ basis, serves as the control of a CNOT operation on the target.

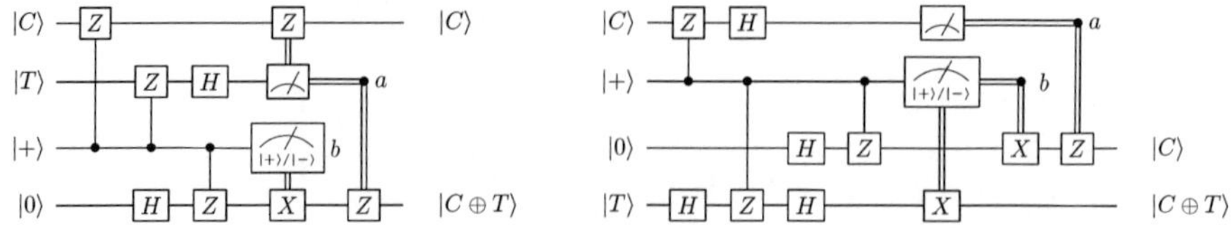

FIGURE 3. Half-memory CNOT gates where the stored qubit is the target (left) and the control (right).

3.3. Direct CNOT gates

If particle H gates are available, we can avoid optical qubit teleportation and save one CZ gate. Figure 4 shows two of these direct CNOT gates. Here, the state of the photon qubits is never completely transferred to the particle.

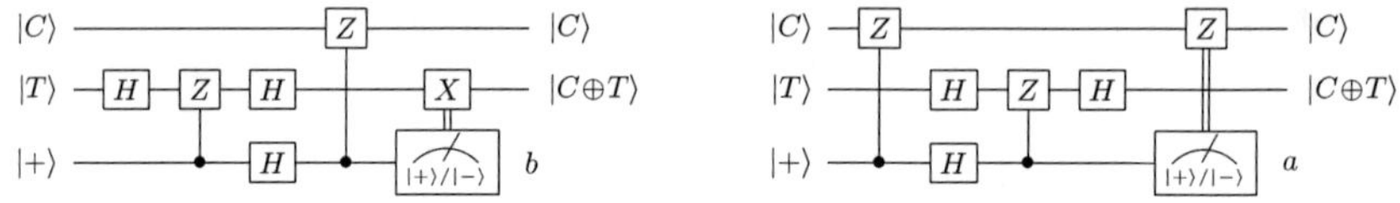

FIGURE 4. Direct CNOT gates with cX (left) and cZ (right) postcorrection.

3.4. Implementation

QI CZ gates are the most delicate part of the scheme. The gates need a coupling mechanism between the particle and the photon. However, QI can work, at a pinch, for partially absorbing objects [7], and couplings weaker than those used in other optical CNOT gate proposals could be enough. The most important challenge is limiting the losses in the QI setup, which can greatly decrease the probability of success of the gates.

ACKNOWLEDGMENTS

This work has been funded by MEC and FEDER grant TEC2007-67429-C02-01.

REFERENCES

1. E. Knill, R. Laflamme and G. J. Milburn, *Nature* **409**, 46–52 (2001).
2. P. G. Kwiat, A. G. White, J. R. Mitchell, O. Nairz, G. Weihs, H. Weinfurter and A. Zeilinger, *Phys. Rev. Lett.* **83**, 4725-4728 (1999).
3. A. Gilchrist, A. G. White and W. J. Munro, *Phys. Rev. A* **66**, 012106 (2002).
4. J. C. García-Escartín and P. Chamorro-Posada, *Quantum Inf. Comput.* **6**, 495–515 (2006).
5. Y. P. Huang and M. G. Moore, *Phys. Rev. A* **77**, 062332 (2008).
6. D. Gottesman and I. L. Chuang, *Nature* **402**, 390-393 (1999).
7. J.-S. Jang, *Phys. Rev. A* **59**, 2322-2329 (1999).

Discrimination of Unitary Transformations and Quantum Algorithms

David Collins

Department of Physical and Environmental Sciences, Mesa State College, Grand Junction, Colorado, USA

Abstract. Quantum algorithms are typically understood in terms of the evolution of a multi-qubit quantum system under a prescribed sequence of unitary transformations. The input to the algorithm prescribes some of the unitary transformations in the sequence with others remaining fixed. For oracle query algorithms, the input determines the oracle unitary transformation. Such algorithms can be regarded as devices for discriminating amongst a set of unitary transformations. The question arises: "Given a set of known oracle unitary transformations, to what extent is it possible to discriminate amongst them?" We investigate this for the Deutsch-Jozsa problem. The task of discriminating amongst the admissible oracle unitary transformations results in an exhaustive collection of algorithms which can solve the problem with certainty.

Keywords: Quantum computing, quantum algorithms, quantum state discrimination, unitary discrimination.
PACS: 03.67Ac, 03.67.-a, 03.65.-w

ORACLE ALGORITHMS AND DISCRIMINATION OF QUANTUM OPERATIONS

Quantum algorithms [1, 2, 3, 4] are usually solve computational problems with the aid of a binary oracle function $f:\{0,1\}^n \mapsto \{0,1\}^m$, which depends on the problem. The simplest case, where $m=1$, occurs in the Deutsch-Jozsa [1] and Grover's [4] algorithms. In quantum algorithms a given oracle, f, is invoked via unitary transformation, $\hat{U}_f$, whose structure depends on the nature of the oracle. For example, in both the Deutsch-Jozsa algorithm [5] and Grover's algorithm the oracle is defined on computational basis states, $|x\rangle \equiv |x_n \dots x_1\rangle$ with $x_i \in \{0,1\}$, to be $\hat{U}_f |x\rangle = (-1)^{f(x)} |x\rangle$ and this is extended linearly to superpositions of computational basis states. The general structure of such oracle algorithms is encapsulated in an algorithm unitary

$$\hat{U}_{\text{alg}} = \hat{V}_M \hat{U}_f \dots \hat{U}_f \hat{V}_1 \hat{U}_f \hat{V}_0 \tag{1}$$

where $\hat{V}_0, \dots, \hat{V}_M$ are oracle independent unitary transformations and the oracle is invoked M times. This is applied to a quantum system in an oracle independent initial state $|\Psi_i\rangle$, giving an oracle dependent final state $|\Psi_f\rangle = \hat{U}_{\text{alg}} |\Psi_i\rangle$, upon which a computational basis measurement is performed. It is important to note that *the input to the algorithm is the oracle unitary* and not the initial state. The output from the algorithm potentially identifies the oracle unitary or a class of oracle unitaries. For example, in Grover's algorithm for searching a database one marked item at location s, $f(x) = 0$ whenever $x \neq s$ and $f(s) = 1$. The standard Grover's algorithm terminates in a compu-

CP1110, *Quantum Communication, Measurement and Computing (QCMC), The Ninth International Conference,* edited by A. Lvovsky

tational basis measurement which yields s with high probability. Since the admissible oracles for this problem can be labeled by s, the algorithm *identifies the input oracle unitary* with high probability. We thus regard the quantum algorithm as a tool for discriminating between classes of admissible input oracle unitaries.

The problem of discriminating between unitary transformations is usually reduced to a quantum state discrimination problem [6, 7]. This considers application of one unitary from a collection of known possible unitaries, $\{\hat{U}_1, \hat{U}_2, \ldots\}$, any of which may be invoked with known probabilities $\{p_1, p_2, \ldots\}$. An (unknown) unitary from this collection is selected and can be applied one or more times to a single quantum system, after which a measurement is performed so as to infer the actual unitary used. We consider the case where the *unitary is used once.* This is converted into a standard state discrimination problem [8] by choosing an initial state, described by a density operator $\hat{\rho}_{\mathrm{i}}$, and applying the given unitary, $\hat{U}_j$, to yield an output state $\hat{\rho}_{\mathrm{f}\,j} = \hat{U}_j \hat{\rho}_{\mathrm{i}} \hat{U}_j^{\dagger}$. The problem of discrimination between the unitaries reduces to discrimination between the states $\{\hat{\rho}_{\mathrm{f}\,1}, \hat{\rho}_{\mathrm{f}\,2}, \ldots\}$. This involves subjecting the system to a POVM with positive operator elements $\{\hat{E}_1, \hat{E}_2, \ldots\}$ that satisfy $\sum_j \hat{E}_j = \hat{I}$ and applying an rule for inferring the state from the measurement outcome (for example, if the measurement yields the outcome associated with $\hat{E}_j$, then the state was $\hat{\rho}_{\mathrm{f}\,j}$). This can be converted to an inference about the unitary used and the task is to find the POVM and the initial density operator which maximizes the probability with which the unitary is correctly inferred.

The central idea of this work is to use techniques for discrimination of quantum states or unitaries to arrive at and assess quantum algorithms. In the remainder of this article we focus on the Deutsch-Jozsa algorithm.

APPLICATION TO THE DEUTSCH-JOZSA ALGORITHM

The Deutsch-Jozsa problem considers functions $f : \{0,1\}^n \mapsto \{0,1\}$ which are required to be in one of two possible classes : *constant or balanced.* A constant function returns the same output for all possible arguments while a balanced function returns 0 for exactly half of the possible inputs and 1 for the other half. The problem is to determine the function class with a minimum number of oracle invocations. A classical algorithm that solves with certainty requires $2^{n-1}+1$ oracle invocations in the worst case [1, 9]. A quantum algorithm for solving this exists [1, 9] and, in its modified form [5], uses the oracle defined on computational basis elements as

$$\hat{U}_f \left|x\right\rangle := (-1)^{f(x)} \left|x\right\rangle . \tag{2}$$

The question that we pose is whether it is possible to arrive at quantum algorithms that solve the Deutsch-Jozsa problem purely by considering the possibility of discriminating between the oracle unitaries for balanced and constant functions, given by Eq. (2).

Since the Deutsch-Jozsa problem requires that a determination of function class rather than the actual function used, the discrimination problem is one of discriminating between the two classes of unitary transformations. The can be conveniently recast as a problem of discriminating between two quantum operations, one corresponding to the two constant functions and the other to the set of all balanced functions. Generally either relevant quantum operation can be represented as $\hat{\rho}_{\mathrm{i}} \mapsto \hat{\rho}_{\mathrm{f}} = \sum_{f \text{ in class}} p_f \hat{U}_f \hat{\rho}_{\mathrm{i}} \hat{U}_f^{\dagger}$ where $\{p_f | f \text{ in class}\}$ are the probabilities with which the various unitaries within each class are applied. The operation for the constant class is independent of these probabilities and performs

$$\hat{\rho}_{\mathrm{i}} \mapsto \hat{\rho}_{\mathrm{fconst}} = \hat{\rho}_{\mathrm{i}}. \tag{3}$$

For the balanced class, we consider the case where p_f are identical for all balanced functions. A detailed calculation shows that the operation is

$$\hat{\rho}_{\mathrm{i}} \mapsto \hat{\rho}_{\mathrm{fbal}} = \frac{1}{N-1}\left(-\hat{\rho}_{\mathrm{i}} + N\sum_{x=0}^{N-1} \hat{P}_x \hat{\rho}_{\mathrm{i}} \hat{P}_x\right) \tag{4}$$

where $N = 2^n$ and $\hat{P}_x := |x\rangle\langle x|$. The problem of discrimination between the two classes of unitaries reduces to that of discrimination between the two density operators, $\hat{\rho}_{\mathrm{fconst}}$ and $\hat{\rho}_{\mathrm{fbal}}$. Conclusive discrimination between these requires a POVM with two elements $\hat{E}_{\mathrm{const}}$ and $\hat{E}_{\mathrm{bal}} = \hat{I} - \hat{E}_{\mathrm{const}}$. If the measurement outcome associated with $\hat{E}_{\mathrm{const}}$ is obtained, then it will be inferred that the state after the application of the operation is $\hat{\rho}_{\mathrm{fconst}}$ and that hence the unitary is one for a constant function; a similar rule applies for the balanced case. The probability with which a correct inference will be made is $\Pr(\text{correct inference}) = p_{\mathrm{const}} \operatorname{Tr}(\hat{\rho}_{\mathrm{fconst}} \hat{E}_{\mathrm{const}}) + p_{\mathrm{bal}} \operatorname{Tr}(\hat{\rho}_{\mathrm{fbal}} \hat{E}_{\mathrm{bal}})$ where p_{const} and $p_{\mathrm{bal}} = 1 - p_{\mathrm{const}}$ are the probabilities with which the function is chosen from the constant or balanced classes respectively. The inference will be correct with certainty for arbitrary p_{const} if $\operatorname{Tr}(\hat{\rho}_{\mathrm{fconst}} \hat{E}_{\mathrm{const}}) = 1$ and $\operatorname{Tr}(\hat{\rho}_{\mathrm{fbal}} \hat{E}_{\mathrm{bal}}) = 1$. The requirement that $\operatorname{Tr}(\hat{\rho}_{\mathrm{fconst}} \hat{E}_{\mathrm{const}}) = 1$ can be shown, via a series of inferences based on the positivity of both $\hat{\rho}_{\mathrm{fconst}}$ and $\hat{E}_{\mathrm{const}}$ and the fact that their eigenvalues are each in the range $[0,1]$, to imply that $\hat{E}_{\mathrm{const}}$ is the identity operator on the support of $\hat{\rho}_{\mathrm{fconst}}$ (i.e. the subspace orthogonal to the kernel) and is zero elsewhere. Applying this to the analogous operators for the balanced case yields the result that the supports of $\hat{\rho}_{\mathrm{fconst}}$ and $\hat{\rho}_{\mathrm{fbal}}$ are orthogonal or, equivalently

$$\hat{\rho}_{\mathrm{fconst}} \hat{\rho}_{\mathrm{fbal}} = \hat{\rho}_{\mathrm{fbal}} \hat{\rho}_{\mathrm{fconst}} = 0. \tag{5}$$

Eqs. (3)-(5) imply that

$$\left[\hat{\rho}_{\mathrm{i}}, \hat{\Lambda}\right] = 0 \quad \text{and} \tag{6}$$

$$N \hat{\Lambda} \hat{\rho}_{\mathrm{i}} = \hat{\rho}_{\mathrm{i}}^2. \tag{7}$$

where $\hat{\Lambda} := \sum_{x=0}^{N-1} \hat{P}_x \hat{\rho}_{\mathrm{i}} \hat{P}_x$ and satisfies the requirements for a density operator. Denote the orthonormal basis in which the two operators $\hat{\rho}_{\mathrm{i}}$ and $\hat{\Lambda}$ can be diagonalized simultaneously by $\{|\phi_j\rangle \,|\, j = 1, \ldots N\}$. Thus $\hat{\rho}_{\mathrm{i}} = \sum_{j=1}^{M} p_j |\phi_j\rangle\langle\phi_j|$ where M is the number of non-zero eigenvalues of $\hat{\rho}_{\mathrm{i}}$ and $\hat{\Lambda} = \sum_{j=1}^{N} \lambda_j |\phi_j\rangle\langle\phi_j|$. Suppose that $p_1 \neq 0$. Then Eqs. (6) and (7) can be shown to imply that

$$\frac{p_1}{N} = p_1 \sum_x |\phi_1(x)|^4 + \sum_{k \neq 1}^{M} p_k \sum_x |\phi_k(x)|^2 |\phi_1(x)|^2. \tag{8}$$

where $\phi_1(x) = \langle x|\phi_1\rangle$. The normalized state $|\phi_1\rangle$ satisfies $\sum_x |\phi_1(x)|^2 = 1$ and a Lagrange multiplier technique demonstrates that, subject to this constraint, $\sum_x |\phi_1(x)|^4 \geqslant 1/N$ with equality attained if and only if $|\phi_1(x)|^2 = 1/N$ for all $x = 0, \ldots N-1$. Thus Eq. (8) implies

$$\frac{p_1}{N} \geqslant \frac{p_1}{N} + \sum_{k \neq 1}^{M} p_k \sum_x |\phi_k(x)|^2 |\phi_1(x)|^2. \tag{9}$$

with equality if and only if $|\phi_1(x)|^2 = 1/N$ for all $x = 0, \ldots N-1$. Since the second term on the right hand side of Eq. (9) is non-negative, the only possibility is that the equality holds and that $|\phi_1(x)|^2 = 1/N$ for all $x = 0, \ldots N-1$. This implies that $p_k = 0$ for $k \neq 1$.

Thus the only possible initial states which discriminates conclusively and correctly with certainty between the two classes of quantum operations for the Deutsch-Jozsa problem are $\hat{\rho}_{\rm i} = |\phi\rangle\langle\phi|$ where

$$|\phi\rangle = \frac{1}{\sqrt{N}} \sum_{x=0}^{N-1} e^{i\theta_x} |x\rangle \tag{10}$$

with θ_x arbitrary real phases. The corresponding POVM elements are $\hat{E}_{\rm const} = |\phi\rangle\langle\phi|$ and $\hat{E}_{\rm bal} = \hat{I} - \hat{E}_{\rm const}$. It is easily verified that this algorithm discriminates with certainty between balanced and constant functions regardless of *a priori* probabilities.

CONCLUSION

We have shown that the techniques of unitary discrimination can be applied to yield an exhaustive collection of algorithms which solve the Deutsch-Jozsa problem with certainty. This suggests that it may be fruitful to investigate unitary discrimination in the context of other algorithms or whenever the set of input states is restricted (such as thermal equilibrium states in NMR).

ACKNOWLEDGMENTS

The author is grateful for support from the Mesa State College Faculty Professional Development Fund.

REFERENCES

1. D. Deutsch, and R. Jozsa, *Proc. R. Soc. Lond. A* **439**, 553 (1992).
2. D. R. Simon, *Proceedings 35th Annual Symposium on Foundations of Computer Science* pp. 116–123 (1994).
3. P. Shor, *SIAM J. Comput.* **26**, 1484 (1997).
4. L. K. Grover, *Phys. Rev. Lett.* **79**, 325 (1997).
5. D. Collins, K. W. Kim, and W. C. Holton, *Phys. Rev. A* **58**, 1633 (1998).
6. A. Acín, *Phys. Rev. Lett.* **87**, 177901 (2001).
7. M. F. Sacchi, *Phys. Rev. A* **71**, 062340 (2005).
8. C. W. Helstrom, *Quantum Detectionand Estimation Theory*, Academic Press, New York, 1976.
9. R. Cleve, A. Ekert, C. Macchiavello, and M. Mosca, *Proc. R. Soc. Lond. A* **454**, 339 (1998).

Generating arbitrary four-qubit decoherence-free states via two singlet states and a partial exchanging device

Yan-Xiao Gong*, Xu-Bo Zou*, Xiao-Ling Niu*, Jian Li†, Yun-Feng Huang* and Guang-Can Guo*

*Key Laboratory of Quantum Information, University of Science and Technology of China, CAS, Hefei, 230026, People's Republic of China
†Department of Physics, Southeast University, Nanjing 210096, People's Republic of China

Abstract. We clarify the original thought for the scheme in our recent paper (Phys. Rev. A, 77, 042317), for preparing arbitrary four-qubit decoherence-free states robust against collective decoherence. Our method needs two singlet states and a tunable partial exchanging device.

Keywords: decoherence-free space, quantum information, collective decoherence
PACS: 03.67.Pp, 03.67.Hk

A significant obstacle to quantum information processing is decoherence induced by uncontrolled coupling between a quantum system and the environment. It may easily destroy the fragile superpositions of quantum states [1]. Several strategies have been devised to deal with decoherence, such as error-correcting codes [2] or dynamical decoupling [3]. However, if the system-environment interaction, no matter how strong, exhibits some symmetry, it is possible to encode logical qubits into a part of the overall Hilbert space, a "decoherence-free space" (DFS) [4, 5, 6], which is inherently immune to decoherence.

Here we consider a particularly useful model of a DFS, where the decoherence mechanism is known as "collective decoherence". In this decoherence model the environment couples identically to each individual qubit in the system. The decoherence-free (DF) states are also invariant under any identical unitary transformation on each component qubit. This property makes the DF states useful for quantum information processing among parties who do not share any reference frame [7].

The dimension of a DFS depends on the number N of the component qubits [4]. For $N = 2$ qubits, there is only one DF state, namely, the singlet state, $|\Psi^-\rangle = (|01\rangle - |10\rangle/\sqrt{2})$. Therefore, it is not sufficient to fully protect one-qubit information. The invariance properties of the singlet state and its immunity against collective decoherence have been demonstrated experimentally in different physical systems [8].

For $N = 4$ cases, the dimension of the DFS is 2, and thus it is sufficient to fully

CP1110, *Quantum Communication, Measurement and Computing (QCMC), The Ninth International Conference,* edited by A. Lvovsky
© 2009 American Institute of Physics 978-0-7354-0647-6/09/$25.00

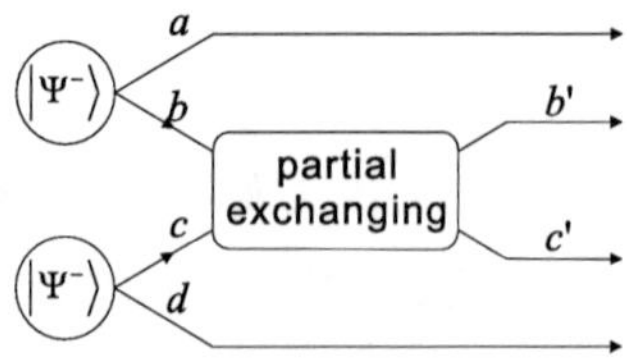

FIGURE 1. Schematic of generating four-qubit decoherence-free states from two singlet states with a partial exchanging device.

protect an arbitrary logical qubit against collective decoherence. The two basis states of the four-qubit DFS, given by Kempe *et al.*[6], are shown as follows,

$$|0\rangle_L = |\Psi^-\rangle \otimes |\Psi^-\rangle = \frac{1}{2}(|01\rangle - |10\rangle)(|01\rangle - |10\rangle), \tag{1}$$

$$|1\rangle_L = \frac{1}{2\sqrt{3}}(2|0011\rangle + 2|1100\rangle - |0101\rangle - |1010\rangle - |0110\rangle - |1001\rangle). \tag{2}$$

Therefore, it is important to prepare any superpositions of the two basis states to encode an arbitrary qubit. Bourennane *et al.* [9] did an experiment to generate the two basis states via a spontaneous parametric down-conversion (SPDC) source and demonstrated the immunity of both states against collective decoherence. In order to prepare superposition states, they could use two sources for the two basis states respectively, and coherently overlap the generated photons. Zou *et al.* [10] proposed another scheme to directly prepare a superposition state from two SPDC sources using only a tunable beam splitter, but only a few particular superposition states could be generated.

In our recent paper [11], we proposed a scheme to generate arbitrary four-photon polarization-entangled decoherence-free states with linear optics. In this paper we aim to clarify the original thought for that scheme and hope to motivate applications in other physical systems.

Our scheme for generating arbitrary four-qubit decoherence states is shown in FIGURE 1. The source states are two singlet states $|\Psi^-\rangle_{ab}$ and $|\Psi^-\rangle_{cd}$. The two qubits in modes b and c are coupled into the key device in our scheme, which is called the partial exchanging device.

Let us first introduce how the partial exchanging device works. Assuming the qubits in modes b and c in pure states $|\psi\rangle_b$ and $|\phi\rangle_c$, respectively, for simplicity, we set the evolution through the device as

$$|\psi\rangle_b|\phi\rangle_c \rightarrow C(\alpha|\psi\rangle_{b'}|\phi\rangle_{c'} + \beta|\phi\rangle_{b'}|\psi\rangle_{c'}), \tag{3}$$

where C is the normalized constant dependent on the initial states and the coefficients α and β are complex numbers.

The operator of the device can be expressed as

$$U = (\alpha+\beta)\Pi_+ + (\alpha-\beta)\Pi_-, \tag{4}$$

where $\Pi_- = |\Psi^-\rangle\langle\Psi^-|$ and $\Pi_+ = I - \Pi_-$ are projectors onto the antisymmetric and symmetric subspaces of the two-qubit space, respectively, and I represents the identity operator. The coefficients of the two projectors are determined by α and β, and therefore by tuning them we can obtain different operations in the form of the Eq. (4). For instance, the square-root SWAP ($\sqrt{\text{SWAP}}$) gate, $U_{\pi/2} = \Pi_+ + i\Pi_-$, which, together with arbitrary single-qubit rotations, is sufficient for universal quantum computing.

Then with the operation of the partial exchanging device, we can write the state evolution of the initial singlet sates as

$$\begin{aligned}
|\Phi\rangle_L &= C_{\text{DFS}}\left(\alpha|\Psi^-\rangle_{ab'}|\Psi^-\rangle_{c'd} + \beta|\Psi^-\rangle_{ac'}|\Psi^-\rangle_{b'd}\right) \\
&= C_{\text{DFS}}\left[\frac{1}{4}(2\alpha+\beta)\left(|0\rangle_a|1\rangle_{b'} - |1\rangle_a|0\rangle_{b'}\right)\left(|0\rangle_{c'}|1\rangle_d - |1\rangle_{c'}|0\rangle_d\right)\right. \\
&\quad + \frac{1}{4}\beta\left(2|0\rangle_a|0\rangle_{b'}|1\rangle_{c'}|1\rangle_d + 2|1\rangle_a|1\rangle_{b'}|0\rangle_{c'}|0\rangle_d - |0\rangle_a|1\rangle_{b'}|0\rangle_{c'}|1\rangle_d\right. \\
&\quad \left.\left. - |1\rangle_a|0\rangle_{b'}|1\rangle_{c'}|0\rangle_d - |0\rangle_a|1\rangle_{b'}|1\rangle_{c'}|0\rangle_d - |1\rangle_a|0\rangle_{b'}|0\rangle_{c'}|1\rangle_d\right)\right] \\
&= C_{\text{DFS}}\left[\left(\alpha + \frac{1}{2}\beta\right)|0\rangle_L + \frac{\sqrt{3}}{2}\beta|1\rangle_L\right], \qquad (5)
\end{aligned}$$

where $|0\rangle_L$ and $|1\rangle_L$ are given by Eqs. (1) and (2), and the normalized constant C_{DFS} is given by

$$C_{\text{DFS}} = \frac{1}{\sqrt{|\alpha|^2 + |\beta|^2 + \text{Re}(\alpha\beta^*)}}. \qquad (6)$$

To obtain an arbitrary four-qubit DF state,

$$|\Psi\rangle_L = \cos\frac{\gamma}{2}|0\rangle_L + \sin\frac{\gamma}{2}e^{i\delta}|1\rangle_L, \qquad (7)$$

where $0 \le \gamma \le \pi$ and $0 \le \delta \le 2\pi$, we need to solve the equations for α and β,

$$C_{\text{DFS}}\left(\alpha + \frac{1}{2}\beta\right) = \cos\frac{\gamma}{2}, \qquad \frac{\sqrt{3}}{2}C_{\text{DFS}}\beta = \sin\frac{\gamma}{2}e^{i\delta}, \qquad (8)$$

together with Eq. (6). The solutions can be given by

$$\begin{cases} \alpha = \frac{\mu}{2(\mu+1)}e^{i(\delta+\nu)} \\ \beta = \frac{1}{2(\mu+1)}e^{i\delta} \end{cases} (\gamma \neq 0), \qquad \begin{cases} \alpha = \frac{1}{2}, \\ \beta = 0 \end{cases} (\gamma = 0) \qquad (9)$$

where,

$$\mu = \left|\frac{1}{2}\left(\sqrt{3}\cot\frac{\gamma}{2}e^{i\delta} - 1\right)\right|, \qquad \nu = \arg\left[\frac{1}{2}\left(\sqrt{3}\cot\frac{\gamma}{2}e^{i\delta} - 1\right)\right]. \qquad (10)$$

Therefore, if we can realize a tunable partial exchanging device, we can prepare arbitrary DF states from two singlet states. In our paper [11] we proposed a possible way to realize the device probabilistically with linear optics and post-selection, which is feasible with existing experimental technology, so that enables the protection of an arbitrary logical qubit against collective decoherence.

In conclusion, we proposed a way to generate arbitrary four-qubit DF states robust against collective decoherence starting from two singlet states. The key building block is a partial exchanging device, which may have other possible applications.

ACKNOWLEDGMENTS

This work was funded by the National Fundamental Research Program (Grant No. 2006CB921907), the National Natural Science Foundation of China (Grant No. 10774139, No. 10674128, No. 60121503 and No. 60621064), the Innovation Funds from the CAS, BaiRen program of CAS and Doctor Foundation of Education Ministry of China (Grant No. 20060358043), Program for NCET and International cooperation program from CAS, and FANEDD of PR China(Grant No. 200729).

REFERENCES

1. W. H. Zurek, *Phys. Today* **44**, 36–44 (1991).
2. P. W. Shor, *Phys. Rev. A* **52**, R2493–R2496 (1995); A. M. Steane, *Phys. Rev. Lett.* **77**, 793–797 (1996); C. H. Bennett, D. P. DiVincenzo, J. A. Smolin, and W. K. Wootters, *Phys. Rev. A* **54**, 3824–3851 (1996); R. Laflamme, C. Miquel, J. P. Paz, and W. H. Zurek, *Phys. Rev. Lett.* **77**, 198–201 (1996).
3. L. Viola, E. Knill, and S. Lloyd, *Phys. Rev. Lett.* **82**, 2417–2421 (1999).
4. P. Zanardi, and M. Rasetti, *Phys. Rev. Lett.* **79**, 3306–3309 (1997);
5. G. M. Palma, K. Suominen, and A. K. Ekert, *Proc. R. Soc. London, Ser. A* **452**, 567–584 (1996); L.-M. Duan, and G.-C. Guo, *Phys. Rev. Lett.* **79**, 1953–1956 (1997); D. A. Lidar, I. L. Chuang, and K. B. Whaley, *Phys. Rev. Lett.* **81**, 2594–2597 (1998); P. Zanardi, *Phys. Rev. A* **60**, R729–R732 (1999); D. Bacon, D. A. Lidar, and K. B. Whaley, *Phys. Rev. A* **60**, 1944–1955 (1999); D. A. Lidar, D. Bacon, J. Kempe, and K. Birgitta Whaley, *Phys. Rev. A* **61**, 052307-1–5 (2000).
6. J. Kempe, D. Bacon, D. A. Lidar, and K. B. Whaley, *Phys. Rev. A* **63**, 042307-1–29 (2001).
7. S. D. Bartlett, T. Rudolph, and R. W. Spekkens, *Phys. Rev. Lett.* **91**, 027901-1–4 (2003); A. Cabello, *Phys. Rev. A* **68**, 042104-1–6 (2003); A. Cabello, *Phys. Rev. Lett.* **91**, 230403-1–4 (2003); J.-C. Boileau, D. Gottesman, R. Laflamme, D. Poulin, and R. W. Spekkens, *Phys. Rev. Lett.* **92**, 017901-1–4 (2004); T.-Y. Chen, J. Zhang, J.-C. Boileau, X.-M. Jin, B. Yang, Q. Zhang, T. Yang, R. Laflamme, and J.-W. Pan, *Phys. Rev. Lett.* **96**, 150504-1–4 (2006); S. D. Bartlett, T. Rudolph, and R. W. Spekkens, *Rev. Mod. Phys.* **79**, 555–609 (2007).
8. P. G. Kwiat, A. J. Berglund, J. B. Altepeter, and A. G. White, *Science* **290**, 498–501 (2000); J. B. Altepeter, P. G. Hadley, S. M. Wendelken, A. J. Berglund, and P. G. Kwiat, *Phys. Rev. Lett.* **92**, 147901-1–4 (2004); D. Kielpinski, V. Meyer, M. A. Rowe, C. A. Sackett, W. M. Itano, C. Monroe, and D. J. Wineland, *Science* **291**, 1013–1015 (2001); E. M. Fortunato, L. Viola, J. Hodges, G. Teklemariam, and D. G. Cory, *New J. Phys.* **4**, 5.1–5.20 (2002); J. E. Ollerenshaw, D. A. Lidar, and L. E. Kay, *Phys. Rev. Lett.* **91**, 217904-1–4 (2003).
9. M. Bourennane, M. Eibl, S. Gaertner, C. Kurtsiefer, A. Cabello, and H. Weinfurter, *Phys. Rev. Lett.* **92**, 107901-1–4 (2004).
10. X.-B. Zou, J. Shu, and G.-C. Guo, *Phys. Rev. A* **73**, 054301-1–4 (2006).
11. Y.-X. Gong, X.-B. Zou, X.-L. Niu, J. Li, Y.-F. Huang, and G.-C. Guo, *Phys. Rev. A* **77**, 042317-1–5 (2008).

Theoretical constraints on implementations of arbitrary single-qubit gates under conservation laws

Tokishiro Karasawa*, Masanao Ozawa† and Kae Nemoto**

*jidai@nii.ac.jp
†ozawa@is.nagoya-u.ac.jp
**nemoto@nii.ac.jp

Abstract. We investigate theoretical limits on the precision of arbitrary single-qubit operations under conservation laws. When a single-qubit gate is implemented by the interaction between the qubit and the control system, the conservation laws on the total system may induce an inevitable imperfection on the qubit gate. We quantify this imperfection of single-qubit gates, employing the gate trace distance between the desired unitary map and a trace preserving completely positive map implemented with the interaction with the control qubits. We present the bounds of the gate accuracy for arbitrary qubit gates under conservation laws of additive quantities.

Keywords: quantum gate, precision limit, conservation law
PACS: 03.67.Lx, 03.65.Yz, 03.65.Ta

INTRODUCTION

One of the most basic elements in quantum information processing is quantum logic gates on a single-qubit. Quantum logic gates can be represented as a unitary operator on the Hilbert space of the qubit; however when we consider an implementation of such operation, it is necessary to introduce an ancilla system as its control system. Hence a single-qubit operation can be implemented through an interaction between the qubit and the ancilla system. During the interaction, while the qubit could decohere due to the noise induced by the interaction with the environment, the implemented gate operation may as well introduce an inevitable imperfection due to the quantum nature of the interaction, which is called the dynamical noise. Unlike the environmental noise, which can in principle be suppressed to arbitrarily small, the dynamical noise may impose a fundamental limitation on the accuracy of a single-qubit gates.

While the dynamical noise has been studied as the quantum effects of the electromagnetic control fields [1, 2, 3, 4], a general approach to estimate the dynamical noise under conservation laws has been independently established [5, 6] based on the Wigner-Araki-Yanase (WAY) theorem [7, 8]. The WAY theorem states that precise and non-disturbing measurements cannot be performed on observables non-commutative with bounded additively conserved quantities. This general approach has been employed to investigate the controlled-NOT gate utilizing the commutation relation between noise operators and conserved quantity, and a dynamical error bound is given to be inversely proportional to the variance of the conserved quantity [5]. A similar approach has been used to obtain an error bound for the Hadamard gate [6]; also see [9, 10]. These bounds have a relevance

CP1110, *Quantum Communication, Measurement and Computing (QCMC), The Ninth International Conference,* edited by A. Lvovsky

to the fault-tolerant error threshold [11] in a way that they provide an estimate of the size of the control system for the fault-tolerant quantum information processing.

Recently, this general approach has been extended to investigate error bounds for quantum-NOT gate using the gate trace distance between the desired unitary map and the completely positive map of the implemented gate operations [12]. Evaluating the lower bound of the gate trance distance, we can obtain error bounds, which could be tighter than ones expected from the previous estimates based on the commutation relation. In this paper, we employ this approach with the gate trace distance to obtain error bounds for arbitrary single-qubit operations.

FORMULATION

A qubit **S** of a spin-1/2 system is supposed to interact with the ancilla system **A** to implement single-qubit gates. We assume that the ancilla system is a multi-qubit system and the composite system $\mathbf{S}+\mathbf{A}$ does not interact with its environment, that is, $\mathbf{S}+\mathbf{A}$ can be considered as an isolated system. It is natural to assume the conservation law of the total angular momentum in the control interaction between **S** and **A**. Thus a possible unitary time evolution operator U of $\mathbf{S}+\mathbf{A}$ satisfies the condition,

$$[U, Z_{\mathbf{S}} \otimes I_{\mathbf{A}} + I_{\mathbf{S}} \otimes Z_{\mathbf{A}}] = 0. \tag{1}$$

Here $Z_{\mathbf{S}} = |0\rangle\langle 0| - |1\rangle\langle 1|$ with a computational basis $\{|0\rangle, |1\rangle\}$, and $Z_{\mathbf{A}}$ is the total angular momentum of the ancilla system, which is given by the total sum of the angular momentum of the spin-1/2 systems in the ancilla system. $I_{\mathbf{S}}$ and $I_{\mathbf{A}}$ represent the identity operators on the qubit and the ancilla system, respectively.

Suppose that the desired gate is an arbitrary unitary operator $U_{\mathbf{S}}$. The output state of the physical implementation of the gate is described by a trace preserving completely positive (TPCP) map $\mathscr{E}_{U,\rho_{\mathbf{A}}}$ defined as

$$\mathscr{E}_{U,\rho_{\mathbf{A}}}(\rho_{\mathbf{S}}) := \mathrm{Tr}_{\mathbf{A}}[U(\rho_{\mathbf{S}} \otimes \rho_{\mathbf{A}})U^{\dagger}], \tag{2}$$

where $\rho_{\mathbf{S}}$ and $\rho_{\mathbf{A}}$ are initial states of **S** and **A**, respectively. $\mathrm{Tr}_{\mathbf{A}}$ stands for the partial trace over the Hilbert space of the ancilla system. If one could perfectly implement the desired gate $U_{\mathbf{S}}$ by the physical implementation, the output state would be given by a unitary map $\mathscr{E}_{U_{\mathbf{S}}}$ defined as

$$\mathscr{E}_{U_{\mathbf{S}}}(\rho_{\mathbf{S}}) := U_{\mathbf{S}}\rho_{\mathbf{S}}U_{\mathbf{S}}^{\dagger}. \tag{3}$$

In order to evaluate the precision limit of the physical implementation of the gate $U_{\mathbf{S}}$, we quantify the difference between the two maps, $\mathscr{E}_{U_{\mathbf{S}}}$ and $\mathscr{E}_{U,\rho_{\mathbf{A}}}$, by the gate trace distance. As the TPCP map $\mathscr{E}_{U,\rho_{\mathbf{A}}}$ is completely characterized by the time evolution operator and the initial state of the ancilla system, one can reduce the difference between the two maps by appropriately choosing U and $\rho_{\mathbf{A}}$. The optimal choice gives the precision limit of the gate implementation. Thus the precision limit is defined by the gate trace distance minimized over all the time evolution operators and the initial states of the ancilla system,

$$\min_{U,\rho_{\mathrm{A}}} \mathscr{D}(\mathscr{E}_{U_{\mathrm{S}}}, \mathscr{E}_{U,\rho_{\mathrm{A}}}). \tag{4}$$

Here $\mathscr{D}(\cdot,\cdot)$ stands for the gate trace distance defined by the output trace distance maximized over all the input states, namely,

$$\mathscr{D}(\mathscr{E}_{U_{\mathrm{S}}}, \mathscr{E}_{U,\rho_{\mathrm{A}}}) := \max_{\rho_{\mathrm{S}}} \frac{1}{2} \mathrm{Tr} \left| \mathscr{E}_{U_{\mathrm{S}}}(\rho_{\mathrm{S}}) - \mathscr{E}_{U,\rho_{\mathrm{A}}}(\rho_{\mathrm{S}}) \right|. \tag{5}$$

RESULTS

We have obtained a lower bound of the minimized gate trace distance,

$$\min_{U,\rho_{\mathrm{A}}} \mathscr{D}(\mathscr{E}_{U_{\mathrm{S}}}, \mathscr{E}_{U,\rho_{\mathrm{A}}}) \geq \frac{1}{2} \left(\sin^2 \frac{\zeta}{2} - \cos \frac{\pi}{N+2} \right) \tag{6}$$

for any ζ with $\cos\zeta \leq -\cos(\pi/N+2)$, and

$$\min_{U,\rho_{\mathrm{A}}} \mathscr{D}(\mathscr{E}_{U_{\mathrm{S}}}, \mathscr{E}_{U,\rho_{\mathrm{A}}}) \geq \frac{1}{2} \frac{1-\cos\frac{\pi}{N+2}}{1+\cos\frac{\pi}{N+2}} \sin^2 \frac{\zeta}{2} \tag{7}$$

for any ζ with $\cos\zeta \geq -\cos(\pi/N+2)$ [13]. Here N is the total number of qubits in the ancilla system, and ζ is the amount of a rotation in a decomposition of U_{S},

$$U_{\mathrm{S}} = \exp[i\gamma] \exp\left[\frac{i\delta Z_{\mathrm{S}}}{2}\right] \exp\left[\frac{i\zeta Y_{\mathrm{S}}}{2}\right] \exp\left[\frac{i\xi Z_{\mathrm{S}}}{2}\right], \tag{8}$$

namely, the rotation angle along the Y-axis. Any unitary operator can be decomposed as Eq.(8). Here $X_{\mathrm{S}} = |0\rangle\langle 1| + |1\rangle\langle 0|$, $Y_{\mathrm{S}} = -i(|0\rangle\langle 1| - |1\rangle\langle 0|)$.

If $\zeta = 0$, the lower bound becomes zero. This is expected because in this case, the desired gate commutes with the conserved quantity. We have not proved the optimality of this lower bound. Yet, for the quantum-NOT gate the optimality has been approximately verified for large ancilla systems [12].

There are useful applications of this bound. One of them is the Hadamard gate which can be decomposed into $H_{\mathrm{S}} = i\exp[iY_{\mathrm{S}}\pi/4]\exp[iZ_{\mathrm{S}}\pi/2]$. (The amount of the rotation along the Y-axis is $\pi/2$.) The general result (7) gives

$$\begin{aligned} \min_{U,\rho_{\mathrm{A}}} \mathscr{D}(\mathscr{E}_{H_{\mathrm{S}}}, \mathscr{E}_{U,\rho_{\mathrm{A}}}) &\geq \frac{1}{4} \frac{1-\cos\frac{\pi}{N+2}}{1+\cos\frac{\pi}{N+2}} \\ &\approx \frac{\pi^2}{16N^2}, \end{aligned} \tag{9}$$

where the approximation holds for a large N. One can calculate the necessary size of the ancilla system to reduce the precision limit below the fault-tolerant error threshold. For example, given an error threshold of 10^{-4}, we require at least 79 qubits to implement the Hadamard gate.

We assumed the total angular momentum as the conserved quantity. However the above argument follows in exactly the same manner for any additively conserved quantities satisfying condition (1). Here the additivity means that the conserved quantity of a composite system is given by the sum of the conserved quantities of the subsystems. One of interesting generalizations of our approach is the case where a control-Hamiltonian satisfies the additive conservation law. A typical and important example of this case is the Jaynes-Cummings Hamiltonian [14], allowing a detuning between the atom and the field,

$$H = \hbar\omega I_{\mathbf{S}} \otimes a^{\dagger}a + \frac{\hbar\omega}{2}\sigma_z \otimes I_{\mathbf{A}} + \hbar g(|e\rangle\langle g| \otimes a + |g\rangle\langle e| \otimes a^{\dagger}). \tag{10}$$

This Hamiltonian has the additively conserved quantity $I_{\mathbf{S}} \otimes a^{\dagger}a + \frac{1}{2}\sigma_z \otimes I_{\mathbf{A}}$. Here a is the annihilation operator of a single mode of the field, and $\sigma_z = |e\rangle\langle e| - |g\rangle\langle g|$ with the excited state $|e\rangle$ and the ground state $|g\rangle$ of the atom qubit. In Ref. [9] the precision limit for the Hadamard gate induced by the above conservation law has been compared with the noise calculated for the above Hamiltonian (10), and concluded to be consistent with the noise induced by the phase fluctuation of the control field.

ACKNOWLEDGMENTS

T.K. acknowledges the Grant-in-Aid for Young Scientists (B). This work was partially supported by the NICT project. M.O.'s research has been supported by the SCOPE project of the MIC and the CREST project of the JST.

REFERENCES

1. J. P. Barnes and W. S. Warren, Phys. Rev. A **60**, 4363 (1999).
2. J. Gea-Banacloche, Phys. Rev. A **65**, 022308 (2002).
3. S. J. van Enk and H. J. Kimble, Quantum Inf. Comput. **2**, 1 (2002).
4. A. Silberfarb and I. H. Deutsch, Phys. Rev. A **69**, 042308 (2004).
5. M. Ozawa, Phys. Rev. Lett. **89,** 057902 (2002).
6. M. Ozawa, Int. J. Quant. Inf. **1**, 569 (2003).
7. E. P. Wigner, Z. Phys. **133**, 101 (1952).
8. H. Araki and M. Yanase, Phys. Rev.**120**, 622 (1960).
9. J. Gea-Banacloche and M. Ozawa, J. Opt. B: Quantum Semiclassical Opt. **7**, S326 (2005).
10. J. Gea-Banacloche and M. Ozawa, Phys. Rev. A **74**, 060301 (2006).
11. M. Nielsen and I. Chuang, *Quantum Computation and Quantum Information* (Cambridge University Press, 2000).
12. T. Karasawa and M. Ozawa, Phys. Rev. A **75**, 032324 (2007).
13. T. Karasawa, M. Ozawa, and K. Nemoto, in preparation.
14. E. T. Jaynes and F. W. Cumming, Proc. IEEE, **51**, 89 (1963).

CHAPTER 14

QUANTUM METROLOGY AND IMAGING

Ghost imaging: from quantum to classical to computational

Jeffrey H. Shapiro* and Baris I. Erkmen*†

*Research Laboratory of Electronics, Massachusetts Institute of Technology
Cambridge, Massachusetts 02139 USA
†Now with the Jet Propulsion Laboratory, Pasadena, California 91109, USA

Abstract. Ghost-imaging experiments correlate the outputs from two photodetectors: a high spatial-resolution (scanning pinhole or CCD camera) detector that measures a field which has not interacted with the object to be imaged, and a bucket (single-pixel) detector that collects a field that has interacted with the object. We review the Gaussian-state framework for ghost imaging setups that rely on biphoton states or pseudothermal states. The understanding derived therefrom leads to two new ghost imaging configurations, employing spatial light modulators to impose controlled spatial incoherence at the source, one of which uses only a single-pixel detector.

INTRODUCTION

Ghost imaging is the acquisition of object information by means of photocurrent correlation measurements. Its first demonstration utilized a biphoton source, thus the image was interpreted as a quantum phenomenon owing to the entanglement of the source photons [1]. Experimental [2],[3] and theoretical [4],[5] work later demonstrated that ghost imaging could be performed with pseudothermal light. Whereas the biphoton requires a quantum description for its photodetection statistics, pseudothermal light can be regarded as a classical electromagnetic wave whose photodetection statistics can be treated via the semiclassical (shot-noise) theory of photodetection [6]. This disparity has sparked interest in the physics of ghost imaging [7]–[10]. Recently [11], we established a Gaussian-state analysis of ghost imaging that unified prior work on biphoton and pseudothermal sources. Our analysis, which we will review below, indicated that ghost-image formation is intrinsically due to classical coherence propagation, with the principal advantage afforded by the biphoton state being high-contrast imagery in the wideband limit. Other recent work [12],[13], however, has offered an alternative explanation for pseudothermal-light ghost imaging: nonlocal two-photon quantum interference. Building from our Gaussian-framework treatment, we will show [14] that ghost imaging can be accomplished with only *one* detector, viz., the bucket detector that collects a single pixel of light which has interacted with the object. As only one light beam and one photodetector are required, this imaging configuration cannot depend on nonlocal two-photon interference. Moreover, it affords background-free imagery in the narrowband limit and a 3D sectioning capability.

CP1110, *Quantum Communication, Measurement and Computing (QCMC), The Ninth International Conference,* edited by A. Lvovsky

GAUSSIAN-STATE GHOST IMAGING

We begin with a review of biphoton and pseudothermal-light lensless ghost imaging, based on [11]. In the biphoton case, shown in Fig. 1, a continuous-wave (cw), type-II phase matched spontaneous parametric downconverter operating at frequency degeneracy (wavelength λ_0, center frequency ω_0) emits a signal $\hat{E}_S(\rho,t)e^{-i\omega_0 t}$ and a reference $\hat{E}_R(\rho,t)e^{-i\omega_0 t}$, that are scalar, positive frequency, paraxial field operators normalized to have units $\sqrt{\text{photons/m}^2\text{s}}$. Here, ρ is the transverse coordinate with respect to each one's optical axis. Both beams undergo paraxial diffraction along their respective optical axes, over an L-m-long free-space path, yielding detection-plane field operators $\hat{E}_1(\rho,t)$ and $\hat{E}_2(\rho,t)$, respectively. At the detection planes, $\hat{E}_1(\rho,t)$ illuminates a quantum-limited pinhole photodetector of area A_1 whose photosensitive region $\rho \in \mathscr{A}_1$ is centered at the transverse coordinate ρ_1, while $\hat{E}_2(\rho,t)$, illuminates an amplitude-transmission mask $T(\rho)$ located immediately in front of a quantum-limited bucket photodetector with photosensitive region $\rho \in \mathscr{A}_2$. The photodetectors are assumed to have identical sub-unity quantum efficiencies and finite electrical bandwidths, but no dark current or thermal noise. The ghost image at transverse location ρ_1 is formed by time-averaging the product of the detector photocurrents to obtain an estimate of the ensemble-average equal-time photocurrent cross-correlation function $C(\rho_1)$ as the pinhole detector is scanned over the plane.[1]

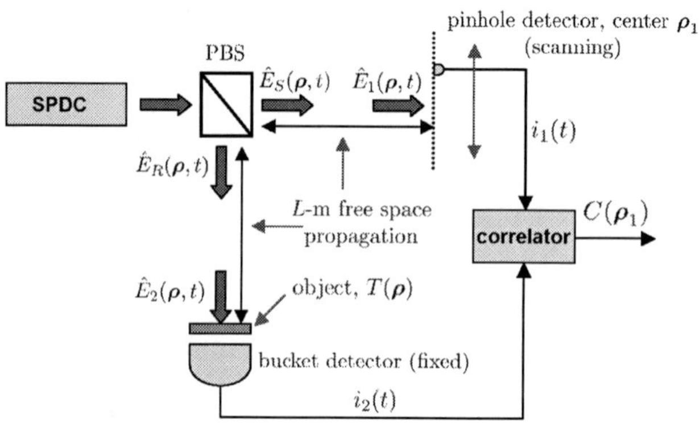

FIGURE 1. Configuration for biphoton lensless ghost imaging. SPDC: spontaneous parametric downconverter. PBS: polarizing beam splitter.

The SPDC source in Fig. 1 produces a jointly Gaussian quantum state that, in its typical low-gain low-flux operating regime, is well approximated within a photodetection coincidence window by a predominant vacuum component plus a weak biphoton component [15]. By assuming a cross-spectrally pure, Gaussian-Schell model for the correlation functions of the SPDC output's jointly Gaussian state, we have obtained the following result for the ghost image cross-correlation function in the far field of that

[1] Our photodetection cross-correlation model is equivalent to photon coincidence counting in the operating regime of interest for the Fig. 1 setup.

spatially incoherent source [11]:

$$C(\rho_1) = q^2\eta^2 A_1 \left(\frac{2P}{\pi a_L^2}\right)^2 \left[\int_{\mathscr{A}_2} d\rho \, |T(\rho))|^2 + \frac{a_0^2}{\sqrt{8\pi}\, PT_0\rho_0^2} \int_{\mathscr{A}_2} d\rho \, e^{-|\rho_1+\rho|^2/\rho_L^2} |T(\rho)|^2\right]. \tag{1}$$

Here: q is the electron charge; η is the detector quantum efficiency; P is the source's photon flux; a_0 and ρ_0 are the source's intensity radius and coherence radius; $a_L = \lambda_0 L/\pi\rho_0$ and $\rho_L = \lambda_0 L/\pi a_0$ are the resulting intensity radius and coherence radius after L m of free-space propagation; and T_0 is the source's coherence time. Equation (1), which assumes that the object transparency lies well within the $\sim a_L$-radius field of view, shows that $C(\rho_1)$ contains an inverted image (second term) of the object's intensity transmission $|T(\rho)|^2$ that has been degraded by convolution with a Gaussian point-spread function of radius ρ_L and embedded in a featureless background (first term). Because $PT_0\rho_0^2/a_0^2 \ll 1$ prevails in the biphoton regime, the background term makes a negligible contribution to $C(\rho_1)$.

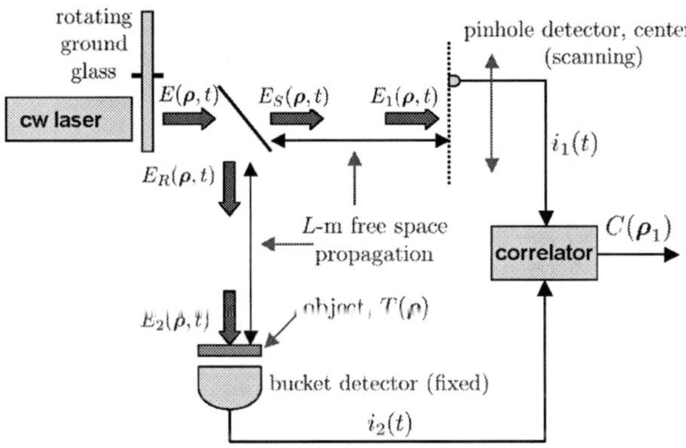

FIGURE 2. Configuration for pseudothermal lensless ghost imaging.

The pseudothermal configuration for lensless ghost imaging, shown in Fig. 2, can be analyzed using classical electromagnetism and shot-noise theory. Here, $E_S(\rho,t)e^{-i\omega_0 t}$ and $E_R(\rho,t)e^{-i\omega_0 t}$ are scalar, positive frequency, classical signal and reference fields that are normalized to photon-units and have center frequency ω_0. They are the outputs from 50/50 beam splitting of $E(\rho,t)$, a cw laser beam that has been transmitted through a rotating ground-glass diffuser. The signal and reference undergo paraxial diffraction over L-m-long free-space paths, yielding measurement-plane fields $E_1(\rho,t)$ and $E_2(\rho,t)$, respectively. The field $E_1(\rho,t)$ illuminates a shot-noise limited pinhole photodetector centered at ρ_1 with sensitive region $\rho \in \mathscr{A}_1$. The field $E_2(\rho,t)$ illuminates an amplitude-transmission mask $T(\rho)$, located immediately in front of a shot-noise limited bucket photodetector with sensitive region $\rho \in \mathscr{A}_2$. The product of the photocurrents from these detectors is time averaged to estimate their ensemble-average cross correlation, $C(\rho_1)$, as the pinhole detector is scanned over the plane. Because pseudothermal light is well modeled as a classical Gaussian random process, we can parallel the development for the biphoton case and show [11],[14] that the the ghost image cross-correlation function in the far field of that spatially incoherent source is given by:

$$C(\rho_1) = q^2\eta^2 A_1 \left(\frac{2P}{\pi a_L^2}\right)^2 \left[\int_{\mathscr{A}_2} d\rho\, |T(\rho))|^2 + \int_{\mathscr{A}_2} d\rho\, e^{-|\rho_1-\rho|^2/\rho_L{}^2} |T(\rho)|^2\right]. \quad (2)$$

Equation (2) shows that the pseudothermal ghost image is erect, with the same spatial resolution as the biphoton ghost image. But the pseudothermal ghost image has a background term that cannot be neglected. Although this background term may be suppressed by employing DC blocks in the paths from the photodetectors to the correlator in Fig. 2, the shot noise and excess noise contributed by the background term affect the resulting signal-to-noise ratio in using a time-average measurement to estimate the desired ensemble-average image [16].

NEW CONFIGURATIONS FOR GHOST IMAGING

With the preceding results in hand, it becomes a simple matter to identify two new configurations for lensless ghost imaging. First, rather than use laser light transmitted through a rotating ground glass as the source of a narrowband, spatially-incoherent $E(\rho,t)$, let us employ the configuration shown in Fig. 3. Here, we transmit a cw laser beam through a spatial light modulator (SLM) whose inputs are chosen to create the desired coherence behavior. Specifically, we assume an idealized SLM consisting of $d \times d$ pixels arranged in a $(2M+1)\times(2M+1)$ array with 100% fill factor within a $D \times D$ opaque pupil, where $D = (2M+1)d$ and $M \gg 1$. We use this SLM to impose a phase $\phi_{nm}(t)$ on the light transmitted through pixel (n,m), with $\{e^{i\phi_{nm}(t)} : -M \leq n,m \leq M\}$ being independent identically-distributed random processes obeying $\langle e^{i\phi_{nm}(t)}\rangle = 0$ and $\langle e^{i([\phi_{nm}(t_2)-\phi_{jk}(t_1)])}\rangle = \delta_{jn}\delta_{km}e^{-|t_2-t_1|/T_0}$, where the coherence time T_0 is long compared to the response times of the photodetectors.

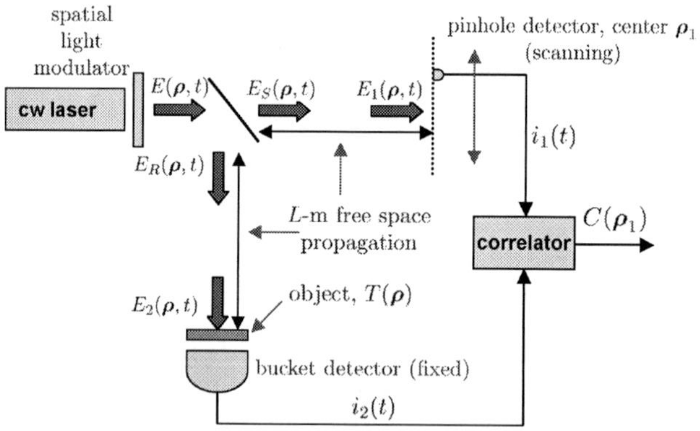

FIGURE 3. Configuration for spatial light modulator lensless ghost imaging.

In this source's far field, the preceding $E(\rho,t)$ leads to $E_1(\rho,t)$ and $E_2(\rho,t)$ that, although not of Gaussian-Schell form, have an intensity width $\sim\lambda_0 L/d$ and a coherence length $\sim\lambda_0 L/D$, behavior which is qualitatively similar to what we saw earlier if we identify $d \approx \rho_0$ and $D \approx a_0$. Furthermore, Central Limit Theorem considerations imply that $E_1(\rho,t)$ and $E_2(\rho,t)$ may be taken to be jointly Gaussian. Hence our Fig. 3 configuration will produce a ghost image of spatial resolution $\lambda_0 L/D$ within a spatial region of width $\lambda_0 L/d$ embedded in a background that can be suppressed with a DC block.

To realize the Fig. 3 ghost imager we could use noise generators to drive the SLM , but it is more interesting, for what will follow, to suppose that deterministic driving functions are employed. What we want at the SLM's output is a narrowband, zero-mean field whose spatial coherence—inferred now from a time average, rather than an ensemble average, because there is no true randomness—is limited to field points within a single pixel. Strong sinusoidal modulation, $\phi_{nm}(t) = \Phi\cos[(\Omega_0 + \Delta\Omega_{nm})t]$, with different $\Delta\Omega_{nm}$ for each pixel will work. This deterministically-modulated source will also yield a ghost image of spatial resolution $\lambda_0 L/D$ within a spatial region of width $\lambda_0 L/d$ embedded in a background that can be suppressed by means of a DC block.

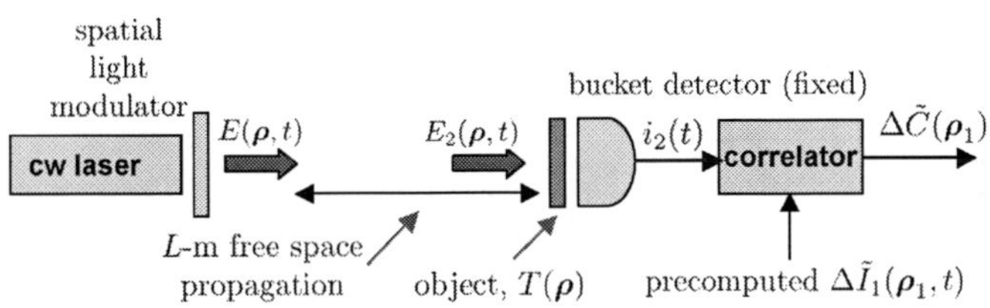

FIGURE 4. Configuration for computational lensless ghost imaging.

At this point, the notion of computational ghost imaging—in which we only use the bucket detector—is easily understood, see Fig. 4. We use deterministic modulation of a cw laser beam to create the field $E_2(\rho,t)$ that illuminates the object transparency, and, as usual, we collect the light that is transmitted through the transparency with a bucket (single-pixel) detector. Knowing the deterministic modulation applied to the original cw laser beam allows us to use diffraction theory to *compute* the intensity pattern, $I_1(\rho_1,t)$, that would have illuminated the pinhole detector in the usual lensless ghost imaging configuration. Furthermore, we can subtract the time average of this intensity, in our computation, and obtain the equivalent of the zero-mean intensity variation, $\Delta I_1(\rho_1,t)$. To distinguish these computed values from actual light-field quantities, we will denote them $\tilde{I}_1(\rho_1,t)$ and $\Delta\tilde{I}_1(\rho_1,t)$, respectively. The time average correlation function, $\Delta\tilde{C}(\rho_1)$, will then be a background-free ghost image—with spatial resolution $\lambda_0 L/D$ over a spatial extent of width $\lambda_0 L/d$—akin to what would be obtained from pseudothermal ghost imaging with $d \approx \rho_0$, $D \approx a_0$, and a DC block applied to the pinhole detector. Now, because only one photodetector has been employed, it is impossible to interpret this computational ghost image as arising from nonlocal two-photon interference.

In addition to obviating the need for a high spatial-resolution detector in ghost-image formation, at the expense of the computational burden associated with the free-space propagation calculation for the field $E(\rho,t)$, computational ghost imaging permits 3D sectioning to be performed. To see that this is so, we return to the Fig. 2 ghost-imaging configuration with pseudothermal light and inquire about its depth of focus,. In other words, how badly is the ghost image blurred if the object is at $z = L$ but the pinhole detector is at $z = L + \Delta L$? This question is easily answered [14]: the ghost-image term from Eq. (2) takes the same form with ρ_L replaced by $\rho_L' \equiv \rho_L\sqrt{1+(\Delta L/k_0\rho_L^2)^2}$ with $k_0 = 2\pi/\lambda_0$, so that the focal region is $|\Delta L| \leq k_0\rho_L^2 = 4L^2/k_0a_0^2$. In the near-field of the pre-diffuser laser beam, the focal region is a very small fraction of the source-to-object path, as reported for the experiments in [12]. This limited depth of focus has the following implications when a range-spread opaque object is imaged in reflection. The pseudothermal ghost imager can only image one focal region at a time. However, because the computational ghost imager can precompute $\Delta\tilde{I}_1(\rho_1, t)$ for a wide range of propagation distances, the same bucket-detector photocurrent can be correlated with many such $\Delta\tilde{I}_1(\rho_1, t)$ to perform 3D sectioning of the object's reflectance. Of course, this sectioning further increases the computational burden, but this burden can be handled off-line, and, for a given SLM and its associated modulation waveforms, the *same* precomputed $\Delta\tilde{I}_1(\rho_1, t)$ can be used for *all* ghost images formed using that system.

CONCLUSIONS

We have reviewed the Gaussian-state theory of biphoton and pseudothermal ghost imaging, and extended it to spatial light modulator and computational ghost imaging.

ACKNOWLEDGMENTS

This work was supported by the U. S. Army Research Office MURI Grant W911NF-05-1-0197, the DARPA Quantum Sensors Program, and the W. M. Keck Foundation for Extreme Quantum Information Theory.

REFERENCES

1. T. B. Pittman, Y. H. Shih, D. V. Strekalov, and A. V. Sergienko, Phys. Rev. A **52,** R3429 (1995).
2. A. Valencia, G. Scarcelli, M. D'Angelo, and Y. Shih, Phys. Rev. Lett. **94,** 063601 (2005).
3. F. Ferri, D. Magatti, A. Gatti, M. Bache, E. Brambilla, and L. A. Lugiato, Phys. Rev. Lett. **94,** 183602 (2005).
4. A. Gatti, E. Brambilla, M. Bache, and L. A. Lugiato, Phys. Rev. A **70,** 013802 (2004).
5. A. Gatti, E. Brambilla, M. Bache, and L. A. Lugiato, Phys. Rev. Lett. **93,** 093602 (2004).
6. L. Mandel and E. Wolf, *Optical Coherence and Quantum Optics* (Cambridge Univ., Cambridge, 1995), chapters. 4, 9, 12.
7. M. D'Angelo, A. Valencia, M. H. Rubin, and Y. Shih, Phys. Rev. A **72,** 013810 (2005).
8. Y. Cai and S.-Y. Zhu, Opt. Lett. **29,** 2716 (2004).
9. Y. Cai and S.-Y. Zhu, Phys. Rev. E **71,** 056607 (2005).
10. G. Scarcelli, V. Berardi, and Y. Shih, Phys. Rev. Lett. **96,** 063602 (2006).
11. B. I. Erkmen and J. H. Shapiro, Phys. Rev. A **77,** 043809 (2008).
12. R. Meyers, K. S. Deacon, and Y. Shih, Phys. Rev. A **77,** 041801(R) (2008).
13. Y. Shih, arXiv:0805.1166 [quant-ph].
14. J. H. Shapiro, arXiv:0807.2614 [quant-ph].
15. F. N. C. Wong, T. H. Kim, and J. H. Shapiro, Laser Physics **16,** 1517 (2006).
16. B. I. Erkmen and J. H. Shapiro, arXiv:0809.4167 [quant-ph].

Quantum metrology with Bose-Einstein condensates

Sergio Boixo*, Animesh Datta†,**, Matthew J. Davis‡, Steven T. Flammia§, Anil Shaji¶, Alexandre B. Tacla¶ and Carlton M. Caves¶

*Institute for Quantum Information, Caltech, Pasadena, California, 91125, USA
†Institute for Mathematical Sciences, 53 Prince's Gate, Imperial College, London, SW7 2PG, UK
**QOLS, The Blackett Laboratory, Imperial College, London, Prince Consort Road, SW7 2BW, UK
‡School of Physical Sciences, University of Queensland, Brisbane, Queensland 4072, Australia
§Perimeter Institute for Theoretical Physics, 31 Caroline Street N., Waterloo, ON N2L 2Y5, Canada
¶Department of Physics and Astronomy, University of New Mexico, Albuquerque, NM, 87131, USA

Abstract. We show how a generalized quantum metrology protocol can be implemented in a two-mode Bose-Einstein condensate of n atoms, achieving a sensitivity that scales better than $1/n$ and approaches $1/n^{3/2}$ for appropriate design of the condensate.

Keywords: quantum metrology, nonlinear interferometry, Bose-Einstein condensate
PACS: 03.65.Ta, 03.75.Nt, 03.65.-w, 03.75.Mn

In a separate paper elsewhere in this volume [1], we showed that two-body couplings between the n qubits that make up the quantum probe in a parameter-estimation scheme can lead to measurement sensitivities that scale as $1/n^{3/2}$ even when the initial state of the probe is unentangled. A Bose-Einstein condensate (BEC) is a physical system in which effective two-body couplings exist between all the atoms in the condensate. We examine in this contribution how a BEC of n atoms can be turned into a quantum probe to measure a parameter with a sensitivity scaling that approaches $1/n^{3/2}$.

The many-body Hamiltonian, in second-quantized notation, for a dilute Bose gas in which the inter-particle spacing is much larger than the scattering length, a, is [2, 3, 4, 5]

$$H = \int d\boldsymbol{r} \left(-\frac{\hbar^2}{2m} \nabla \hat{\psi}^\dagger \cdot \nabla \hat{\psi} + V(\boldsymbol{r}) \hat{\psi}^\dagger \hat{\psi} + \frac{g}{2} \hat{\psi}^\dagger \hat{\psi}^\dagger \hat{\psi} \hat{\psi} \right), \tag{1}$$

where $\hat{\psi}^\dagger$ and ψ are creation and annihilation field operators that obey bosonic commutation relations. The gas is in a trapping potential $V(\boldsymbol{r})$ and the coupling constant $g = 4\pi\hbar^2 a/m$. In a BEC at zero temperature almost all the atoms are in the ground state; we can write $\hat{\psi}^\dagger(\boldsymbol{r}) \simeq \psi_n(\boldsymbol{r})\hat{a}^\dagger$, where $\psi_n(\boldsymbol{r})$ is the single-particle ground-state wave function and $\hat{a}^\dagger$ is operator that creates atoms with this wave function.

So far we have assumed that all the atoms in the BEC are in the same atomic state, but to use the atoms as components of a quantum probe, we want them to be two-level systems or qubits. We thus consider two-mode BECs in which the atoms can occupy one of two internal states, labeled $|1\rangle$ and $|2\rangle$. These two states are typically hyperfine levels of the atoms. In practice, the atoms are cooled to form the BEC while they are all in the same internal state, and then an external field is used to drive transitions between the two levels to achieve the desired coherent superposition of the two levels. We assume that

CP1110, *Quantum Communication, Measurement and Computing (QCMC), The Ninth International Conference,* edited by A. Lvovsky

the collisions between the atoms are elastic. We also assume that the the second internal state is chosen so that it sees the same trapping potential $V(\boldsymbol{r})$; therefore, for short times, the wave functions for atoms in both states $|1\rangle$ and $|2\rangle$ can be approximated by the same wave function $\psi_n(\boldsymbol{r})$. Using these assumptions can write down the Hamiltonian for the two-mode BEC as

$$\begin{aligned} H &= \int d\boldsymbol{r}\Big[-\frac{\hbar^2}{2m}|\nabla\psi_n(\boldsymbol{r})|^2(\hat{a}_1^\dagger\hat{a}_1+\hat{a}_2^\dagger\hat{a}_2)+V(\boldsymbol{r})|\psi_n(\boldsymbol{r})|^2(\hat{a}_1^\dagger\hat{a}_1+\hat{a}_2^\dagger\hat{a}_2) \\ &+\frac{g_{11}}{2}|\psi_n(\boldsymbol{r})|^4\hat{a}_1^\dagger\hat{a}_1^\dagger\hat{a}_1\hat{a}_1+\frac{g_{22}}{2}|\psi_n(\boldsymbol{r})|^4\hat{a}_2^\dagger\hat{a}_2^\dagger\hat{a}_2\hat{a}_2+g_{12}|\psi_n(\boldsymbol{r})|^4\hat{a}_1^\dagger\hat{a}_2^\dagger\hat{a}_1\hat{a}_2\Big]\,. \end{aligned} \quad (2)$$

To make the connection to the generalized quantum metrology protocols described in [1, 6, 7], we define the following two operators, $\hat{n}\equiv\hat{a}_1^\dagger\hat{a}_1+\hat{a}_2^\dagger\hat{a}_2$ and $\hat{J}_z\equiv(\hat{a}_1^\dagger\hat{a}_1-\hat{a}_2^\dagger\hat{a}_2)/2$. The operator $\hat{n}$ is simply the number operator; it gives the total number of atoms in the BEC, which is the relevant metrological resource. Treating $\hat{n}$ as a c-number, the Hamiltonian becomes

$$H=H_0+\gamma_1 n\hat{J}_z\eta+\gamma_2\hat{J}_z^2\eta\ , \quad (3)$$

where $(g_{11}-g_{22})/2\equiv\gamma_1$, $(g_{11}+g_{22})/2-g_{12}\equiv\gamma_2$ and $\eta=\int d\boldsymbol{r}|\psi_n(\boldsymbol{r})|^4$.

The first term, H_0, of the Hamiltonian (3) can be ignored, because it acts identically on atoms in states $|1\rangle$ and $|2\rangle$. Since the initial states of the probe that we consider are of the form $(\alpha|1\rangle+\sqrt{1-\alpha^2}|2\rangle)^{\otimes n}$, the effect of H_0 is only to produce an overall phase in the time evolved state of the probe. In the next two terms, we have both nJ_z and J_z^2 couplings that suggest that we might be able to measure the constants γ_1 and γ_2 with an accuracy that scales as $1/n^{3/2}$ [1, 6, 7]. This assumes that the factor η has no dependence on n. In fact, η is inversely proportional to the effective volume occupied by the ground-state wave function. Adding more atoms to a harmonically trapped BEC spreads out the wave function because of the repulsive scattering, thereby reducing η as n increases. To pin down the exact scaling of the measurement accuracy with n, we need to know how η behaves as a function of n.

Scaling of the measurement uncertainty

Since we first create a BEC of n atoms all in hyperfine state $|1\rangle$, before putting them in a superposition of states $|1\rangle$ and $|2\rangle$, we can focus on the dependence of η on n for a single-mode BEC of atoms in state $|1\rangle$. An obvious strategy to suppress the n dependence of η is to constrain the BEC within a hard-walled trap so that it cannot expand as more atoms are added. BECs effectively confined to two or one dimensions and held in power-law trapping potentials along the effective dimensions are the sort that are found in real experiments. Thus we look at the dependence of η on n for a BEC in d dimensions, referred to as *longitudinal* (L) dimensions, and held in a trap along those dimensions of the form $V_L(\boldsymbol{r})=\frac{1}{2}kr^q$, where q is a positive integer. There are $D=3-d$ tightly confined, *transverse* (T) dimensions along which we take the trapping potential to be harmonic, i.e., $V_T(\boldsymbol{\rho})=\frac{1}{2}m\omega_T^2\rho^2$. The bare, one-atom ground-state half-widths along the longitudinal and transverse dimensions are denoted by $r_0 \equiv (\hbar^2/mk)^{1/(q+2)}$ and $\rho_0 \equiv (\hbar/2m\omega_T)^{1/2}$, where r_0 is an approximate expression that scales properly with the atomic and trap parameters.

We define two critical atom numbers, n_L and n_T, as the number of atoms in the BEC that makes the scattering energy comparable to the longitudinal and transverse kinetic energies:

$$n_L \equiv \frac{r_0}{a}\left(\frac{\rho_0}{r_0}\right)^D, \qquad n_T = \frac{\rho_0}{a}\left(\frac{r_0}{\rho_0}\right)^{d(q+2)/q}. \tag{4}$$

Here a is the scattering length of the atoms in state $|1\rangle$.

When $n \ll n_L$, the scattering energy can be ignored, and the ground-state wave function is simply the ground-state solution for the trapping potential $V_L(\boldsymbol{r}) + V_T(\boldsymbol{\rho})$, i.e., a product wave function independent of n, $\psi_n(\boldsymbol{r},\boldsymbol{\rho}) = \phi(\boldsymbol{r})\chi(\boldsymbol{\rho})$. In this case η is a constant and the uncertainty in an estimate of γ_1 or γ_2 scales as $1/n^{3/2}$.

We are mainly interested in the regime of atom numbers satisfying $n_L \ll n \ll n_T$. For a 1D trap, typical trap parameters can make $n_L \sim$ 1–10, so it only takes a few atoms to make the scattering energy important along the longitudinal direction. As long as $n \ll n_T$, however, we can approximate the ground-state wave function to be a product, $\psi_n(\boldsymbol{r},\boldsymbol{\rho}) = \phi_n(\boldsymbol{r})\chi(\boldsymbol{\rho})$, where $\chi(\boldsymbol{\rho})$ is the ground-state, Gaussian wave function for the harmonic transverse trapping potential, and $\phi_n(\boldsymbol{r})$ is the ground-state solution for a d-dimensional, time-independent Gross-Pitaevskii equation that includes the longitudinal trapping potential $V_L(\boldsymbol{r})$ and the scattering term. For a product wave function, we have $\eta = \eta_L\eta_T$, where $\eta_T = \int d\boldsymbol{\rho}|\chi(\boldsymbol{\rho})|^4 = 1/(2\sqrt{\pi}\rho_0)^D$. To find η_L we assume that $n \gg n_L$ and solve the Gross-Pitaevskii equation in the Thomas-Fermi approximation, which ignores the longitudinal kinetic energy. Using the Thomas-Fermi wave function along the longitudinal dimensions and its normalization condition, we get

$$\eta = \eta_T\eta_L = (\alpha_{q,d}/\rho_0^D r_0^d)(n_L/n)^{d/(d+q)}, \tag{5}$$

where $\alpha_{q,d}$ is a geometrical factor of order unity that depends on q and d but not on n. Numerical computation of η using the full, 3D Gross-Pitaevskii equation indicates that this expression for η is quite accurate in the regime $n_L \ll n \ll n_T$, in spite of the several approximations that went into obtaining it [8].

For n between n_L and n_T, the effective n-dependence of the last two terms in Eq. (3) is $n^{2-d/(d+q)} = n^{(d+2q)/(d+q)}$, and the measurement uncertainties in γ_1 and γ_2 scale as

$$\delta\gamma_{1,2} \sim \frac{1}{n^{\xi}}, \quad \text{where} \quad \xi = \frac{d+3q}{2(d+q)}. \tag{6}$$

We find that a 2D BEC ($d = 2$) in a harmonic trap ($q = 2$) matches the $1/n$ scaling and a 1D BEC in a harmonic trap betters this scaling. Two-D and 1D BECs in harder traps ($q > 2$) perform even better with respect to the scaling of the measurement uncertainty.

If $n > n_T$ we cannot meaningfully consider the BEC to be lower dimensional. For a BEC in a 3D spherically symmetric trap, $n_L = n_T = r_0/a$ and $\xi = (3+3q)/(6+2q)$, which becomes $\xi = 9/10$ for a 3D harmonic trap. This is worse than the $1/n$ scaling, but better than the shot-noise-limited scaling of $1/\sqrt{n}$. If the trap is anisotropic and the

BEC is cigar-shaped, then assuming that the trap is harmonic along the two transverse dimensions, we find $\xi = (4q+1)/(4q+2)$, which reduces to $\xi = 9/10$ if the trap is harmonic along the longitudinal dimension as well. Thus we see that for doing better than the $1/n$ scaling when $n > n_T$, we have to use harder traps in all three dimensions.

A good candidate for implementing the generalized metrology protocol is a Bose-Einstein condensate made of rubidium (^{87}Rb) atoms [9]. Typically, the $|F=1; M_F = -1\rangle \equiv |1\rangle$ is trapped and cooled to the condensation point. Once the atoms in $|1\rangle$ have formed in the condensate ground state, a two-photon drive can be applied that couples the $|1\rangle$ state to the $|F=1; M_F=-1\rangle \equiv |2\rangle$ state. The s-wave scattering lengths for the three processes, $|1\rangle|1\rangle \to |1\rangle|1\rangle$, $|2\rangle|2\rangle \to |2\rangle|2\rangle$, $|1\rangle|2\rangle \to |1\rangle|2\rangle$ are nearly degenerate for ^{87}Rb, with the ratios $\{a_{22} : a_{12} : a_{11}\} = \{0.97 : 1 : 1.03\}$. This means that γ_2 is very close to zero for ^{87}Rb, and so we realize the generalized quantum metrology protocol with just the $\gamma_1 n\eta J_z$ coupling. We look at the practical considerations and limitations of a realistic metrology experiment using a BEC in [8].

The quantity that is measured in our metrology protocol is essentially a constant scattering length. Estimating a constant using sophisticated quantum measurement schemes is sometimes interesting, but in this only as a proof of principle. For our proposal using a BEC, one possibility is to work around a broad Feshbach resonance that makes the scattering lengths sensitive to external magnetic fields. We might then be able to use our scheme for high-precision magnetometry.

Acknowledgments: This work was supported in part by the US Office of Naval Research (Grant No. N00014-07-1-0304), the Australian Research Council's Discovery Projects funding scheme (Project No. DP0343094), EPSRC Grant No. EP/C546237/1 and the NSF under grant PHY-0803371. STF was supported by the Perimeter Institute for Theoretical Physics; research at Perimeter is supported by the Government of Canada through Industry Canada and by the Province of Ontario through the Ministry of Research & Innovation.

REFERENCES

1. S. Boixo, A. Datta, M. J. Davis, S. T. Flammia, A. Shaji, A. B. Tacla, and C. M. Caves, "Quantum metrology from an information theory persepective," in *QCMC-2008 proceedings*, 2008.
2. A. J. Leggett, *Rev. Mod. Phys.* **73**, 307 (2001).
3. A. L. Fetter, "Theory of a dilute low-temperature trapped Bose Einstein condensate," in *Bose-Einstein Condensation in Atomic Gases: Proceedings of the International School of Physics "Enrico Fermi", Varenna on Lake Como 1998*, edited by C. E. W. M. Inguscio, S. Stringari, IOS Press, 1999, p. 201.
4. F. Dalfovo, S. Giorgini, L. P. Pitaevskii, and S. Stringari, *Rev. Mod. Phys.* **71**, 463 (1999).
5. J. Rogel-Salazar, S. Choi, G. H. C. New, and K. Burnett, *Journal of Optics B: Quantum and Semiclassical Optics* **6**, R33–R59 (2004).
6. S. Boixo, S. T. Flammia, C. M. Caves, and J. Geremia, *Phys. Rev. Lett.* **98**, 090401–4 (2007).
7. S. Boixo, A. Datta, S. T. Flammia, A. Shaji, E. Bagan, and C. M. Caves, *Phys. Rev. A* **77**, 012317 (2008).
8. S. Boixo, A. Datta, M. J. Davis, S. T. Flammia, A. Shaji, A. B. Tacla, and C. M. Caves, Quantum-limited metrology and bose-einstein condensates (2008), (In preparation).
9. J. E. Williams, *The Preparation of Topological Modes in a Strongly-Coupled Two-Component Bose-Einstein Condensate*, Ph.D. thesis, University of Colorado (1999).

Quantum metrology from an information theory perspective

Sergio Boixo*, Animesh Datta†,**, Matthew J. Davis‡, Steven T. Flammia§, Anil Shaji¶, Alexandre B. Tacla¶ and Carlton M. Caves¶

*Institute for Quantum Information, Caltech, Pasadena, California, 91125, USA
†Institute for Mathematical Sciences, 53 Prince's Gate, Imperial College, London, SW7 2PG, UK
**QOLS, The Blackett Laboratory, Imperial College, London, Prince Consort Road, SW7 2BW, UK
‡School of Physical Sciences, University of Queensland, Brisbane, Queensland 4072, Australia
§Perimeter Institute for Theoretical Physics, 31 Caroline Street N., Waterloo, ON N2L 2Y5, Canada
¶Department of Physics and Astronomy, University of New Mexico, Albuquerque, NM, 87131, USA

Abstract. Questions about quantum limits on measurement precision were once viewed from the perspective of how to reduce or avoid the effects of quantum noise. With the advent of quantum information science came a paradigm shift to proving rigorous bounds on measurement precision. These bounds have been interpreted as saying, first, that the best achievable sensitivity scales as $1/n$, where n is the number of particles one has available for a measurement and, second, that the only way to achieve this Heisenberg-limited sensitivity is to use quantum entanglement. We review these results and show that using quadratic couplings of n particles to a parameter to be estimated, one can achieve sensitivities that scale as $1/n^2$ if one uses entanglement, but even in the absence of any entanglement at any time during the measurement protocol, one can achieve a super-Heisenberg scaling of $1/n^{3/2}$.

Keywords: quantum metrology, nonlinear interferometry, Bose-Einstein condensate
PACS: 03.65.Ta, 03.75.Nt, 03.65.-w, 03.75.Mn

In quantum metrology, the description "Heisenberg-limited scaling" refers to the best possible scaling of the measurement uncertainty with the resources put into a measurement. The phrase arises not from Heisenberg uncertainty relations, but from uncertainty relations of the Mandelstam-Tamm type [1], $\delta\gamma\langle\Delta^2 K\rangle^{1/2} \geq \frac{1}{2}$, in units with $\hbar = 1$. The uncertainty $\delta\gamma$ in a parameter γ that, in part, determines the state of a quantum system is related to the standard deviation of the operator K that generates translations of the state along a path parameterized by γ. A sequence of logical and mathematical steps is needed to provide a rigorous connection between the problem of measurement uncertainty in quantum metrology and uncertainty relations of the Mandelstam-Tamm type. The pioneering work of Helstrom [2], Holevo [3], Braunstein, Caves and Milburn [4, 5] and others laid out and elucidated these steps.

The discussion in this Paper is restricted to single-parameter estimation. The first step in estimating the value of a parameter is to identify an elementary physical system that is sensitive to changes in the parameter. One or more of these elementary units makes up the measuring device or *probe*. We expect the measurement uncertainty to depend on the initial state of the quantum probe, its evolution, and the measurement made on the probe to extract information about the parameter. The *quantum Cramér-Rao bound* quantifies the idea that the optimal measurement uncertainty is inversely proportional to

CP1110, *Quantum Communication, Measurement and Computing (QCMC), The Ninth International Conference,* edited by A. Lvovsky

the change in the state of the probe corresponding to small changes in the value of the parameter [2, 3, 4, 5]:

$$(\delta\gamma)^2 \geq \frac{1}{(ds_{\rm DO}/d\gamma)^2} = \frac{1}{\mathfrak{I}(\gamma,t)} \geq \frac{1}{2\langle\Delta^2 K(\gamma,t)\rangle^{1/2}} \geq \frac{1}{\|K(\gamma,t)\|}\,. \tag{1}$$

Here $ds_{\rm DO}$ denotes a metric in the space of density operators of the probe, $\mathfrak{I}(\gamma,t)$ is the quantum Fisher information and the Hermitian generator $K(\gamma,t)$ is defined by $\partial\rho/\partial\gamma = -i\big[K(\gamma,t),\rho(\gamma,t)\big]$. The uncertainty in determining γ is quantified by the units-corrected, root-mean-square deviation of one's estimate of the parameter, $\gamma_{\rm est}$, from the true value γ: $\delta\gamma = \langle(\gamma_{\rm est}/|\partial\langle\gamma_{\rm est}\rangle/\partial\gamma| - \gamma)^2\rangle^{1/2}$. The last but one inequality in Eq. (1) is a rigorous statement of the Mandelstam-Tamm uncertainty relation. The last inequality in Eq. (1) is obtained by noting that the variance of a Hermitian operator is bounded from above by $\langle\Delta^2 K\rangle \leq \|K\|^2/4$, where $\|\cdot\|$ denotes a particular semi-norm of a Hermitian operator, defined as the difference between its largest and smallest eigenvalues.

The number n of elementary units of the probe can be regarded as the most significant resource that goes into a measurement scheme. To see the dependence of the bound on $\delta\gamma$ on n and to understand what "Heisenberg-limited scaling" means, we view quantum metrology from the perspective of quantum information theory using the language of quantum circuits in the next section.

QUANTUM CIRCUITS FOR METROLOGY PROTOCOLS

We follow Giovannetti, Lloyd, and Maccone [6] in using quantum circuits to describe and analyze metrology protocols. From this perspective, we first look at a couple of well-known measurement schemes that were considered in [6]—Ramsey interferometry and cat-state interferometry—with the aim of generalizing these circuits to new protocols that were introduced in [7, 8]. In these initial discussions of Ramsey interferometry and cat-state interferometry, we assume that the probe units are qubits. The quantization axis is taken to be along the z-direction of a Bloch-sphere representation, with the basis states along this direction denoted as $|0\rangle$ and $|1\rangle$.

Ramsey and cat-state interferometry

A typical Ramsey interferometer, such as the one in [9], can be represented by the following quantum circuit.

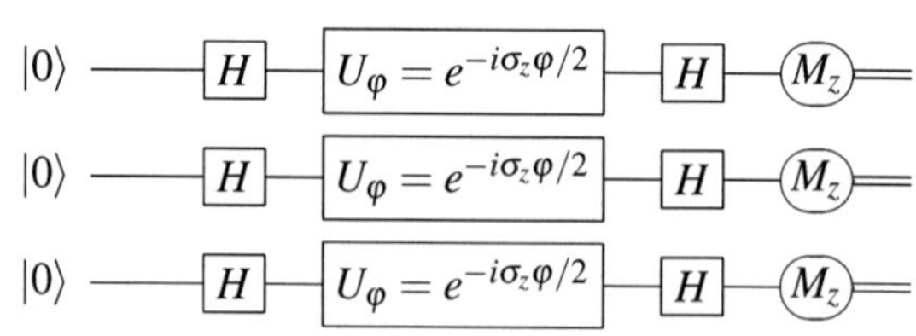

In this measurement protocol, each of the n qubits that make up the probe evolves independently. Here and in the other quantum circuits depicted in this section, we use $n = 3$ probe units as an example. All the qubits are initialized in the state $|0\rangle$, which could represent the ground state in Ramsey interferometry using atoms. The Hadamard gate H puts each of the qubits in an equal superposition of the two basis states, $(|0\rangle + |1\rangle)/\sqrt{2}$. The parameter-dependent evolution of the quantum probe is generated by the Hamiltonian

$$H_{\text{Ramsey}} = \gamma \sum_{j=1}^{n} \sigma_{z;j}/2 = \gamma J_z\,, \tag{2}$$

where $\sigma_{z;j}$ denotes the σ_z operator acting on the jth probe qubit and J_z is the z component of the "total angular momentum" for all the qubits.

After evolution under this Hamiltonian for a time t and the remaining Hadamard gates, the state of the probe qubits just before the readout is $\cos(\varphi/2)|0\rangle + \sin(\varphi/2)|1\rangle$, where $\varphi = \gamma t$. Each qubit is measured along the z-direction. This leads to a measured signal $\langle J_z\rangle \equiv \langle\sum_{j=1}^{n}\sigma_{z;j}/2\rangle = n\cos\varphi/2$, with variance $\langle\Delta^2 J_z\rangle = n\langle\Delta^2\sigma_z\rangle/4 = n\sin^2\varphi/4$. The uncertainty in the estimate of γ from the measured signal is $\delta\gamma_{\text{Ramsey}} = \langle\Delta^2 J_z\rangle^{1/2}/|d\langle J_z\rangle/d\gamma| = 1/t\sqrt{n}$.

For the Ramsey Hamiltonian (2), the generator of translations in γ is tJ_z. Thus, according to the quantum Cramér-Rao bound (1), the measurement uncertainty is bounded from below by $\delta\gamma \geq 1/t\|J_z\| = 1/tn\|\sigma_z/2\| = 1/tn$. The Ramsey interferometer described here does not achieve the best measurement uncertainty given by the quantum Cramér-Rao bound. The Hamiltonian H_{Ramsey} that governs the evolution of the probe qubits is fixed by the choice of physical systems that are the qubits. Given a choice of probe qubits, however, we still have the freedom to choose an optimal initial state for the probe and an optimal measurement of the qubits to minimize the measurement uncertainty. It turns out that the best possible scaling for the measurement uncertainty can be achieved if the probe is initialized in an entangled, "Schrödinger-cat" state.

The quantum circuit that employs an initial Schrödinger-cat state is the following.

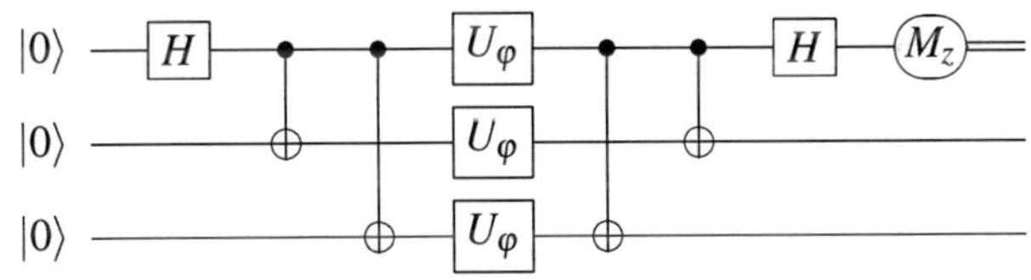

The Hadamard gate on the first qubit, followed by the controlled-NOT gates to the remaining qubits, initializes the probe in the state $|\text{cat}\rangle = \big(|0\ldots0\rangle + |1\ldots1\rangle\big)/\sqrt{2}$. This state is often referred to as the cat state because when the number of qubits is large, it is a superposition of two macroscopically distinct states. The probe qubits evolve under the same Hamiltonian (2) as in Ramsey interferometry. The parameter-dependent evolution of the probe for a duration t changes the probe state to $\big(e^{-in\varphi/2}|0\ldots0\rangle + e^{in\varphi/2}|1\ldots1\rangle\big)/\sqrt{2}$.

After the parameter-dependent evolution, one way to handle the readout, depicted in the circuit above, is to subject the qubits to a sequence of gates that kick the phases picked up by the two components of the cat state into amplitudes on the first qubit, so that the state of the probe just before readout is $[\cos(n\varphi/2)\,|0\rangle + \sin(n\varphi/2)\,|1\rangle] \otimes |0\rangle^{\otimes(n-1)}$. The readout of the probe can then be performed by measuring the σ_z operator on the first qubit. This leads to a measured signal and variance given by $\langle \sigma_{z;1} \rangle = \cos n\varphi$ and $\langle \Delta^2 \sigma_{z;1} \rangle = \sin^2 n\varphi$. The frequency of the γ-dependent fringe in cat-state interferometry is n times greater than the frequency of the signal in ordinary Ramsey interferometry. This leads to an enhanced sensitivity in the estimate of γ in cat-state interferometry: $\delta\gamma_{\text{cat}} = \langle \Delta^2 \sigma_{z;1} \rangle^{1/2} / |d\langle \sigma_{z;1} \rangle / d\gamma| = 1/(tn)$

We can put our interferometry circuits in a general setting by considering the case in which the probe units are arbitrary systems and the Hamiltonian of the probe is of the form $H_{\text{linear}} = \gamma h_{\text{linear}} = \gamma \sum_{j=1}^{n} h_j$. Here the operators h_j denote identical couplings to the probe units; the use of independent couplings to the parameter is the source of our appelation "linear" for this Hamiltonian. The generator of translations in γ is $K(\gamma, t) = t h_{\text{linear}}$, so the quantum Cramér-Rao bound (1) on the uncertainty in a determination of γ takes the form

$$\delta\gamma \geq \frac{1}{t\|h_{\text{linear}}\|} = \frac{1}{tn(\Lambda - \lambda)} \tag{3}$$

where Λ and λ are the largest and smallest eigenvalues respectively of the single-unit operators h_j.

The quantum circuit for a measurement protocol of this sort is the following.

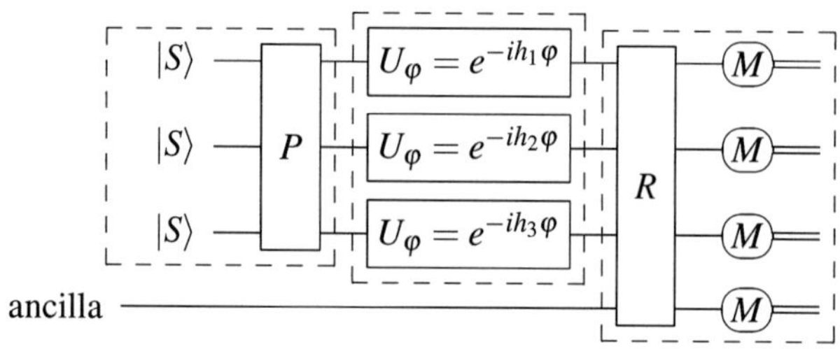

The dashed boxes highlight the three stages of this protocol: probe preparation, dynamics, and readout. All the probe units begin in a standard state $|S\rangle$. The arbitrary unitary operator P can then prepare any initial state as input to the dynamics. In the dynamics stage, the gates U_φ imprint information about the parameter on the probe. The final readout stage includes an arbitrary unitary interaction R among the probe units and with an arbitrary ancilla system. This unitary followed by measurements on each subsystem in a standard basis can be used to perform any quantum measurement. The quantum Cramér-Rao bound (3) applies to all circuits of the above form. Indeed, the bound actually applies to somewhat more general situations in which the unitary operator R is interleaved with the gate dynamics and the results of ancilla measurements are fed back onto the probe [7].

If the preparation unitary P is omitted from the circuit, making the input to the dynamics a product state, then the uncertainty in the generator of γ displacements is bounded by $\langle \Delta^2 K \rangle^{1/2} \leq t\sqrt{n}(\Lambda - \lambda)/2$. The resulting bound on measurement uncertainty, from Eq. (1), is $\delta\gamma \geq 1/t\sqrt{n}(\Lambda - \lambda) \equiv \delta\gamma_{\text{QNL}}$. This bound, a general form of that for Ramsey interferometry, is called the *quantum noise limit* or the *shot-noise limit*.

One can achieve the Cramér-Rao bound (3) by operating the circuit in a way that takes advantage of entangled input states. The preparation operator is chosen to take the initial product of standard states to the state $(|\Lambda,\ldots,\Lambda\rangle + |\lambda,\ldots,\lambda\rangle)/\sqrt{2}$, where $|\Lambda\rangle$ and $|\lambda\rangle$ are the eigenstates of h corresponding to its largest and smallest eigenvalues. In the dynamics stage, this "cat-like" initial state is subject to a period of parameter-dependent evolution that changes it to $(e^{-in\Lambda\varphi}|\Lambda,\ldots,\Lambda\rangle + e^{in\lambda\varphi}|\lambda,\ldots,\lambda\rangle)/\sqrt{2}$. The readout process kicks back the differential phase shift into an amplitude information, which produces fringes with frequency proportional to $n(\Lambda-\lambda)$, thus achieving the optimal measurement uncertainty, $\delta\gamma_{\mathrm{HL}} = 1/tn(\Lambda-\lambda)$, of the Cramér-Rao bound (3). This optimal measurement uncertainty, a general form of that for cat-state interferometry, is often called the *Heisenberg limit*.

The general quantum metrology scheme considered in this subsection indicates that probe preparation buys an enhancement of $1/\sqrt{n}$ over the case where the probe qubits are initialized in a product state. Readout has already been optimized to take advantage of this entangled input, so we conclude that when the parameter-dependent dynamics acts independently on the probe qubits, Heisenberg-limited scaling is indeed the $1/n$ scaling. The one remaining way of exploring whether the $1/n$ scaling can be improved is to consider more general dynamics [10, 11, 7, 12, 8, 13, 14].

Heisenberg-limited metrology with nonlinear Hamiltonians

A generalized quantum metrology scheme in which the dynamics of the probe is generated by a Hamiltonian that includes all k-body couplings between the probe qubits was first considered in [7]. This nonlinear coupling Hamiltonian has the form $H_{\mathrm{nonlinear}} = \gamma h_{\mathrm{nonlinear}} = \gamma(\sum_{j=1}^{n} h_j)^k = \gamma\sum_{j_1,\ldots,j_k=1}^{n} h_{j_1}h_{j_2}\cdots h_{j_k}$. The generator of translations in γ is $K(\gamma,t) = th_{\mathrm{nonlinear}}$, so the quantum Cramér-Rao bound for this dynamics is $\delta\gamma \geq 1/tn^k(\Lambda_{\max}-\Lambda_{\min})$, where $\Lambda_{\max}$ and $\Lambda_{\min}$ are functions of Λ and λ, the largest and smallest eigenvalues, respectively, of the single-unit operators h_j. For instance, if both Λ and λ are positive, then $\Lambda_{\max} = \Lambda^k$ and $\Lambda_{\min} = \lambda^k$ for all values of k. The other possible signs of Λ and λ are discussed in [8]; they all lead to a scaling $1/n^k$. The quantum circuit for metrology with nonlinear Hamiltonians has the following form.

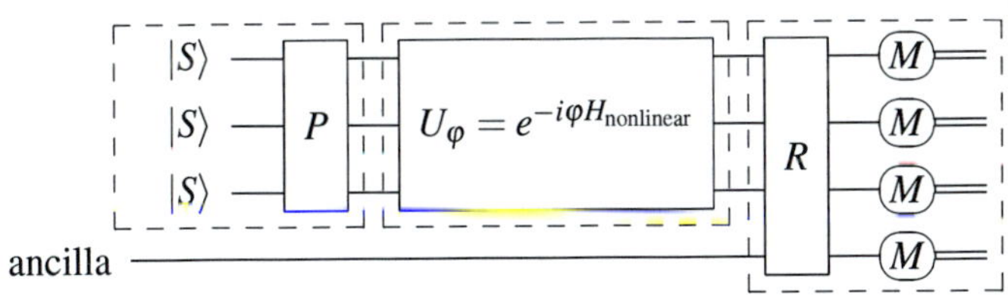

The only difference between this circuit and the one for linear quantum metrology is in the dynamics stage, where the gate that imprints information about the parameter on the probe can involve multi-unit coupling to the probe units.

To achieve the $1/n^k$ scaling made available by using a nonlinear Hamiltonian, the probe units have to be initialized in an entangled state that is very much like a cat state. Experimental limitations up till now have precluded making such cat-like states for large numbers of systems. To avoid this difficulty, Boixo *et al.* [8] analyzed the performance of quantum metrology protocols employing nonlinear Hamiltonians when the initial state of the probe is a product state. In this case the optimal measurement uncertainty scales as $\delta\gamma \sim 1/tn^{k-1/2}$. The factors multiplying this scaling depend on the particular nonlinear coupling Hamiltonian [8]. It is noteworthy that the optimal $1/tn^{k-1/2}$ sensitivity can be achieved using product measurements. The key point is that for a k-body coupling Hamiltonian, the use of a product-state input costs only a factor of $\sim \sqrt{n}$ relative to the optimal possible sensitivity. The quantum noise limit and the Heisenberg limit of linear metrology are a special case of this $\sqrt{n}$ loss of sensitivity when using input product states as opposed to an optimal entangled state.

With two-body couplings and an initial product state for the probe, a measurement uncertainty scaling as $1/n^{3/2}$ is possible [15]. Since two-body couplings between all probe units is not an especially onerous requirement for a probe system, the prospect of improving upon the $1/n$ Heisenberg scaling motivates us to investigate candidate systems for such metrology schemes. In a separate paper in this volume [16], we consider a Bose-Einstein condensate as such a candidate system, with the aim of developing a detailed, realistic, and viable proposal for an experiment that achieves better than $1/n$ scaling for the measurement uncertainty in quantum single-parameter estimation.

Acknowledgments: This work was supported in part by the US Office of Naval Research (Grant No. N00014-07-1-0304), the Australian Research Council's Discovery Projects funding scheme (Project No. DP0343094), EPSRC Grant No. EP/C546237/1, and the NSF under grant PHY-0803371. STF was supported by the Perimeter Institute for Theoretical Physics; research at Perimeter is supported by the Government of Canada through Industry Canada and by the Province of Ontario through the Ministry of Research & Innovation.

REFERENCES

1. I. E. Mandelstam, and I. Tamm, *J. Phys. (USSR)* **9**, 249 (1945).
2. C. W. Helstrom, *Quantum detection and estimation theory*, vol. 123 of *Mathematics in science and engineering*, Academic Press, New York, 1976, 1 edn.
3. A. S. Holevo, *Probabilistic and statistical aspects of quantum theory*, vol. 1 of *North-Holland series in statistics and Probability theory*, North-Holland, Amsterdam, 1982, 1 edn.
4. S. L. Braunstein, and C. M. Caves, *Phys. Rev. Lett* **72**, 3439–3443 (1994).
5. S. L. Braunstein, C. M. Caves, and G. J. Milburn, *Ann. Phys.* **247**, 135–173 (1996).
6. V. Giovannetti, S. Lloyd, and L. Maccone, *Phys. Rev. Lett.* **96**, 010401 (2006).
7. S. Boixo, S. T. Flammia, C. M. Caves, and J. Geremia, *Phys. Rev. Lett.* **98**, 090401–4 (2007).
8. S. Boixo, A. Datta, S. T. Flammia, A. Shaji, E. Bagan, and C. M. Caves, *Phys. Rev. A* **77**, 012317 (2008).
9. S. Gleyzes, S. Kuhr, C. Guerlin, J. Bernu, S. Deleglise, U. B. Hoff, M. Brune, J.-M. Raimond, and S. Haroche, *Nature* **446**, 297–300 (2007), ISSN 0028-0836.
10. A. Luis, *Phys. Lett. A* **329**, 8–13 (2004).
11. J. Beltran, and A. Luis, *Phys. Rev. A* **72**, 045801 (2005).
12. A. M. Rey, L. Jiang, and M. D. Lukin, *Phys. Rev. A* **76**, 053617–8 (2007).
13. S. Choi, and B. Sundaram, *Phys. Rev. A* **77**, 053613 (2008).
14. M. J. Woolley, G. J. Milburn, and C. M. Caves, *eprint, 0804.4540* (2008).
15. S. Boixo, A. Datta, M. J. Davis, S. T. Flammia, A. Shaji, and C. M. Caves, *Phys. Rev. Lett.* **101**, 040403 (2008).
16. S. Boixo, A. Datta, M. J. Davis, S. T. Flammia, A. Shaji, A. B. Tacla, and C. M. Caves, "Quantum metrology with Bose-Einstein Condensates," in *QCMC-2008 proceedings*, AIP, 2008.

Improved resolution in imaging through quantum post-selection

Vittorio Giovannetti*, Seth Lloyd†, Lorenzo Maccone** and Jeffrey H. Shapiro‡

*NEST-CNR-INFM & Scuola Normale Superiore, Piazza dei Cavalieri 7, I-56126, Pisa, Italy.
†MIT, Dept. of Mechanical Engineering, 77 Mass. Ave., Cambridge, MA 02139, USA.
**QUIT, Dip. Fisica "A. Volta", Univ. Pavia, via Bassi 6, I-27100 Pavia, Italy.
‡MIT, Research Laboratory of Electronics, 77 Mass. Ave., Cambridge, MA 02139, USA.

Abstract. We show how we can obtain improved resolution when imaging an object by filtering "high-resolution" quantum states from impinging classical radiation. The method allows one to beat the Rayleigh diffraction bound that is intrinsic in any imaging apparatus. It is based on using focused classical illumination sources and N-photon coincidence detection.

Keywords: Imaging, Rayleigh bound, diffraction, quantum optics
PACS: 42.50.-p,03.67.-a,42.50.Dv,42.30.Va

When performing optical imaging, the finite size of the entrance pupil typically imposes a limitation on the resolution available in the resulting image. Due to diffraction, any point in the object plane is typically mapped into an Airy disk in the image plane: details that are smaller than the radius of this disk cannot be resolved (Rayleigh diffraction bound). Quantum effects have been used to overcome this limitation [1, 2]. However, quantum effects—such as entanglement and squeezing—are very sensitive to noise and imperfections, so that conventional quantum-enhancement methods are quite impractical. Thus, a recent trend is to start from noise-insensitive classical light sources and to filter the appropriate high-resolution quantum states at the detection stage [3]. A particularly simple scheme is the one we presented in Ref. [4]. Here we briefly review one of the results presented in that paper and give some new simulated examples of its application.

The main idea of the quantum imaging scheme of [4] is to constrain all the photons to leave the object from the same point and to be detected at the same point. This technique achieves an enhancement in resolution of the order of $\sqrt{N}$, with N being the number of detected photons. Among all possible localized states that can be used to illuminate the object, the most appealing is an incoherent mixture of coherent states focused on a small area on the object, namely the state

$$\sigma = \int \frac{d^2\vec{r}_o}{\mathscr{A}} |\alpha\rangle_{\vec{r}_o}\langle\alpha| \text{ with } |\alpha\rangle_{\vec{r}_o} \equiv e^{\alpha b^\dagger(\vec{r}_o) - \alpha^* b(\vec{r}_o)}|0\rangle, \tag{1}$$

where $\vec{r}_o$ is the transverse vector indicating the points in the object plane of area $\mathscr{A}$ and where $|\alpha\rangle_{\vec{r}_o}$ is a coherent state for the mode b localized around position $\vec{r}_o$, its transverse spread being Δk_t and its bandwidth $\Delta\omega$, namely

$$b(\vec{r}_o) \equiv \int \frac{d\omega}{\sqrt{\Delta\omega}} \int \frac{d^2\vec{k}_t}{\sqrt{\pi\Delta k_t^2}} \, e^{i\vec{k}_t\cdot\vec{r}_o} a(\vec{k}) \;, \tag{2}$$

CP1110, *Quantum Communication, Measurement and Computing (QCMC), The Ninth International Conference,* edited by A. Lvovsky

with $a(\vec{k})$ being the annihilation operator of the mode associated with the wave-vector $\vec{k}$.

At the image plane we perform N-photon coincidence detection, namely we collect only the instances when N photons arrive within a time Δt all in a single photosensitive element of surface area $\mathscr{S}$. From conventional photodetection theory [5] we find that the probability of such an event is

$$P_N(\vec{r}_i) \simeq \frac{(\eta \mathscr{S} c \Delta t)^N}{N!} \langle \left[E_i^{(-)}(\vec{r}_i,t)\right]^N \left[E_i^{(+)}(\vec{r}_i,t)\right]^N \rangle \,, \tag{3}$$

where η is the detector's quantum efficiency and the positive part of the electromagnetic field at the transverse position $\vec{r}_i$ in the image plane is given by

$$E_i^{(+)}(\vec{r}_i,t) = \int d^3\vec{k}\, \mathscr{E}_i^{(+)}(\vec{r}_i;\vec{k})\, e^{-ikt/c} a(\vec{k}) \,, \tag{4}$$

with $\mathscr{E}_i^{(+)}(\vec{r}_i;\vec{k})$ being the solution to the Helmholtz equation for the field at that plane.

We will consider a very simple imaging apparatus, namely a circular thin lens of radius R and focal length f placed such that the image plane is in focus with respect to the object, namely $\frac{1}{D_o} + \frac{1}{D_i} = \frac{1}{f}$, where $D_{o,i}$ are the object-lens and image-lens distances, respectively. For monochromatic light in the paraxial approximation (i.e., when the transverse component k_t of the electromagnetic field is negligible, $k_t \ll k$), conventional imaging theory [6] yields the image-plane field as the following function of the object-plane field:

$$\mathscr{E}_i^{(+)}(\vec{r}_i;\vec{k}) = \int \frac{d^2\vec{r}_o}{\mathscr{A}}\, A(\vec{r}_o)\, h(\vec{r}_i,\vec{r}_o)\, \mathscr{E}_o^{(+)}(\vec{r}_o;\vec{k}) \,, \tag{5}$$

where $A(\vec{r}_o)$ is the object transfer function, and $h(\vec{r}_i,\vec{r}_o)$ is the imaging apparatus transfer function which, for a focused thin lens is

$$h(\vec{r}_i,\vec{r}_o) = \frac{R^2 k^2 \mathscr{A}}{4\pi D_o D_i} e^{i\vartheta} \mathrm{somb}(R\, k\, |\vec{r}_o + \vec{r}_i/m|/D_o), \tag{6}$$

where $\mathrm{somb}(x) \equiv 2J_1(x)/x$ (with J_1 being the first-order Bessel function), $m \equiv D_i/D_o$ is the image magnification factor, and where the phase factor ϑ can be compensated using standard techniques [6]. Now we can put all the ingredients together, namely we calculate the expectation value of Eq. (3) using the input state σ of Eq. (1) and the electromagnetic field calculated through Eq. (5). If we assume that the input state is focused on the object into an area which is small compared to the Rayleigh limit of the lens (namely $D_o/R \gg k/\Delta k_t$), we find (see [4] for details)

$$P_N(\vec{r}_i) \simeq \xi^N \frac{|\alpha|^{2N}}{N!} \int \frac{d^2\vec{r}_o}{\mathscr{A}}\, |\tilde{A}(\vec{r}_o)|^{2N} |h(\vec{r}_i,\vec{r}_o)|^{2N}, \text{ where } \xi \equiv \eta \frac{\Delta\omega \Delta t \mathscr{S}}{\pi \Delta k_t^2 \mathscr{A}^2} \,, \tag{7}$$

where $\tilde{A}$ is the convolution of the object transfer function A with the somb function arising from the finite spot size of the illuminating source on the object (which has a finite transverse spatial bandwidth Δk_t), namely

$$\tilde{A}(\vec{r}_o) \equiv \int \frac{d^2\vec{r}_o}{\mathscr{A}} A(\vec{r}) F_{\Delta k_t}(|\vec{r}_o - \vec{r}|) \,,\ F_{\Delta k_t}(x) \equiv \pi \Delta k_t^2 \mathscr{A}\, \mathrm{somb}(\Delta k_t\, x). \tag{8}$$

If the spot size is sufficiently small (i.e., for large transverse wave-vector spread Δk_t), A and $\tilde{A}$ do not differ very much, because $X^2 \mathrm{somb}(X|\vec{x}|) \to 4\pi\delta^{(2)}(\vec{x})$ for $X \to \infty$ [6]. Then Eq. (7) describes the image, reconstructed through the N-photon coincidence detection events. For $A \sim \tilde{A}$, the object-image point-spread function is essentially given by the factor $|h(\vec{r}_i, \vec{r}_o)|^{2N}$.

This result must be compared with conventional incoherent imaging, where the intensity of the light at the image plane is detected when the object is entirely illuminated with quasimonochromatic spatially-incoherent light. In this case, Eq. (3) for $N = 1$ gives

$$P_1(\vec{r}_i) = \frac{\eta \mathscr{S} c\Delta t}{2\pi k^2 \mathscr{A}} I_o \int \frac{d^2\vec{r}_o}{\mathscr{A}} \left| A(\vec{r}_o)\, h(\vec{r}_i, \vec{r}_o) \right|^2, \tag{9}$$

where $I_o = \langle E_o^{(-)} E_o^{(+)} \rangle$ is the field intensity at the object. Here the object-image point-spread function is given by the factor $|h(\vec{r}_i, \vec{r}_o)|^2$. Each point at the object plane is mapped at the image plane into a ring (Airy disk) of radius $x_R \simeq 0.61 \times 2\pi m D_o/(kR)$, so that details that are smaller than x_R cannot be discerned at the image: x_R is the Rayleigh bound of the imaging apparatus.

The resolution improvement of our method is essentially due to the fact that the point-spread function h^N in Eq. (7) is a much narrower function than h, so that Eq. (7) yields more accurate images than Eq. (9). An example of reconstructed images is given in Fig. 1, which shows that our method also works for complex grey-scale images (in addition to the black-and-white examples given in Ref. [4]).

The resolution improvement can be roughly estimated by gauging by how much the "width" of the point-spread function h is reduced by taking its Nth power. When imaging a point object, the Rayleigh diffraction implies that about 84% of the light intensity falls in the first ring of the Airy diffraction pattern given by the $\mathrm{somb}(x) = 2J_1(x)/x$ function, i.e., about 84% of the radial integral of $\mathrm{somb}^2(|\vec{x}|)$ is located inside a ring of radius x_R. By calculating the ratio between the Rayleigh diffraction bound and the radius $r(N)$ that contains 84% of the radial integral of the $\mathrm{somb}^{2N}(|\vec{x}|)$, we see that this quantity goes like $1/\sqrt{N}$, which gives the scaling with N of the resolution enhancement. This should be taken only as a rough estimate, as $r(N)$ is also the radial dimension of a point-like object imaged using the classical post-processing strategy of taking the Nth power of the classically reconstructed image from Eq. (9). For images more complex than a single point, the actual resolution enhancement will also depend on the parameter Δk_t of the illumination source. Note that the $1/\sqrt{N}$ scaling can also be achieved through the classical strategy of measuring the arrival position of each photon at the image plane and calculating the average arrival position. This may not always be possible (as for example in the case of lithography or film photography). Finally, note that it is possible to further improve the resolution, up to the $1/N$ conjectured upper bound for imaging, by requiring that all the N photons coherently traverse the lens at the same position [4].

REFERENCES

1. M. I. Kolobov, Rev. Mod. Phys. **71**, 1539 (1999).
2. L. A. Lugiato, A. Gatti and E. Brambilla, J. Opt. B: Quantum Semiclass. Opt. **4,** S176 (2002).

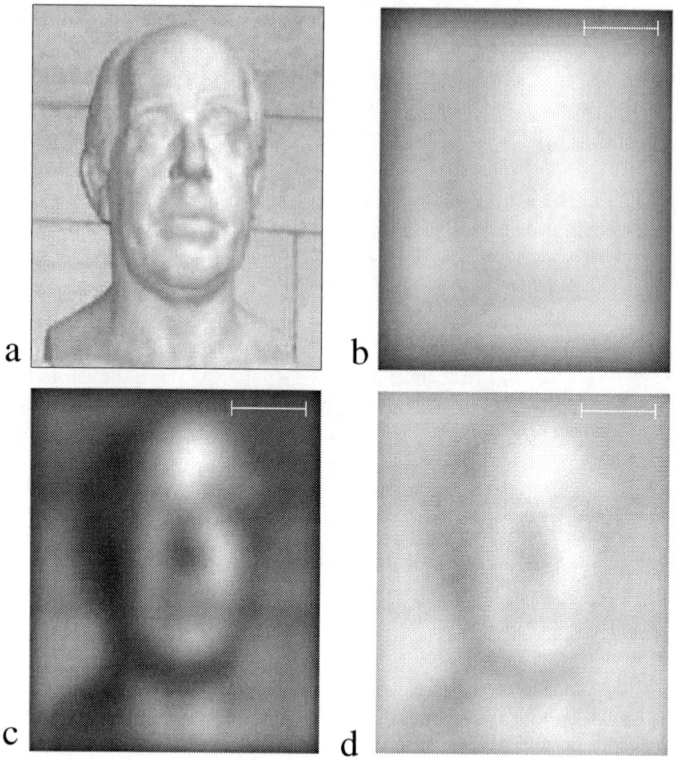

FIGURE 1. (a) Object transfer function A that is imaged (the bust of N. Bohr in the town hall of Copenhagen). (b) Incoherent imaging of the previous object through Eq. (9). The image is completely smeared out since here the parameters are chosen so that the Rayleigh bound x_R (segment on the top right of the figure) is of the same order of magnitude as the object's details. The image parameters are $D_o/R = 400$, $k = 6000L$, $\Delta k_t = 600L$, and the unit of length L is given by the image width. (c) Quantum imaging through Eq. (7) with $N = 5$: many of the details of the image are recovered as the resolution in the reconstruction is increased. Notice also that the contrast of the image is increased due to the exponent N that appears in the image transfer function $|\tilde{A}|^{2N}$ in Eq. (7). (d) The Nth root of the image in (c): it is a classical post-processing of the high resolution image to reduce the effect of the increased contrast. (These images come from simulated data.)

3. P. R. Hemmer, *et al.*, Phys. Rev. Lett. **96,** 163603 (2006); A. Muthukrishnan, M. O. Scully, and M. S. Zubairy, J. Opt. B: Quantum Semiclass. Opt. **6,** S575 (2004); D. Korobkin, and E. Yablonovitch, Opt. Eng. **41,** 1729 (2002); C. Thiel, *et al.*, quant-ph/070.1024 (2007); S. J. Bentley and R. W. Boyd, Opt. Exp. **12,** 5735 (2004); M. Zhang, *et al.*, Phys. Lett. A in press, quant-ph/0612060 (2006); K. Wang and D.-Z. Cao, Phys. Rev. A **70,** 041801(R) (2004); Y. Shih, IEEE J. Sel. Topics in Quantum Electron. **13,** 1016 (2007); E. Yablonovitch and R.B. Vrijen, Opt. Eng. **38,** 334 (1999); A. Pe'er, *et al.*, Opt. Express **12,** 6600 (2004); B. I. Erkmen and J. H. Shapiro, Phys. Rev. A **77,** 043809 (2008); M. I. Kolobov (Ed.) *Quantum Imaging*, (Springer, New York, 2006).
4. V. Giovannetti, S. Lloyd, L. Maccone, and J. H. Shapiro, arXiv:0804.2875 [quant-ph] (2008).
5. L. Mandel and E. Wolf *Optical Coherence and Quantum Optics* (Cambridge Univ. Press, Cambridge, 1995).
6. J. W. Goodman, *Introduction to Fourier optics* (McGraw-Hill, New York, 1988).

LADAR resolution improvement using squeezed-vacuum injection

Zachary Dutton* and Jeffrey H. Shapiro†

*BBN Technologies, 10 Moulton St., Cambridge, MA 02138
†Research Laboratory of Electronics, Massachusetts Institute of Technology

Abstract. We evaluate the signal-to-noise ratio (SNR) dependence of LADAR spatial resolution, and the improvement afforded by squeezed-vacuum injection at the receiver. We define spatial resolution as the ability to distinguish two point targets from a single point target. For speckle returns and a soft-aperture homodyne receiver, the resolution scales as $\mathrm{SNR}^{-3/10}$ at high SNR, and exhibits a threshold SNR below which targets are unresolvable. Injecting a 15 dB squeezed vacuum improves the resolution by a factor of about 2 at high SNR, and provides approximately 3.5 dB of threshold extension, which permits some classically-unresolvable targets to become distinguishable.

Keywords: LADAR, remote sensing, squeezed vacuum, spatial resolution
PACS: 42.30.Va, 42.50.Lc

INTRODUCTION

Conventional laser detection and ranging (LADAR) systems have spatial resolutions that are restricted by the Rayleigh limit $0.61\lambda/R$, where R is their entrance-pupil radius, and their signal-to-noise ratio (SNR). An open question is whether quantum resources can provide improved spatial resolution. We shall address that question for the particular case of a soft-aperture, homodyne-reception system. We begin with a classical system—a coherent-state transmitter with a conventional homodyne receiver—to establish a rigorous baseline. We then introduce augment the classical sensor by injecting squeezed-vacuum light to ameliorate the loss of high spatial-frequency information that is incurred at the soft aperture. Both systems are shown in Fig. 1. For their analysis, we will rely on decision theory, viz., we will define spatial resolution to be the minimum angular separation between two speckle-return point targets at which they can be distinguished from a single speckle-return point target within a specified error-probability bound.

DETECTION THEORY FORMALISM

We will treat spatial resolution as a problem of binary hypothesis testing in which it is equally likely for there to be one or two point targets at range L from the LADAR. We will illuminate these targets with coherent-state (laser) light, and we assume that each point target imposes an independent fully-developed laser speckle on its return, as is typical for operation at $\sim 1\ \mu$m wavelengths. For simplicity, we consider just one transverse dimension, but our results easily generalize to 2D. The classical target-return

CP1110, *Quantum Communication, Measurement and Computing (QCMC), The Ninth International Conference,* edited by A. Lvovsky

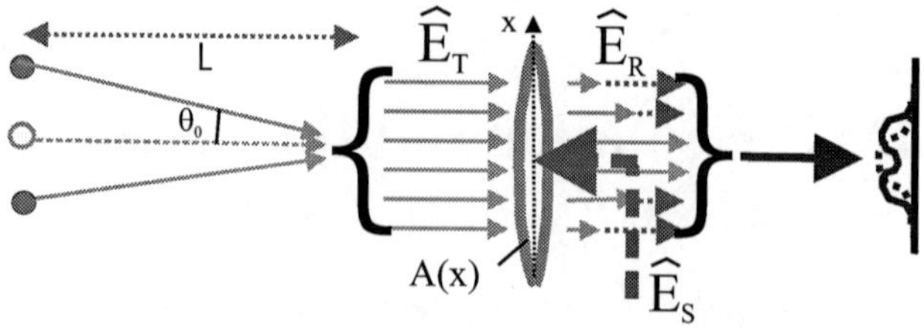

FIGURE 1. LADAR system configurations. A coherent-state transmitter (not shown) results in point-target returns from either one target (open circle) or two targets (filled circles). The target return, $\hat{E}_T(x,t)$, traverses a soft aperture with transmissivity $A(x)$. Without squeezed-vacuum injection (SVI), vacuum-state quantum noise mixes with the target return emerging from the soft aperture, and the resulting total field, $\hat{E}_R(x,t)$, is mixed with a strong local-oscillator (LO) field (not shown) and homodyne detected on a detector array. With SVI, a squeezed-vacuum field, $\hat{E}_S(x,t)$, replaces the standard vacuum-state injection, and $\hat{E}_R(x,t)$ is imaged onto the detector array, as before, with homodyne reception.

field envelopes, just before the receiver's entrance pupil, are thus

$$E_T^{(1)}(x,t) = v_0 C_1 e^{ikx^2/2L} \text{ and } E_T^{(2)}(x,t) = C_2\left[v_+ e^{ik(x-L\theta_0)^2/2L} + v_- e^{ik(x+L\theta_0)^2/2L}\right], \quad (1)$$

for the one-target ($E_T^{(1)}$) and two-target ($E_T^{(2)}$) cases, respectively.[1] Here: C_1, C_2 are constants chosen to ensure that the average power transmitted through the receiver's soft aperture is P_r regardless of which hypothesis is true; $k = (2\pi)/\lambda$ is the wave number; and v_0, v_+, v_- are independent, identically-distributed, zero-mean, unity-variance, isotropic complex-valued Gaussian random variables that represent the speckle.

The target return is transmitted through a soft aperture with transmissivity $A(x) = e^{-2x^2/R^2}$, that, because of the spatial Fourier transform relation in Fraunhofer diffraction, suppresses high spatial-frequency content in the resulting detector-plane field. Without SVI, there are increasing amounts of vacuum-state quantum noise injected in the increasingly attenuating regions of this soft aperture, noise that is needed to ensure that $\hat{E}_R(x,t)$ has the proper free-field commutator bracket. With SVI, a squeezed-vacuum field, $\hat{E}_S(x,t)$, is reflected from the soft-aperture—according to the beam splitter relation $\hat{E}_R(x,t) = A(x)\hat{E}_T(x,t) + \sqrt{1-|A(x)|^2}\hat{E}_S(x,t)$—so that one quadrature of $\hat{E}_R(x,t)$ will see attenuated quantum noise, hence improved local SNR, at points of soft-aperture attenuation of $\hat{E}_T(x,t)$. With or without SVI, the lens shown in Fig. 1 casts a minified, degraded image of the target-plane field on a detector array that has been set up for homodyne detection.

The minimum error probability decision rule for processing the data obtained from the homodyne array makes use of a sufficient statistic vector $\mathbf{r} = [\, r_+ \;\; r_- \;\; r_0 \,]^T$, where r_+ (r_-) is the sum (difference) of the overlap integrals of the detected field with the ideal noiseless signals produced by point targets at $\theta = \pm\theta_0$, and r_0 the overlap with a target

[1] Conditioned on knowledge of the speckle variables and the true hypothesis (one target or two targets), these classical fields are the coherent-state eigenfunctions of the field operator $\hat{E}_T(x,t)$, shown in Fig. 1, over the observation time.

at $\theta = 0$. Conditioned on the true hypothesis, $\mathbf{r}$ is zero-mean Gaussian random vector with covariance matrices Λ_n, for $n = 1$ or 2 targets with

$$\begin{aligned}
\Lambda_1 &= \begin{bmatrix} 2\mathrm{SNR}h_1^2 + h_{02}^+ & \sqrt{2}h_1(\mathrm{SNR}h_0+1) & 0 \\ \sqrt{2}h_1(\mathrm{SNR}h_0+1) & (\mathrm{SNR}h_0+1)h_0 & 0 \\ 0 & 0 & h_{02}^- \end{bmatrix} \\
\Lambda_2 &= \begin{bmatrix} h_{02}^+[\mathrm{SNR}h_{02}^+/2+1] & \sqrt{2}h_1[\mathrm{SNR}h_{02}^+/2+1] & 0 \\ \sqrt{2}h_1[\mathrm{SNR}h_{02}^+/2+1] & \mathrm{SNR}h_1^2 + h_0 & 0 \\ 0 & 0 & h_{02}^-[\mathrm{SNR}h_{02}^-/2+1] \end{bmatrix} \\
h_0 &= \frac{2}{\sqrt{\pi}R}\int dx \frac{e^{-x^2/R^2}}{D(x)}, \quad h_1 = \frac{2}{\sqrt{\pi}R}\int dx \frac{e^{-x^2/R^2}\cos(kx\theta_0)}{D(x)}, \\
h_2 &= \frac{2}{\sqrt{\pi}R}\int dx \frac{e^{-x^2/R^2}\cos(2kx\theta_0)}{D(x)}, \quad D(x) = e^{-r}(1-|A(x)|^2) + |A(x)|^2. \qquad (2)
\end{aligned}$$

Here $h_{02}^{\pm} = h_0 \pm h_2$ and $\mathrm{SNR} = P_r T/\hbar\omega$.[2] In the $D(x)$ expression, $r \geq 0$ is the squeeze parameter, with $r = 0$ corresponding to conventional (no SVI) operation. The standard likelihood-ratio test for Gaussian random vectors [1] now gives

$$\frac{1}{2}\mathbf{r}^T[(\Lambda_1)^{-1} - (\Lambda_2)^{-1}]\mathbf{r} \ \underset{\text{decide one target}}{\overset{\text{decide two targets}}{\gtrless}} \ \ln\left(\frac{|\Lambda_2|^{1/2}}{|\Lambda_1|^{1/2}}\right), \qquad (3)$$

where $|\Lambda_n|$ is the determinant of Λ_n.

SPATIAL RESOLUTION

To assess the classical resolution limit we generated 10,000 sufficient-statistic vectors each for the one-target and two-target hypotheses with $r = 0$, i.e., no SVI. We then applied the decision rule (3) and used the frequency of decision errors to estimate the true error probability, P_E, as a function of the target spacing θ_0 and SNR. Figure 2(a) plots P_E versus θ_0 for several SNR values, where angle is measured in units of the Rayleigh angle, $\theta_0^{(\mathrm{Ray})} = 0.6\,\lambda/R$. We have chosen to say that two targets are resolved resolution $P_E \leq 0.03$. Note that for SNR=17 dB, the minimum value of P_E, as θ_0 increases without bound, exceeds 0.03, implying that the system has no resolution capability at this SNR. Resolution versus SNR, for the classical case, is plotted as circles in Fig. 2(b). We see that there is a threshold SNR of 20.5 dB below which $\theta_0/\theta_0^{(Ray)}$ diverges, because $P_E > 0.03$. Also, at high SNR we find a slow resolution improvement with SNR. To understand this high-SNR behavior, we computed the Bhattacharyya error bounds [2]

[2] In this SNR formula, T is the post-detection integration time, $\hbar\omega$ is the photon energy at the operating frequency $\omega = 2\pi c/\lambda$, and unity quantum efficiency has been assumed.

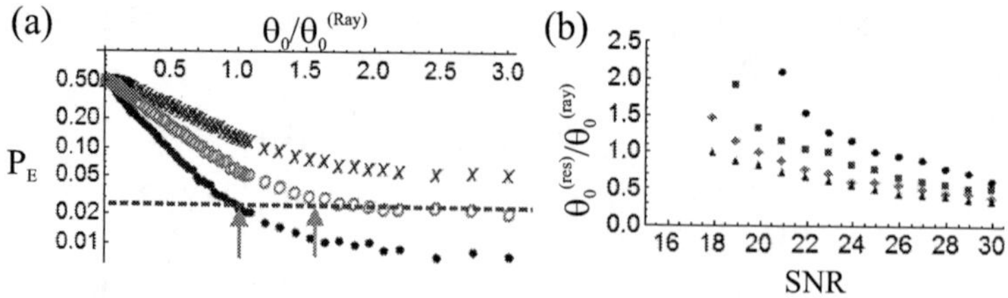

FIGURE 2. (a) Error probability P_E without SVI versus normalized angular separation for SNR = 17 dB (red, crosses), 22 dB (green, open circles), and 27 dB (blue, filled circles). The dashed line is our $P_E = 0.03$ resolution threshold, and the arrows indicate the resolutions for SNR = 22 and 27 dB, i.e., where their error probabilities cross this threshold. (b) Resolution versus SNR. Filled black circles are for no SVI, red squares are for 5 dB SVI, green diamonds are for 10 dB SVI, and blue triangles are for 15 dB SVI.

associated with the covariance matrices (2) and took their asymptotic behavior in the limit $\theta_0 \ll \theta_0^{(\mathrm{Ray})}$ and SNR $\to \infty$. This high-SNR analysis showed $\theta_0^{(\mathrm{res})} \propto \mathrm{SNR}^{-3/10}$.

Resolution versus the classical-system $\mathrm{SNR} = P_r T/\hbar\omega$ is also plotted in Fig. 2(b) when varying amounts of SVI are employed, [3] Here we see a horizontal shift of the curves with increasing amounts squeezing, corresponding to an effective boost in SNR that results in improved spatial resolution. At high SNR, this improvement is approximately a factor of 2 times for 15 dB SVI squeezing. At low SNR, the threshold shifts, giving an SNR range for which the SVI system resolves targets that a classical system cannot.

CONCLUSIONS

We have compared LADAR spatial resolution capabilities for conventional and SVI homodyne receivers. For both there is an SNR threshold below which targets are unresolvable, and high-SNR resolution proportional to $\mathrm{SNR}^{-3/10}$. When SVI is employed in a soft aperture system we get an increase in effective SNR, resulting in modest improvements in resolution for high SNR and dramatic improvements for SNR near threshold. We have assumed unit detector quantum efficiency, but we note the SVI performance degradation incurred with sub-unity efficiency can be ameliorated by phase-sensitive amplification [3].

REFERENCES

1. H. L. Van Trees, *Detection, Estimation, and Modulation Theory, Part I*, Wiley, New York, 1968.
2. H. Chernoff, Ann. Math. Stat. **23,** 493–507 (1952).
3. S.-K. Choi, M. Vasilyev, and P. Kumar, Phys. Rev. Lett. **83**, 1938–1941 (1999); erratum: **84,** 1361 (2000).

---

[3] These results were obtained by the same Monte Carlo procedure described above for the classical case.

CHAPTER 15

QUANTUM PROCESSES

Quantum Stochastic Processes

William Joseph Spring

Quantum Information & Probability Group,
School of Computer Science, University of Hertfordshire,
Hatfield, Herts AL10 9AB, UK
j.spring@herts.ac.uk

Abstract.

We consider quantum analogues of n-parameter stochastic processes, associated integrals and martingale properties extending classical results obtained in [1, 2, 3], and quantum results in [4, 5, 6, 7, 8, 9, 10].

Keywords: Quantum Stochastic Process, Isometry, Quantum Stochastic Integral, Martingale
PACS: 02.30.Cj, 02.30.Lt, 02.30.Rz, 02.30.Tb, 02.50.CW, 02.50.Ey, 02.30.-d, 02.30.Fk

INTRODUCTION

In modeling a classical stochastic process $X = (X_i)$ the concept of a stochastic base $(\Omega, \mathscr{F}, \{\mathscr{F}_i\}_{i\in I}, \mathscr{P}, I)$ is introduced, in which $(\Omega, \mathscr{F}, \mathscr{P})$ forms a complete probability space, $\{\mathscr{F}_i\}_{i\in I}$ denotes a filtration and I denotes an associated index set. The filtration $\{\mathscr{F}_i\}_{i\in I}$ consists of an increasing family of sub σ - fields of $\mathscr{F}$, is right continuous and contains, for each $\mathscr{F}_i$ in the filtration, all $\mathscr{P}$ - null sets. X is said to be an adapted process in the sense that for all $i \in I$, X_i is $\mathscr{F}_i$ measurable. [11].

For quantum stochastic processes $Q = (Q_z)$ one may develop a quantum stochastic base $(\mathscr{H}, \mathscr{C}, \{\mathscr{C}_z\}_{z\in\mathscr{R}_+}, m, \mathscr{R}_+)$, in which $(\mathscr{H}, \mathscr{C}, m)$ forms a non-commutative probability gage space [12], $\mathscr{H}$ denotes a Hilbert Space, $\mathscr{C}$ a von Neumann algebra and g a gage. The 'nest of algebras' $\{\mathscr{C}_z\}_{z\in\mathscr{R}_+}$ is referred to as a filtration of von Neumann subalgebras of $\mathscr{C}$. Their union generates the von Neumann $\mathscr{C}$ and satisfies a left continuity conditionand and their intersection right continuity conditions, [4, 5, 9, 11]. The filtrations may be realised at either the operator or Hilbert space level, [5]. A quantum process Q is said to be a $\mathscr{C}$ adapted process if for all $z \in \mathscr{R}_+$, $Q_z \in \mathscr{C}_z$.

THE CLIFFORD STOCHASTIC BASE

For the Clifford construction the Hilbert space $\mathscr{H}$ takes the form $\mathscr{F}(L^2(\mathfrak{h}))$ denoting Fermi-Fock space in which $\mathfrak{h} = L^2(\mathbb{R}^n_+)$ is the one particle anti-symmetric fermion space with underlying parameter space $\mathbb{R}^n_+$ with $n \in \mathbb{N}^+$. The von Neumann algebra $\mathscr{C}$ in our base is generated by elements of the form $\psi(f) = a^*(f) + a(f)$ with real valued $f \in L^2_{loc}(\mathbb{R}^n_+)$, in which a^* and a denote creation and annihilation operators respectively [13]. Restricting attention to elements of the form $\chi_{R_z} g \in L^2(\mathbb{R}^n_+)$ with R_z the n-dimensional closed cube in $\mathbb{R}^n_+$, with infimum at the origin and supremum at z,

CP1110, *Quantum Communication, Measurement and Computing (QCMC), The Ninth International Conference,* edited by A. Lvovsky

a filtration $\{\mathscr{C}_z\}_{z\in\mathbb{R}^3_+}$ may be constructed. Resulting conditional expectation operators $m(\ |\mathscr{C}_z)$ may be extended to non commutative $L^p(\mathscr{C})$ spaces generated via the gage $m(|a|) = \langle\Omega||a^p|^{1/p}|\Omega\rangle$. $L^\infty(\mathscr{C})$ is defined to be $\mathscr{C}$ with the usual operator norm, [14]. m we note, is faithful, central and normal, [15].

The operator ψ is a self adjoint operator satisfying the canonical anticommutation relation $\{\psi(f),\psi(g)\} = 2\langle f|g\rangle\mathbb{I}$. For von Neumann algebras $\mathscr{A}$ generated by $\psi(f)$ as f varies in $\mathscr{H}_\mathscr{A}$ and $\mathscr{B}$ generated by $\psi(g)$ as g varies in $\mathscr{H}_\mathscr{B}$ we note that for $\mathscr{H}_\mathscr{A}$ orthogonal to $\mathscr{H}_\mathscr{B}$ $m(ab) = m(a)m(b)$ for all $a\in\mathscr{A}$, and $b\in\mathscr{B}$, [16]. $\mathscr{A}$ and $\mathscr{B}$ are said to be independent and $m(ab) = m(a)m(b)$ for a and b as above, is described as the independence property of m. It may be shown that the process $(\psi_z(f))_{z\in\mathbb{R}^n_+} = (\psi(\chi_{R_z}f)_{z\in\mathbb{R}^n_+}$ is a martingale process in $\mathscr{C}$ and so may be employed as an integrator for quantum analogues of the classical *Ito* and the *Wong and Zakai* integrals. We defer to [5, 6, 7, 8, 9] and references therin for further details.

Definition Let $X = (X_z)$ denote an $L^2(\mathscr{C})$ process. We say that X is a martingale if $m(X_{z''}|\mathscr{C}_{z'}) = X_{z'}$ for $z'\prec z''$. A martingale is said to be centred if its gage is zero.

PARAMETER BASE

We describe the parameter base for the case $\mathbb{R}^3_+$ and extend to the general case. Let $z=(t,u,v), z'=(t',u',v'), z''=(t'',u'',v'')$ and $z'''=(t''',u''',v''')$. $z\prec z'$ is used to indicate that $t\le t', u\le u'$ and $v\le v'$. We write $z' \overset{\wedge}{\wedge} z''$(relative to R_z) to indicate that $t'\ge t> t''', u'<u\le u''', v'\ge v$ and $v'''\ge v$ or; $u''\ge u>u', v''<v\le v', t'\ge t$ and $t''\ge t$ or; $v'''\ge v>v'', t'''<t\le t'', u''\ge u$ and $u'''\ge u$. Finally, we write $z' \overset{\wedge}{\wedge} z'' \overset{\wedge}{\wedge} z'''$ (relative to R_z) to indicate that $t'\ge t>t''$ and $t''', u''\ge u>u'$ and u''', and; $v'''\ge v>v'$ and v''.

We write $A \overset{\wedge}{\wedge} B$ to denote the set $\{(z',z'')\,|\,z' \overset{\wedge}{\wedge} z''\}$ with $A\cap B=\emptyset$. Likewise we write $A \overset{\wedge}{\wedge} B \overset{\wedge}{\wedge} C$ to denote $\{(z',z'',z''')\,|\,z' \overset{\wedge}{\wedge} z'' \overset{\wedge}{\wedge} z'''\}$ with A,B and C mutually exclusive. These 'increments' on R_z lead to new type 2 and type 3 integrals extending previous results found in [4, 5, 6, 7, 8, 9].

Let $a\in\mathscr{C}_z$. A type 1 elementary process is one of the form $a\chi_{\Delta_z}$ with $inf\Delta_z = z$. So Δ_z is an increment forward of $a\in\mathscr{C}_z$. A type 2 elementary process is of the form $a\chi_{\Delta_{z_1}}\chi_{\Delta_{z_2}}$ with $\Delta_{z_1} \overset{\wedge}{\wedge} \Delta_{z_2}$ and each forward increments with respect to $a\in\mathscr{C}_z$. A type 3 elementary process is one of the form $a\chi_{\Delta_{z_1}}\chi_{\Delta_{z_2}}\chi_{\Delta_{z_3}}$ with $\Delta_{z_1} \overset{\wedge}{\wedge} \Delta_{z_2} \overset{\wedge}{\wedge} \Delta_{z_3}$ again with each an increment forward of the parameter space upon which $a\in\mathscr{C}_z$ is based. Type 1, 2 and 3 simple processes are then formed through summing similar types of elementary process. This discussion can be extended to the general case where we define elementary r adapted processes as follows.

Definition A $\mathscr{C}$ - valued elementary r adapted process is of the form

$$h:\mathbb{R}^n_+\times\ldots\mathbb{R}^n_+ \longrightarrow L^2(\mathscr{C}) \text{ by } (z_1,\ldots,z_r)\mapsto h(z_1,\ldots,z_r) = a\prod_{i=1}^{r}\chi_{\Delta_i}(z_i)$$

with $\Delta_1 \overset{\wedge}{\wedge} \cdots \overset{\wedge}{\wedge} \Delta_r$ and $a\in\mathscr{C}_{inf\Delta_1\vee\ldots\vee inf\Delta_r}$

QUANTUM STOCHASTIC INTEGRALS

We consider the case of Clifford stochastic integrals for $\mathfrak{h} = L^2(\mathbb{R}^n_+)$ with $n \in \mathbb{N}^+$.

Definition For h a $\mathscr{C}$ - valued elementary r adapted process we define the type r integral $\mathscr{S}_r$ of h over R_z to be

$$\mathscr{S}_r(h,z,f_1,\dots,f_r) = \int_{R_z} \dots \int_{R_z} h(z_1,\dots,z_r)d\psi_{z_1}(f_1)\dots d\psi_{z_1}(f_r) = a\prod_{i=1}^{r}\psi(\chi_{\Delta_i \cap R_z} f_i)$$

For $r = 1$ type 1 quantum stochastic analogues of the classical Itô integral result [1], and for $r = 2$ quantum analogues of the Wong Zakai integral are produced, [5, 6]. For simple adapted processes we extend the above definition by linearity.

Properties

It may be shown that each type r integral satisfies isometry conditions of the form $\| \mathscr{S}_r(h,z,f_1,\dots,f_r)\|_2^2 = \int_{R_z} \dots \int_{R_z} \| h\|_2^2 \prod_{j=1}^{r} |f_j|^2 dz_1 dz_2 \dots dz_r$. and are orthogonal to each of the the other types of integral [10]. The integrals are also centred martingales.

Theorem The type r integrals $(\mathscr{S}_r(h,z,f_1,\dots,f_r))$ form centred martingales.

Proof. Let $X_{z'} = a\psi(\chi_{\Delta_1})\dots\psi(\chi_{\Delta_r})$ denote a type r integral with $1 \le r \le n$. We want to show that $m(X_{z'}|\mathscr{C}_z) = X_z$ and that $m(X_z) = 0$. Both of these properties follow by independence of the gage. For the centred property we note that each of the increments χ_{Δ_i} with $1 \le i \le r$ have a parameter space mutually exclusive to each other, and also to the parameter space from which $a \in \mathscr{C}_{inf\Delta_1 \vee \dots \vee inf\Delta_r}$ is generated. It follows that $\exists \Delta_j$ with $1 \le j \le r$ such that $|m(a\prod_{i=1}^{r}\chi_{\Delta_i}(z_i))| = |m(a\prod_{i=1, i\neq j}^{r}\chi_{\Delta_i}(z_i)).m(\chi_{\Delta_j}(z_j))| = 0$ by independence of the gage m, [16].

For the martingale property one can work with either the conditional expectation operator $D^{-1}\Gamma(\chi_{R_z})D$ for D the duality transform [4], or employ properties of the gage [5, 15, 16]. For $X \in L^1(\mathscr{C}), z \in \mathbb{R}^n_+$ and $c \in \mathscr{C}_z$ the conditional expectation operator $m(\ |\mathscr{C}_z) \in L^1(\mathscr{C})$ is unique, and satisfies $m(cm(X|\mathscr{C}_z)) = m(cX)$ [5].

$$\begin{aligned} m(X_{z'}|\mathscr{C}_z) - X_z &= m(X_{z'} - X_z|\mathscr{C}_z) \\ &= m(a(\prod_{i=1}^{r}\psi(\chi_{\Delta_i \cap R'_z \cap R_{z'}} f_i) \pm \Sigma \prod_{\substack{i=1 \\ k+k'=r}}^{k} \psi(\chi_{\Delta_i \cap R_z} f_i)\prod_{j=1}^{k'}\psi(\chi_{\Delta_j \cap R'_z \cap R_{z'}} f_j))|\mathscr{C}_z) \end{aligned}$$

Since $m(m(X_{z'} - X_z|\mathscr{C}_z)) = m(\mathbb{I}_{\mathscr{C}_z}(X_{z'} - X_z)) = 0$ by independence of the gage, and the conditional expectation operator is idempotent, and self adjoint on the ψ, it follows, since the gage is faithful, that $m(X_{z'} - X_z|\mathscr{C}_z)) = m(X_{z'}|\mathscr{C}_z)) - X_z = 0$.

□

Completion

The type r integrals discussed above extend via the isometry property to completions in $L^2(\mathscr{C})$ where the isometry, orthogonality and centred martingale properties continue to hold. Further research [17] has been carried out in this and related areas the details of which will appear elsewhere.

ACKNOWLEDGMENTS

Many thanks to all at QCMC08 for their warm hospitality and stimulating discussions.

REFERENCES

1. Kiyosi Ito, *Memoirs of the American Mathematical Society*, **4**, 1-51, 1951.
2. E. Wong and M. Zakai Z. *Wahrscheinlichkeitstheorie und Verw. Gebiete*, **29**, 109-122, 1974.
3. P. Imkeller, *Stochastic Processes and their Applications*, **20**, 1-40, 1985.
4. C. Barnett, R. F. Streater and I. F. Wilde, *J. Funct. Analysis*, **52** (1), 19-47, 1982.
5. William J. Spring and Ivan F. Wilde, *Reports on Mathematical Physics*, **42**, 389-399, 1998.
6. William J. Spring and Ivan F. Wilde, *Reports on Mathematical Physics*,**49**, 63-76, 2002.
7. William J. Spring and Ivan F. Wilde, *Quantum Probability and White Noise Analysis*, **15**, 243-253, 2003.
8. William J. Spring, *Quantum Communication, Measurement and Computing*, **8**, 409-412, American Institute of Physics, 2006.
9. William J. Spring, 28th Quantum Probability and Related Topics, CIMAT, Guanajuato, Mexico, 2nd September - 8th September, 2007.
10. William J. Spring, Fifth International Conference of Applied Mathematics and Computing, Plovdiv, Bulgaria, 12th August - 18th August, 2008. Submitted to Int. J. Pure and App. Math.2008.
11. P. E. Kopp, *Martingales and Stochastic Integrals*, Cambridge University Press, 1984.
12. I. E. Segal and R. A. Kunze, *Integrals and Operators*, Grundlehren der mathematischen Wissenschaften, Springer-Verlag, 1978.
13. Ola Bratteli and Derek W. Robinson *Operator Algebras and Quantum Statistical Mechanics II*, Texts and Monographs in Physics, Springer-Verlag, 1981.
14. M. Terp, *L^p Spaces Associated with Von Neumann Algebras*, Lic. Scient. Thesis, University of Odense, 1981.
15. I. Segal, Ann. of Math,**57**, 401-457, 1953.
16. L. Gross, J. Funct. Analysis,**10**,52-109, 1972.
17. William J. Spring, *(In Preparation)*, 2008.

Coherent-State Quantum Process Tomography

Mirko Lobino,[1] Dmitry Korystov,[1] Connor Kupchak,[1] Eden Figueroa,[1] Barry C. Sanders,[1] and A. I. Lvovsky[1]

[1] *Institute for Quantum Information Science, University of Calgary, Calgary, Alberta T2N 1N4, Canada*

Abstract. The technologies of quantum information and quantum control are rapidly improving, but full exploitation of their capabilities requires complete characterization and assessment of processes that occur within quantum devices. We present a method for characterizing, with arbitrarily high accuracy, any quantum optical process. Our protocol recovers complete knowledge of the process by studying, via homodyne tomography, its effect on a set of coherent states, i.e. classical fields produced by common laser sources. We demonstrate the capability of our protocol by evaluating and experimentally verifying the effect of a test process on squeezed vacuum.

Keywords: Quantum process tomography
PACS: 03.65.Wj

A quantum process $\mathscr{E}$ can be represented by a completely positive, trace-preserving linear map (superoperator) on the linear space $\mathbb{L}(\mathbb{H})$ of all density matrices over Hilbert space $\mathbb{H}$. It can be expressed as a rank-4 tensor $\mathscr{E}_{lk}^{nm}$ that relates the matrix elements of the output $\mathscr{E}(\hat{\rho})$ and input $\hat{\rho}$ states in some basis:

$$[\mathscr{E}(\hat{\rho})]_{lk} = \sum_{nm} \mathscr{E}_{lk}^{nm} \rho_{nm}, \tag{1}$$

where summation is from 1 to $\dim\mathbb{H}$.

Characterization of a process (known as quantum process tomography, or QPT) means finding all components of the superoperator tensor. It can be implemented by determining the output state for each of the $(\dim\mathbb{H})^2$ elements of a spanning set of $\mathbb{L}(\mathbb{H})$. Such a direct approach to QPT [1] was experimentally realized on one qubit teleportation [2], the Hamiltonian evolution of vibrational states of atoms in an optical lattice [3], on a two-qubit controlled-NOT gate [4, 5] and Bell-state filter [6]. As an alternative, ancilla-assisted QPT exploits an isomorphism between processes and states [7] and has been used to characterize a controlled-NOT gate [8] and a general single qubit gate [9, 10].

Existing QPT methods suffer from a number of shortcomings. As a rule, they require an unwieldy set of input states and measurements, preventing scalability and restricting accessible systems to very low dimension. In optics, QPT has been applied to processes on one or two dual-rail qubits, with post-selection based on photon coincidences projecting the input and output states onto these qubit subspaces. This approach cannot provide complete information about a state or a process because optical losses, imperfect sources, detector dark counts, and other imperfections lead to departure from the qubit subspaces. In post-selected tomography, such phenomena are simply eliminated from the analysis.

CP1110, *Quantum Communication, Measurement and Computing (QCMC), The Ninth International Conference,* edited by A. Lvovsky

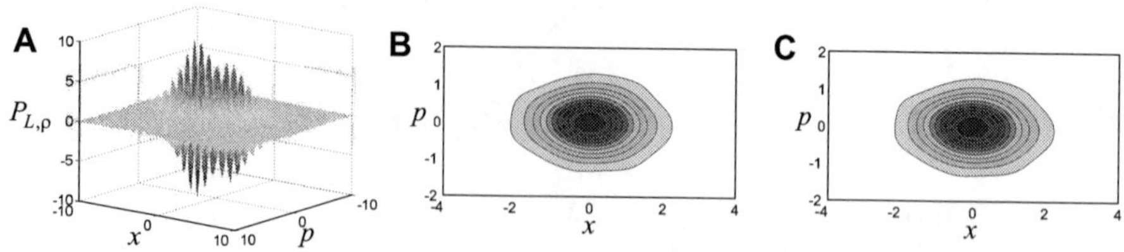

FIGURE 1. (**A**) Regularized Glauber-Sudarshan decomposition of the squeezed state. (**B**) and (**C**) Wigner representations of, respectively, the original and approximated squeezed vacuum states.

Here we introduce a scheme that enables *complete* characterization of a general quantum-optical process [11]. We use optical homodyne tomography to obtain full information on the process across all photon number sectors and also the coherence between sectors. As inputs, we use only coherent states that are readily available from a laser source, which makes the method easily scaled up.

Our method is based on the fact that any density matrix $\hat{\rho}$ can be represented as a sum of coherent states' density matrice [12, 13]:

$$\hat{\rho} = 2\int P_\rho(\alpha)|\alpha\rangle\langle\alpha|d^2\alpha, \tag{2}$$

where $P_\rho(\alpha)$ is the state's Glauber-Sudarshan P function, α is the coherent state with mean position and momentum observables $(x,p) = (\sqrt{2}\mathrm{Re}\,\alpha, \sqrt{2}\mathrm{Im}\,\alpha)$. We use the convention $[\hat{x},\hat{p}] = i$ and integration is performed over the entire phase space. Therefore, if we know the effect $|\alpha\rangle\langle\alpha| \mapsto \hat{\rho}(\alpha) = \mathscr{E}(|\alpha\rangle\langle\alpha|)$ of the process on all coherent states, we can predict its effect upon state $\hat{\rho}$:

$$\mathscr{E}(\hat{\rho}) = 2\int P_\rho(\alpha)\hat{\rho}(\alpha)\,d^2\alpha. \tag{3}$$

An obstacle to direct application of this approach is posed by singular behavior of the function $P_\rho(\alpha)$. Fortunately, this can be overcome by applying a theorem proven by Klauder [14]: for any bounded operator $\hat{\rho}$ there exists an operator $\hat{\rho}_L$ with continuous and rapidly decreasing P function arbitrarily close to $\hat{\rho}$ in the trace-class norm.

As an example, we applied the Klauder approximation to squeezed vacuum that has a noise reduction in the squeezed quadrature of -1.58 dB and excess noise in the orthogonal quadrature of 2.91 dB. As evidenced by Fig. 1, this state, whose P function is highly singular, can be well approximated by a state with a reguilar Glauber-Sudarshan fucntion. The two states exhibit a quantum fidelity of more than 0.9999.

Although the above method permits finding the process output for an arbitrary input state, it requires one first to determine the input state's P function. This step can be avoided by calculating the process superoperator in the Fock basis, so the output can be found from the input density matrix according to Eq. 1. To this end, we express the Glauber-Sudarshan function as

$$P_\rho(\alpha) = \sum_{mn} \rho_{nm} P_{nm}(\alpha), \tag{4}$$

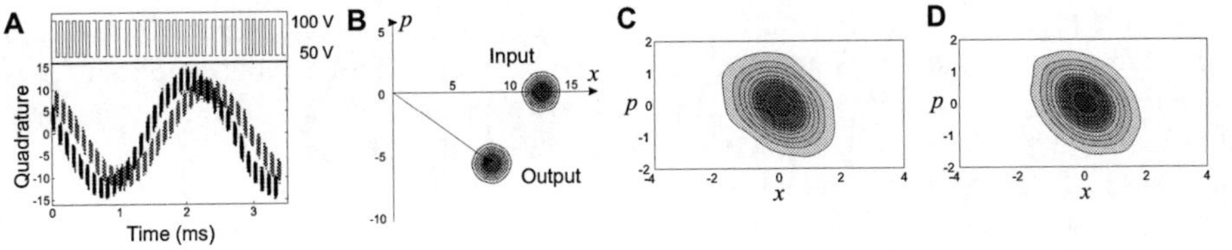

FIGURE 2. (**A**) Time-dependent quadrature values acquired from homodyne detection of a coherent state with input $\alpha = 8.3$. Black dots correspond to the state before the process; red dots, after the process. The top curve shows the EOM driving voltage. (**B**) The Wigner function of the coherent state before and after the process. (**C**) and (**D**) Wigner representations of the measured output squeezed state compared to the one obtained from process tomography.

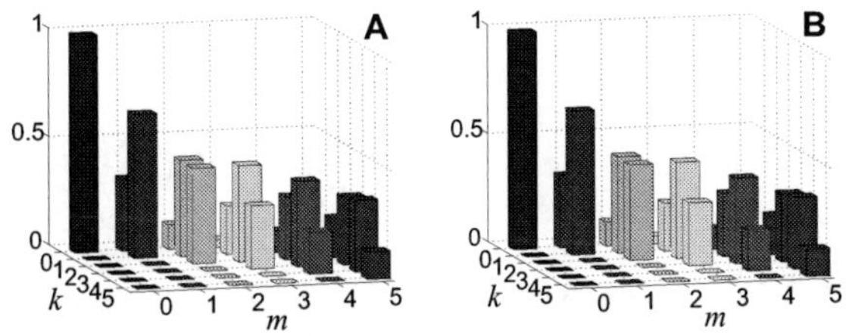

FIGURE 3. The "diagonal" values of the superoperator $\mathcal{E}_{kk}^{mm}$. (**A**): as obtained in the experiment. (**B**): theoretical model.

where $P_{nm}(\alpha)$ is the P function of the operator $|n\rangle\langle m|$. We now replace these functions by their regularized versions $P_{L,nm}(\alpha)$ and rewrite Eq. 3 as

$$\mathcal{E}(\hat{\rho}) = 2\sum_{nm} \rho_{nm} \int P_{L,nm}(\alpha)\hat{\rho}(\alpha)\, d^2\alpha, \tag{5}$$

from which we determine the process superoperator as

$$\mathcal{E}_{lk}^{nm} = 2\int P_{L,nm}(\alpha)\rho_{lk}(\alpha)\, d^2\alpha. \tag{6}$$

A simplification arises for phase-symmetric processes, in which there is no phase coherence between the "processing unit" and input states. In this case, if two inputs $\hat{\rho}$ and $\hat{\rho}_1$ are different by an optical phase shift $\hat{U}(\varphi)$, the states $\mathcal{E}(\hat{\rho})$ and $\mathcal{E}(\hat{\rho}_1)$ will differ by the same phase shift:

$$\mathcal{E}[\hat{U}(\varphi)\hat{\rho}\hat{U}^\dagger(\varphi)] = \hat{U}(\varphi)\mathcal{E}(\hat{\rho})\hat{U}^\dagger(\varphi). \tag{7}$$

Then, if we know the effect of the process on a coherent state $|\alpha\rangle$, we also know what happens to $|\alpha e^{i\varphi}\rangle$.

In order to test this procedure, we have implemented a test process using an electro optical modulator (EOM) followed by a polarizer. The field experienced minimal distortion when a bias voltage $V_1 = 100$ V was applied to the EOM. Switching the voltage to

$V_2 = 50$ V produced birefringence, and thus losses at the polarizer, along with a phase shift. We reconstructed the input and output states at 11 different input amplitude levels between $\alpha_1 = 0$ and $\alpha_{11} = 10.9$. In order to keep track of the relative phase shift, the EOM voltage was switched between V_1 and V_2 every 100 μs (Fig. 2A, top) while the phase of the local oscillator was linearly scanned by a piezoelectric transducer at 100 Hz. The homodyne photocurrent was recorded with an oscilloscope (Fig. 2A) and 50,000 phase and quadrature values for both the input and output states were sampled and used to calculate density matrices by likelihood maximization [15, 16] (Fig. 2B). The output state reconstruction showed a phase shift of 36° and a loss of 34 % with respect to the input state. The experimental data have been used to determine the process superoperator tensor with the phase symmetry assumption in Eq. 7. The elements $\mathscr{E}_{kk}^{mm}$ of the tensor in the photon number basis are plotted in Fig. 3A. A theoretical model for the process tensor has been calculated using the Bernoulli transformation to account for a lossy channel and a phase shift superoperator (Fig. 3B). This calculated tensor bears close resemblance to the experimental result.

For additional verification, we applied this result to predict the effect of the process on the squeezed vacuum state described in the previous section. The state before (Fig. 1B) and after (Fig. 2C) the process was reconstructed using homodyne tomography as described above. By applying the process superoperator to the input squeezed state, we also determine the prediction for the process output (Fig. 2D). The maximum (minimum) quadrature noise variance amounted to 2.19(−1.07) dB for the measured state, and 2.15(−0.95) dB for the prediction, corresponding to a quantum fidelity of 0.994.

REFERENCES

1. J. F. Poyatos, J. I. Cirac, P. Zoller, *Phys. Rev. Lett.* **78**, 390 (1997).
2. M. A. Nielsen, E. Knill, R. Laflamme, *Nature* **396**, 52 (1998).
3. S. H. Myrskog, J. K. Fox, M. W. Mitchell, A. M. Steinberg, *Phys. Rev. A* **72**, 013615 (2005).
4. J. L. O'Brien *et al., Phys. Rev. Lett.* **93**, 080502 (2004).
5. A. M. Childs, I. L. Chuang, D. W. Lueng *Phys. Rev. A* **64**, 012314 (2001).
6. M. W. Mitchell, C. W. Ellenor, S. Schneider, A. M. Steinberg *Phys. Rev. Lett.* **91**, 120402 (2003).
7. G. M. D'Ariano, P. Lo Presti, *Phys. Rev. Lett.* **86**, 4195 (2001).
8. M. Riebe *et al., Phys. Rev. Lett.* **97**, 220407 (2006).
9. J. B. Altepeter *et al., Phys. Rev. Lett.* **90**, 193601 (2003).
10. F. De Martini, A. Mazzei, M. Ricci, G. M. D'Ariano, *Phys. Rev. A* **67**, 062307 (2003).
11. M. Lobino *et al.*, Science **322**, 563 (2008)
12. R. J. Glauber, *Phys. Rev. Lett.* **10**, 84 (1963).
13. E. C. G. Sudarshan, *Phys. Rev. Lett.* **10**, 227 (1963).
14. J. R. Klauder, *Phys. Rev. Lett.* **16**, 534 (1966).
15. A. I. Lvovsky, *J. Opt. B: Q. Semiclass. Opt.* **6**, S556 (2004).
16. J. Řeháček, Z. Hradil, E. Knill, A. I. Lvovsky, *Phys. Rev. A* **75**, 042108 (2007).

This work was supported by NSERC, CIFAR, iCORE, AIF, CFI and Quantum*Works*. A. I. L. is a CIFAR scholar, B. C. S. is a CIFAR associate. Figures are partially reproduced from Ref. [11]

Characterization of quantum processes with a faulty Bell-state analyzer

M. Mohseni* and A. T. Rezakhani†

*Department of Chemistry and Chemical Biology, Harvard University, 12 Oxford St., Cambridge, MA 02138, USA
Research Laboratory of Electronics, Massachusetts Institute of Technology, 77 Massachusetts Avenue, Cambridge, MA 02139, USA
†Center for Quantum Information Science and Technology, University of Southern California, Los Angeles, CA 90089, USA

Abstract. Accuracy of quantum process estimation is limited to noisy operations in the preparations and measurements devices. Such imperfections may lead to inaccuracies in the process reconstruction due to the fact that standard data analysis techniques presume ideal devices. Here, we provide a scheme for estimation of completely positive quantum maps acting on qubit systems using calibrated faulty state generator and analyzer. Specifically, we present a generalization of "Direct Characterization of Quantum Dynamics" [M. Mohseni and D. A. Lidar, **97**, 170501 (2006)] with an imperfect Bell-state analyzer.

Keywords: Quantum Process Tomography, Quantum Maps, Bell-state Measurements
PACS: 03.65.Wj, 03.67.-a, 03.67.Pp

Quantum process tomography provides a general experimental procedure for estimating dynamics of a system which has an unknown interaction with its embedding environment [1, 2, 3, 4]. In these methods the full information is obtained by a complete set of experimental settings associated with the set of required input states and non-commuting measurements. In recent developments [3, 4, 5, 6, 7, 8], it has been demonstrated that the minimal number of required experimental settings can indeed be substantially reduced by using degrees of freedom of auxiliary quantum systems correlated with the system of interest. In particular, Direct Characterization of Quantum Dynamics (DCQD) [4, 5, 6] has been proposed as a scheme for measuring certain superoperator parameters with a single Bell-state analyzer. The construction of the full information about the dynamical process is then possible via preparation of a set of mutually unbiased input states over a subspace of the total Hilbert space of the principal system and an ancilla [5]. The DCQD approach has been originally developed with the assumptions of ideal (i.e., error-free) quantum state preparation, measurement, and ancilla channels. However, in a realistic estimation process, due to decoherence, limited preparation/measurement accuracies, or other imperfections, certain errors may occur hindering the overall procedure. In this work, we generalize the DCQD approach to the cases in which the preparations and measurements are realized with known systematic faulty operations.

According to the DCQD scheme two quantum systems A and B are prepared in certain forms of Bell correlated states before an unknown quantum operation is applied on system A, and next a Bell-state measurement in performed on the combined system [4, 5] (see Table 1). Much progress have been made in manipulation of two-body correlations

CP1110, *Quantum Communication, Measurement and Computing (QCMC), The Ninth International Conference,* edited by A. Lvovsky

TABLE 1. Input states and measurements for direct characterization of single-qubit χ, where $\mathscr{E}(\rho) = \sum_{mn}\chi_{mn}(t)\sigma_m^A\rho\sigma_n^A$, and $\mathscr{E}$, is a completely positive quantum dynamical map acting on the system A. The positive matrix $\chi = [\chi_{mn}]$ encompasses all information about the dynamics relative to the Pauli operator basis. Here $|\Phi_\alpha^+\rangle = \alpha|00\rangle + \beta|11\rangle$, $|\Phi_\alpha^+\rangle_{x(y)} = \alpha|++\rangle_{x(y)} + \beta|--\rangle_{x(y)}$ where $|\alpha| \neq |\beta| \neq 0$ and $\mathrm{Im}(\bar{\alpha}\beta) \neq 0$, and $\{|0\rangle,|1\rangle\}$, $\{|\pm\rangle_x\}$, $\{|\pm\rangle_y\}$ are eigenstates of the Pauli operators σ_z, σ_x, and σ_y. P_{Φ^+} is the projector on the Bell state $|\Phi^+\rangle$, and similarly for the other projectors. See Refs. [7, 9]

input state	Bell-state analyzer	χ_{mn}
$\|\Phi^+\rangle$	$P_{\Psi^\pm}, P_{\Phi^\pm}$	$\chi_{00}, \chi_{11}, \chi_{22}, \chi_{33}$
$\|\Phi_\alpha^+\rangle$	$P_{\Phi^+} \pm P_{\Phi^-}, P_{\Psi^+} \pm P_{\Psi^-}$	χ_{03}, χ_{12}
$\|\Phi_\alpha^+\rangle_x$	$P_{\Phi^+} \pm P_{\Psi^+}, P_{\Phi^-} \pm P_{\Psi^-}$	χ_{01}, χ_{23}
$\|\Phi_\alpha^+\rangle_y$	$P_{\Phi^+} \pm P_{\Psi^-}, P_{\Phi^-} \pm P_{\Psi^+}$	χ_{02}, χ_{13}

in different physical systems, however, in almost all of the cases high fidelity entangled Bell-state preparations and measurements are still lacking. To address this issue, we note that any experimentally known two-body correlation, as initial preparation or final measurement, can be described by an ideal Bell-state preparation and measurement followed by appropriately chosen completely-positive (CP) fault maps acting on the combined system. Note that any other quantum map acting on ancilla system B can be included in either preparation or the measurement CP fault maps. This naturally provides a generalization of the DCQD scheme to the cases of calibrated faulty preparations, measurement, and ancilla channels. For simplicity, in this work we concentrate on the qubit systems; however, generalization to qudit systems with d being a power of a prime is straightforward according to the construction of Ref. [5].

Let us consider a maximally entangled prepared state between the qubits A (the system of interest) and B (the ancilla) as $|\Phi^+\rangle_{AB} = (|00\rangle + |11\rangle)_{AB}/\sqrt{2}$. We first apply a *known* quantum fault map $\mathscr{E}^{(\mathrm{i})}$ to qubit A and B: $\mathscr{E}^{(\mathrm{i})}(\rho) = \sum_{pqrs}\chi_{pqrs}^{(\mathrm{i})}\sigma_p^A\sigma_q^B\rho\sigma_r^B\sigma_s^A$, where $\rho = |\Phi^+\rangle\langle\Phi^+|$ and $\{\sigma_0 \equiv I, \sigma_1, \sigma_2, \sigma_3\}$ are the identity and Pauli operator basis for qubits. Next, we apply the unknown quantum map $\mathscr{E}$: $\mathscr{E}[\mathscr{E}^{(\mathrm{i})}(\rho)] = \sum_{mn}\chi_{mn}\sigma_m^A(\sum_{pqrs}\chi_{pqrs}^{(\mathrm{i})}\sigma_p^A\sigma_q^B\rho\sigma_r^B\sigma_s^A)\sigma_n^A$. Finally we apply a *known* quantum fault map before the Bell-state measurement. (Note that in this approach any error on the ancilla channel can be absorbed into either $\mathscr{E}^{(\mathrm{f})}$ or $\mathscr{E}^{(\mathrm{i})}$). The total map acting on the combined system AB is $\mathscr{E}^{(\mathrm{T})} = \mathscr{E}^{(\mathrm{f})} \circ \mathscr{E} \circ \mathscr{E}^{(\mathrm{i})}$, given by:

$$\mathscr{E}^{(\mathrm{T})}(\rho) = \sum_{mnpp'qq'rr'ss'}\chi_{mn}\chi_{pqrs}^{(\mathrm{i})}\chi_{p'q'r's'}^{(\mathrm{f})}(\sigma_{p'}^A\sigma_{q'}^B\sigma_m^A\sigma_p^A\sigma_q^B)\rho(\sigma_r^B\sigma_s^A\sigma_n^A\sigma_{r'}^B\sigma_{s'}^A), \quad (1)$$

where the parameters $\chi_{pqrs}^{(\mathrm{i})}$ and $\chi_{p'q'r's'}^{(\mathrm{f})}$ are known (from calibration of the operational/systematic errors in the preparation and measurement devices). Now by defining $\omega_{mnp's'} = (-1)^{(\delta_{m0}-1)(\delta_{p'0}-1)\delta_{mp'}+(\delta_{n0}-1)(\delta_{s'0}-1)\delta_{ns'}}$ and

$$\widetilde{\rho}_{mn} = \sum_{pp'qq'rr'ss'}\omega_{mnp's'}\chi_{pqrs}^{(\mathrm{i})}\chi_{p'q'r's'}^{(\mathrm{f})}(\sigma_{p'}^A\sigma_{q'}^B\sigma_p^A\sigma_q^B)\rho(\sigma_r^B\sigma_s^A\sigma_{r'}^B\sigma_{s'}^A),$$

Eq. (1) can be written as follows: [1]

$$\mathscr{E}^{(\mathrm{T})}(\rho) = \sum_{mn} \chi_{mn} \sigma_m^A \widetilde{\rho}_{mn} \sigma_n^A.$$

By construction, the parameters $\chi^{(\mathrm{i})}_{pqrs}$ and $\chi^{(\mathrm{f})}_{p'q'r's'}$ are all *a priori* known; the matrices $\widetilde{\rho}_{mn}$ which are functions of $\chi^{(\mathrm{i})}$, $\chi^{(\mathrm{f})}$, and the initial state ρ, are also known. Therefore, in order to develop a generalized DCQD scheme for the systems with faulty Bell-state preparation (BSP) and measurement (BSM), we need to consider a set of modified (input) states, $\widetilde{\rho}_{mn}$, rather than a pure Bell-state-like inputs. Expanding $\widetilde{\rho}_{mn}$ in the Bell basis yields

$$\widetilde{\rho}_{mn} = \sum_{kk'} \lambda_{mn}^{kk'} P^{kk'},$$

where $\lambda_{mn}^{kk'} = \mathrm{Tr}[P^{k'k}\widetilde{\rho}_{mn}]$, $P^{kk'} = |\mathscr{B}_k\rangle\langle\mathscr{B}_{k'}|$, and $|\mathscr{B}_k\rangle$ for $k = 0,1,2,3$ correspond to the Bell-states $|\Phi^+\rangle$, $|\Psi^+\rangle$, $|\Psi^-\rangle$, and $|\Phi^-\rangle$, respectively, where $|\Phi^{\pm}\rangle = (|00\rangle \pm |11\rangle)/\sqrt{2}$, $|\Psi^{\pm}\rangle = (|01\rangle \pm |10\rangle)/\sqrt{2}$. (Henceforth throughout the paper, superscripts refer to Bell basis and subscripts refer to Pauli operator basis.) The $\lambda_{mn}^{kk'}$ is a known function of $\omega_{mnp's'}$, $\chi^{(\mathrm{f})}$, $\chi^{(\mathrm{i})}$, and ρ. The overall output state can be rewritten as:

$$\mathscr{E}^{(\mathrm{T})}(\rho) = \sum_{kk'mn} \lambda_{mn}^{kk'} \chi_{mn} \sigma_m^A P^{kk'} \sigma_n^A.$$

We now apply the standard DCQD data analysis to read the matrix elements of $\chi^{(\mathrm{T})}$ (representing $\mathscr{E}^{(\mathrm{T})}$). After performing a BSM, i.e., $\{P^{jj}\}_{j=0}^3$, we obtain the Bell-state $|\mathscr{B}_j\rangle$ with the probability:

$$\mathrm{Tr}[P^{jj}\mathscr{E}^{(\mathrm{T})}(\rho)] = \sum_{kk'mn} \Lambda^{(j)}_{kk',mn} \chi_{mn},$$

where $\Lambda^{(j)}_{kk',mn} = \lambda_{mn}^{kk'} \mathrm{Tr}[P^{jj}\sigma_m^A P^{kk'} \sigma_n^A]$. Similar set of equations for the inputs $\{\rho^{(i)}\}_{i=0}^3$ of the standard DCQD can be written. We represent all of these equations in a compact vector form as $|\chi^{(\mathrm{T})}\rangle = \Lambda|\chi\rangle$, where the $\Lambda(\chi^{(\mathrm{i})}, \chi^{(\mathrm{f})}, \{\rho^{(i)}\})$ matrix contains full information about all faulty experimental configurations. Given $\chi^{(\mathrm{i})}$, $\chi^{(\mathrm{f})}$, and the set of inputs $\{\rho^{(i)}\}$ for the standard DCQD, one can calculate the Λ matrix. The standard DCQD experimental data analysis also determines $|\chi^{(\mathrm{T})}\rangle$. Now, *if* the Λ matrix is *invertible*, one can obtain χ matrix by an inversion: $|\chi\rangle = \Lambda^{-1}|\chi^{(\mathrm{T})}\rangle$. Invertibility of the Λ matrix, i.e., $\det\Lambda \neq 0$, depends on the input states $\{\rho^{(i)}\}$ and the fault operations $\chi^{(\mathrm{i})}$ and $\chi^{(\mathrm{f})}$. It may happen that the Λ matrix becomes ill-conditioned [10] for a specific set of input states (for some given noise operations $\chi^{(\mathrm{i})}$ and $\chi^{(\mathrm{f})}$). That is, even small errors, whether operational, stochastic, or round-off, in estimation of $\chi^{(\mathrm{T})}$ can be amplified after multiplication by Λ^{-1}. This in turn can render the estimation of χ — the sought after unknown map $\mathscr{E}$ — completely unreliable. To minimize the statistical errors in this case, the input states should be chosen such that $\det\Lambda$ becomes as far from zero as possible.

[1] Here we have used the commutation relation: $\sigma_a\sigma_b = (-1)^{(\delta_{a0}-1)(\delta_{b0}-1)\delta_{ab}}\sigma_b\sigma_a$.

That is, given $\chi^{(\mathrm{i})}$ and $\chi^{(\mathrm{f})}$, we need to maximize $\det\Lambda$ to obtain $\{\rho_{\mathrm{opt}}^{(i)}\}$, the optimal set of input states (optimal in the sense of statistical errors). The generalization of this approach to higher-dimensional systems is straightforward by introduction of generalized Pauli operators [5].

Summary.—We have provided a theoretical scheme for utilizing auxiliary quantum correlations to perform process estimation tasks with faulty quantum operations. Our proposed method may have near-term applications to a variety of realistic quantum systems or devices, such as in trapped ions, liquid-state NMR, optical lattices, and entangled pairs of photons with the current state-of-the-technology.

Acknowledgments.—We thank A. Aspuru-Guzik, J.T. Barreiro, P. G. Kwiat, D. A. Lidar, A. M. Steinberg for helpful discussions. This work was supported by Faculty of Arts and Sciences of Harvard University, and Natural Sciences and Engineering Research Council of Canada (NSERC) to MM, and the USC Center for Quantum Information Science and Technology, *i*CORE, MITACS, and PIMS to AR.

REFERENCES

1. M. Paris and J. Rehacek (Ed.), *Quantum State Estimation* (Springer-Verlag, Berlin, 2004).
2. M. A. Nielsen and I. L. Chuang, *Quantum Computation and Quantum Information* (Cambridge University Press, Cambridge, UK, 2001).
3. J. B. Altepeter *et al.*, Phys. Rev. Lett. **90**, 193601 (2003).
4. M. Mohseni and D. A. Lidar, Phys. Rev. Lett. **97**, 170501 (2006).
5. M. Mohseni and D. A. Lidar, Phys. Rev. A **75**, 062331 (2007).
6. Z. Wang *et al.*, Phys. Rev. A **75**, 044304 (2007); W. T. Liu *et al.*, *ibid.* **77**, 032328 (2008).
7. M. Mohseni, A. T. Rezakhani, and D. A. Lidar, Phys. Rev. A **77**, 032322 (2008).
8. A. Bendersky, F. Pastawski, and J. P. Paz, Phys. Rev. Lett. **100**, 190403 (2008).
9. M. Mohseni, A. T. Rezakhani, and A. Aspuru-Guzik, Phys. Rev. A **77**, 042320 (2008).
10. C. D. Meyer, *Matrix Analysis and Applied Linear Algebra* (SIAM, Philadelphia, 2000).

Process Characterization of Experimental Photonic One-Way Quantum Computing

Yuuki Tokunaga*,†, Shin Kuwashiro*, Takashi Yamamoto*, Masato Koashi* and Nobuyuki Imoto*

**Division of Materials Physics, Graduate school of Engineering Science, Osaka University, Toyonaka, Osaka 560-8531, Japan*
CREST Photonic Quantum Information Project, 4-1-8 Honmachi, Kawaguchi, Saitama 331-0012, Japan
†NTT Information Sharing Platform Laboratories, NTT Corporation, 3-9-11 Midori-cho, Musashino, Tokyo 180-8585, Japan

Abstract. We demonstrate process characterization of basic gates of an optical cluster-state quantum computing. We present complete process tomography of a single-qubit gate and evaluate the bounds on the process fidelity of a two-qubit gate.

Keywords: optical one-way quantum computing, process characterization
PACS: 03.67.Bg, 03.67.Lx, 42.50.Ex

INTRODUCTION

Cluster states are important entangled states for quantum information processing because they are used as a resource for one-way quantum computing (QC), in which computation proceeds by a sequence of single-qubit measurements with classical feedforward [1].

In this work, we report process characterization of a photonic one-way quantum computing. We present complete process tomography of a single-qubit gate using maximum-likelihood estimation, which is used for constructing physical process matrix from the experimental data. Obtaining full process tomography for two-qubit gates requires a large number of measurements [2], but we can evaluate the bounds on the process fidelity using fewer number of measurements than that of the full process tomography [3]. We also present the bounds on the process fidelity for an important two-qubit interacting gate, controlled-phase (C-Phase) gate.

We use four-photon cluster states, e.g., $|C_4\rangle = \frac{1}{2}(|HHHH\rangle + |HHVV\rangle + |VVHH\rangle - |VVVV\rangle)$, as a resource for one-way quantum computing. Here, $|H\rangle$ ($|V\rangle$) represents the state of a photon with horizontal (vertical) polarization. Our scheme for preparing $|C_4\rangle$ (Fig. 1) is a slight modification of the scheme in [4]. It is constructed from an entangled photon pair, two single photons, polarizing beam splitters (PBSs), half-wave plates (HWPs), and photon detectors.

CP1110, *Quantum Communication, Measurement and Computing (QCMC), The Ninth International Conference,* edited by A. Lvovsky

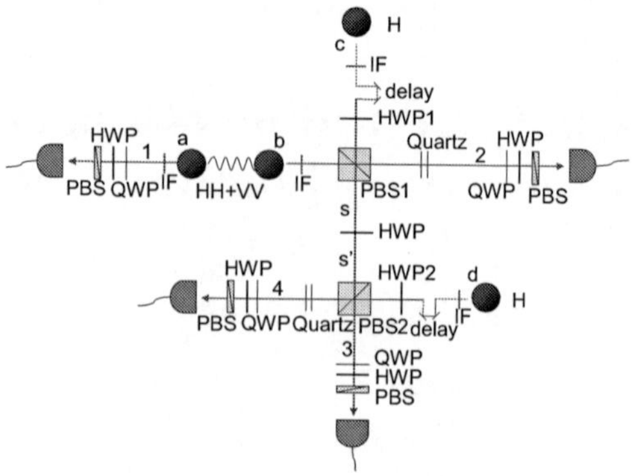

FIGURE 1. Experimental setup for preparing four-photon cluster states. $|C_4\rangle$.

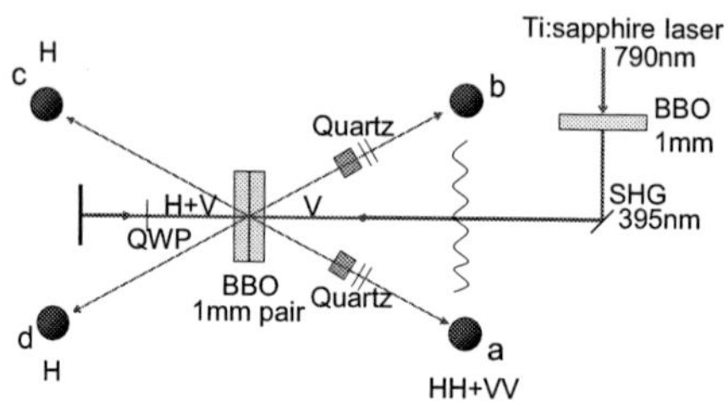

FIGURE 2. Experimental setup for preparing two single photons and an entangled photon pair.

EXPERIMENTAL SETUP

We use PDC for the preparation of the entangled photon pair [5] and two single photons (Fig. 2). An ultraviolet pulse with a central wavelength of 395 nm, an average power of 220 mW from a frequency-doubled mode-locked Ti:sapphire laser (wavelength, 790 nm; pulse width, 140 fs; repetition rate, 76 MHz) pumps a pair of 1mm-thick BBO (β-Barium Borate, type-I) crystal for PDC. The group delay is compensated by thick quartz crystals to erase the information on the origin (the first or the second BBO) of the photon pairs. The relative phase between H and V polarizations is adjusted by a pair of thin quartz crystals. The typical two-fold coincidence rate of entangled photon pairs is around 2500/s and the visibility is $\sim 97\%$. The temporal overlap is adjusted by moving mirrors on motorized stages (delays in Fig. 1). The thin quartz crystal pairs in modes 2 and 4 are placed to compensate additional phase shifts. The spectral filtering is achieved with narrow bandwidth interference filters (IFs) with bandwidth of 2.7 nm (FWHM). Photon detectors (silicon avalanche photodiodes) are placed after single-mode optical fibers to select a single spatial mode to ensure a high visibility. Polarization correlations are recorded by coincidence counting among four photon detectors for various angles of quarter-wave plates (QWPs) and HWPs. The typical four-fold coincidence rate is around 100 per hour.

We obtain lower bounds on the fidelity F of the produced state to $|C_4\rangle$ using methods with fewer measurement settings [6] compared to the method for obtaining the exact fidelity. In this experiment, we measured four settings $XXZZ$, $ZZXX$, $YYZZ$, and $ZZYY$. We obtain a lower bound on fidelity $F \geq 0.860 \pm 0.015$. from four-fold coincidence probabilities for the four settings.

ONE-WAY QUANTUM COMPUTATION

We demonstrate one-way quantum computing using the produced four-photon cluster state. Here we report characterization of the process of basic gate operations.

Single-qubit rotation. The quantum circuit in Fig. 3 (a) shows a simple implementation of a single-qubit rotation. This implementation is basically similar to [7, 8]. Input state $|\psi_{\text{in}}\rangle$ is prepared by measuring qubit 1. Qubits 2 and 3 are measured in the basis $B'(\alpha)$ and $B(\beta)$, respectively, where $B'(\theta) = \{\frac{|+\rangle + e^{-i\theta}|-\rangle}{\sqrt{2}}, \frac{|+\rangle - e^{-i\theta}|-\rangle}{\sqrt{2}}\}$ and $B(\theta) = \{\frac{|0\rangle + e^{-i\theta}|1\rangle}{\sqrt{2}}, \frac{|0\rangle - e^{-i\theta}|1\rangle}{\sqrt{2}}\}$. Here $|\pm\rangle = (|0\rangle \pm |1\rangle)/\sqrt{2}$ and we take $\{|0\rangle \equiv |H\rangle, |1\rangle \equiv |V\rangle\}$ as a standard basis. The gate proceeds as in Fig. 3 (a) and we obtain the outcome $|\psi_{\text{out}}\rangle = R_X(\beta)R_Z(\alpha)|\psi_{\text{in}}\rangle$, where $R_Z(\theta) = \exp(-i\sigma_Z/2)$ and $R_X(\theta) = \exp(-i\sigma_X/2)$. In this experiment, the feed-forwarded Pauli operations are substituted by the appropriate reassignment of measurement bases for qubits 1 and 4. We prepared four kinds of input states $|H\rangle$, $|V\rangle$, $|+\rangle$, or $|R\rangle$, chose $\alpha = \beta = \pi/2$, and performed tomographic measurements on the output states for each input state. From the obtained experimental data, the process matrix χ [9, 2, 10] is calculated by using maximum-liklihood estimation. The obtained matrix is shown in Fig. 4 and the process fidelity is calculated as $F_p = \text{Tr}(\chi_{ideal}\chi) = 0.904$.

Two-qubit gate. The quantum circuit in Fig. 3(b) shows an implementation of a two-qubit gate. Input state $|\psi_{\text{in}}\rangle$ is prepared by rotating the HWP1 and HWP2 in Fig. 1. Qubits 2 and 3 are measured in the basis $B(\alpha)$ and $B(\beta)$. The gate proceeds as in Fig. 3 (b) and we obtain the output state $|\psi_{\text{out}}\rangle = (R_Z(\alpha) \otimes R_Z(\beta))\text{CZ}|\psi_{\text{in}}\rangle$ on qubits 1 and 4. Here, CZ (C-Phase) operation is defined as $|j\rangle|k\rangle \mapsto (-1)^{jk}|j\rangle|k\rangle$, where $j,k \in 0,1$. We prepared eight kinds of input states $|\psi_{\text{in}}\rangle = |+\rangle|+\rangle$, $|+\rangle|-\rangle$,

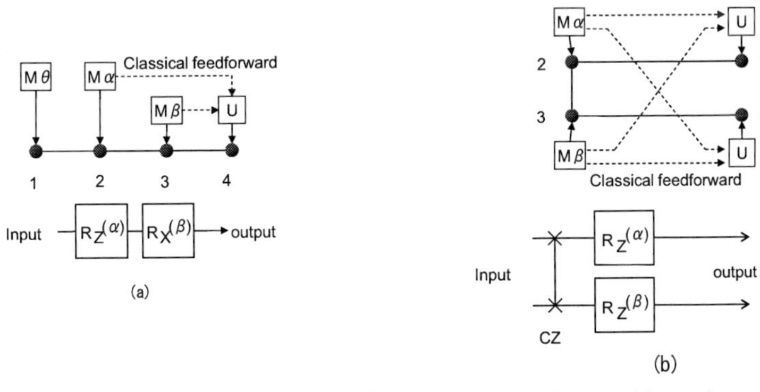

FIGURE 3. Physical implementation and the quantum circuit of (a) a single-qubit rotation and (b) a two-qubit gate.

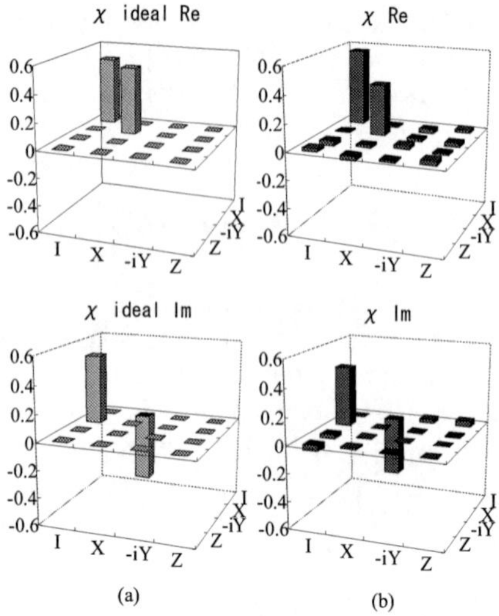

FIGURE 4. Process matrix χ of single-qubit rotation $R_X(\pi/2)R_Z(\pi/2)$. (a) Ideal matrix. (b) Maximum-likelihood experimental matrix.

$|-\rangle|+\rangle$, $|-\rangle|-\rangle$, $|R\rangle|R\rangle$, $|R\rangle|L\rangle$, $|L\rangle|R\rangle$, or $|L\rangle|L\rangle$. We chose $\alpha = \beta = 0$. In this case, the performed two-qubit gate is just the C-Phase gate. We obtain the bounds on the process fidelity of the C-Phase gate using the method of Ref. [3]. The obtained bounds are $0.813 \pm 0.014 \leq F_p \leq 0.892 \pm 0.021$.

SUMMARY

In summary, we demonstrate process characterization of basic gates of optical cluster-state quantum computation. We present complete process tomography of a single-qubit gate and evaluate the bounds on the process fidelity of a two-qubit gate.

REFERENCES

1. R. Raussendorf and H. J. Briegel, Phys. Rev. Lett. **86**, 5188 (2001).
2. A. G. White, A. Gilchrist, G. J. Pryde, J. L. O'Brien, M. J. Bremner, and N. K. Langford, J. Opt. Soc. Am. B **24**, 172 (2007).
3. H. F. Hofmann, Phys. Rev. Lett. **94**, 160504 (2005).
4. Y. Tokunaga, T. Yamamoto, M. Koashi, and N. Imoto, Phys. Rev. A **71**, 030301(R) (2005).
5. P. G. Kwiat, E. Waks, A. G. White, I. Appelbaum, and P. H. Eberhard, Phys. Rev. A **60**, R773 (1999).
6. Y. Tokunaga, T. Yamamoto, M. Koashi, and N. Imoto, Phys. Rev. A **74**, 020301(R) (2006).
7. P. Walther *et al.*, Nature **434**, 169 (2005).
8. R. Prevedel *et al.*, Nature **445**, 65 (2007).
9. I. Chuang and M. Nielsen, J. Mod. Opt. **44**, 732 (1997).
10. J. L. O'Brien, G. J. Pryde, A. Gilchrist, D. F. V. James, N. K. Langford, T. C. Ralph, and A. G. White, Phys. Rev. Lett. **93**, 080502 (2004).

Linear optics quantum process tomography

J. B. Altepeter*, E. R. Jeffrey†, M. Medic** and P. Kumar‡,**

*Center for Photonic Communication and Computing, EECS Department, Northwestern University, Evanston, IL 60208, USA
†Leiden Institute of Physics, Niels Bohrweg 2, 2333 CA Leiden, Netherlands
**Department of Physics and Astronomy, Northwestern University, Evanston, IL 60208, USA
‡Center for Photonics Communication and Computating, EECS Department, Northwestern University, Evanston, IL 60208, USA

Abstract. Linear optics quantum computation (LOQC) eliminates the need for a strongly nonlinear photon-photon interaction to create entangling gate operation, instead replacing it with the fundamental nonlinearity of quantum measurement. The LOQC gates rely on nondeterministic performance conditioned on detection or nondetection of events in ancillary modes, and are normally characterized in the multiple photon basis in which they are designed to operate. By assuming that the underlying physical structure of an LOQC gate is linear, and therefore describable entirely by it's action on single-photon input states, it is possible to exponentially reduce the number of parameters necessary to characterize that gate.

Keywords: process, tomography, LOQC
PACS: 42.50.Dv, 03.67.Mn, 42.50.Ex, 03.67.-a

INTRODUCTION

Linear optics quantum computation (LOQC) [1] requires only linear optics and photon detection, but any successful LOQC implementation will require the construction and accurate characterization of the basic building block of any quantum information processor—the quantum gate. Controlled-NOT (CNOT) gates have been experimentally constructed and, by extending the principles of quantum state tomography, have been completely characterized via process tomography [2, 3, 4].

Quantum process tomography of a two-qubit gate presents several challenges. A large number of measurement settings (> 255) are necessary to completely characterize an arbitrary, unknown process. Even the most general two-qubit process may not be general enough to account for systematic errors or background noise whose states extend into a larger Hilbert space than that of the process being characterized (such as those generated when too many photons are used as input states during the measurement—a serious possibility when pulsed sources of photon pairs are used as inputs). For certain types of processes, it may be possible to mitigate these problems by tailoring the process tomography to the type of process, as investigated here using the first experimental implementation of a telecom-band (1550-nm) CNOT gate [5].

The CNOT gate allows the state of one qubit to be conditionally changed depending on the state of a second qubit. For LOQC gates, this photon-photon interaction occurs not through a nonlinear medium, but through an effective nonlinearity conditioned on the detection or lack of detection of photons in ancillary modes [1]. Normally, these additional modes are not used in a process tomography, since they determine when a

CP1110, *Quantum Communication, Measurement and Computing (QCMC), The Ninth International Conference,* edited by A. Lvovsky

gate succeeded, rather than the behavior of the gate itself. By including these ancilla, it becomes obvious that any LOQC process is linear, thus dramatically reducing the size of the Hilbert space that must be spanned by a best-fit algorithm. Here we describe the use of a simplified version of this method to estimate the errors in an experimental CNOT characterization due to multiple input pairs. We then present a parameter counting argument which suggests that it may be possible to characterize linear optical gates using fewer measurements than previously thought, in addition to permitting the use of non-single-photon or non-single-photon-pair inputs for characterizing a gate that is designed for use with more precisely prepared input states.

MULTIPLE INPUT PAIRS AND CNOT PERFORMANCE

This approach was used to analyze the first experimental implementation of a telecom-band CNOT gate. By encoding quantum information in the telecom band, it is feasible to distribute qubits over long distances using the existing telecommunications infrastructure. Figure 1 shows a schematic of the measured CNOT gate [5].

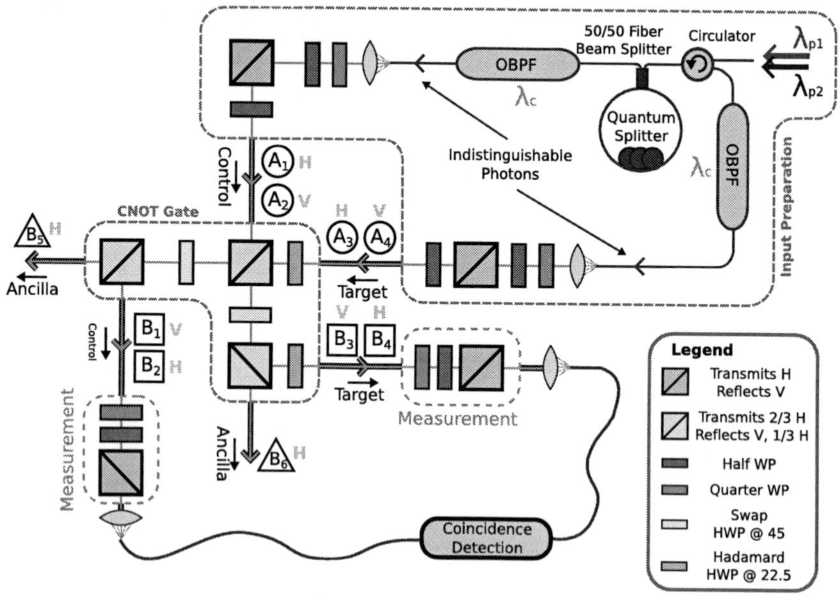

FIGURE 1. Experimental schematic of a Sagnac-loop based source of identical photon pairs used as the input to a linear optical CNOT gate. Logical input modes are marked by A_i while logical output modes (squares) and ancillary output modes (triangles) are denoted by B_i.

A copolarized dual-frequency pump enters a fiber Sagnac loop composed of a 50/50 fiber coupler, a piece of dispersion-shifted fiber (DSF), and a fiber polarization controller (FPC). The total peak pump power P is equally split into two dual-frequency, counter-propagating pump pulses of power $P/2$. Inside the DSF, two pump photons of different frequencies annihilate to produce a pair of energy-degenerate daughter photons at their mean frequency. When the counter propogating powers are balanced, the clockwise and

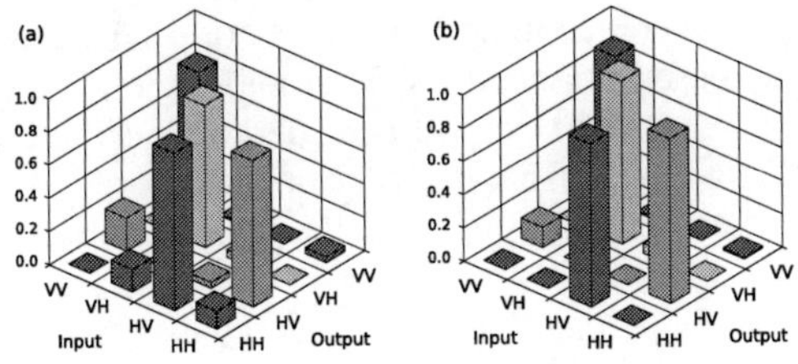

State	F	T	S_L
HH+VV	0.80(2)	0.43(6)	0.42(4)
HH-VV	0.75(3)	0.30(5)	0.53(4)
HV+VH	0.75(3)	0.34(6)	0.53(4)
HV-VH	0.80(2)	0.29(4)	0.49(3)

State	F	T	S_L
HH+VV	0.94(2)	0.88(6)	0.07(5)
HH-VV	0.92(3)	0.75(10)	0.17(7)
HV+VH	0.95(3)	0.85(10)	0.10(7)
HV-VH	0.92(3)	0.75(10)	0.17(7)

FIGURE 2. (a) Experimentally measured truth table for the CNOT gate in the logical basis (data represents coincidence counts minus accidental counts due to Raman noise and detector dark counts). Also shown is the Bell-state characterization for the same gate. Fidelity (F), tangle ($T \equiv \text{Concurrence}^2$), and linear entropy ($S_L \equiv \frac{4}{3}\left\{1-\text{Tr}(\rho^2)\right\}$) are given for all CNOT-generated Bell states. (b) The same data is shown after subtracting accidental coincidences due to multiple photon-pair production [5].

counterclockwise pumps scatter copolarized photon pairs with equal probability. When these photon-pair amplitudes meet on the 50/50 fiber coupler with an appropriately chosen phase, they deterministically split via the reverse Hong-Ou-Mandel effect [6]. These identical photon pairs act as the input to the CNOT gate.

The CNOT gate, composed entirely of linear optical elements, can be modeled as a two-photon gate with four inputs and four outputs:

$$A_1^\dagger A_3^\dagger \rightarrow B_1^\dagger B_3^\dagger \qquad A_2^\dagger A_3^\dagger \rightarrow B_2^\dagger B_4^\dagger \qquad A_1^\dagger A_4^\dagger \rightarrow B_1^\dagger B_4^\dagger \qquad A_2^\dagger A_4^\dagger \rightarrow B_2^\dagger B_3^\dagger, \tag{1}$$

or as a single-photon gate with four inputs and six outputs (see Figure 1):

$$\begin{aligned} A_1^\dagger &\rightarrow \sqrt{\tfrac{1}{3}}\left(B_1^\dagger + B_4^\dagger - B_3^\dagger\right) & A_3^\dagger &\rightarrow \sqrt{\tfrac{1}{3}}\left(B_1^\dagger - B_4^\dagger - B_6^\dagger\right) \\ A_2^\dagger &\rightarrow \sqrt{\tfrac{1}{3}}\left(-\sqrt{2}B_5^\dagger - B_2^\dagger\right) & A_4^\dagger &\rightarrow \sqrt{\tfrac{1}{3}}\left(B_1^\dagger + B_3^\dagger + B_6^\dagger\right). \end{aligned} \tag{2}$$

This equivalent characterization allows additional inferences to be made about the gate's behavior when input states of more than two photons are used (e.g., multiple photon pairs). By fitting our CNOT's experimental data to the closest pure process of the type shown in Figure 2, we were able to estimate the CNOT gate's preformance for single photon-pair inputs (high pump powers in the experiment resulted in a 10% probability of multiple photon-pair inputs). Figure 2 shows the CNOT characterization results both before and after this correction.

LINEAR OPTICS PROCESS TOMOGRAPHY

While successful at estimating the errors from multiple-photon-pair inputs to a particular CNOT gate, this technique has the potential to be extended to a more general method for analyzing quantum gates where the underlying behavior can be assumed to be linear. In other words, completely describing a gate's action on single-photon inputs is sufficient to completely describe that gate. Here we consider via a simple parameter counting argument what gains can be made from this type of analysis.

Consider an LOQC gate consisting of N photonic polarization-encoded qubits. The legal input and output states for this gate exist in a Hilbert space spanned by 2^N orthonormal basis vectors, which may in general undergo loss and/or decoherence as part of the gate. This quantum process is therefore completely characterized by $(2^N)^4 = 2^{4N}$ independent parameters [7]. Consider the same process as characterized by it's behavior on single-photon inputs. There are $2N$ single-photon inputs and $2N$ single-photon outputs to this gate. There are, therefore, $(2N)^4$ parameters which completely describe this process—an exponential reduction using only the assumption that the gate's physical components have no appreciable photon-photon interactions.

Note that these simple characterizations *completely* describe only lossless gates, in the sense that the results of a process tomography cannot be used to determine the gate's operation on non-standard inputs (e.g., multiple-photon-pair inputs to a CNOT gate). A lossy gate will leak photons into unmeasured and uncharacterized modes, such as the ancillary modes B_5 and B_6 in Figure 1. It is adequate characterization of these loss modes, and their relative phase with the measured output modes, which enable complete characterization. In general, a single loss mode exists for every input mode (although these loss modes may in general be superpositions of many spatial paths or polarization states). Together these $2N$ general loss modes can be described by an orthonormal basis of $2N$ modes. Even if we completely model the input and output behavior of each of these generalized loss modes, it is still only necessary to use $(4N)^4$ parameters to completely model the gate under the assumption of linearity, a model which will now predict the gate's behavior under *any combination of input photons*.

In showing this exponential reduction in parameters, we have made one additional implicit assumption: that the gate operation is not conditional on photon detection in ancillary (loss) modes. Because every conditional configuration would have to be independently characterized, this would increase the number of parameters necessary to characterize a gate by a factor of 2^D, where D is the number of conditional detections. Finally, it is an open question whether a useful set of *experimental measurements* can be constructed for an arbitrary gate which exploits this exponential reduction in parameters.

REFERENCES

1. E. Knill, R. Laflamme, and G. J. Milburn, *Nature* **409**, 46 (2001).
2. J. L. O'Brien, G. J. Pryde, A. G. White, T. C. Ralph, and D. Branning, *Nature* **426**, 264 (2003).
3. R. Okamoto, H. F. Hofmann, S. Takeuchi, and K. Sasaki, *Physical Review Letters* **95**, 210506 (2005), `arXiv:quant-ph/0506263`.
4. S. Gasparoni, J.-W. Pan, P. Walther, T. Rudolph, and A. Zeilinger, *Physical Review Letters* **93**, 020504 (2004), `arXiv:quant-ph/0404107`.
5. J. Chen, J. B. Altepeter, M. Medic, K. F. Lee, B. Gokden, R. H. Hadfield, S. W. Nam, and P. Kumar, *Physical Review Letters* **100**, 133603 (2008), `0802.1441`.
6. J. Chen, K. F. Lee, and P. Kumar, *Physical Review A* **76**, 031804 (2007).
7. M. A. Nielsen, and I. L. Chuang, *Quantum Computation and Quantum Information*, Cambridge University Press, Cambridge, UK, 2000.

Author Index

A

B

C

D

E

F

G

H

I

J

K

L

M

N

O

P

R

S